URBAN SURFACE WATER MANAGEMENT

URBAN SURFACE WATER MANAGEMENT

Stuart G. Walesh
Valparaiso University
Valparaiso, Indiana

A WILEY-INTERSCIENCE PUBLICATION
JOHN WILEY & SONS, INC.
New York • Chichester • Brisbane • Toronto • Singapore

Library of Congress Cataloging in Publication Data:

Walesh, S. G.
Urban surface water management.

"A Wiley-Interscience publication."
Includes bibliographies and index.
1. Urban hydrology. 2. Urban runoff. I. Title.
TC409.W325 1989 628'.2 88–27704
ISBN 0–471–83719–9

Printed in the United States of America

10 9 8 7 6 5 4 3 2

To Jerrie

CONTENTS

Preface **xiii**

1 Fundamentals of Urban Surface Water Management **1**

1.1 Definition of Terms, 2
1.2 Objectives of Urban Surface Water Management, 4
1.3 Flooding: A Natural and Certain Phenomenon, 6
1.4 How Traditional Floodplain Management and Stormwater Management Are Related, 7
1.5 Statistics, 11
1.6 Means by Which Floodwater May Enter a Structure, 14
1.7 History of Urban Drainage, 15
1.8 History of Floodplain Management, 17
1.9 Components of the Urban Surface Water System, 19
1.10 Approaches to Control of the Quantity of Urban Runoff, 20
1.11 Emergency and Convenience Systems, 31
1.12 Role of Natural Features in Surface Water Management, 34
1.13 Channel Erosion and Sedimentation in Urbanizing Areas, 35
1.14 Floodplain Regulations, 38
1.15 Flood Insurance Studies, 42
1.16 Surface Water Management in Geologically Unique Areas, 46
References, 47

2 **The Hydrologic Cycle in the Urban Environment** 53

2.1 The Hydrologic Cycle, 53
2.2 Impact of Urbanization on the Quantity of Runoff, 57
2.3 Impact of Urbanization on the Quality of Runoff, 67
2.4 The Sequential Nature of Erosion/Sedimentation and Flooding Problems, 73
References, 74

3 **Techniques for Hydrologic Analyses** 77

3.1 Hydrologic Data Types, Sources, and Uses, 78
3.2 Watershed Delineation, 82
3.3 Hydrologic Techniques: An Overview, 86
3.4 Statistical Analyses for Determining Peak Discharge and Volume, 90
3.5 Regional Methods for Determining Peak Discharge, 93
3.6 Transfer Method for Determining Peak Discharge, 95
3.7 Rainfall–Runoff Methods: Introduction, 97
3.8 Rational Method for Determining Peak Discharge, 100
3.9 Modified Rational Method for Determining Discharge and Volume, 105
3.10 Soil Conservation Service TR55 Method for Determining Discharge and Volume, 112
3.11 British Road Research Laboratory Method for Determining Discharge and Volume, 124
3.12 Other Hydrologic Methods, 132
References, 133

4 **Floodplain Hydraulics** 139

4.1 Floodplain Map and Flood Stage and Streambed Profiles, 139
4.2 Fundamentals of Open Channel Hydraulics, 142
4.3 Data Needs and Uses: Overview, 145
4.4 Acquisition of Channel–Floodplain Geometry and Roughness Data, 149
4.5 Acquisition of Hydraulic Structure Geometry Data, 158
4.6 Performing Flood Stage Profile Calculations with Computer Programs, 159
4.7 Sensitivity Studies, 160
4.8 Floodway Determination, 162
References, 165

5 Stormwater Facility Hydraulics **167**

5.1 Hydraulic Analysis and Design Guidelines, 168
5.2 Force Exerted on a Person by Moving Floodwater, 175
5.3 Hydraulic Design of a Culvert, 178
5.4 Determining Culvert Flow for Given Headwater and Tailwater Conditions, 184
5.5 Flow Capacity of an Urban Street, 188
5.6 Storage Capacity of an Urban Street, 192
5.7 Development of Stage–Discharge Relationship for a Detention/Retention Outlet Control Structure, 195
5.8 Other Hydraulic Methods, 199
References, 200

6 Computation of Average Annual Monetary Flood Damage **203**

6.1 Types of Flood Losses, 203
6.2 The Uses of Average Annual Monetary Damage, 205
6.3 Calculation Procedure, 207
6.4 Determining the Average Annual Monetary Benefit of Alternative Flood Damage Mitigation Management Measures, 213
References, 215

7 Nonpoint-Source Pollution Load Techniques **217**

7.1 Definition and Essential Features of Nonpoint-Source Pollution, 218
7.2 Nonpoint-Source Pollution Load Techniques: An Overview, 219
7.3 Unit Load Method, 220
7.4 Preliminary Screening Procedure, 223
7.5 Universal Soil Loss Equation, 227
7.6 Concentration Times Flow Method, 232
7.7 Constituent Rating Curve–Flow Duration Curve Method, 236
7.8 Other Nonpoint-Source Pollution Load Techniques, 242
References, 242

8 Planning and Designing Detention/Retention Facilities **245**

8.1 Definition of Terms, 245
8.2 Planning and Design Procedure: An Overview, 246
8.3 Step 1: Collect and Analyze Watershed Data, 247

8.4 Step 2: Identify Potential Sites, 254
8.5 Step 3: Select the Design Discharge–Probability Condition, 257
8.6 Step 4: Perform Preliminary Hydrologic Design, 262
8.7 Step 5: Accommodate Lateral and Vertical Constraints, 265
8.8 Step 6: Perform Refined Hydrologic–Hydraulic Design, 269
8.9 Step 7: Finalize the Design, 287
8.10 Step 8: Begin Implementation, 287
8.11 Multiple Detention/Retention Facilities, 290
8.12 Negative Aspects and Misuse of Detention/Retention Facilities, 294
References, 295

9 Sedimentation Basin Design 297

9.1 Overview, 297
9.2 Components of a Sedimentation Basin, 299
9.3 Design Procedure: Overview, 301
9.4 Step 1: Collect and Analyze Watershed Data, 301
9.5 Step 2: Identify Potential Sites, 302
9.6 Step 3: Select Design Recurrence Intervals, 304
9.7 Step 4: Select Design Particle Size and Settling Efficiency, 304
9.8 Step 5: Determine Design Flows and Volume of the Settling Zone, 307
9.9 Step 6: Design the Inlet Zone, 308
9.10 Step 7: Determine Settling Zone Geometry, 308
9.11 Step 8: Determine Volume of the Sediment Storage Zone, 309
9.12 Step 9: Design the Outlet Zone, 310
References, 313

10 Computer Modeling 315

10.1 System and Modeling Concepts, 316
10.2 Nature of Watershed Digital Computer Models, 319
10.3 Model Selection Guidelines, 325
10.4 Introduction to Selected Computer Models, 328
10.5 Use of Two or More Computer Models on One Project, 335
10.6 Calibration: Key to Credibility in Modeling, 337
10.7 Applications of Computer Models: Examples, 361
10.8 Costs and Benefits of Computer Modeling, 385
10.9 Modeling: Maintaining Perspective, 386
References, 387

11 Management Measures **391**

11.1 Definition of Structural and Nonstructural Management Measures, 392
11.2 Structural Measures, 394
11.3 Nonstructural Measures, 416
References, 448

12 Preparation of a Master Plan **453**

12.1 Definition of and Need for Master Planning, 453
12.2 Fluctuating Interest in Master Planning, 456
12.3 Master Planning Principles, 459
12.4 The Planning Process: An Overview, 465
12.5 Step 1: Establish Objectives and Standards, 466
12.6 Step 2: Conduct Inventory, 469
12.7 Step 3: Analyze Data and Prepare Forecasts, 471
12.8 Step 4: Formulate Alternatives, 473
12.9 Step 5: Compare Alternatives and Select the Recommended Plan, 480
12.10 Step 6: Prepare Plan Implementation Program, 481
12.11 Step 7: Implement the Plan, 492
12.12 Economics of Master Planning, 493
References, 495

Appendices

A Glossary **497**

B Input Data on Coding Form for ILUDRAIN Example **503**

C Partial Output for ILUDRAIN Example **507**

Index **513**

PREFACE

Urban Surface Water Management shows how state-of-the-art engineering tools and techniques have been and can be used to manage the quantity and quality of urban stormwater runoff. The book's focus is on planning and designing facilities and systems to prevent or to control flooding, erosion, sedimentation, and nonpoint-source pollution. The pragmatic application of state-of-the-art methodologies reflects, in part, my urban water resources engineering experience while employed in the public and private sectors.

The state of the art of urban surface water management is far ahead of the "state of the practice." As a group, engineers, planners, and public officials are using or are benefiting from the use of only a small fraction of what is known. For example, surface water systems are often designed and built without even considering or using state-of-the-art concepts or tools, such as the emergency and convenience system approach, watershed master planning, computer modeling, and multipurpose flood control–water quality enhancement–recreation facility possibilities. Certainly, more research and development are needed, but the public would benefit by closing the gap between what we know how to do and what we do. This book is intended to help close the gap between the state of the art and the state of the practice.

Under the umbrella of closing the gap, several convictions influenced the organization and content of this book. First, rather than being a problem, surface water can be a valuable aesthetic and other resource in urban areas if we plan and design those areas to be compatible with the hydrologic cycle. Second, existing or potential surface water quantity and quality problems—flood control, erosion, sedimentation, nonpoint-source pollution—should be addressed simultaneously in planning and design because causes of and solutions to quantity and quality problems are inextricably tied to each other and to the hydrologic cycle. Third, surface water management is usually of interest to a community's citizens and elected officials only when a disaster occurs. Accordingly, high-quality technical work coupled with effective public education and interaction programs are needed to raise the profile of surface water management. Fourth, much more emphasis must be placed on preventive rather than remedial efforts; on planning instead of fixing.

AUDIENCE

This book was written primarily for civil engineers and planners employed in one of two sectors: (1) those employed in consulting firms and doing planning and design of urban surface water facilities for public and private clients, and (2) civil engineers and planners working for local, regional, state, and federal government units and agencies and having responsibilities for the planning, design, operation, and maintenance of urban surface water systems and for review of plans and designs prepared by others. For this principal audience, I have assumed that the reader has a basic knowledge of subjects such as hydraulics, hydrology, sanitary engineering, soil mechanics, chemistry, computers, and economics typically covered during an undergraduate civil engineering program. Extensive references cited in the text and listed at the end of each chapter facilitate further study by the technically prepared reader.

Significant portions of this book are intended to be of value for reference or textbook purposes to faculty in civil engineering and planning programs and to their senior and graduate-level students. Academics may be interested in learning more about the application of state-of-the-art tools and techniques to management of the quantity and quality of urban surface water. The upper-level undergraduate student and the graduate student may benefit from seeing how fundamentals such as hydraulics and hydrology are integrated and used in engineering practice. Students typically exposed to simplified hypothetical examples may welcome and learn from the complex, real systems and situations frequently used in the book. Furthermore, undergraduate and graduate students may benefit from seeing how the professional must address both technical and nontechnical factors and considerations if planning, analysis, and design are to be fruitful in terms of action and implementation.

Local to federal government administrators and elected officials having policymaking authority and administrative responsibilities may also find portions of the book useful. Examples are Chapter 1, "Fundamentals of Urban Surface Water Management," which defines key terms and discusses many and varied concepts, and Chapter 12, "Preparation of a Master Plan," which offers a comprehensive and largely nontechnical treatment of master planning.

Finally, *Urban Surface Water Management* may be valuable as a reference for residential, commercial, and industrial land developers. This sector of our economy is increasingly expected to understand and meet the ever-rising expectations and requirements of government and society in all aspects of land development, including management of the quantity and quality of surface water.

ORGANIZATION AND CONTENT

Subject matter in the book is arranged to flow from fundamentals through engineering analysis and design methodologies, concluding with application and integration. Chapter 1, "Fundamentals of Urban Surface Water Management," and Chapter 2, "The Hydrologic Cycle in the Urban Environment," focus on fundamentals in that objectives are presented, terms are defined, concepts are discussed, historical perspective is provided, and the consequences of no or inadequate management are illustrated.

Chapters 3 through 7 present many and varied engineering analysis and design methodologies. The range and depth of topics covered in this middle portion of the book is suggested by the titles of Chapters 3 through 7, which, are, respectively; "Techniques for Hydrologic Analyses," "Floodplain Hydraulics," "Stormwater Facility Hydraulics," "Computation of Average Annual Monetary Flood Damage," and "Nonpoint-Source Pollution Load Techniques." In almost all cases where methodologies are presented, more than one analysis or design technique is offered. Many worked examples are provided, most of which are drawn from or based on actual projects.

Beginning with Chapter 8 and extending through Chapter 12, the last chapter, application and synthesis are emphasized. Technical material presented in earlier chapters is utilized and nontechnical and other considerations are introduced. The planning and design of detention/retention facilities and of sedimentation basins are covered, in Chapters 8 and 9 titled, respectively, "Planning and Designing Detention/Retention Facilities" and "Sedimentation Basin Design." Chapter 10, "Computer Modeling," provides a comprehensive and in-depth treatment of modeling in urban surface water management, with emphasis on computer modeling as being a practical and often the only way to utilize fully the technology available in urban surface water management. Chapter 11, "Management Measures," provides a comprehensive summary of structural and nonstructural measures available for managing the quantity and quality of urban surface water. The salient planning and design characteristics of each alternative measure are presented. Chapter 12, "Preparation of a Master Plan," concludes the book by presenting a proven process for preparing urban surface water master plans that are intended to be implemented. Planning is chronologically one of the first steps to be undertaken in urban surface water management. However, the master planning process is discussed at the end of this book because effective planning requires understanding of fundamentals and the ability to use tools and techniques all of which are presented in the preceding chapters.

ACKNOWLEDGMENTS

Material presented in this book was developed over a two-decade period beginning in 1967 when I began a three-year assignment teaching water-related courses in the Department of Civil Engineering at Valparaiso University. Other content was developed during the period 1974 through 1978 when I served as a Visiting Instructor and then Adjunct Associate Professor of Civil and Environmental Engineering at the University of Wisconsin–Madison. Refinements were made as a result of my participation as an instructor in the American Society of Civil Engineers Continuing Education Program for a decade beginning in 1978. Additional materials, reflecting primarily applications, were developed during the period 1970 to 1978, while I was on the staff of the Southeastern Wisconsin Regional Planning Commission and from 1978 to 1985 when I was employed by Donohue and Associates, Inc., a civil engineering consulting firm. Independent consulting assignments beginning in 1985 added further to the experience base of the book.

In addition to the cited references included in this book, I have received a wealth of ideas and much information from and have been influenced by numerous colleagues associated with the preceding and other organizations and in various professional circles. My debt to other professionals is suggested, in part, by the extensive lists of cited references that appear at the ends of chapters. But two civil engineer colleagues warrant specific mention. I am particularly indebted to Mr. Daniel H. Lau, PE and Mr. Gary E. Raasch, PE, with whom I worked closely for many years. They helped to maintain a balance between our enthusiastic desire to apply new, state-of-the-art approaches and the need to resolve our clients' problems on time and within budget.

I also fondly and thankfully acknowledge former teachers who directly influenced my early studies and indirectly influenced my later work and the writing of this book. I am especially indebted to Dr. A. Sami El-Naggar, my undergraduate civil engineering professor at Valparaiso University, who enthusiastically introduced me to study, research, and practice in the field of water resources engineering; to Dr. John C. Geyer, who encouraged and supported early graduate work at The Johns Hopkins University; and to Dr. Peter L. Monkmeyer, who as my advisor at the University of Wisconsin–Madison exemplified discipline and competence in research and teaching.

Book writing labor and logistics are almost overwhelming. I gratefully acknowledge the help of Ms. Kathleen F. Marlowe, my assistant in the College of Engineering at Valparaiso University, who supervised and did some of the transcription of early chapter drafts and coordinated various supplemental tasks. Ms. Susan Price enthusiastically and artistically prepared all the figures. Her contribution to the appearance and readability of the illustrations in the book is appreciated. I also acknowledge the assistance of Mr. Aaron K. Smith formerly of the Moellering Library at Valparaiso University, who so effectively and quickly provided or obtained reference materials. Jill, my daughter, did large amounts of word processing and her help is happily noted. Finally, my wife, Jerrie, did most of the word processing required for numerous revisions and refinements and, as always, provided total support.

STUART G. WALESH

Valparaiso, Indiana
January 1989

URBAN SURFACE WATER MANAGEMENT

1

FUNDAMENTALS OF URBAN SURFACE WATER MANAGEMENT

This chapter presents concepts, definitions, and other background information used in the planning, design, operation, and maintenance phases of urban surface water management (SWM). By defining key words and describing certain concepts and ideas, this chapter establishes much of the nomenclature used throughout the book.

Words and terms commonly used in the profession, such as hydrology, hydraulics, channel, floodplain and flooding, are defined in this chapter. The concept of flooding as a natural and certain phenomenon is presented. Distinctions and similarities between what have been traditionally termed stormwater management and floodplain management are discussed. The replacement concept and terminology of surface water management is introduced.

Selected statistical concepts and risk equations suited to urban SWM are presented and illustrated with examples. The vulnerability to flooding of typical urban residential and other structures is described, followed by a review of the history of urban drainage and floodplain management. Components of a typical urban surface water system are described and the system itself is contrasted with other urban utilities and facilities.

The fundamentally different conveyance-oriented and storage-oriented approaches to urban SWM are defined and contrasted. The emergency system–convenience system concept applicable to the analysis and design of an urban SWM system is introduced. The potential integration of natural landscape features into the urban SWM system is discussed. Floodplain regulation fundamentals, including the two-district approach, are discussed, followed by a review of the history of the National Flood Insurance Program and its current applicability. The chapter concludes with a brief discussion of urban SWM in unique geologic areas.

1.1 DEFINITION OF TERMS

Hydrology. Analysis of physical behavior of water from its occurrence as precipitation through its movement on or beneath the earth's surface; to its entry into sewers, streams, lakes, and the oceans; and its return to the atmosphere. Simply stated, hydrology is the study of spatial and temporal changes in water volumes and discharge rates.

Hydraulics. Analysis of the physical behavior of water as it flows over the land surface; on streets and parking lots; within sewers and stream channels and on natural floodplains; under and over bridges, culverts, and dams; and through lakes and impoundments. Simply stated, hydraulics is the study of flow paths, velocities, and stages.

Channel. Long and narrow, continuous, low-lying area as shown in Figure 1.1 that, except in time of flood, contains all of the flow of a river, stream, or other waterway.

Floodplain. Wide, flat to gently sloping area contiguous with and usually lying on both sides of the channel as shown in Figure 1.1. The floodplain is typically very wide relative to the channel. The floodplain is more subtle, that is, more difficult to observe than the channel, particularly in developing or developed urban areas, where the topographic discontinuity between the floodplain and adjacent higher ground is masked by urban development.

Flooding. Temporary inundation of all or part of the floodplain along a well-defined channel or temporary localized inundation occurring when surface water runoff moves via surface flow, swales, channels, and sewers toward well-defined channels. Flooding is not necessarily synonymous with flooding problems. For example, occasional inundation of a properly planned floodplain park and recreation area will cause temporary inconvenience contrasted with the disruption, damage, and sometimes death that result when unprotected residential, commercial, or industrial development are placed on floodplains. Flooding may even have positive results. For example, Egyptian farmers depended on flooding of the Nile River to restore fertile soil [Geological Society of America (GSA), 1978].

With respect to the meteorological phenomena that cause floods, three types may be identified. The first is a large amount of rainfall, occurring at a relatively low intensity over a long period of time and on a large area. An example is Hurricane Agnes, which resulted in extensive death, damage, and disruption in the eastern United States in 1972. The second type of flood is that caused by high-intensity thunderstorms occurring over small areas. Examples include the Rapid City, South Dakota, flood of 1972 and

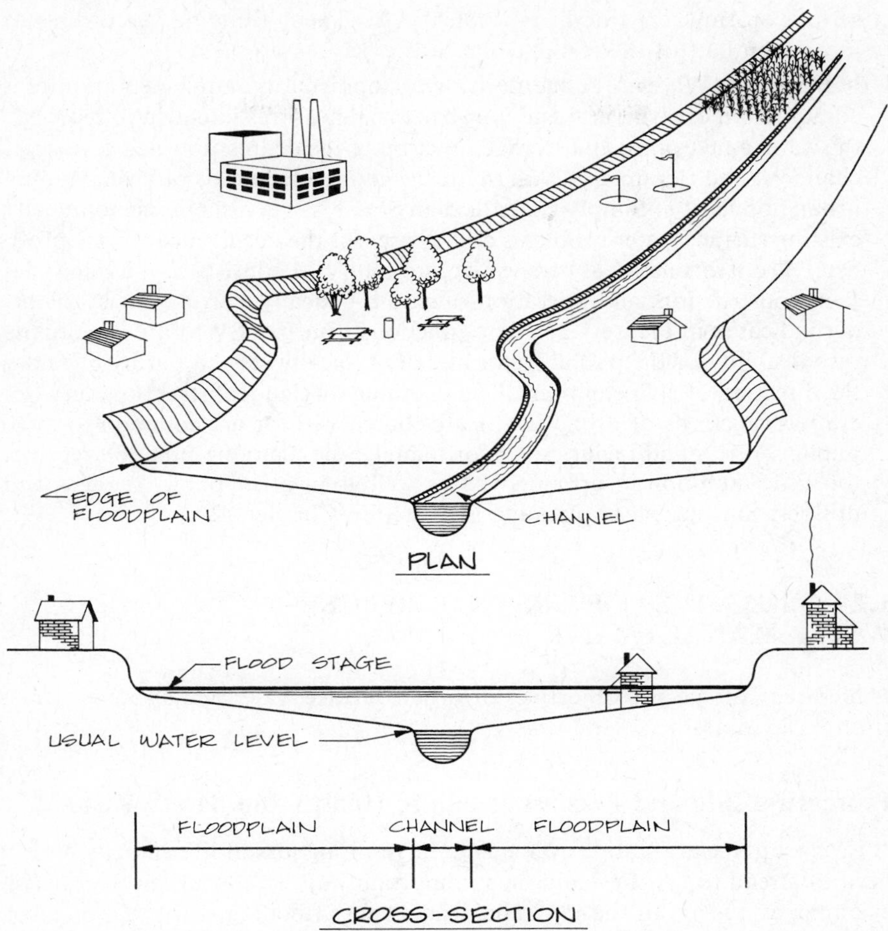

FIGURE 1.1 Channel and floodplain

the Big Thompson, Colorado, flood of 1976. The third type of flood is that caused by rapid snowmelt, possibly in combination with rainfall and ice jams. This type of flood is typical of that which occurs on large watersheds in the midwest (GSA, 1978).

Flooding Problem. Disruption to community affairs, damage to property and facilities, and danger to human life and health that occur when land use is incompatible with the hydrologic-hydraulic system. Flooding problems are widespread throughout the United States. For example, responses by 325 public agencies to a 1980 American Public Works Association Survey (1980) indicated that nearly half of the responding communities had serious stormwater flooding problems. Basement flooding, according to more than half of the respondents, was a serious problem.

Urban Surface Water Management. Development and implementation of a combination of structural and nonstructural measures intended to reconcile the water conveyance and storage function of the depressions, lakes, swales, channels, and floodplains with the space and related needs of an expanding urban population. Simply stated, urban SWM is everything done to remedy existing surface water problems and to prevent the occurrence of new problems. From a functional perspective, urban SWM consists of planning, design, construction, and operation functions—ideally, carried out in the order indicated in Figure 1.2. These functions of urban SWM are usually the responsibility of the public sector and are typically carried out by or under the direction of civil engineers. The planning, design, construction, and operation functions of urban SWM are shared with or are common to most public services and facilities. Unfortunately, the planning function receives too little attention in urban SWM as well as in other public services and utilities. Surface water planning is treated in Chapter 12.

1.2 OBJECTIVES OF URBAN SURFACE WATER MANAGEMENT

A hierarchy of specific objectives of urban surface water management, listed from basic to aesthetic, follows.

Protecting Life and Lessening Public Health and Safety Risks

A review of recent flooding events and the resulting loss of life emphasizes the need to attend to this first and most important objective of urban SWM. For example, as a result of the 1972 Hurricane Agnes flooding, 122 lives were lost in 12 midatlantic states. The 1972 flood in Rapid City, South Dakota, caused 238 deaths. In 1976, 139 people died as a result of the Big Thompson flood in Colorado (GSA, 1978). The July 1977 flood in Johnstown, Pennsylvania, affected 7300 families and left 70 persons dead (Mendenhall, 1986). Finally, the September 1977 flood in Kansas City, Missouri and Kansas, took 25 lives (Hauth et al., 1981).

It is interesting to note that a 1907 flood in Rapid City, South Dakota,

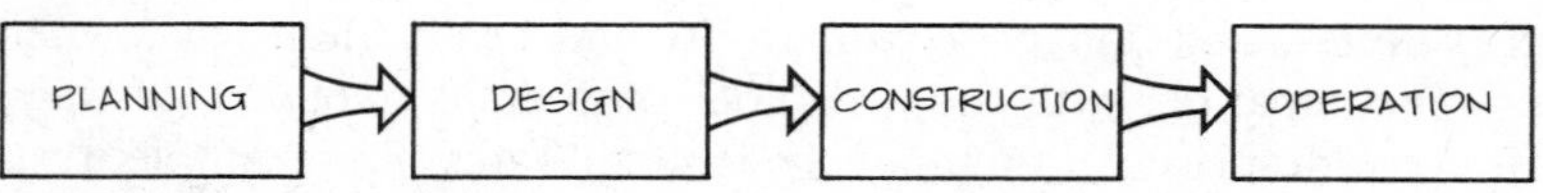

FIGURE 1.2 Functions of urban surface water management

approximated, in terms of depth and areal extent, the devastating 1972 flood. Furthermore, some "farsighted and thoughtful individual" plotted the limits of the 1907 flood. Unfortunately, the historic flood records and maps were not used to prevent the subsequent occurrence of flood-prone development. This set the stage for the devastation that occurred 65 years later (Swanson, 1982).

Although historic accounts of floods clearly illustrate the threat to human life, the public health threat is less apparent. Nevertheless, floodwaters can transport potentially harmful substances and objects. Examples are toxic materials from industrial operations and pathogenic bacteria from on-site waste disposal systems, sanitary and combined sewer systems, and sewage treatment plants. Potentially harmful flood-borne objects include drums and tanks, debris from buildings and other structures destroyed by the flood, trees, and other large buoyant materials. These substances and materials can be carried from their sources and brought in contact with, and cause harm to, area residents.

Besides causing physiological harm, the occurrence and potential occurrence of floods can adversely affect the psychological well being of people living in flood-prone areas. Owners and tenants of flood-prone buildings and properties are subject to stress because of the need to be in a constant state of readiness. This is particularly true in some smaller, urban watersheds, where flooding can occur, with little warning, almost any time of the year. People are faced with the difficult task of cleaning sand, silt, and other material and debris from their homes and places of business. Even after cleanup and repair have been completed, residual odors and other reminders of the recent flooding impose an additional psychological stress on the occupants of flood-prone property.

Reducing Risks of Monetary Damage to Private and Public Property

Flood damage to private and public property ranges from minor occurrences of strictly local interest to large, catastrophic events that attract national or international attention. Accordingly, the latter are better documented. For example, Hurricane Agnes caused $3.5 billion of damage in the eastern United States in 1972 (GSA, 1978). The 1976 Big Thompson flood in Colorado destroyed 252 homes and 16 businesses and 4 years and $50 million were required to restore the canyon (*Milwaukee Journal,* 1981). The 1977 flooding in Kansas City, Missouri and Kansas, caused over $80 million in damage (Hauth et al., 1981). About 400 homes were completely destroyed and 340 businesses destroyed or severely damaged as a result of the 1977 flood in Johnstown, Pennsylvania (Mendenhall, 1986). Almost a century earlier in Johnstown, Pennsylvania, 7000 lives were lost in a disastrous flood caused by failure of a dam in 1889 (Mead, 1950).

Minimizing Disruption of Community Affairs

Next to protecting life and property, urban SWM seeks to minimize the usually annoying and sometimes costly disruption caused by flooding in urban areas. This disruption often extends well beyond the flooded areas and affects many people not residing or working in the flooded areas. Manufacturing and business activities are interrupted not only during the flood period but also after the flood during the cleanup and repair period. In the public sector, routine operations are curtailed as government personnel provide relief to affected areas. Floods may result in the temporary closing of streets, highways, and railroads, further disrupting normal urban activities.

Protecting the Quality of Surface and Groundwater

The Nationwide Urban Runoff Program (NURP), which is discussed in Chapter 2, concluded that stormwater runoff from urban areas generally contains significant quantities of potential pollutants. Examples are heavy metals, coliform bacteria, nutrients, oxygen-demanding substances, and total suspended solids (Athayde et al., 1983). These substances can have an adverse impact on surface and groundwater resources.

Enhancement of the Quality of Life in Urban Areas

As urbanization proceeds, consideration should be given to retaining natural drainage features, such as swales, channels, floodplains, ponds, and wetlands, and incorporating them into the surface water system. Preservation, conservation, or even restoration of these features and of their recreational, aesthetic, ecological, and cultural values can greatly enhance the quality of life in urban areas (Walesh, 1973, 1976). From the preservation–conservation–restoration perspective, the management of surface waters in urbanizing areas can be viewed more as an opportunity than a problem.

1.3 FLOODING: A NATURAL AND CERTAIN PHENOMENON

Most of the time, rivers, streams, and other waterways are passive and tranquil and the water is contained within their channels. Occasionally—approximately every 2 years for natural channels (Wolman and Leopold, 1957)—rainfall, snowmelt, and antecedent moisture and other conditions develop such that the channels do not contain the resulting flows and flooding occurs. Flooding is a natural and certain phenomenon, although its time of occurrence is generally unpredictable. The floodplain is as much a part of the creek, stream, or river as is the channel—it is just used less often.

Unfortunately, the public press tends to treat flooding as being abnormal.

For example, during times of flooding, rivers and streams are often treated as animate objects with destructive tendencies. Adjectives used in news accounts to describe streams in a flood condition include "rampaging," "devastating," "raging," "relentless," "overflowing," and "destructive." The public reads or hears that people were "forced from their homes" as "floodwaters . . . poured . . . raised . . . fanned out. . . ."

This popularized but inaccurate portrayal of flooding and flood problems is unfortunate because it masks the fundamental cause of flooding problems: failure by individuals, groups, and communities to recognize that flooding is natural and certain, and to act accordingly. Natural flooding can be prevented only by the construction of massive and costly flood control works, such as reservoirs, channels, diversions, dikes, and floodwalls. In the absence of these major public facilities, flooding of floodplains and other areas is inevitable. T.S. Eliot, in his poem "The Dry Salvages," describes the inevitability of flooding this way:

> I do not know much about gods; but I think that the river is a strong brown god—sullen, untamed and intractable, Patient to some degree, at first recognized as a frontier; Useful, untrustworthy, as a conveyor of commerce; Then only a problem confronting the builder of bridges. The problem once solved, the brown god is almost forgotten By the dwellers in cities—ever, however, implacable, Keeping his seasons and rages, destroyer, reminder Of what men choose to forget . . .

1.4 HOW TRADITIONAL FLOODPLAIN MANAGEMENT AND STORMWATER MANAGEMENT ARE RELATED

Floodplain management and stormwater management evolved and, in many communities, continue to occur as largely separate activities. Much of the separation between floodplain management and stormwater management has little substance and tends to detract from the effectiveness of both. Distinctions that have developed between floodplain management and stormwater management are reviewed here. Significant similarities are discussed, as is the value of emphasizing the similarities in developing urban SWM programs (Walesh, 1982).

Distinctions between Floodplain and Stormwater Management

Intramunicipal versus Intermunicipal and Small versus Large Watersheds. The distinction between floodplain management and stormwater management has often been determined by a combination of scale and relationship to intermunicipal boundaries. For example, if the tributary watershed lies wholly or largely within a given community, the necessary remedial or preventive action to resolve or prevent the occurrence of flooding problems can be taken unilaterally. Such situations are typically and popularly called stormwater management. In contrast, if the watershed is large, intermunicipal cooperation is required and the resulting effort is sometimes referred to as floodplain management.

Floodplain Management Is Required by Regulations and Encouraged by Funding. Statutes in many states (Bloomgren, 1982), reinforced with Federal Emergency Management Agency requirements applicable to all states, make floodplain management mandatory. That is, some minimum level of floodplain management is required—there is no choice. In contrast, with few exceptions, such as Pennsylvania (Commonwealth of Pennsylvania, 1978) and Minnesota (State of Minnesota, 1982), the decision to have a comprehensive stormwater management program is a local option. Furthermore, if significant federal funds are involved, the effort is often called floodplain management. Examples are FEMA-funded Flood Insurance Studies (FIS) and Corps of Engineers–funded small projects. In contrast, if primarily local funds are expended, the project is often referred to as stormwater management.

Watershed Orientation of Stormwater Management. When stormwater management plans are prepared, they tend to be done on a watershed basis and usually reflect projected future land use. In contrast, floodplain management tends to be disjointed with respect to watersheds. For example, state-mandated floodplain regulations tend to be implemented on a community-by-community, reach-by-reach basis rather than on a watershed basis. The result is often a series of hydrologic–hydraulic engineering studies executed by different consulting firms or government agencies over a period of years using widely varying methodologies. Repetitive, costly engineering may be required and the results may be inconsistent. Furthermore, FISs usually do not include the affect of future urbanization of the watershed. This FIS deficiency is most critical on small urbanizing watersheds, where the urbanization process can significantly, and in a short period of time, increase flood flows and stages.

Growing Quantity–Quality Orientation of Stormwater Management. In recent years, stormwater management has been increasingly directed at the quality as well as the traditional quantity aspects of stormwater runoff. The growing concern with quality has focused on nonpoint-source pollutants, primarily suspended solids and adsorbed substances. In contrast, water quality is usually not considered in floodplain management.

Higher Level of Protection in Floodplain Management. In many states, a new single-family residence constructed in a floodplain fringe must be placed on fill with the finished ground grade around the house at least 1 foot (ft) above the 1 percent (100-year-recurrence-interval) flood stage, and the first floor must be at least 2 ft above the 1 percent flood stage. In contrast, a new single-family residence located outside the floodplain in the same community

is usually served by a storm sewer designed to carry a 50 percent (2-year) to 10 percent (10-year) flow.

Street planimetric layouts, street cross sections, and site grades are rarely designed to provide protection against large runoff events. Therefore, whereas 1 percent protection is usually explicitly designed into floodplain management projects, it is infrequently explicitly designed into stormwater management projects.

Similarities Between Floodplain and Stormwater Management

Life and Safety Objectives. Floodplain management and stormwater management share the primary objective of prevention and mitigation of death, damage, and disruption attributable to surface water flooding. A flood victim gets just ''as wet'' regardless of whether the floodplain management or stormwater management project is deficient in a given community. Engineers, planners, and other professionals and public officials often forget the primary and common objectives of floodplain management and stormwater management.

Public Interest Tends to Be Intense, but Brief and Infrequent. Popular, citizen interest—and in some cases the interest of public officials—in urban flood control is typically strong during and immediately after a damaging flood. The rise and fall of public interest often parallels the rise and fall of the floodwaters. This highly variable public response, perception, and concern stands in contrast with the more constant public attitudes often exhibited toward deficiencies and problems with other utilities and facilities, such as sanitary sewerage, water supply systems, and streets and highways. These services probably receive more continuous attention because they are more visible, are in almost continuous operation or use, and often have continuous revenue-producing mechanisms to support them.

Similar Structural and Nonstructural Measures. The available preventive and remedial measures commonly used in both floodplain management and stormwater management are similar. Furthermore, they have evolved from a single purpose–single means approach to a multiple purpose–multiple means approach (White, 1969). For example, storage facilities are used in floodplain management, where they are usually referred to as flood control reservoirs and often serve other purposes, such as recreation. Storage facilities are also used in stormwater management, where they are often referred to as detention/retention facilities, and sometimes serve recreation, aesthetic, and other functions. As another example, public acquisition of flood-prone land is used in both floodplain management, where it typically takes the form of long and narrow riverine parkways, and in stormwater management, where it often takes the form of local neighborhood parks and recreation areas.

Identical Hydrologic–Hydraulic Processes and Engineering Tools. The underlying hydrologic–hydraulic processes and the engineering tools and techniques used to analyze the processes and to design remedial and preventive measures are the same. For example, the Manning equation can be used to compute flood stage profiles along the Mississippi River and depth of flow in a roadside swale.

Parts of the Same Watershed System. Small watersheds, typically of concern in stormwater management, contribute runoff to the riverine areas, typically of concern in floodplain management. More specifically, urbanization on the small watersheds can increase flood stages and floodplain width in the receiving major stream. Similarly, flood stages on the major streams can affect flood stages in tributary watersheds.

Diminishing Distinctions between Floodplain and Stormwater Management

1. Consider use of the terminology "surface water management" (SWM) to combine the best elements of floodplain management and stormwater management. This convention is adopted in this book.
2. Continue to adjust the National Flood Insurance Program so that there is more of a watershed orientation. For example, have one contractor or engineering consultant do all the FISs in a given watershed simultaneously. Furthermore, include the option of future condition floodplain delineation in FISs for use in floodplain regulations, particularly in small urbanizing watersheds.
3. Consider use of watershed-wide "flood flow mapping" (Debo and Lumb, 1977). That is, perform watershed-wide hydrologic analyses based on future land use to obtain design or regulatory flows throughout the watershed. Perform the more costly hydraulic analyses needed for floodplain mapping as needs dictate and funds are available.
4. Join with neighboring communities in the same watershed to finance and prepare SWM plans addressing both the quantity and quality of stormwater runoff. These programs are particularly important if communities already have flooding problems and anticipate additional urbanization. Allocate the cost of preparing the programs on the basis of factors such as land areas, imperviousness, population, or assessed value.
5. If a FIS is being done in the community by a consulting firm or government agency under contract to FEMA, the community should consider retaining the firm or agency to develop a SWM plan simultaneously or subsequently. The FIS should be seen as the beginning of the plan—not the end. The FIS alone, particularly the floodplain delineation aspect,

often identifies a community problem that previously was ill or poorly defined, without offering solutions.

6. Stormwater systems in newly developing areas should be designed in accordance with the emergency–convenience (major–minor) system concept discussed later in this chapter. With this approach, sewers convey runoff from frequent storm events (e.g., 2-year or 50 percent), and streets convey runoff from severe, infrequent runoff events (e.g., 100-year or 1 percent). Surface and subsurface detention facilities may often be readily integrated into the emergency–convenience system approach to achieve cost savings and multipurpose objectives.
7. Insist on state-of-the-art approaches. The traditional floodplain management and stormwater management fields have advanced significantly in the past few decades. Advances occurred in terms of analysis and design approaches and in terms of the availability of proven structural and nonstructural remedial and preventive measures. As communities move to a coordinated SWM approach, they have an opportunity to utilize and benefit from state-of-the-art tools and techniques.

In summary, the traditional distinctions between floodplain management and stormwater management seem to be more matters of scale and funding than of engineering substance.

1.5 STATISTICS

Within the SWM field, statistical analysis has two broad areas of application. One is use by professionals while working with other professionals in quantifying the magnitude of hydrologic–hydraulic–water quality events and quantifying the severity of flooding and pollution problems. Statistics provide a precise, shorthand means of communication within professional circles.

The second general use of statistical analysis is communication with, and education of, elected officials and the public at large. Fundamental statistical parameters can be used to explain the relative severity of floods to the public and to describe levels of protection provided by alternative management measures. For example, the public can understand the relative difference between a 1 percent and a 10 percent flood event and can use the difference to help decide if the marginal cost of going from 10 percent protection to 1 percent protection is worth the perceived marginal benefit.

A few selected concepts and equations used in the application of statistical analysis to urban SWM are presented in this section. Detailed treatments of statistical analysis are presented elsewhere (e.g., Bedient and Huber, 1988; Chow et al., 1988; Linsley et al., 1982; McCuen, 1979; U.S. Water Resources Council, 1981; Viessman et al., 1977).

Probability and Recurrence Interval Concepts

As noted earlier, flooding is a natural and certain phenomenon that is unpredictable only in that the exact time of occurrence of flood of a given magnitude cannot be predetermined. However, the probability of occurrence of a flood of given magnitude in a given year—annual probability—or given number of years is amenable to statistical analysis.

Such analysis requires a quantitative means of stating the severity of a flood. Probability is one means of quantifying flood severity. A 1 percent flood is a flood that has a 1 percent probability of being reached or exceeded in a given year. Similarly, a 10 percent flood has a 10 percent probability of being reached or exceeded in a given year.

Flood severity can also be expressed in terms of the recurrence interval stated in years. The recurrence interval (T) in years is related to probability (P) in percent by the following relationship:

$$T = \frac{1}{P} \times 100$$

For example, a 1 percent flood is one that will be reached or exceeded on the average of once every 100 years.

Casual use of the recurrence interval concept, in contrast with the probability concept, can lead to a serious misunderstanding by the public at large. For example, the recurrence interval approach often is incorrectly interpreted to mean that there is a fixed interval between floods of given magnitude. The public may incorrectly assume that if there has been a 100-year flood, another 100 years will pass before the next occurrence of a 100-year flood.

In contrast, the use of probability to define the magnitude of the flood tends to focus attention on the current, the next, or some other specified year. Thus the probability approach is less likely to be misunderstood and to portray the flood risk more accurately.

Risk Equations

The multiplication theorem of the statistics can be used to derive the equation

$$R = 1 - (1 - P)^N$$

where R is the risk that a flood with the annual probability P will be reached or exceeded at least once in N years (Linsley et al., 1982; Viessman et al., 1977). The relationship represented by this risk equation can be depicted graphically (McCuen, 1979). As illustrated in Table 1.1, the equation can be used to demonstrate to the public the risk associated with buying or building a home or other structure in a flood-prone area.

TABLE 1.1 Probability of Flooding as a Function of Vertical Position in the Floodplain

Position in Floodplain—First Floor Coincident With:	Percent Probability of First Floor or More Damage in Any Year	Percent Probability of First Floor or More Damage One or More Times before 25-Year Mortgage Is Paid
5-year recurrence interval flood stage	20.0	99.6
10-year recurrence interval flood stage	10.0	92.8
25-year recurrence interval flood stage	4.0	64.0
50-year recurrence interval flood stage	2.0	40.0
100-year recurrence interval flood stage	1.0	22.2

The multiplication theorem can also be used to derive the equation

$$R = \frac{(N!)\,(PI)\,(1 - P)^{N-I}}{I(N - I)!}$$

where R is the risk that flood with an annual probability P will be reached or exceeded exactly I times in N years. The equation is called the binomial expression (U.S. Water Resources Council, 1981). For example, to calculate the risk that a flood with an annual probability of 10 percent will occur exactly two times during the next 5 years, substitute $P = 0.1$, $I = 2$, and $N = 5$. This yields $R = 0.073$; therefore, there is a 7.3 percent risk that a 10 percent or 10-year flood will occur exactly two times during the next 5 years.

Numerous other special-purpose risk equations can be derived. For example, the multiplication and addition theorems of statistics can be used to derive the equation

$$R = 1 - (1 - P)^N - (NP)\,(1 - P)^{N-1}$$

where R is the risk that a flood with an annual probability P will be reached or exceeded two or more times in N years (U.S. Water Resources Council, 1981). For example, to determine the risk that a flood having an annual probability of 10 percent will be reached or exceeded two or more times during the next 5 years, substitute $P = 0.1$ and $N = 5$ in the equation and calculate R

= 0.081. Therefore, there is an 8.1 percent risk that a 10 percent or 10-year flood will occur two or more times in the next 5 years.

Responding to the Skeptics

Elected officials and the public are more likely to respond positively to the use of statistical analysis in making SWM decisions if they recently experienced a major flood. In the absence of such a flood, some elected officials and certain segments of the public may express skepticism and disbelief over the possibility of ever experiencing floods as large as the 2 percent (50-year recurrence interval) or 1 percent (100-year recurrence interval).

In response, it is possible to cite historic floods that have reached or exceeded such levels. For example, rainfall responsible for the June 9, 1972 flood in Rapid City, South Dakota, totaled 25 inches (in.) and included a 6-hour period in which 15 in. fell. The peak measured flow was 50,000 cubic feet per second (ft^3/sec) and was determined to have a recurrence interval of roughly 250 years (Swanson, 1982). Similarly, flows recorded at 10 gauging stations in the Kansas City, Missouri and Kansas, area were equal to or in excess of the 100-year recurrence interval discharge during the September 12–13, 1977 floods. The flood was caused by back-to-back record-setting rainstorms (Hauth et al., 1981). On August 6, 1986, 6.8 in. of rainfall occurred over a 24-hour period in the vicinity of the Kinnickinnic watershed in Milwaukee, Wisconsin, with 5.24 and 5.73 in. occurring over, respectively 2- and 3-hour segments of the storm. All portions of this rainfall event, from 30 minutes to 24 hours, were in excess of 100-year amounts. As a result, the 18.1-square mile (mi^2) Kinnickinnic River watershed generated a peak flow of about 10,000 ft^3/sec, estimated to have a 500-year recurrence interval [Southeastern Wisconsin Regional Planning Commission (SEWRPC), 1986].

1.6 MEANS BY WHICH FLOODWATER MAY ENTER A STRUCTURE

SWM objectives include protecting people and property. Basic to achieving this is knowing how floodwaters may enter a structure. Preventing direct damage to structures and their contents is a primary concern in urban areas.

As suggested by Figure 1.3, the typical urban structure is a virtual "sieve" with respect to the means by which floodwaters may enter. Various surface and subsurface points of entry exist. The number and variety of entry points increases with rising flood stages in proximity to a particular building. One typical and generally preferable variation on the structure presented in Figure 1.3 for illustration purposes is having downspouts discharge to the ground surface. Other variations include foundation drains connected to a storm sewer or connected to a sump from which water is pumped to the ground surface at some point away from the structure.

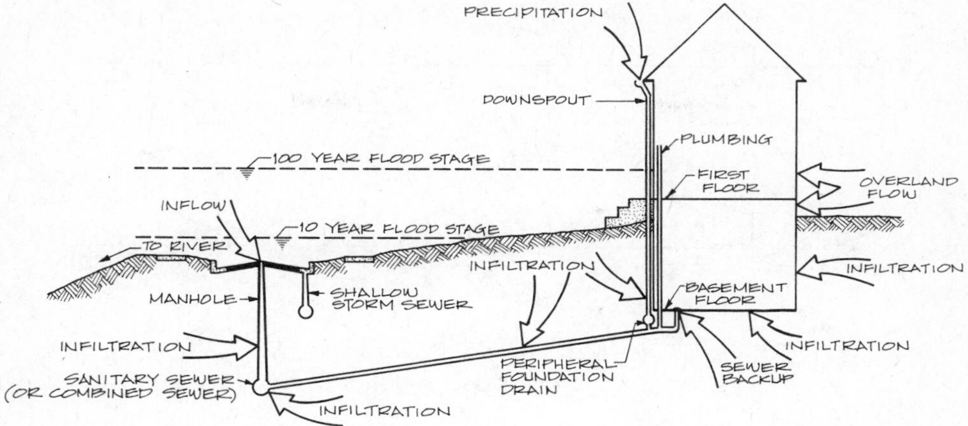

FIGURE 1.3 Means by which floodwaters may enter a structure (*Source:* Adapted from Southeastern Wisconsin Regional Planning Commission, 1976)

Consideration of the surface and subsurface means by which damaging floodwater can enter a structure gives rise to the concepts of a primary flooding zone and a secondary flooding zone. As illustrated in Figure 1.4, the primary flooding zone includes structures that incur damage because they are surrounded by or contiguous with areas experiencing surface flooding. In contrast, the secondary flooding zone, although outside the area of surface flooding, may still incur damage because of the hydraulic connections between the secondary flooding zone and the primary flooding zone. Examples of such hydraulic connections are sanitary and combined sewer systems. The structures in the secondary flooding zone have their lowest floors below flood stage and therefore are subject to flooding if there are subsurface hydraulic connections between the primary flooding zone and the secondary flood zone.

1.7 HISTORY OF URBAN DRAINAGE

Subsurface masonry storm drains were constructed at least as early as 3000 years ago. Extending well into the nineteenth century, sewers and drains in Europe were intended primarily for stormwater conveyance. There were no sanitary sewers; human and other wastes were deposited in courtyards and on streets. The discharge of fecal matter and other human wastes into storm sewers and drains was generally prohibited. With the realization that infectious diseases such as typhoid fever, dysentery, and cholera were waterborne, planning and construction of combined sewer systems began in large European and U.S. cities in the mid-nineteenth century. Construction of combined sewer systems subsequently ceased early in this century, primarily because surface

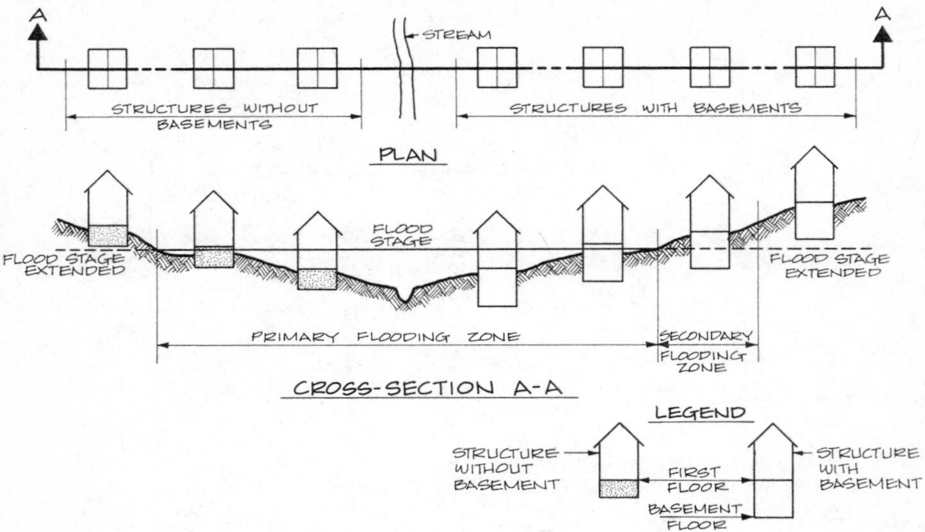

FIGURE 1.4 Primary and secondary flooding zones (*Source:* Adapted from Southeastern Wisconsin Regional Planning Commission, 1976)

water pollution attributed to combined sewer overflows (Fair and Geyer, 1963; Metcalf & Eddy, 1972). However, some of the original systems and associated pollution and basement flooding problems remain to this day. Planning, engineering, and construction are under way in many large cities to mitigate the combined sewer overflow problems.

Despite the provision of separate sewer systems and advances in controlling municipal and industrial point source pollution, surface water pollution problems remain. As a result, interest in control of pollution caused by strictly stormwater runoff grew in the 1970s and 1980s. The major impetus for the nonpoint-source pollution control was the 1972 amendments to the Federal Water Pollution Control Act, which led to the preparation of nonpoint-source pollution control plans for major metropolitan areas throughout the United States.

Detention/retention (D/R) facilities, possibly operating in series with sedimentation basins, are one means of managing nonpoint-source pollution. The historic development of D/R facilities is discussed later in this chapter.

In summary, the first sewers were for stormwater control, not for conveyance of sanitary wastes. Combined sewers first began to appear in the nineteenth century to control the spread of waterborne diseases. Because of the combined sewer overflow problem, separate sewers became the policy in the United States beginning early in the twentieth century. Now that the infectious disease problem is largely solved, there is growing concern with pollution of surface water by stormwater runoff.

1.8 HISTORY OF FLOODPLAIN MANAGEMENT

Native American and then European settlers tended to settle on or near floodplains. Creeks and rivers provided water supply, wastewater disposal, water power, fishing, transportation of people and goods, and strategic military advantages (Vollmer, 1970). Most of these factors that originally motivated development near rivers are no longer of major importance to cities because of modern technology. However, a significant portion of our nation's urban development remains flood-prone because those factors were once very important.

Urban lands cover approximately 3.5 percent of the area of the United States and 1 percent floodplains comprise about 7 percent of the total land. Whereas a small portion of the existing urban land is already located on floodplains, there is little validity to the argument that floodplains are needed for development because of land scarcity (Goddard, 1973).

Floodplain management has undergone a significant change in the United States during the twentieth century. Table 1.2 summarizes the three-phase evolution of floodplain management in the United States (White, 1969). The first

TABLE 1.2 Changing Strategies of Floodplain Management in the United States

	Phase		
	(1) Single Purpose–Single Means	(2) Single Purpose–Multiple Means	(3) Multiple Purpose–Multiple Means
Objective	Protect existing development and/or permit new development	Protect existing development and/or permit new development	Protect or remove existing development, and capitalize on recreational, scenic, ecologic, and cultural values of floodplains
Means of achieving	Structural	Structural and nonstructural	Structural and nonstructural
Milestones and/or examples	Miami Conservacy District—1913 Flood Control Act—1917 TVA—1933 Flood Control Act—1936	TVA—1950–1960s Federal Task Force—1966 ASCE Task Force—1962	Denver Urban Drainage and Flood Control District—1969 S.E. Wisconsin Regional Planning Commission Watershed Plans—1966

and longest period, extending from early in the century to about 1960, embodied what has been called the single purpose–single means approach. The single purpose was to protect existing floodplain development and facilitate additional floodplain development. The single means was construction of major facilities, such as dams, dikes, floodwalls, and channel improvements.

Milestone events during the single purpose–single means period were the 1913 founding of the Miami Conservancy District in Ohio, which marked one of the first times that floodplain management was done in a comprehensive fashion on a large (400-mi^2) basis. The Miami Conservancy District also was one of the first uses of reservoirs for flood control; they had previously been used for navigation, irrigation, and floating logs (Burges, 1979; Leuba, 1971). Another milestone was the U.S. Congress's 1917 authorization of levee construction on the lower Mississippi River (Thomas, 1979) and the 1933 formation of the Tennessee Valley Authority (TVA), which encompasses 41,000 mi^2 and was the first federal activity in multiple-purpose water resources planning on a watershed basis (Burges, 1979; Weathers, 1972). A final important milestone during the single purpose–single means period of floodplain management in the United States was the 1936 passage of the Flood Control Act by the U.S. Congress, which marked the beginning of a national flood control policy (Thomas, 1979).

Despite the efforts carried out under the largely single purpose–single means approach, over $9 billion had been spent on major river flood control as of the early 1960s and monetary damage continued to rise (U.S. House of Representatives, 1966; U.S. Water Resources Council, 1968). In response, the American Society of Civil Engineers established the Task Force on Floodplain Regulations, which in 1962 published guidelines for the development of floodplain regulations. These guidelines set forth a comprehensive approach recommending the application of nonstructural as well as structural measures to achieve the primary single-purpose objective of floodplain development [American Society of Civil Engineers, (ASCE), 1962].

In 1966, the Task Force on Federal Flood Control Policy recommended explicit consideration of nonstructural measures (U.S. House of Representatives, 1966). TVA had already begun a floodplain regulation program in 1953 and in 1960 began a program in which both structural and nonstructural measures were considered as part of their floodplain management activities (Weathers, 1972). So by the late 1960s, progressive floodplain management was employing a single purpose–multiple means approach. The single purpose continued to be protecting existing development of floodplains and permitting additional development. Multiple means included structural as well as nonstructural measures.

The third and final phase of floodplain management, referred to as the multiple purpose–multiple means approach, began in the 1970s. A collection of papers published in 1969 by Dougal in effect argued for a multiple purpose–multiple means without using that particular terminology. Multiple purposes included protecting existing development, discouraging new development, and emphasizing and capitalizing on the recreational, scenic, ecologic, and cultural values of floodplains (Walesh, 1973, 1976). Means of achieving the multiple objectives continued to be a combination of structural and nonstructural measures. Regional examples of the multiple purpose–multiple means approach include the watershed-oriented floodplain management efforts of the Denver area Urban Drainage and Flood Control District (Tucker and DeGroot, 1975) and the Southeastern Wisconsin Regional Planning Commission (e.g., SEWRPC, 1976). The multiple purpose–multiple means approach is now clearly the strategy of the federal government (Federal Emergency Management Agency, 1986).

In summary, today's riverine flood problems can be traced back to historic development needs. Although most of those needs are no longer valid, many flooding problems remain because of the original floodplain siting of communities. Floodplain management evolved from the original single purpose–single means philosophy early in this century into the present multiple purpose–multiple means approach.

1.9 COMPONENTS OF THE URBAN SURFACE WATER SYSTEM

An urban surface water system consists of various integrated components, each of which is intended to perform one or more functions in controlling the quantity and quality of stormwater runoff (ASCE, 1969). Figure 1.5 is a hypothetical urbanized watershed illustrating most components of the surface water control system, with the exception of the channel and floodplain, which are illustrated in Figure 1.1. System components shown in Figure 1.5 are listed in Table 1.3 together with a brief description of their function and additional comments.

One characteristic of the surface water system is that, for a variety of reasons, many of the components are not readily apparent to the casual observer. Some of the components, such as storm sewers and subsurface detention facilities, are underground and therefore out of sight. Some of the components, such as inlets and catch basins, are small and easily overlooked. Other components, such as carefully designed detention basins and conveyance swales, tend to conform with natural topography and are not readily apparent. To a large extent, components of the surface water system are visible or noticed only when they malfunction or are alleged to malfunction.

Another somewhat unique characteristic of the urban surface water system is that it functions infrequently, that is, immediately after rainfall or snowmelt events. This contrasts with most other municipal and private utilities and facilities, such as the sanitary sewerage system, the water supply system, the electri-

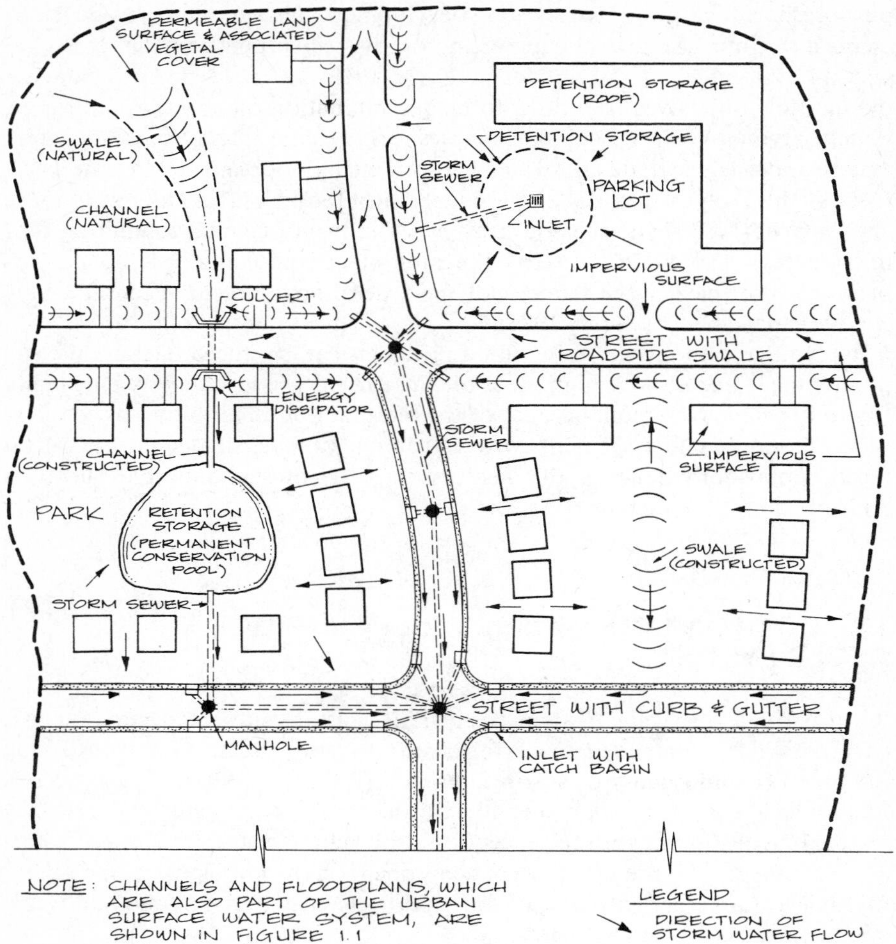

FIGURE 1.5 Components of an urban surface water system

cal system, the natural gas system, and streets and highways, all of which are in essentially continuous use or service.

1.10 APPROACHES TO CONTROL OF THE QUANTITY OF URBAN RUNOFF

The state of the art of surface water management has developed to the point where there are two fundamentally different approaches to controlling the quantity, and to some extent the quality, of stormwater runoff. Using a ''be-

TABLE 1.3 Components of an Urban Surface Water System

Component	Surface Water Control Function	Comment
Permeable land surface and associated vegetal cover	Permit interception and infiltration and provide for surface runoff.	Includes land surfaces in natural or rural areas on the fringes of urban or urbanizing areas as well as the permeable land surfaces—lawns, parks, and other open space—in urban and urbanizing areas.
Swale and open channel	Receive, concentrate, and transmit surface runoff from the land surface to other subsurface components of the stormwater system.	Swales and open channels may be natural, constructed, or a combination.
Culvert	Provide for the passage of stormwater flow beneath highways, streets, and private drives and driveways.	
Roadways with roadside ditches or curb and gutter	a. Provide, during minor runoff events, for the collection of surface runoff and its conveyance, through stormwater inlets, to subsurface conduits so as to minimize disruptive ponding on roadways. b. Provide, during major runoff events, for temporary storage of stormwater in and conveyance of stormwater through urban areas to avoid serious stormwater inundation problems.	
Parking lots, rooftops, and other impervious surfaces	a. Provide, during minor runoff events, for the collection and conveyance of stormwater	

(continued)

TABLE 1.3 (Continued)

Component	Surface Water Control Function	Comment
	through stormwater inlets and downspouts to surface and subsurface channels and conduits so as to minimize disruptive ponding. b. Provide, during major runoff events, for the collection, conveyance, and temporary storage of stormwater to minimize serious stormwater inundation problems.	
Inlet	Provide, primarily during minor runoff events, for transition between the surface and subsurface components of a stormwater control system.	There are three basic types of inlets: a. Gutter or grated inlets located in the gutter surface near a curb or in depressions in areas such as parking lots. b. Curb opening inlets located in the vertical face of a curb. c. Combination inlets used in a street or roadway having a curb and gutter.
Catch basin	Provide, by means of a sump, for the retention of sediment and other heavy debris that enters an inlet to prevent potentially troublesome sediment and debris accumulation in storm sewers and pollution of receiving waters.	
Storm sewer	a. Provide, during minor runoff events, for the collection and conveyance of stormwater to	

TABLE 1.3 (Continued)

Component	Surface Water Control Function	Comment
	minimize disruptive surface ponding in streets, parking lots, and other low areas. b. Provide, during major runoff events, for the collection and conveyance of a portion of stormwater so that, in combination with the storage and conveyance capability of surface components of the system, serious stormwater inundation problems will be avoided.	
Manhole	a. Provide for the intersection of storm sewers having different grade, elevation, direction, or size. b. Permit access to storm sewers for inspection, maintenance, repair, and system expansion.	
Detention facility	a. Provide, in a normally dry area or enclosure, for the temporary storage of stormwater runoff for subsequent slow release to downstream channels or storm sewers, thus minimizing disruption and damage in downstream areas during both minor and major events. b. Provide for the settling of sediment and other suspended mate-	A detention facility is normally dry, or contains very little water, and is designed to fill only during runoff events. Examples of detention storage include natural swales provided with crosswise earthern berms as control structures, excavated or constructed ponds, underground tanks, and rooftop storage. Detention storage areas are often multi-

(*continued*)

TABLE 1.3 (Continued)

Component	Surface Water Control Function	Comment
	rial in runoff, thus reducing the load of potential pollutants on receiving streams.	purpose facilities in that they can serve recreational functions and have aesthetic value.
Retention facility	a. Provide, in a reservoir that normally contains a substantial volume of water at a predetermined conservation pool level, for the temporary storage of additional stormwater runoff for subsequent slow release to downstream channels or storm sewers, thus minimizing disruption and damage in downstream areas during both minor and major runoff. b. Provide for the settling of sediment and other suspended material in runoff, thus reducing the load of potential pollutants on receiving streams.	The volume of water normally maintained in a retention storage facility serves recreational and aesthetic functions. Examples of retention storage reservoirs include permanent ponds in residential developments and in public park and open space areas.
Sedimentation basin	Trap suspended solids, suspended and buoyant debris, and adsorbed or absorbed potential pollutants which are carried by stormwater runoff. Sedimentation basins are usually designed for minor events, because overall, these events transport the largest quantity of solids, debris, and associated pollutants.	A sedimentation basin may be part of a multipurpose detention/retention facility, in which case it is usually located upstream of the detention/retention portion of the facility and is sometimes followed by a natural or developed wetland.

TABLE 1.3 (Continued)

Component	Surface Water Control Function	Comment
Energy dissipator	Prevent erosion of channel material and prevent structural damage such as undermining of paved channel inverts.	Examples of energy dissipators include drop structures along improved channels and stilling basins at culvert outlets.

Sources: Adapted from American Society of Civil Engineers (1969) and Urban Land Institute (1975).

fore and after'' format, Figure 1.6 illustrates selected characteristics of the two available approaches to surface water management.

Conveyance-Oriented Approach

The first of the two approaches is the more traditional conveyance-oriented surface water system. Systems designed in accordance with this approach provide for the collection of stormwater runoff, followed by the immediate and rapid conveyance of the stormwater from the collection area to the discharge point to minimize damage and disruption within the collection area. Principal components of conveyance-oriented stormwater systems are culverts, storm sewers, and channels supplemented with inlets and catch basins.

Storage-Oriented Approach

A potentially effective, less common approach to surface water control is the storage-oriented system. Its function is to provide for the temporary storage of stormwater runoff at or near the point of origin, with subsequent slow release to downstream storm sewers or channels. This approach minimizes damage and disruption both within and downstream of the site. One or more detention/retention facilities are the principal elements in a storage-oriented system. These principal elements are often supplemented with conveyance facilities, such as culverts, storm sewers, inlets, and catch basins.

Comparison of Features

Selected characteristics of conveyance-oriented and storage-oriented surface water control systems are presented for comparison purposes in Table 1.4. A

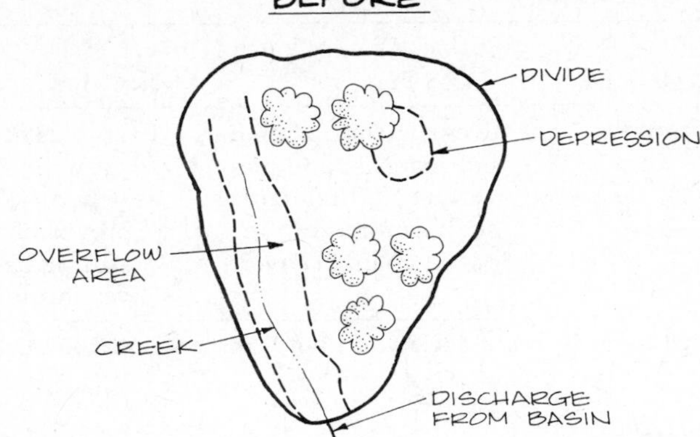

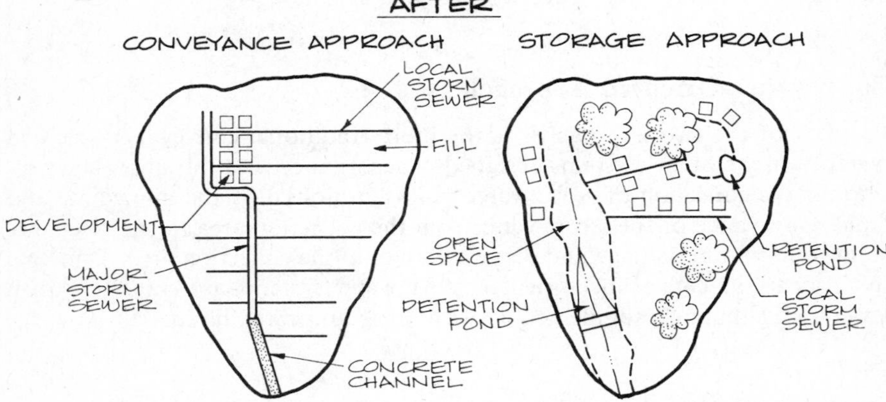

FIGURE 1.6 Conveyance and storage approaches to surface water management

principal advantage of the traditional conveyance-oriented approach is applicability to both existing and newly developing urban areas, contrasted with the storage-oriented approach, which is more difficult to retrofit into already developed areas. Other advantages of the conveyance-oriented approach are rapid removal of stormwater from the service area, minimal operation and maintenance requirements and costs, and accepted analysis and design procedures. Principal advantages of the storage-oriented approach are possible cost reductions in newly developing urban areas, prevention of downstream adverse flooding and pollution associated with stormwater runoff, and potential for multiple-purpose uses.

Neither the conveyance-oriented approach nor the storage-oriented approach is inherently better. Both approaches should be considered, at least when a project or development is at the conceptual level.

TABLE 1.4 Comparison of Conveyance and Storage Approaches

Characteristic or Feature	Approach: Conveyance-Oriented	Approach: Storage-Oriented
Function	Provide for the collection of stormwater runoff in the service areas and the rapid conveyance of stormwater from that area to minimize disruptive and possibly damaging surface ponding in streets and low-lying areas and possible inundation of residential and other sites and structures.	Provide for the temporary storage of stormwater runoff in the service area for subsequent slow release to downstream channels or storm sewers, thus minimizing disruption and damage within and downstream of the service area and reducing the required size and therefore cost of any constructed downstream conveyance facilities.
Components		
Principal	Open channels and storm sewers.	Surface or subsurface detention storage facilities and surface retention storage facilities.
Secondary	Inlets and catch basins. Culverts. Energy dissipators.	Open channels and storm sewers. Inlets and catch basins. Culverts. Energy dissipators. Inlet and outlet works and/or pumping facilities.
Applicability	Suitable for installation in existing and in newly developing urban areas.	Most suitable for incorporation in newly developing urban areas but may be used in existing urban areas if suitable surface or subsurface sites are available.
Downstream impact		
Quantity	Tends to significantly increase, relative to predevelopment conditions, downstream discharges, stages, and areas of inundation.	May be designed to cause no significant increase, relative to predevelopment conditions, in downstream discharges, stages, and areas of inundation. Decreased discharges, stages, and area of inundation are possible.

(*continued*)

TABLE 1.4 (Continued)

Characteristic or Feature	Approach	
	Conveyance-Oriented	Storage-Oriented
Quality	Transports suspended material and other potential pollutants to downstream areas.	Provides for removal, by the natural settling process, of sediment and other suspended material, thus reducing the pollution loading on receiving waters. Provide opportunity for physical–chemical treatment, such as filtration, disinfection, coagulation–flocculation, and swirl concentration.
Multipurpose capability	Storm sewers serve only a stormwater collection and conveyance function. Open channels can provide a basis for development of linear park and open space areas.	Quantity control. Quality control. Recreation. Aesthetic. Water supply. Groundwater recharge.
Operation and maintenance	Minimal periodic cleaning.	Pump and/or inlet–outlet control operation and maintenance. Sediment and debris removal. Weed and insect control.
Impact on sanitary sewer system	Surcharging of storm sewers accompanied by inundation of streets and roadways may result in infiltration of stormwater from storm sewers to adjacent sanitary sewers through manholes. Flow in excess of stormwater channel capacity may also result in surface inundation and inflow to sanitary sewers.	Runoff volumes in excess of available storage volume and runoff rates in excess of the capacity of tributary storm sewers and channels accompanied by inundation of streets and roadways may result in infiltration of stormwater from storm sewers to adjacent sanitary sewers and inflow of stormwater into sanitary sewers through manholes.
Hazards	Minimal hazard associated with storm sewers. High velocities in improved open channels may pose safety hazard, particularly to children.	Minimal hazard associated with subsurface and rooftop storage, but surface storage, particularly retention basins, may pose a safety hazard, especially to children.

TABLE 1.4 (Continued)

Characteristic or Feature	Approach	
	Conveyance-Oriented	Storage-Oriented
Hydrologic–hydraulic analysis and design procedure	Requires determination only of the peak rate of flow associated with a specified recurrence interval.	Requires determination of both a peak rate and a volume of inflow associated with a specified recurrence interval and an estimate of allowable outflow rate and design of pumps or control works to satisfy the discharge conditions.

Source: Adapted from Southeastern Wisconsin Regional Planning Commission (1976).

The conveyance- and storage-oriented approaches to surface water management are not necessarily mutually exclusive within the same hydrologic–hydraulic system. Depending on the circumstances, the two approaches may be compatible, and integrated use of the two approaches may lead to a more optimum surface water control system. One example of the joint use of the conveyance- and storage-oriented approaches is to rely primarily on conveyance-oriented facilities in one portion of a watershed and storage-oriented facilities in another portion. Another example of the combined use of the two approaches is to use conveyance-oriented facilities for the convenience system and storage-oriented facilities for the emergency system. The latter approach is illustrated by a rehabilitated surface water system in Skokie, Illinois (Walesh and Schoeffmann, 1984), where the preexisting combined sewers are the convenience system and new street surface detention and underground tank detention constitute the emergency system.

Historic Development of the Storage-Oriented Approach

The original motivation for using the newer storage-oriented approach over the traditional conveyance-oriented approach apparently was that the former offered cost advantages. Most documented examples of the cost advantage of the storage-oriented approach over the conveyance-oriented approach relate to newly developing areas (e.g., Poertner, 1974). More recently, however, there have been situations in which already developed areas are being retrofitted with a storage-oriented system at significantly less cost than that of a traditional conveyance-oriented system (e.g., Fujita, no date; Walesh and Schoeffmann, 1984).

A complete comparison of conveyance-oriented and storage-oriented systems for a particular location must consider other costs and benefits. For example, reduction in developable land with the storage-oriented system is a cost, and increased land values for areas contiguous to detention/retention facilities is a benefit. Cost analyses must be conducted on a case-by-case basis. Documented case studies and experience suggest that detention/retention facilities should be at least considered for controlling the quantity of stormwater runoff because of the potential for cost savings.

After initial use of detention/retention facilities for the single purpose of controlling the quantity of stormwater runoff, detention/retention facilities found increased use as multiple-purpose developments. In addition to their primary surface water control function, facilities were designed to provide, or be part of, sites for recreation including such activities as fishing, boating, tennis, jogging, ski touring, sledding, and field sports. Well-planned, well-designed, and well-operated detention/retention facilities were also found to have aesthetic value for contiguous and nearby residential areas.

In addition to the obvious erosion and sedimentation problems often associated with urbanization, it became increasingly apparent in the 1970s that urban stormwater runoff contributes a significant part of some of the pollutants finding their way to surface waters. For example, an early study conducted in Durham, North Carolina, compared the quality of urban runoff with that of secondary municipal sewage treatment effluent on the basis of weight per unit area per year (Colston, 1974). On an annual basis, the urban runoff contributed 91 percent of the chemical oxygen demand, 89 percent of ultimate biochemical oxygen demand, and 99 percent of the suspended solids. Many Public Law 92-500 208 studies also concluded that urban stormwater runoff was a major contributor of pollutants to surface waters.

It appeared as though controlling the quality of runoff—at least the first and most important increment of control—should focus on suspended solids, partly because erosion and sedimentation are problems in urbanizing areas. Furthermore, studies in the 1970s (e.g., Betz Environmental Engineers, 1976; Davis, 1979; Whipple, 1979) indicated that many pollutants, such as phosphorus, pesticides, heavy metals, and bacteria, are carried by soil particles. Therefore, successful erosion and sedimentation control was also likely to lead to significant control of other pollutants.

Many measures were suggested for controlling urban area nonpoint-source pollution in general and erosion and sedimentation in particular. The use of detention/retention was one of these measures. The state of the art of using detention/retention facilities to control the quality of urban stormwater runoff is still in its infancy.

In summary, detention/retention facilities are being increasingly used for controlling the quantity of runoff because of the cost advantages and because of their recreation and aesthetic values. They are also being increasingly designed to accomplish a third function of controlling the quality of water (e.g., Raasch, 1982, Urbonas and Roesner, 1986).

The evolution of using the storage-oriented approach in surface water management is summarized in Figure 1.7. Beginning with the single quantity control function, detention/retention facilities have evolved so that they now can

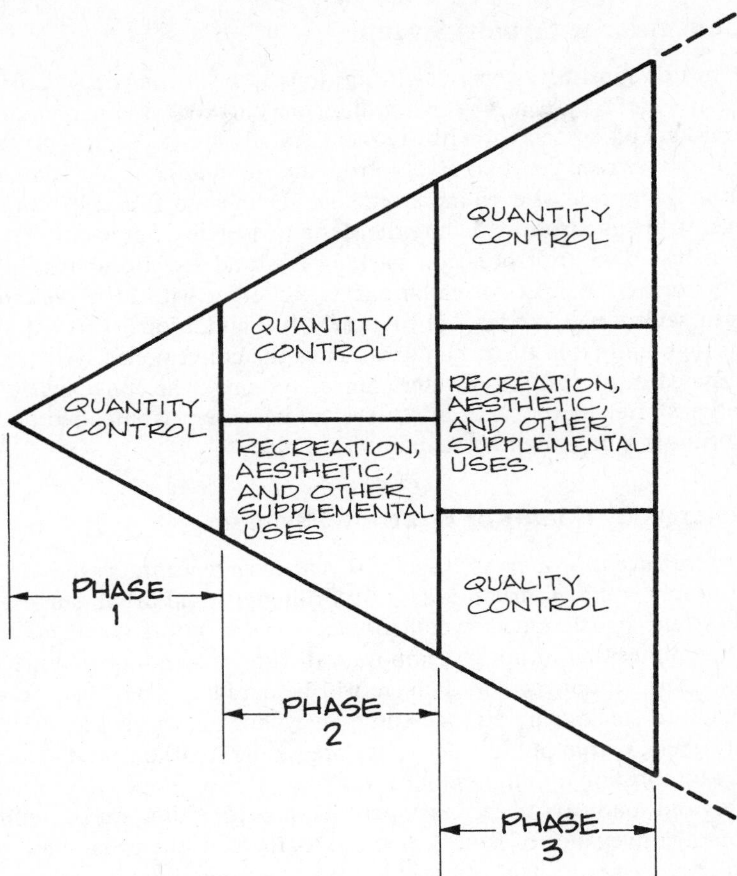

FIGURE 1.7 Historic development and use of detention/retention facilities

serve three compatible functions: quantity control; recreation, aesthetic, and other supplemental uses; and quality control (e.g., McCuen et al., 1983).

1.11 EMERGENCY AND CONVENIENCE SYSTEMS

The surface water control system may be thought of as two systems, one functionally and physically superimposed on the other. One system, the convenience or "minor" system, contains components that accommodate frequent, small runoff events. The other system, the emergency or "major" or overflow system, consists of components that control infrequent but major runoff events. Although many of the components are common to both the convenience and emergency systems, their relative importance in the two systems varies significantly.

The Convenience (Minor) System

Surface water control systems have traditionally been designed to convey all the design runoff without street flooding, parking lot or other ponding, or basement backup associated with frequent, small runoff events—up to about the 5- or 10-year recurrence interval—from an urban area with no damage and little or no disruption or even inconvenience. Peak runoff rates expected from such events are incorporated into the design. System components are sized such that water will not pond on parking lots and streets, almost all water reaching stormwater inlets or catch basins will enter with little ponding, and the storm sewers will, under full or nearly full conditions, carry the design flow through and from the urban area. Thus the convenience system is explicitly designed into most stormwater control systems. The convenience system is sometimes referred to as the minor system because it is intended to accommodate runoff from small or ''minor'' rainfalls.

The Emergency (Major or Overflow) System

Major runoff events—such as 50- or 100-year recurrence interval events—will also inevitably occur in urban areas. Accordingly, some urban surface water control systems are designed to control major event runoff rates and volumes in such a manner that although temporary disruptions and inconvenience will occur, widespread danger and damage will be avoided. This is accomplished by allowing for temporary storage and conveyance of stormwater on parking lots and streets, within public open space areas, and in other suitable low-lying areas; by establishing building grades well above street grades; and by designing streets and roadways to serve as open channels providing for the temporary storage and conveyance of runoff as it moves through the urban area toward a safe discharge point. The emergency system is sometimes called the major system because it is designed to control runoff from ''major'' rainfall events. Sometimes the emergency system is referred to as the overflow system because it is the system that begins to function when the capacity of the convenience system is exceeded and it overflows.

Most surface water control systems, however, are not explicitly designed to accommodate major runoff events. Nevertheless, major runoff events occur and the emergency system will, by default, function during such events with sometimes catastrophic damage and disruption.

Combined Convenience and Emergency System

The ideal surface water system is planned and designed to include both the emergency and convenience systems in anticipation of the inevitable occurrence of both major and minor runoff events. In a combination system, essentially complete control of minor runoff events is achieved to minimize disruption and damage during smaller, frequently occurring rainfall events.

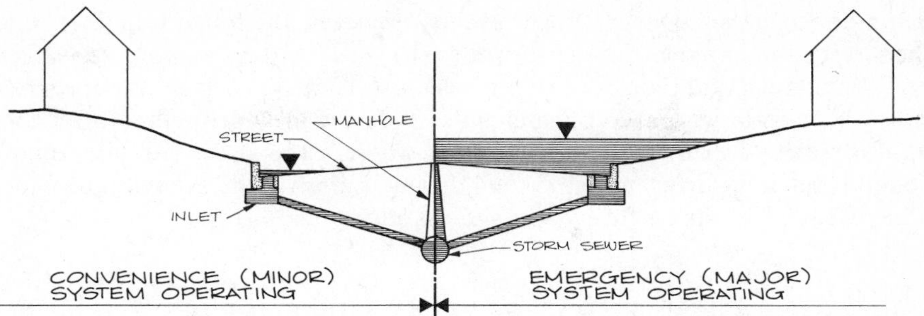

FIGURE 1.8 Emergency and convenience system along an urban street

Emergency components of the system are designed to accept some temporary disruption and inconvenience during relatively infrequent events, but to prevent widespread damage and disruption during such events. (Jones, 1967; Theil, 1977; Urban Land Institute, 1975).

Figure 1.8 illustrates the emergency and convenience system concept applied to a typical urban street cross section. Similarly, Figure 1.9 illustrates the emergency and convenience system applied to a channel–floodplain passing through an urban area.

Components in the surface water system can be examined from the perspective of whether or not they function, how they function, and the relative importance of their functioning under both convenience and emergency conditions. Consider, for example, stormwater inlets located along the curbs and gutters of city streets. For minor runoff events, such inlets are normally designed to pass essentially all the discharge conveyed to them, but under major events should be expected to intercept only a small portion of the flow moving along the gutter. Thus, whereas inlets are key elements in a convenience system, they are of little importance in an emergency system.

In contrast, streets are graded longitudinally and laterally to provide, during minor runoff events, for rapid runoff of stormwater to curbs and inlets or

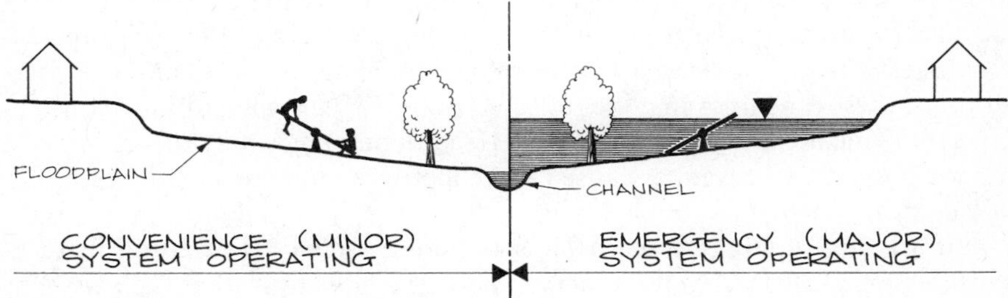

FIGURE 1.9 Emergency and convenience system along a channel and floodplain

to roadside ditches. During major events, however, the longitudinal slope of the streets and the relative elevation of the streets and contiguous residences and commercial and industrial structures must be designed such that the street functions as a large, paved open channel or reservoir which temporarily conveys or stores stormwater runoff. Thus, whereas the street is one of many components in a surface water system during minor runoff events, it becomes a key element in the surface water system during major events.

1.12 ROLE OF NATURAL FEATURES IN SURFACE WATER MANAGEMENT

Most of the components of the typical urban surface water control system are not natural features of the landscape; that is, they are constructed during the urban development process. Examples of constructed components are culverts, inlets, catch basins, detention/retention facilities, concrete-lined channels, and storm sewers.

Some of the components of the urban surface water system are natural or represent minor modifications to natural features. Examples include swales and open channels, ponds and low-lying areas, and the overall spatial orientation of the surface water system.

One potential advantage of retaining some of the natural features for incorporation into the surface water system of a newly developing area is minimizing the capital cost of the system. Considering just capital expenditures for system components, for example, it may cost less to retain a natural channel and swale as part of the surface water system than to construct a lined channel or storm sewer along the alignment of the natural drainageway. Similarly, modifying an existing low-lying area to serve as a detention/retention facility is likely to cost less than excavating and constructing a new storage facility. There are, of course, other costs to be considered in the planning and design of urban development, including the revenues that might be lost when there are fewer building sites to sell as a result of not being able to develop those areas containing natural swales and natural depressions.

Another advantage to retaining some of the natural features and incorporating them in a surface water system is the opportunity to preserve open space and natural areas in the urban environment. Such areas often have recreational, aesthetic, ecological, and cultural values (Walesh, 1973, 1976). Preservation of these amenities may be reflected in the higher value of land contiguous with or near the open space area. Higher land values may offset, at least in part, the reduced revenues caused by reduction in the areal extent of buildable land.

A third and final reason to consider retaining some of the natural features is protection of surface water quality. There are indications that the nature of the urban surface water control system may significantly influence the quality of streams, ponds, and lakes. For example, pipe conveyance systems have been shown to deliver significantly more suspended solids, total phosphorus, and lead to surface waters than do systems draining similar residential areas but relying primarily on natural or natural-like turf-lined swales (Athayde et al., 1983; Wisconsin Department of Natural Resources, 1978b).

In summary, in the planning and design of urban development in general, and the supporting surface water control system in particular, it is prudent to consider the natural surface water system features as the point of departure. Possible revenue loss associated with reductions in buildable areas as a result of incorporation of some of the natural land features must be considered. Some of these revenue losses, however, may be offset by the surface water system cost savings, possible land value enhancement as a result of proximity to or availability of open space and recreational areas, and water quality protection.

1.13 CHANNEL EROSION AND SEDIMENTATION IN URBANIZING AREAS

The Equilibrium Concept

Each stream seeks a state of macroequilibrium with the flow regime it carries, that is, a condition of relatively constant grade and length. Complete or microequilibrium does not exist in that a stream is continuously exhibiting small localized changes in channel location, width, depth, and grade (Kinori and Mevorach, 1984; Linder, 1976; Linsley et al., 1982; Petersen, 1986).

A stream can be put in a state of disequilibrium by any of several external changes. These changes are usually brought about by urbanization. An exception is the disequilibrium that may occur as a result of a major flood event or severe drought.

Some development-related changes are made directly to streams, for example, shortening a channel by cutting across meanders, widening or deepening a channel to increase its flood-carrying capacity, lowering or raising a channel at the point where it is tributary to another channel or flows into a lake or reservoir, and construction of detention/retention facilities. Other changes, such as various forms of urban development and the attendant increase in impervious surface, are made directly on or to the tributary watershed. These watershed modifications typically increase total runoff volumes, increase peak runoff rates, and diminish low flows which place the stream in a state of disequilibrium.

Years or even a few decades may be required before a new state of equilibrium is achieved. During that time, and as a result of the initial disequilibrium, the stream undergoes many, sometimes very significant changes. Direct effects of disequilibrium include changes in channel width and depth, an increase or decrease in the extent of meandering and of braiding patterns, and channel bottom erosion or sedimentation. Indirect or secondary effects include filling in or undercutting in or near culverts, bridges, and other hydraulic structures; undercutting of adjacent roadways, buildings, sewers, water lines, and other utilities; tipping and falling trees; gradual loss of public and private land; and unsightly erosion and sedimentation.

The channel disequilibrium phenomenon and resulting adverse effects is analogous to the coastal disequilibrium phenomenon and resulting effects. In the case of a coastal environment, disequilibrium may be caused by the construction of a pier or offshore breakwater to protect a harbor from waves and sedimentation. Sedimentation and beach expansion usually occur on the up-current side of the pier or breakwater, and beach erosion occurs on the down-current side.

Effects of Channel Shortening

Yearke (1971) reports on the results of shortening a New Hampshire river by replacing a meander loop with a section of new channel. At a point in the watershed where the drainage area was 38 mi^2, an approximately 3000-ft-long, 50- to 75-ft-wide channel having a slope of 52 ft per mile was shortened 850 ft and straightened.

As a result of the disequilibrium, severe erosion began to occur at the upstream end of the modified reach (18 ft of vertical drop in 4 years) and immediately upstream of the modified reach. Sedimentation occurred downstream of the modified reach. These erosion and sedimentation effects are illustrated in Figure 1.10. Major problems caused by the stream's reaction to the disequilibrium were unsightly erosion and deposition.

One way of explaining how the disequilibrium brought about by the channel shortening caused erosion is to consider stream velocity. The initial shortening

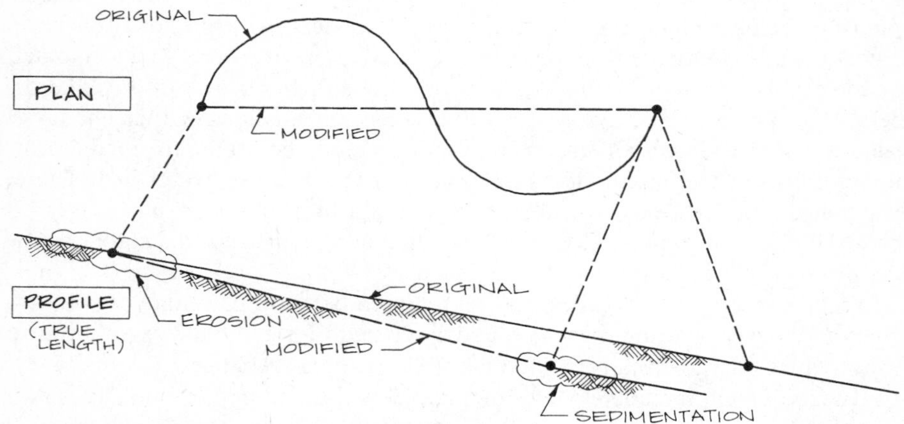

FIGURE 1.10 Erosion and sedimentation effects caused by channel shortening

of the channel increased its slope and thus stream velocity, which increased the stream's capacity to transport sediment (Linder, 1976).

Effects of Channel Widening

Linder (1976) describes the progression of channel adjustment processes that typically occur when a channel is put into disequilibrium by widening to increase its flood flow carrying capacity. The effects, most of which are adverse, are illustrated in Figure 1.11.

The enlarged cross section at the upstream end of the widened reach causes an abrupt decrease in stream velocity. This induces sedimentation in the reach, with the greatest deposition occurring near the upstream end. Gradually, the stream develops a narrower, meandering channel through the deposits.

The enlarged cross section produces a drawdown curve effect which begins at the upstream end of the widened reach and extends upstream. Resulting increased velocities cause erosion of the upstream natural channel, and the erosion progresses upstream.

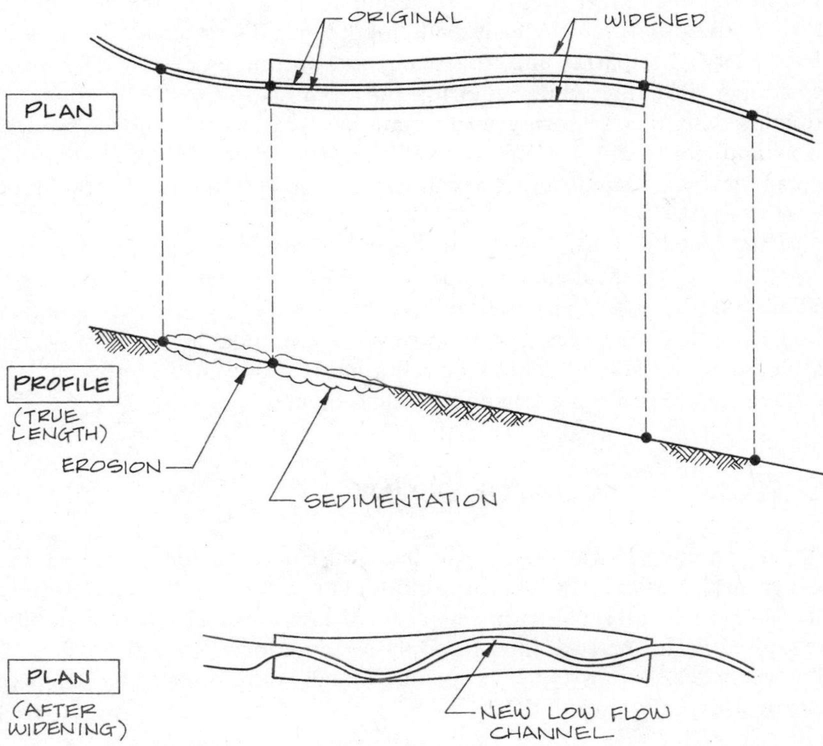

FIGURE 1.11 Erosion and sedimentation effects caused by channel widening

Effects of Watershed Urbanization

Robinson (1976) studied eight watersheds in the Baltimore, Maryland–Washington, D.C., area, each having drainage areas of about 1 mi^2. Three watersheds were rural, two were being urbanized, and three were urbanized.

Many effects of urbanization on channel morphology were observed. The urban streams had channel cross-section areas approximately twice those of rural streams, and width–depth ratios about 1.7 times those of rural streams. The urban channels had coarser channel bottom material than rural streams, that is, less silt and sand and more cobbles.

In the urban streams, a smaller portion of the channel width carried flow during low-flow periods than in the rural streams. Also, as would be expected, the urban stream channels contained a greater accumulation of debris. Most of the preceding urbanization effects are illustrated in Figure 1.12.

Preventive and Remedial Measures

Some undesirable channel changes can be prevented by planning and designing special features or structures prior to development. Detention facilities can be used to maintain predevelopment flow regimes. Composite channels may be used to simulate natural channels and floodplains. Drop structures provide localized energy dissipation and control channel bottom grades. Widely spaced levees control flooding while requiring no modification to the channel and floodplains. Channel armoring using materials such as reinforced concrete, gabions, and riprap can prevent erosion. High-flow cutoff channels or floodways can be constructed across meanders to carry flood flows at reduced stages (Linder, 1976).

Once channel and erosion problems have developed, a remedial or corrective approach may be taken. The specific techniques that may be used in a remedial mode are essentially the same as those available for use in a preventive mode. For a given stream reach, however, the number and type of options available at this stage will usually be less than if a preventive approach had been taken, and costs will probably be much higher.

1.14 FLOODPLAIN REGULATIONS

Floodplain regulations can be an effective nonstructural floodplain management measure, particularly where the floodplain has not yet been developed for urban or other flood vulnerable uses. As of 1980, over 17,000 communities in the United States had adopted floodplain regulations (Kusler, 1982).

Floodplain regulations typically include one or more of the following regulatory measures (Evenson, 1979):

1. A zoning ordinance that regulates the use of private land in the public interest. This approach uses districts or overlay provisions to regulate

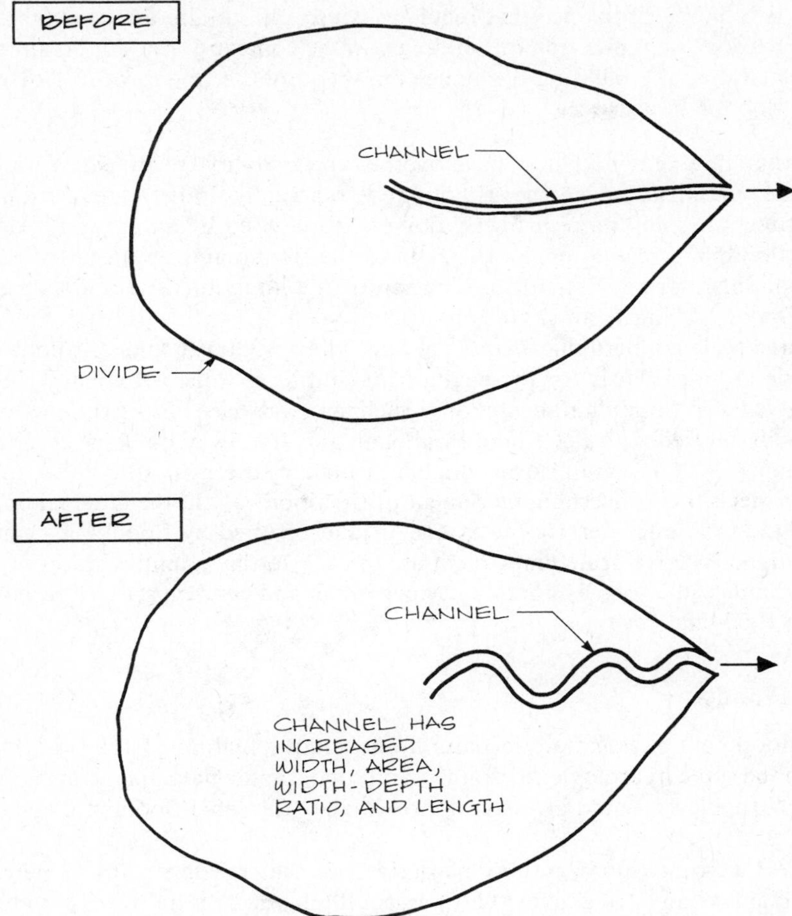

FIGURE 1.12 Channel changes caused by watershed urbanization

land use, type of structures, and placement of structures. For example, the zoning ordinance may establish floodway and floodplain fringe districts overlaid to other districts, such as residential, commercial, and industrial.

2. A land division ordinance that regulates the division, conveyance, and marketing of land. In the context of floodplain regulation, a land division ordinance may require dedication of the floodplain for public park use.
3. A sanitary or health ordinance that regulates facilities such as on-site water supply and septic tank systems. With respect to floodplain regulation, the sanitary or health ordinance may prohibit septic tanks in the floodplains.

4. A building ordinance that regulates construction, alteration, repair, extension, or conversion of buildings. In the context of the floodplain regulations, the building ordinance may require the floodproofing of new floodplain structures.

Although an engineer may have overall responsibility for preparing floodplain regulations, the engineer is not likely to actually draft the regulations—that function being more properly done or supervised by an attorney. However, floodplain regulations must relate to the floodplain, the floodway, and the floodplain fringe. The determination of the limits of the floodplain and floodway is primarily an engineering function.

Large-scale topographic maps, such as 1 in. = 200 ft and 2-ft contour, provide the ideal basis for the engineering and legal work culminating in an effective set of floodplain regulations. Such maps are used during the engineering analysis needed to determine floodplain and floodway limits as described in Chapter 4. They are also invaluable in making the hydraulic and nonhydraulic decisions concerning placement of the floodway. Large-scale maps enable landowners to identify the extent of land affected by floodplain regulations. Finally, large-scale maps facilitate the day-to-day administration of the floodplain regulations by community personnel and can be useful in responding to legal challenges.

The Floodway

The floodplain is essentially a natural and unique feature of the landscape. Given the same hydrologic, hydraulic, and topographic data and information, various engineers would arrive at approximately the same floodplain delineation.

The 1 percent floodway may be defined as that portion of the 1 percent floodplain required to convey the 1 percent discharge "safely". The floodway includes all of the channel plus that portion of the floodplain not suited, because of hazard to life and property, for human habitation. Figure 1.13 is a schematic representation of the floodplain, floodway, and floodplain fringe. Although floodway determination is based on engineering considerations, nontechnical factors enter into the establishment of floodway limits. Therefore, many floodway configurations are possible for a given river reach. Floodplain delineation and floodway determination are discussed in Chapter 4.

Use of the Floodplain and Floodway in Floodplain Regulations

Using the floodplain and floodway limits, the floodplain area is typically partitioned into a floodway district and a floodplain fringe district, as illustrated in Figure 1.13. Permitted uses in the floodway typically include agriculture, park and open space, restricted parking, and storage. Within the floodplain

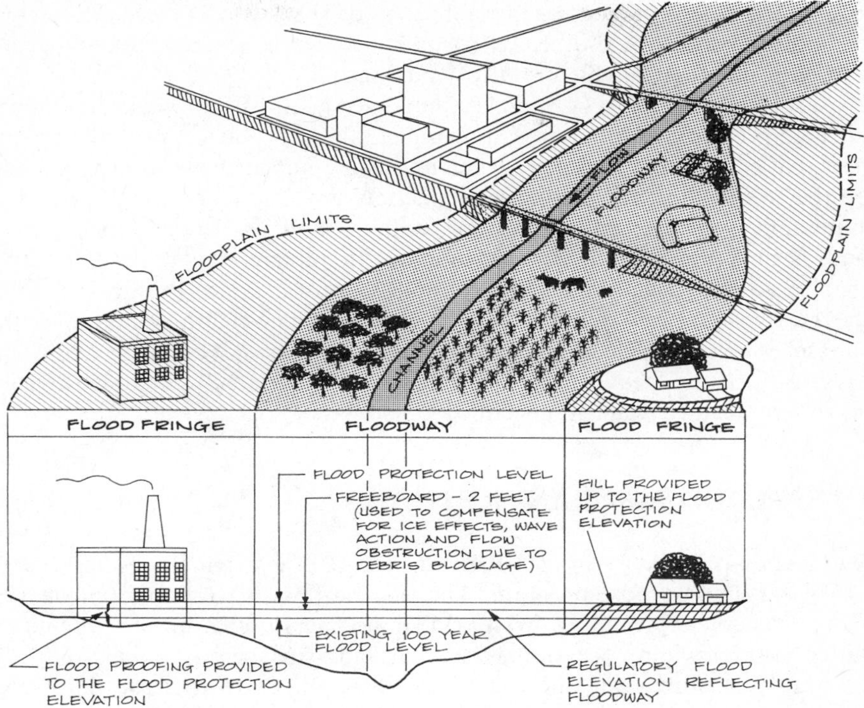

FIGURE 1.13 Floodplain, floodway and floodplain fringe

fringe, almost any use is permitted provided that it is flood protected and otherwise meets the provisions of underlying zoning districts, such as residential, commercial, and conservancy.

Another option is to use a single floodplain district. This approach has the advantage of not requiring a floodway determination and tends to discourage floodplain development. The single floodplain district may be very well suited for watersheds in which there is concern over the hydrologic effects of developing floodplain fringes. However, the single floodplain district approach may be cumbersome to administer if a hydraulic effects analysis is required for each proposed structure or development in the floodplain.

Floodplain Regulations and the Courts

A survey (Kusler, 1982) of legal tests of floodplain regulations concludes that "judicial support for floodplain regulations was overwhelming" during the period 1970 to 1980. More specifically, the survey found that when legal problems occurred, they tended to be procedural, not conceptual or substantive. Assuming that a regulating government unit or agency has sufficient statutory power, the courts judge the reasonableness of regulations on: acceptable use of police power, providing equal treatment to similarly situated landowners, and permitting reasonable use of private land.

According to the survey, judicially endorsed goals of floodplain regulations include: preventing land uses that aggravate flood problems; prohibiting land uses that cause other nuisances, such as placing unprotected hazardous waste on the floodplains; preventing victimization and fraud; reducing the cost of community services; and promoting suitable land use. Regulating agencies have been supported in their right to adopt floodplain regulations more strict than those established under the National Flood Insurance Program. Finally, some courts have said that private landowners have no right to damage natural suitability or capability of lands. Refer to Sheaffer et al. (1982) for additional discussion of the legal aspects of floodplain regulations, as well as the legal aspects of surface water in general.

1.15 FLOOD INSURANCE STUDIES

The National Flood Insurance Program (NFIP), as originally established in 1968, is a dual-purpose program. The first purpose is to spread out or level federal emergency funding by subsidizing some flood insurance premiums rather than providing occasional, irregular, and site-specific massive aid. The second, more-long-term purpose of the program, is to increase the use of floodplain regulations throughout the United States. The NFIP is, in effect, a form of federal land use control in floodplains.

Historic Overview

The federal government established the NFIP because of growing (on an average annual basis) monetary flood losses. In addition, the private insurance industry has traditionally considered flood insurance based on actuarially determined premiums to be too costly to property owners and therefore not a feasible business venture.

The NFIP began as the National Flood Insurance Act of 1968. The program was under the direction of the U.S. Housing and Urban Development (HUD) and made federally subsidized flood insurance available to property owners if a community made a commitment to floodplain management. Upon request of a community, HUD established a rate map and made insurance available—some at subsidized and some at actuarially determined rates—through local insurance agents. This optional program failed. For example, only 29 policies were sold in Rapid City, South Dakota, prior to the 1972 flood in which 236 lives were lost and for which $100 million of federal assistance was provided (GSA, 1978). The probable reason for failure of the initial stage of the program was lack of incentive for participation, other than the possibility of monetary compensation for flood damage. Property owners apparently did not appreciate the monetary risk associated with floodplain sites.

The program was modified with the Flood Disaster Protection Act of 1973. The modified program was similar to the initial program, with two important exceptions. First, limits of insurance coverage were increased. Second, the community was required to participate if any federal financial assistance was to be provided in the event of flooding, and property owners were required to participate if they obtained a mortgage from a federally insured lending institution.

Amendments to the Flood Disaster Protection Act in 1977 softened federal sanctions imposed on a nonparticipating community. For example, a federally insured lending institution could grant a mortgage on a floodplain structure in a community not participating in the NFIP provided that the lending institution informed the buyer of three things: (1) the structure was in the floodplain, (2) flood insurance was not available, and (3) the owner would not be eligible for federal aid if the area were declared a federal disaster area. In addition, federal relief and rehabilitation funds were not to be provided to a community that did not participate in the NFIP.

Coastal zone provisions of the NFIP in 1981 designated certain coastal zones as V, for "high-velocity wave zones." These are 100-year-recurrence-interval flood-prone areas. If a property owner wants to build or substantially improve a structure in a V zone and qualify for flood insurance, plans must be approved by a registered engineer or architect. This change in the program was motivated by the realization that the NFIP was paying too much in claims on coastal zone structures (ASCE, 1981).

The program was modified again in 1983 to reduce insurance coverage on basements. More specifically, flood insurance no longer covered improvements in basements. Examples of uninsurable items are paneling and carpeting. Insured items include sump pumps, oil tanks, furnaces, and freezers. In addition, damage caused by sewer backup was no longer covered unless the building was simultaneously flooded by surface water (Illinois Department of Transportation, 1984).

Emergency and Regular Programs

A community enters the emergency phase of the NFIP by applying to the Federal Insurance Administration (FIA) of the Federal Emergency Management Agency (FEMA) and by documenting the existence of adequate land use control measures or by adopting acceptable measures. During the emergency phase, the FIA conducts, at no cost to the community, a flood insurance study that culminates in a flood insurance report described later. Once the report and accompanying maps are accepted by the community and the FIA, the community is moved into the regular phase of the NFIP. The regular phase of the program remains in effect unless the community fails to enforce land use regulations (Sheaffer et al., 1982).

Available Coverage

Table 1.5 summarizes the type of insurance coverage available and the premiums charged. Emergency program coverage and rates are available for buildings and their contents existing before the community enters the regular program. Under the regular program, actuarially determined rates apply for additional coverage on buildings eligible under the emergency program and for all coverage on other buildings (Illinois Department of Transportation, 1984; Sheaffer et al., 1982).

There are about 2 million policyholders in 17,500 communities in the United States, with a total of about $100 billion of insurance in force. There are an estimated 6 to 8 million people living in flood-prone areas and, therefore, 4 to 6 million are living in uninsured buildings (Association of State Floodplain Managers, 1983; Mendenhall, 1986).

Flood Insurance Report

Flood insurance studies (FIS) are prepared by consultants or government agencies under contract to the FIA. The typical FIS report describes the area studied and briefly discusses hydrologic–hydraulic methods. The report includes 10-, 50-, 100-, and 500-year-recurrence-interval flood stage profiles, floodplain maps showing the 100- and 500-year-recurrence-interval floodplain limits, and a 100-year-recurrence-interval floodway. The 100- and 500-year-recurrence-interval floodplain limits and the floodway are shown on the Flood Boundary and Floodway Map (FBFW). Also included is the Flood Insurance Rate Map (FIRM), which defines flood hazard areas and indicates relative risks.

The two principal uses of an FIS report correspond to the two principal purposes of the NFIP. First, local insurance agents indirectly use FIS report data to establish the insurance premium per $100 of protection based on horizontal and vertical location of a property and building type. Second, the community uses 100-year-recurrence floodplain and floodway data to develop, adopt, and implement floodplain regulations.

Features of the National Flood Insurance Program

An important positive feature of the NFIP is the mandatory floodplain regulations. These regulations provide the basis for land use controls in riverine areas. Experience indicates that such land use controls are a very effective preventive measure. Another positive feature of the NFIP is the certainty of postflood financial assistance via insurance benefits, contrasted with the earlier uncertainties associated with the vagaries of federal disaster assistance. A

TABLE 1.5 Flood Insurance Coverage Available under the National Flood Insurance Program as of 1986[a]

	Coverage Available	
Coverage On:	Emergency Program[b]	Regular Program[c]
Building		
Single family	$ 35,000	$185,000
Other residential	100,000	250,000
Small business	100,000	250,000
Churches and other properties	100,000	200,000
Contents		
Residential	10,000	60,000
Small business	100,000	300,000
Churches and other properties	100,000	200,000

Source: Adapted from Mendenhall (1986); National Archives (1986).

[a]Flood insurance is available in Alaska, Hawaii, Guam, and the U.S. Virgin Islands, but protection limits on buildings are different.

[b]The emergency program is for buildings constructed before a community enters the regular flood insurance program.

[c]Total coverage available, that is, includes emergency program coverage.

third important positive feature is that flood insurance provides an equitable transition for property owners caught in the changing approach to floodplain management. For example, the homeowner who purchases floodplain property prior to delineation of floodplains and the imposition of floodplain regulations is eligible for a federal subsidy on the basic insurance coverage on the building and contents.

A potential negative feature of the NFIP is that delineation of a floodplain and a floodway may be viewed as an invitation to fill and develop the floodplain fringe. In certain riverine areas, filling of the fringe may result in degradation and destruction of natural resource amenities; increases in downstream flood discharges, stages, and damage; and an implicit commitment to costly future channel improvements. Another negative aspect of the NFIP is its civil division rather than watershed orientation. The civil division orientation gives rise to redundant and therefore costly engineering work and the possibility of inconsistent hydrologic–hydraulic data for identical points on a stream system. Another potential problem with the NFIP is the use of existing-watershed-condition hydrologic–hydraulic data for floodplain regulation purposes. This may result in residential, commercial, and industrial structures becoming flood-prone in the future even though they are constructed today in conformance with floodplain regulations reflecting existing flood hazards (Walesh, 1979).

Subrogation

A condition of a community's participation in the NFIP is adoption and enforcement of floodplain regulations. Serious enforcement-related problems have and will continue to arise as the federal government administers the NFIP. The FIA may suspend a community from the program if the community fails to adopt or enforce floodplain regulations.

The federal government, as insurer, has the option of invoking subrogation. This means that the federal government pays flood insurance claims and then seeks to recover funds from the individuals or organizations responsible for the flood damage. Defendants in subrogation actions may include engineering or surveying firms that allegedly incorrectly certified that buildings were placed at proper elevations with respect to the 100-year-recurrence-interval flood level. Defendants may also include organizations that have placed fill and development in floodplain areas contrary to local regulations (Sokolove, 1982).

1.16 SURFACE WATER MANAGEMENT IN GEOLOGICALLY UNIQUE AREAS

Certain areas of the United States, by virtue of unique geologic and other conditions, have unusual surface water management challenges. These unique locations include arid areas and areas of karst topography, each of which is discussed briefly here. Much of the material presented in this book is applicable to such areas. However, surface water management in certain areas will also require efforts beyond the scope of this book.

Arid Areas

Surface water management in arid areas must recognize special conditions. A study (James et al., 1980) of flooding hazards in Utah, and by implication, other arid areas indicates that cloud-burst floods dominate in upland areas, and one consequence is rapid rise of floodwater and high velocities. Large volumes of sediment and debris accompany floodwaters. As a result, blockage of hydraulic structures and channels and shallow flooding are common. Erosion and sedimentation problems are aggravated by development. Most damage tends to be incurred by agricultural land and public facilities—little residential and commercial damage occurs.

Floodplain mapping is very difficult in alluvial fans and other flat areas because of occasional shifting of the stream channels and resulting changes in topography. Alluvial fan flooding is shallow and the principal problem is water, sediment, and debris entering basements. Alluvial fans are important groundwater recharge areas and surface water management efforts must not interfere with this. Federal restrictions on basements in flood insurance zones are not always realistic because of shallow flooding.

The unique flooding and flood problem characteristics of arid areas can be addressed, in part, by emphasizing watershed-wide land treatment to reduce erosion and minimize landslides and by use of debris traps and basins. Upstream storage reservoirs on larger streams can be used to contain snowmelt floods. Consideration should be given to the use of diversions and berms and floodproofing of individual structures to provide protection to scattered groups of buildings. Inspection and maintenance of debris basins, channels, and other hydraulic structures are particularly important. Finally, emergency action plans might be particularly effective because of the flashy nature of surface water runoff in arid areas.

Karst Topography

Recent population shifts to the sunbelt, in combination with the fact that one-half of soluble rock in the United States is located in the sunbelt, is increasing the risk of sinkhole flooding and collapse. A possible solution is the concept of a "sinkhole floodplain" (Kemmerly, 1981). Under this approach, sinkhole floodplains would be delineated in FIS for use in local land use regulations and for establishment of flood insurance premiums.

REFERENCES

American Public Works Association, *Stormwater Update,* No. 3. APWA, Institute for Water Resources, Chicago, IL, 1980.

American Society of Civil Engineers (ASCE), Task Force on Floodplain Regulations, "Guide for the Development of Floodplain Regulations." *J. Hydraul. Div., Am. Soc. Civ. Eng.* **88**(HY5), 73–119 (1962).

American Society of Civil Engineers (ASCE) and Water Pollution Control Federation, *Design and Construction of Sanitary and Storm Sewers,* ASCE Man. Rep. Eng. Pract. No. 37. ASCE/WPCF, New York, 1969.

American Society of Civil Engineers (ASCE), "Engineering the Non-Structural." *Civ. Eng. (N.Y.)* March, p. 114 (1981).

Association of State Floodplain Managers, *News and Views,* Vol. I, No. 2. ASFM, Madison, WI, 1983.

Athayde, D. N., P. E. Shelley, E. D. Driscoll, D. Gaboury, and G. Boyd, *Results of the Nationwide Urban Runoff Program,* Exec. Summ. U.S. Environ. Prot. Agency, Washington, DC, December 1983.

Bedient, P. B. and W. C. Huber, *Hydrology and Floodplain Analysis,* Chapter 3, Addison-Wesley, New York, 1988.

Betz Environmental Engineers, *Planning Methodologies for Analysis of Land Use/Water Quality Relationships,* Tech. Appendix. U.S. Environ. Prot. Agency, Washington, DC, 1976.

Bloomgren, P. A., *Strengthening State Floodplain Management,* Spec. Publ. No. 3. National Hazards Research and Information Center, University of Colorado, Boulder, 1982.

Burges, S. J., "Water Resource Systems Planning in the U.S.A.: 1776-1976." *J. Water Resour. Plann. Manage. Div., Am. Soc. Civ. Eng.* **105**(WR1), 91–111 (1979).

Chow, V. T., D. R. Maidment, and L. W. Mays, *Applied Hydrology,* Chapters 11 and 12, McGraw-Hill, New York, 1988.

Colston, N. V., Jr., *Characterization and Treatment of Urban Land Runoff,* Environ. Prot. Technol. Ser., EPA-670/2-74-096. U.S. Environ. Prot. Agency, Washington, DC, 1974.

Commonwealth of Pennsylvania, *Storm Water Management Act,* P.L. 864, NO. 167. Harrisburg, PA, 1978.

Davis, E. M., *Maximum Utilization of Water Resources in a Planned Community—Bacterial Characteristics of Stormwaters in Developing Areas,* Environ. Prot. Technol. Ser., EPA-600/2-79-050f. U.S. Environ. Prot. Agency, Washington, DC, 1979.

Debo, T. N., and M. Lumb, "Flood Flow Mapping—A Cost Effective Method of Preparing Hydrologic Studies for Flood Mapping." *Hydrocomp Simul. Network Newsl.* **9**(4) (1977).

Dougal, M. D. (Ed.), *Flood Plain Management: Iowa's Experience,* Iowa State University Press, Ames, IA, 1969.

Evenson, P. E., *Tailoring Floodland Regulations to Meet Community Needs.* University of Wisconsin Extension, Institute on Floodplain Management, Madison, May 1979.

Fair, G. M., and J. C. Geyer, *Water Supply and Waste Water Disposal,* pp. 5–8. Wiley, New York, 1963.

Federal Emergency Management Agency, Interagency Task Force on Floodplain Management, *A Unified National Program for Floodplain Management.* FEMA, Washington, DC, March 1986.

Fujita, S., "Experimental Sewer System for Reduction of Urban Storm Runoff" (no date).

Geological Society of America (GSA), Committee on Geology and Public Policy, *Floods and People: A Geological Perspective.* GSA, Boulder, CO., October 1978.

Goddard, J. E., *An Evaluation of Urban Flood Plains,* Tech. Memo. No. 19. ASCE Urban Water Resources Research Program, New York, December 1973.

Hauth, L. D., W. J. Carswell, Jr., and E. H. Chin, "Floods in Kansas City, Missouri and Kansas, September 12-13, 1977." *Geol. Surv. Prof. Pap. (U.S.)* **1169** (1981).

Illinois Department of Transportation, Division of Water Resources, *Flood Insurance.* IDT, Springfield, April 1, 1984.

James, L. D., D. T. Larson, D. H. Hoggan, and T. F. Glover, "Floodplain Management Needs Peculiar to Arid Climates." *Water Resour. Bull.* **16**(6), 1020–1029 (1980).

Jones, D. E., Jr., "Urban Hydrology—A Redirection." *Civ. Eng. (N.Y.)* August, pp. 58–62 (1967).

Kemmerly, P., "The Need for Recognition and Implementation of a Sink Hole—Floodplain Hazard Designation in Urban Karst Terrains." *Environ. Geol. (N.Y.)* pp. 281–292 (1981).

Kinori, B. Z., and J. Mevorach, *Manual of Surface Drainage Engineering,* Vol. II. Am. Elsevier, New York, 1984.

Kusler, J. A., *Floodplain Regulations and the Courts, 1970-1981,* Spec. Publ. No. 5. Natural Hazards Research and Applications Information Center, University of Colorado, Boulder, 1982.

Leuba, C. J., *A Road to Creativity—Arthur Morgan—Engineer, Educator, Administrator.* Christopher Publishing House, North Quincy, MA, 1971.

Linder, W. M., "Designing for Sediment Transport." *Water Spectrum* Spring-Summer, pp. 36–43 (1976).

Linsley, R. K., Jr., M. A. Kohler, and J. L. H. Paulhus, *Hydrology for Engineers, Basis for Planning,* 3rd ed. McGraw-Hill, New York, 1982.

McCuen, R. H., "Statistical Terminology: Definitions and Interpretation for Flood Peak Estimation." *Water Resour. Bull.* **15**(4), 1106–1116 (1979).

McCuen, R. H., S. G. Walesh, and W. J. Rawls, "Control of Urban Stormwater Runoff by Detention and Retention." *Misc. Publ.—U.S., Dep. Agric.* **1428,** 1–75 (1983).

Mead, D. W., *Hydrology—The Fundamental Basis of Hydraulic Engineering,* 2nd ed., Chapter 20. McGraw-Hill, New York, 1950.

Mendenhall, R. S., "A Primer on Flood Insurance." *J. Am. Insurance* **62,** 22–26 (1986).

Metcalf & Eddy, *Wastewater Engineering,* pp. 2–3. McGraw-Hill, New York, 1972.

Milwaukee Journal, "Flood Taught Deadly Lesson," October 6 (1981).

National Archives, 44 Code of Federal Regulations, Emergency Management and Assistance, National Flood Insurance Program, October 1, 1986.

Petersen, M. S., *River Engineering.* Prentice-Hall, Englewood Cliffs, NJ, 1986.

Poertner, H. G., *Practices in Detention of Urban Storm Water Runoff,* Spec. Rep. No. 43. Am. Public Works Assoc., Chicago, IL, 1974.

Raasch, G. E., "Urban Stormwater Control Project in an Ecologically Sensitive Area." *Proc. Int. Symp. Urban Hydrol., Hydraul., Sediment Control, 1982* (1982).

Robinson, A. M., "The Effects of Urbanization on Stream Channel Morphology." *Proc. Natl. Symp. Urban Hydrol., Hydraul., Sediment Control, 1976* pp. B-27–B-39 (1976).

Sheaffer, J. R., K. R. Wright, W. C. Taggart, and R. M. Wright, *Urban Storm Drainage Management,* Chapter 2. Dekker, New York, 1982.

Sokolove, R. D., *Subrogation—A Federal Regulatory Tool to Enforce Floodplain Management.* Presented at the American Society of Civil Engineers National Spring Convention, Las Vegas, NV, April 26-30, 1982.

Southeastern Wisconsin Regional Planning Commission (SEWRPC), *A Comprehensive Plan for the Menomonee River Watershed,* Plann. Rep. No. 26. SEWRPC, Waukesha, WI, October 1976.

Southeastern Wisconsin Regional Planning Commission (SEWRPC), "Major Storm Event Causes Flooding and Local Stormwater Drainage Problems in Southeastern Wisconsin." *Newsletter* **26**(6), 1–19 (1986).

State of Minnesota, "Surface Water Management Law." *Laws of Minnesota,* Chapter 509. St. Paul, MN, 1982.

Swanson, L. F., "Floodplain Management Provides Recreation Opportunities." *Am. Public Works Assoc., Rep.* January, pp. 10–11 (1982).

Theil, P. E., "Urban Drainage Design for New Development." *Conf. Proc.—Res. Program Abatement Munic. Pollut. Provis. Can-Ont. Agreement Great Lakes Water Qual.* **5** (1977).

Thomas, F. H., "The Federal Government's Evolving View of Floodplain Management." *Rice Univ. Stud.* **65**(1), Winter (1979).

Tucker, L. S., and W. G. DeGroot, *Regional Floodplain Management.* Presented at the American Society of Civil Engineers National Convention, November 1975.

Urban Land Institute, American Society of Civil Engineers, and National Association of Home Builders, *Residential Storm Water Management—Objectives, Principles and Design Considerations.* ULI/ASCE/NAHB, Washington, DC, 1975.

Urbonas, B. and L. A. Roesner, (Eds.), *Urban Runoff Quality—Impact and Quality Enhancement Technology,* Proceedings of an Engineering Foundation Conference, American Society of Civil Engineers, New York, 1986.

U.S. House of Representatives, *A Unified National Program for Managing Flood Losses.* Report on the Task Force on Federal Flood Control Policy to the Committee on Public Works, House Doc. No. 465, 89th Congress, Second Session, August 1966.

U.S. Water Resources Council, *The Nation's Water Resources,* Chapter 2. USWRC, Washington, DC, 1968.

U.S. Water Resources Council, *Guidelines for Determining Flood Flow Frequency,* Bull. No. 17B. Hydrol. Comm., USWRC, Washington, DC, September 1981 (revised).

Viessman, W., Jr., J. W. Knapp, G. L. Lewis, and T. E. Harbaugh, *Introduction to Hydrology,* 2nd ed., Chapter 5. Dun-Donnelley, New York, 1977.

Vollmer, A. H., "Urban Planning and Development of River Cities." *J. Urban Plann. Dev. Div., Am. Soc. Civ. Eng.* **96**(UP1), 59–64 (1970).

Walesh, S. G., "Floodland Management: The Environmental Corridor Concept." *Hydraul. Eng. Environ.—Proc. 21st Annu. Hydraul. Div. Spec. Conf., Am. Soc. Civ. Eng.,* 105–111, 1973.

Walesh, S. G., "Floodplain Management: The Environmental Corridor Concept." *Tech. Rec., Southeast. Wis. Reg. Plann. Comm.* **3**(6), 1–13 (1976).

Walesh, S. G., "Flood Insurance Program: Regional Agency's Perspective." *J. Water Resour. Plann. Manage. Div., Am. Soc. Civ. Eng.* **105**(WR2), 243–257 (1979).

Walesh, S. G., "Floodplain Management and Stormwater Management: Integrating Them at the Local Level." Presented at the Workshop of Innovations in Floodplain Management: Cost Effective Approaches, Association of State Floodplain Managers, Madison, WI, June 7-9, 1982.

Walesh, S. G., and M. L. Schoeffmann, *Surface and Subsurface Detention in Developed Urban Areas: A Case Study.* Presented at the American Society of Civil Engineers Water Resources Planning and Management Division Conference, Baltimore, MD, May 1984.

Weathers, J. W., *TVA's Experience in Floodplain Management.* Presented at the American Society of Civil Engineers Annual and National Environmental Engineering Meeting, Houston, TX, October 1972.

Whipple, W., Jr., "Dual-Purpose Detention Basins." *J. Water Resour. Plann. Manage. Div., Am. Soc. Civ. Eng.* **105**(WR2), 403–412 (1979).

White, G., *Strategies of American Water Management*. Univ. of Michigan Press, Ann Arbor, 1969.

Wisconsin Department of Natural Resources, *Floodplain Regulation Administration Manual.* WDNR, Madison, WI, January 1978a.

Wisconsin Department of Natural Resources, University of Wisconsin System Water Resources Center, and Southeastern Wisconsin Regional Planning Commission, *Menomonee River Pilot Watershed Study—Summary Pilot Watershed Report.* WDNR, Madison, WI, January 1978b.

Wolman, M. G., and L. B. Leopold, "River Floodplains: Some Observations on Their Formation." *Geol. Surv. Prof. Pap. (U.S.)* **282-C** (1957).

Yearke, L. W., "River Erosion Due to Channel Relocation." *Civ. Eng. (N.Y.)* August, pp. 39–40 (1971).

2

THE HYDROLOGIC CYCLE IN THE URBAN ENVIRONMENT

In its most elemental form, surface water management (SWM) seeks to integrate urban development into the hydrologic cycle to satisfy the space and other needs of the community while accommodating the surface water conveyance and storage function of the natural environment. Processes comprising the hydrologic cycle are reviewed in this chapter with emphasis on the significance of those processes in urban SWM.

Various analytical and other means are then used to illustrate the potential significant impact of urbanization on the quantity and quality of stormwater runoff. In the absence of a SWM program, these impacts can lead to disruption, damage, and death. The chapter concludes with a discussion of the sequential nature of erosion–sedimentation problems and flooding problems.

2.1 THE HYDROLOGIC CYCLE

The hydrologic cycle is the continuous, unsteady circulation of the water resource from the atmosphere to and under the land surface and, by various processes, back to the atmosphere. As suggested by Figure 2.1, which shows the hydrologic cycle in schematic form, the hydrologic cycle is dynamic in that the quantity and quality of water at a particular location may vary greatly with time. Temporal variations may occur in the atmosphere, on the land surface, in surface waters, and in the groundwaters of an area. Within the hydrologic cycle, water may appear in all three of its states: solid, liquid, and gas.

As shown in Figure 2.1, the hydrologic cycle consists of various unsteady processes occurring in the atmosphere and on and beneath the earth's surface.

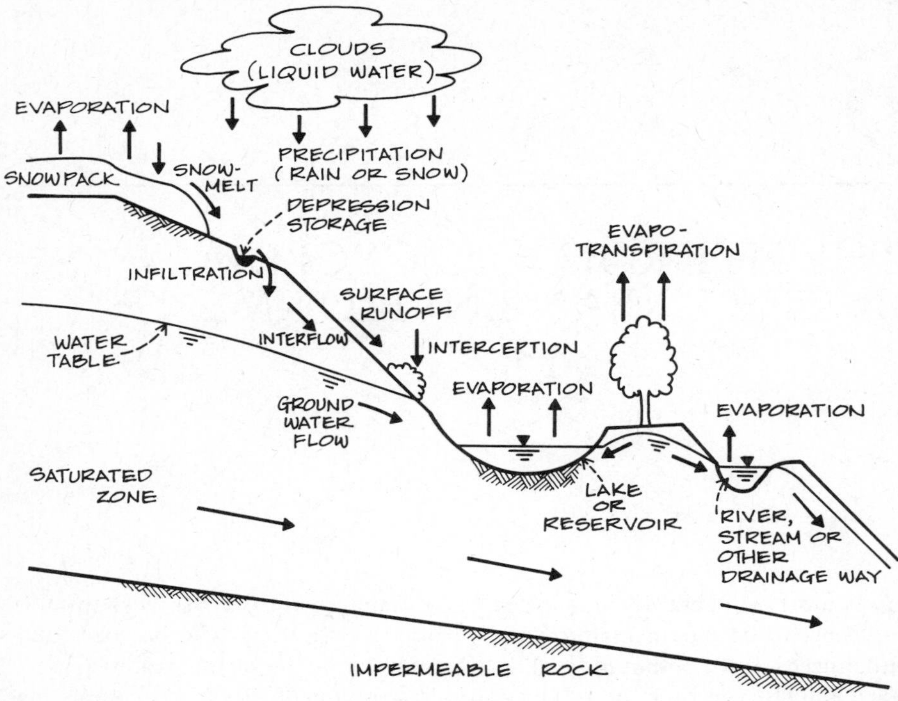

FIGURE 2.1 Hydrologic cycle

Each important process is described briefly in subsequent sections. Emphasis is placed on factors influencing each of the processes and the process's significance in the planning, design, and operation of urban SWM systems. For a detailed description of processes comprising the hydrologic cycle, the reader is referred to various monographs and textbooks on the subject (e.g., Bedient and Huber, 1988; Chow et al., 1988; Kibler, 1982; Linsley et al., 1982; Viessman et al., 1977).

Precipitation

Precipitation occurs as rain, sleet, hail, and snow. Annual amounts of total precipitation are unpredictable and highly variable, ranging from 5 to 100 in. for various locations in the United States. In a sense, precipitation is the most important process in a hydrologic cycle because it is the "driving force" providing water which must be accommodated in the urban environment.

Interception

The amount of precipitation that wets and adheres to aboveground objects—primarily vegetation—until it is evaporated back into the atmosphere is called interception. The annual amount of interception in a particular area is affected by factors such as the amount and type of precipitation, the extent and type of vegetation, and wind. In forested areas, interception may be as much as 25 percent of the annual precipitation. Interception is not likely to be an important process in urban SWM programs.

Depression Storage

This process is defined as the amount of total precipitation detained in and evaporated from depressions on the land surface. Depression storage is water that does not run off or infiltrate. Depression storage is affected by surface type and slope and the factors influencing evaporation. For a given rainfall event, common nominal values of depression storage are 0.1 in. of precipitation for impervious surfaces such as asphalt and concrete and 0.2 in. for pervious surfaces such as clay or sand. Because of its small magnitude, depression storage is not likely to be important in urban SWM investigations.

Snowmelt

Snowmelt can be thought of as precipitation that is temporarily stored as snowpack, much of which eventually melts and moves on or into the ground. Snowmelt is affected by solar radiation, conductive and convective transfer of heat from the overlying air, conduction of heat from the underlying soil, and heat supplied by precipitation falling directly on snowpack. Snowmelt is not likely to be significant in determining the quantity of runoff from "small" watersheds, but it is very likely to be important in determining the quantity of runoff from "large" watersheds. Snowmelt may also be important in determining the quality of spring runoff from watersheds of all sizes.

Infiltration

Infiltration is defined as the passage of water through the air–soil interface. Infiltration rates are affected by factors such as time since the rainfall event began, soil porosity and permeability, antecedent soil moisture conditions, and presence of vegetation. Infiltration is a very important process in urban SWM and, therefore, essentially all hydrologic methods explicitly account for infiltration. Urbanization usually decreases infiltration with a resulting increase in runoff volume and discharge.

Evaporation and Transpiration

Evaporation is the process whereby water is transformed from the liquid or solid state into the gaseous state. Transpiration is the mechanism whereby water moves up through vegetation and is then evaporated. Evapotranspiration rates are affected by factors such as temperature, wind, vapor pressure, differences, plant characteristics, and availability of soil moisture. Although evapotranspiration is of a very little practical significance during precipitation events, evapotranspiration is very important in preparing hydrologic budgets for watersheds, lakes, or reservoirs.

Surface Runoff

The process whereby water moves to the ground surface to a natural or constructed channel, a lake or reservoir, or other receiving waters is called surface runoff and is sometimes also referred to as overland flow. Surface runoff is affected by other processes in the hydrologic cycle, such as precipitation and infiltration, plus factors such as imperviousness and land slope. Surface runoff determines the quantity of stormwater that must be locally managed and affects the quantity of potential pollutants transported to the receiving waters. In the absence of SWM programs, urbanization usually dramatically increases surface runoff volumes and rates.

Interflow

Interflow, sometimes referred to as subsurface flow, is the process whereby water moves essentially laterally beneath the land surface but above the groundwater table. Interflow occurs until water enters a surface water channel, drainageway, lake, or other impoundment; enters the groundwater table; crosses back over the air–soil–water interface; or is evapotranspired. Interflow is affected by the same factors as those for surface runoff. Interflow is rarely explicitly analyzed—it is usually considered part of the surface runoff. Surface runoff, interflow, and precipitation falling directly on water bodies are sometimes lumped together and called direct runoff.

Groundwater Flow

Groundwater flow, sometimes referred to as base flow, is water moving laterally beneath the water table toward and into natural or artificial channels, lakes, and other receiving waters. Unlike most other processes in the hydrologic cycle, groundwater flow is essentially a continuous process. It maintains flows in natural and man-made drainageways and impoundments. Urbanization usually decreases groundwater flow. Groundwater discharge through the sidewalls of newly constructed turf or other earth-lined open channels may cause slumping of channel sidewalls. In a related matter, channel deepening and widening or sewer construction may lower the local groundwater table and damage building foundations, particularly in organic soils.

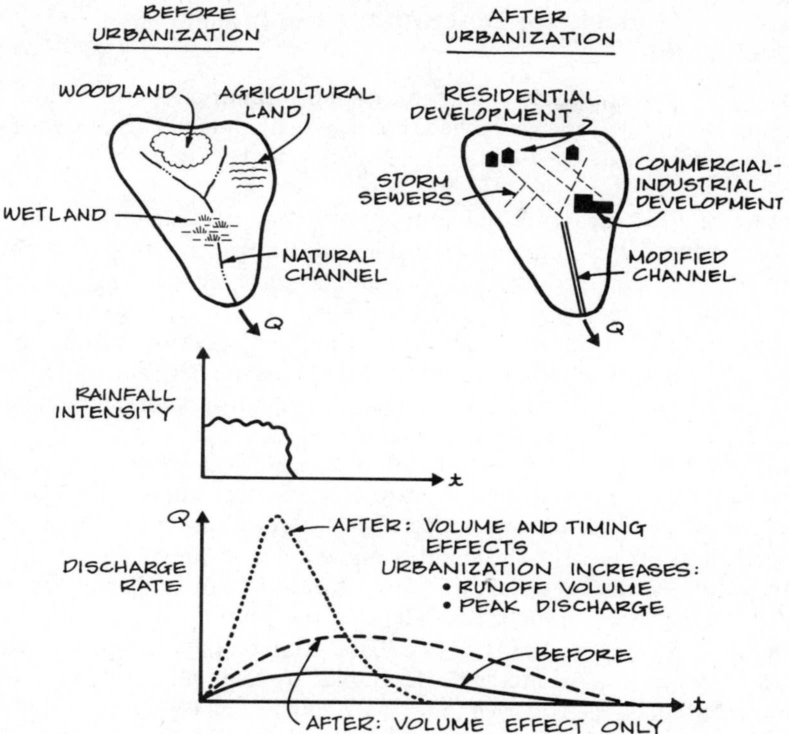

FIGURE 2.2 Typical hydrologic–hydraulic impact of urbanization in the absence of a surface water management program

2.2 IMPACT OF URBANIZATION ON THE QUANTITY OF RUNOFF

As illustrated in Figure 2.2, the usual hydrologic–hydraulic effects of urbanization are an increase in the volume of direct runoff, accompanied by a corresponding decrease in the volume of the base flow, and a decrease in runoff time. The usual impacts of these two effects in the absence of mitigating SWM measures are larger flow rates, higher stages, and expanded areas of inundation.

There are, however, exceptions. For example, a computer modeling investigation (Malcom, 1978) concluded that the gradual placement of culverts in a watershed as a part of the development of the street and highway system and resulting formation of detention areas immediately upstream of the culverts had the accidental but favorable effect of reducing peak flows throughout the watershed.

Techniques Available for Determining the Impact of Urbanization

The ideal way to illustrate the impact of urbanization is long-term monitoring of rainfall and runoff for a watershed that has been gradually urbanized. Such measurements are rare and certainly not likely to be available for a given watershed.

A more realistic means of illustrating the impact of urbanization on the quantity of runoff is to monitor two watersheds similar in critical characteristics, such as size, shape, soil type, slope, and meteorology, but with one catchment being in agricultural or other undeveloped use and the other catchment being urban. Another approach to quantifying the impact of urbanization is application of empirical or theoretical tools and techniques ranging from the simple approaches, such as the rational method, to more complex methodologies, such as a hydrologic–hydraulic computer model.

Examples of the preceding techniques are presented in subsequent sections of this chapter. The primary reason for presenting the examples is to emphasize the dramatic, often adverse impact that urbanization, in the absence of a SWM program, can have on the quantity and quality of runoff. The secondary purpose of presenting the examples is to introduce analytic techniques some of which are discussed in more detail later in this book.

Many other studies of the effect of urbanization on the quantity of surface water runoff appear in the literature. Examples are Leopold's (1968) summary of empirical studies, the American Society of Civil Engineer's reports (ASCE, 1969a, 1974), a computer modeling analysis by James and Lumb (1975), Hollis's (1975) study of the effect of urbanization on floods of different recurrence intervals, and reports of engineering investigations (e.g., Donohue & Associates, 1984).

Impact of Urbanization on Peak Flows from a Small Catchment Using the Rational Method

A hypothetical watershed before and after urban development is presented in Figure 2.3 along with hydrologic–hydraulic parameters needed to apply the rational method (ASCE, 1969b). On the assumption that the watershed is located in Boston, Massachusetts, for purposes of assigning intensity–duration–frequency relationships, the 2-, 5-, and 10-year recurrence interval discharges were computed for the before and after conditions. Hydrologic–hydraulic parameters used in the calculations are summarized in Table 2.1 together with the resulting 2-, 5-, and 10-year recurrence interval discharges obtained using the rational method. Peak flows after development are seen to be over five times peak flows before development.

Assume that a concrete storm sewer at a slope of 0.01 ft per foot and flowing full is to be provided to carry a 10-year-recurrence-interval design discharge from the watershed to the receiving stream. A storm sewer at least

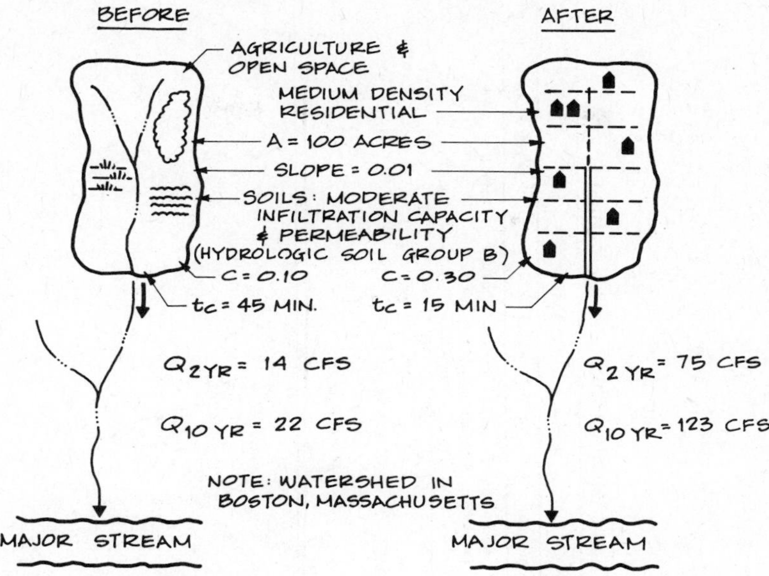

FIGURE 2.3 Effect of urbanization on peak discharges from a small watershed based on the rational method

36 in. in diameter would be required for the before condition. For the after development condition, the storm sewer would have to be at least 66 in. in diameter to carry the larger flows resulting from urbanization of the watershed.

In summary, the impact of urbanization on a small watershed can be illustrated easily using the rational method. The impact can be measured in terms of relative increase in flows or in incremental increase in the size of storm sewers or other conveyance facilities required to accommodate the increased flow.

Hydrologic–Hydraulic–Damage Effects of Urbanization Using a Computer Model

One approach to determining the consequences of urbanization changes on flood flows, stages, and damage involves partitioning the watershed land surface into floodplain and nonfloodplain areas as illustrated in Figure 2.4. Various combinations of floodplain and nonfloodplain development can be hypothesized, as suggested by Figure 2.5.

This conceptual approach was developed and used to assess the impact of seven combinations of floodplain and nonfloodplain urbanization on a 136-mi^2 urbanizing watershed in southeastern Wisconsin (Walesh and Videkovich, 1978). A hydrologic–hydraulic–flood damage model consisting of a combina-

TABLE 2.1 Effect of Urbanization on Peak Discharge from a Small Watershed Based on the Rational Method

	Watershed Characteristics						Rainfall–Runoff Analysis[a]								
							2-Year Recurrence Interval			5-Year Recurrence Interval			10-Year Recurrence Interval		
Watershed Condition	Area (acres)	Slope (ft/ft)	Soil Type	Land Use	Runoff Coeffi-cient	Time of Concentra-tion (min)	Rainfall Intensity (in./hr)	Peak Discharge (ft^3/sec)	Relative Increase in Discharge	Rainfal Intensity (in./hr)	Peak Discharge (ft^3/sec)	Relative Increase in Discharge	Rainfall Intensity (in./hr)	Peak Discharge (ft^3/sec)	Relative Increase in Discharge
Before development	100	0.01	Moderate infiltration capacity and permeability	Agriculture and open space	0.1	45	1.4	14	—	1.9	19	—	2.2	22	—
After development	Same	Same	Same	Medium-density residential	0.3	15	2.5	75	5.4	3.5	105	5.5	4.1	123	5.6

Source: Rainfall intensity–duration–frequency data from American Society of Civil Engineers (1969*b*)

[a]Analysis assumes that the watershed is in Boston, Massachusetts.

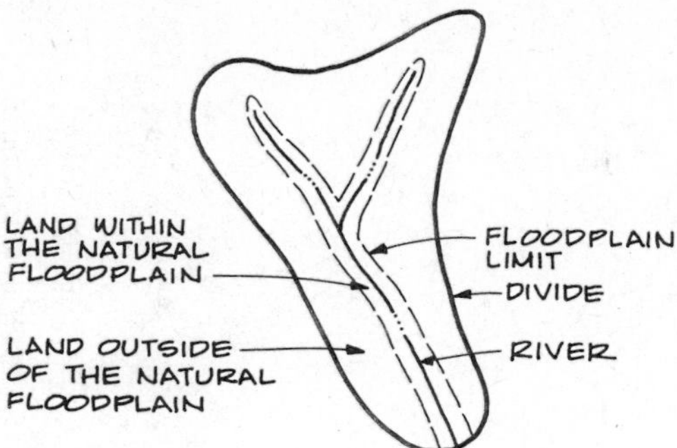

FIGURE 2.4 Floodplain and nonfloodplain areas of a watershed (*Source:* Adapted from Walesh and Videkovich, 1978)

tion of continuous process and steady-state computer programs was used as the analytic tool. The seven floodplain and nonfloodplain development conditions are presented schematically in Figure 2.6.

One hundred-year-recurrence-interval flood flows for nine locations and all seven conditions are presented in Figure 2.7. Discharge–frequency relations for location b in Figure 2.7 are shown on Figure 2.8.

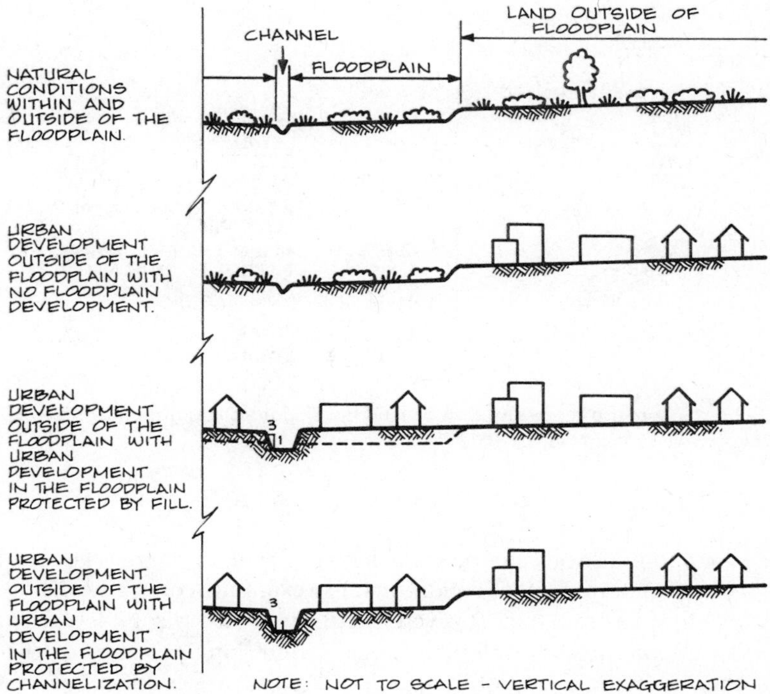

FIGURE 2.5 Development in floodplain and nonfloodplain areas of a watershed (*Source:* Adapted from Walesh and Videkovich, 1978)

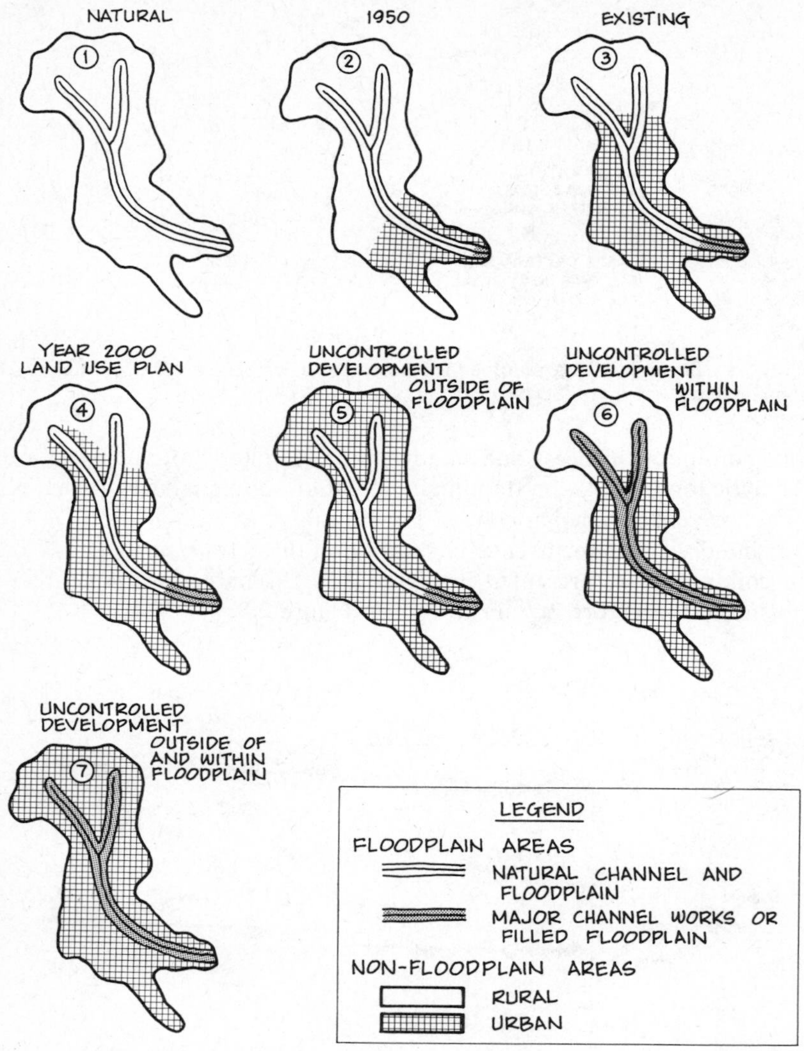

FIGURE 2.6 Schematic representation of land development conditions (*Source:* Adapted from Walesh and Videkovich, 1978)

Numerous comparisons can be made between peak discharges at each location for the seven development conditions. For example, compare existing conditions (condition 3) to the most severe situation, which is complete urbanization of floodplain and nonfloodplain areas (condition 7). Ratios of the 100-year discharge under condition 7 to 100-year discharges under condition 3

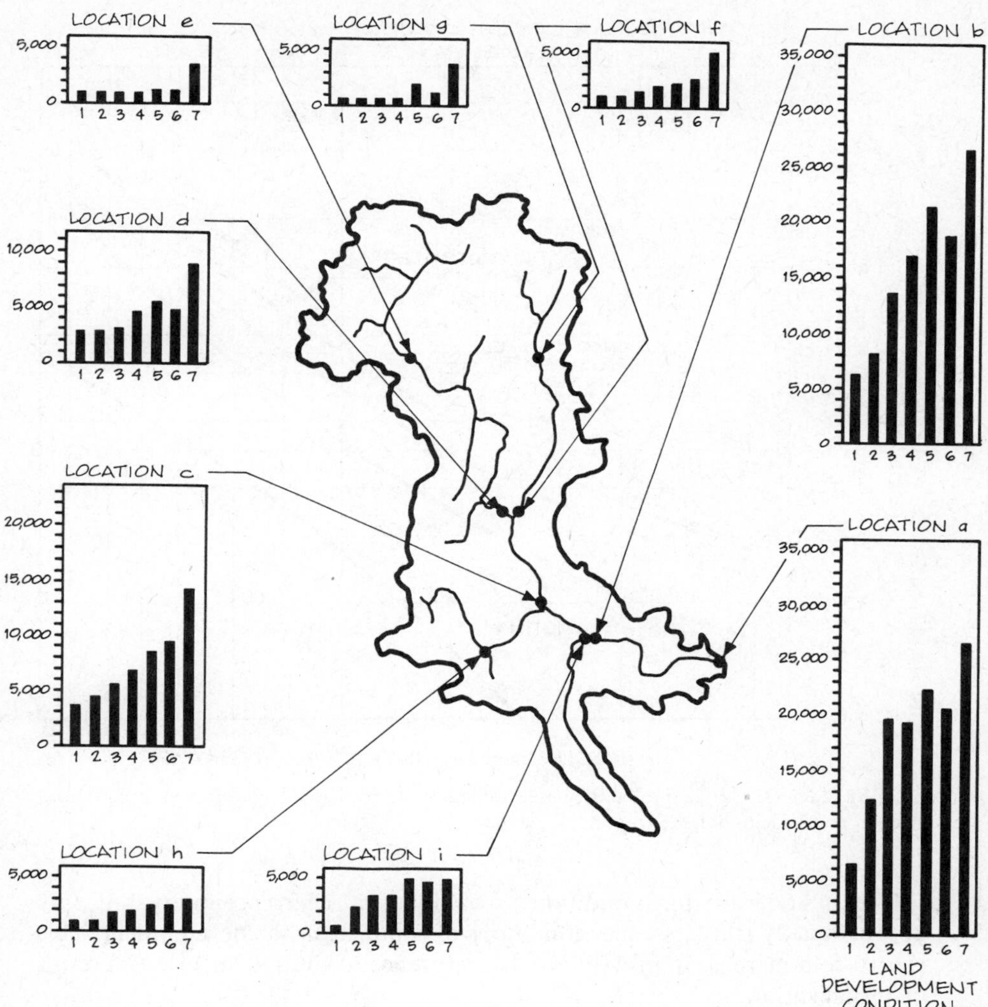

FIGURE 2.7 Effect of land development conditions on 100-year discharges (*Source:* Adapted from Walesh and Videkovich, 1978)

range from 1.4 (locations a and j) to 6.4 (location g), the median value being 2.

Examination of the discharge–frequency relationships presented in Figure 2.8 for conditions 3 and 7 at location b near the mouth of the watershed indicates that complete urbanization will almost double the existing-condition 100-year discharge of 13,500 ft^3/sec. Or, stated differently, the peak flow of 13,500 ft^3/sec, which has an expected average recurrence interval of about

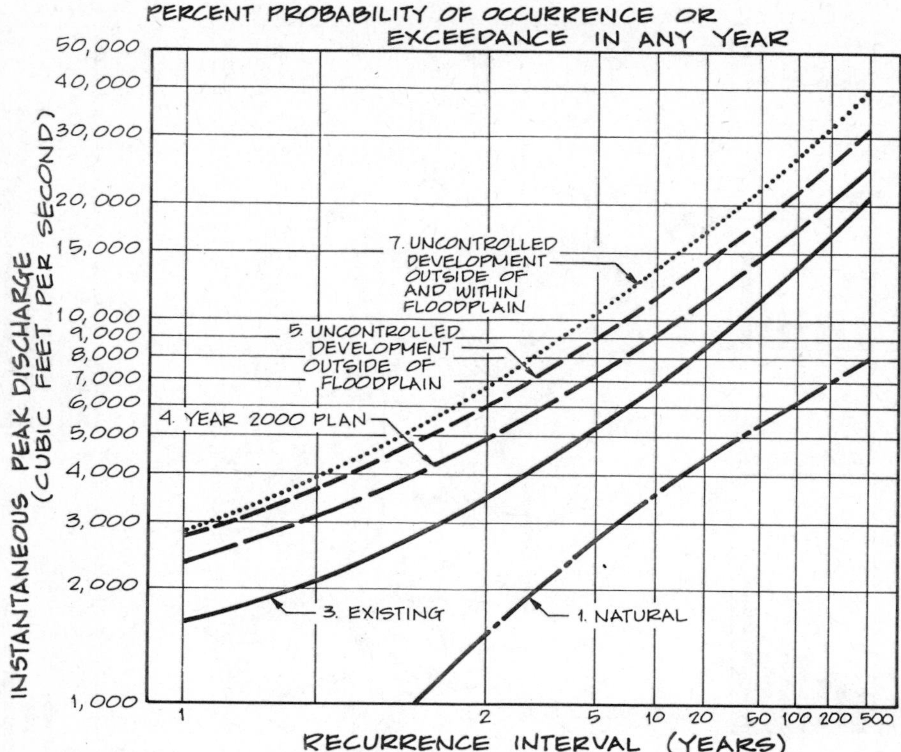

FIGURE 2.8 Discharge–probability relationships for location b (*Source:* Adapted from Walesh and Videkovich, 1978)

100 years under existing conditions, would have a recurrence interval of only approximately 10 years under conditions of complete urban development. This is a 10-fold increase in probability of occurrence of the stated flow as a result of urbanization.

Average annual monetary flood damage offers a second means of quantifying the impact of urban development on flood problems. Average annual monetary damages for four reaches and five development conditions are presented graphically in Figure 2.9. In all cases, the flood damage computations are based on the assumption that no additional flood-prone development will be constructed in the floodplains, or if additional floodplain development occurs, as would be probable under conditions 6 and 7, the assumption is that the structures would be floodproofed or otherwise protected against flood damage. Furthermore, monetary flood damage calculations assume that complete floodplain development occurs in all reaches upstream of the region in question. The combination of uncontrolled development outside of and within the floodplain areas (condition 7) may be expected to increase average annual

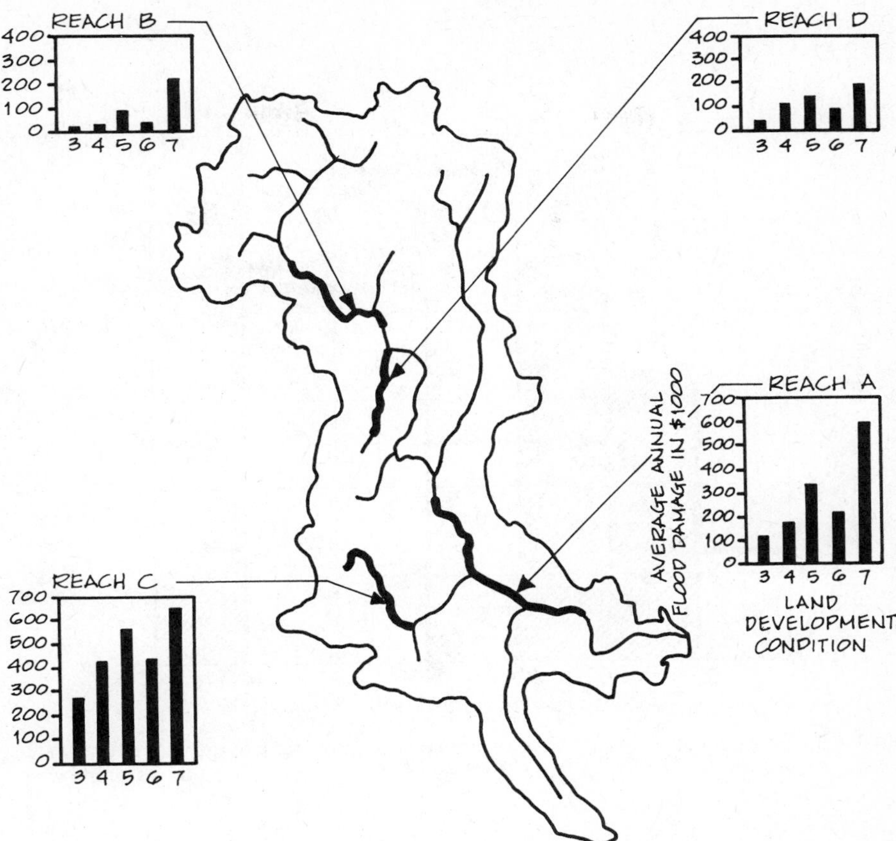

FIGURE 2.9 Effect of land development conditions on average annual flood damage (*Source:* Adapted from Walesh and Videkovich, 1978).

flood damage relative to existing conditions (condition 3) by factors of 2.4 to 8.5 for the four flood-prone reaches.

A similar computer modeling analysis was conducted on a 44-mi^2 watershed in California which was 20 percent urbanized at the time of the investigation (James, 1965). Existing and complete urbanization conditions were simulated. Modeling predicted that the discharge at the watershed outlet having a recurrence interval of 100 years under existing conditions would have a 2-year recurrence discharge under complete urbanization conditions. In other words, in the absence of a SWM program, urbanization would convert a 100-year discharge to a 2-year recurrence interval discharge.

In summary, computer models can be used to quantify the impact of urbanization. Impacts can be presented in a variety of ways, including increase in peak flows and increase in monetary flood damage.

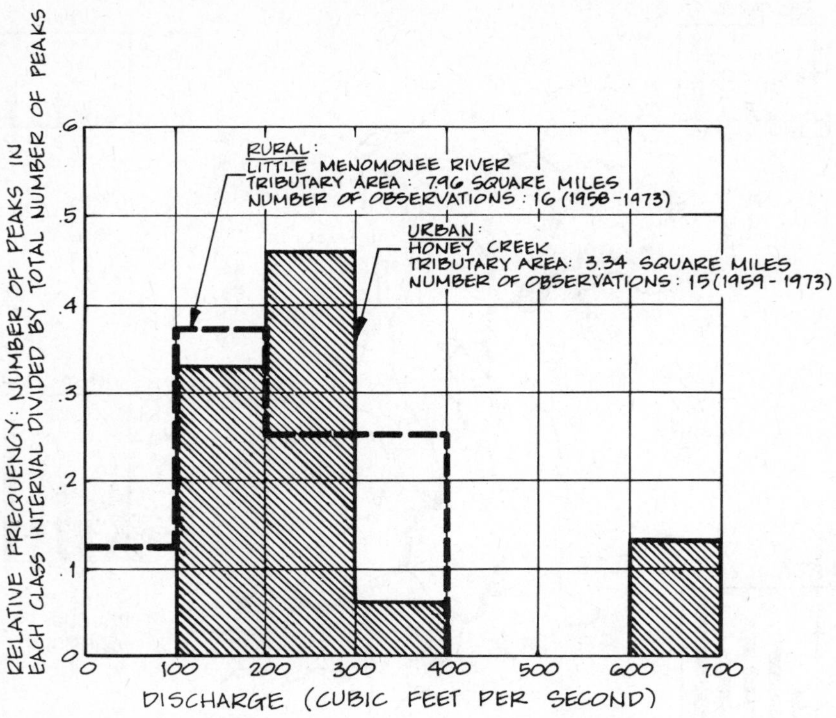

FIGURE 2.10 Frequency distributions of annual peak discharges for urban and rural watersheds (*Source:* Adapted from Southeastern Wisconsin Regional Planning Commission, 1976)

Effect of Urbanization in Terms of Recorded Peak Discharges

A frequency distribution of annual instantaneous peak discharges for a rural 7.96-mi^2 watershed in southeastern Wisconsin is shown in Figure 2.10. Superimposed on the distribution for the rural watershed is a similar distribution for a 3.34-mi^2 urban watershed in the same area. Because of their proximity, the two watersheds are exposed to similar meteorologic conditions.

The frequency distributions are very similar in shape even though the urban watershed has an area less than half that of the rural watershed. Using annual instantaneous peak discharges as an indicator, the much smaller area of the urban watershed relative to the rural watershed is offset by the hydrologic–hydraulic effects of urbanization. That is, the urban watershed, although smaller, produces annual instantaneous peak discharges similar to the rural watershed because the urban watershed produces greater runoff volumes and shorter runoff times for similar meteorologic events.

Impact of Urbanization on Local Meteorologic Conditions

The preceding illustrations of ways to analyze the impact of urbanization assume that meteorologic conditions are unchanged and all adverse hydrologic–hydraulic effects are caused by urban development. Recent evidence indicates that urbanization of large areas can alter local meteorology. Furthermore, these alterations tend to increase rainfall amounts, thus potentially further exacerbating the typical adverse effect of urbanization.

For example, rainfall monitoring reveals that the frequency and severity of flooding increased in the Chicago metropolitan area over the past five decades (Changnon, 1984). There are 76 percent more 1-in. rains in the city than in the surrounding rural areas. One reason is the increase in summer rainfall caused by additional particulate matter in the atmosphere, which provide more nuclei for drop formation. Other reasons for increased frequency and severity of flooding are additional heating from the urban centers, which increases rising air, cooling, and condensation; additional moisture from some industrial processes; and mechanical mixing of air due to urban structures. Other examples of increased rainfall caused by urban development are cited by Changnon (1976) and Detwyler and Marcus (1972).

2.3 IMPACT OF URBANIZATION ON THE QUALITY OF RUNOFF

Urbanization influences the quality of stormwater runoff because of two effects. First, the many and varied activities occurring in an urban area generate, expose, or otherwise make available at or near land surface many different types of potential pollutants. This first effect is accentuated if the urban area is undergoing development. As illustrated in Figure 2.11, substances having the potential to cause toxic pollution, organic pollution, nutrient pollution, pathogenic pollution, and sediment pollution are present or generated in urban areas. Also, substances having an adverse aesthetic impact may be produced in an urban setting. Generating processes or activities include motor vehicle operation, leaf fall, on-site waste disposal systems, excessive application of chemical and organic fertilizers and pesticides, human littering, careless material storage and handling, poor property maintenance, construction and demolition activity, animal droppings, pavement disintegration, and application of deicing compounds and sand.

The second mechanism by which urbanization adversely affects the quality of stormwater runoff is the more efficient hydrologic–hydraulic system. As noted previously and as illustrated conceptually in Figure 2.2, urbanization typically increases the volume of surface water runoff and decreases runoff times, resulting in increases in velocities and flow rates. The discharge of larger volumes of stormwater runoff from the land surface at higher discharge rates

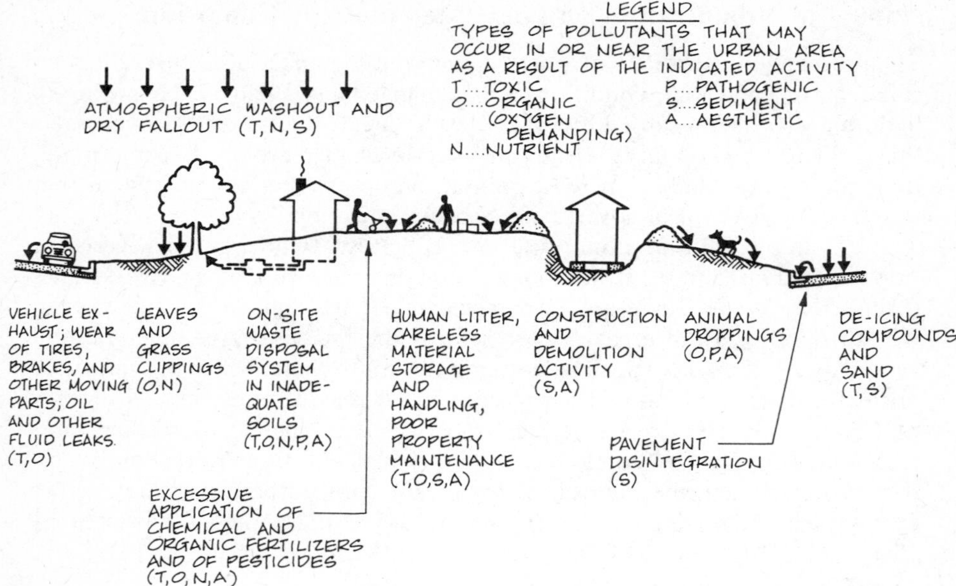

FIGURE 2.11 Impact of urbanization on the availability of potential pollutants on or near the land surface

and greater velocities provides a more effective means for transporting potential polluting materials from the land surface to the receiving waters.

Introductory treatments of nonpoint-source pollution may be found in the literature (e.g., Torno et al., 1986; Sheaffer et al., 1982; Whipple et al., 1983). Simplified, manual methods for estimating nonpoint-source pollution loads are presented in Chapter 7.

Techniques Available for Determining the Impact of Urbanization

As with the quantity of runoff, the water quality effects of urbanization may be illustrated by the monitoring of rural and urban watersheds having similar size, shape, soil type, slope, and meteorology. Another approach for illustrating the impact of urbanization on the quality of stormwater runoff is comparison of unit nonpoint-source pollutant loads. Unit loads may be expressed in various ways, including pounds per acre per year, and the values may be drawn from numerous studies involving monitoring of the quantity and quality of runoff from rural and urban areas. Computer models as well as simpler desktop methods such as those presented in Chapter 7 may also be used to illustrate the impact of urbanization on the quality of stormwater runoff.

Examples based on the preceding techniques are used in the following sections of this chapter to demonstrate the potential significant and adverse water quality effect in the absence of a SWM program. The examples also serve to introduce methods of analysis which are discussed in more detail in later chapters.

Numerous additional investigations of the effect of urbanization on the quality of stormwater runoff appear in the literature. Examples are Donigian's (1977) literature summary of unit loads of potential pollutants as a function of land uses, Bowlby and Ponce's (1977) monitoring study of the effect of recreation home development on surface water quality. Cherkauer's (1977) field study of the effect of urban lakes on downstream water quality, Wilber and Hunter's (1977) field investigation of heavy metal enrichment in river sediments as a result of urbanization, Eisen's (1977) monitoring study of the impact of urbanization on the quality of shallow groundwater, and Klein's (1979) study of stormwater quality impairment caused by urban development.

Urbanization Impact Determined by Monitoring of Adjacent Different Watersheds

Rainfall, stormwater runoff, and selected quality constituents were monitored during and after rainfall events occurring on a rural watershed and an urban watershed located in proximity to each other in east central Wisconsin (Cherkauer, 1975). The two watersheds had similar topography, soil type, geology, and meteorology. However, as indicated in Table 2.2, one watershed was predominantly urban and undergoing additional urban development, and the other was predominantly rural.

Also presented in Table 2.2 are peak flow, total runoff volume, and total sediment yield for the two watersheds as a result of receiving 0.87 in. of rainfall in 5 hours after 7 days with no rainfall. Although subjected to the same meteorologic conditions, the urban watershed responded with much greater channel flow depth, peak flow, and unit sediment yield than did the rural watershed. The higher sediment yield of the urban watershed may be attributed, in part, to active construction.

Impact of Urban Development on Stormwater Quality as Determined by Fish Kills

Several severe fish kills occurred over a two-decade period on the James River downstream of Springfield, Missouri. Rainstorms preceded all fish kills, but not all rainstorms resulted in fish kills. Various hypotheses were presented as to the cause of the fish kills, including the impact of stormwater runoff, bypass of sewage at the municipal sewage treatment plant, resuspension of organic material deposited immediately downstream of the wastewater treatment plant, and poor-quality spring water discharging into the river.

Continuous monitoring revealed the principal cause to be dissolved oxygen

TABLE 2.2 Runoff Response of Rural and Urban Watersheds Subjected to the Same Rainfall Event

Characteristic	Urban	Rural (Agriculture)	Ratio: Urban/Rural
Location	North of Milwaukee, Wisconsin—similar topography, soils, geology, and meteorology		
Area (m^2)	2.90	3.74	0.77
Urban land (%)	65.1	5.7	11.4
Under construction (%)	1.5	0.0	—
Point sources	None	None	—
Drainage system	Roadside ditches connected to large storm sewers, which are in turn connected to a concrete-lined channel.	Some channels were dredged and straightened for agricultural purposes.	—
Rainfall event	October 6, 1987 beginning at 0530—0.87 in. in 5 hours after 7 days with no rain		—
Peak flow conditions in principal channel	Three-fourths bank full	No significant rise above base flows	—
Peak flow (ft^3/sec)	16.8	0.064	262
Suspended sediment[a] (lb/acre)	2.56	0.0071	361

Source: Adapted from Cherkauer (1975).

[a]Between 0530 on October 6 and 0900 on October 8.

depletion in the James River. Furthermore, the oxygen depletion was caused by oxygen-demanding substances carried into the James River as a result of stormwater runoff from Springfield. The wash-off of organic materials was significant only when sufficient time occurred between rainstorms for organic materials to build up on the land surface (Barrett, 1971).

Nationwide Urban Runoff Program

Conducted from 1978 to 1983, the Nationwide Urban Runoff Program (NURP) investigated the extent to which urban runoff was causing water quality problems. Assuming that urbanization had some adverse effect on the quality of stormwater runoff and on the receiving surface and groundwater, the second purpose of NURP was to test the effectiveness of alternative control measures (Athayde et al., 1983).

NURP's principal conclusions (quoted from Athayde et al., 1983) concerning the quality of urban runoff are:

1. Heavy metals (especially copper, lead, and zinc) are by far the most prevalent priority pollutant constituents found in urban runoff. End-of-pipe concentrations exceed EPA ambient water quality criteria and drinking water standards in many instances. Some of the metals are present often enough and in high enough concentrations to be potential threats to beneficial uses.
2. The organic priority pollutants were detected less frequently than the heavy metals and at lower concentrations.
3. Coliform bacteria are present at high levels in urban runoff and can be expected to exceed EPA water quality criteria during and immediately after storm events in many surface waters, even those providing high degrees of dilution.
4. Nutrients are generally present in urban runoff, but with a few individual site exceptions, concentrations do not appear to be high in comparison with other possible discharges to receiving water bodies.
5. Oxygen demanding substances are present in urban runoff at concentrations approximating those in secondary treatment plant discharges. If dissolved oxygen problems are present in receiving waters of interest, consideration of urban runoff controls as well as advanced waste treatment appears to be warranted.
6. Total suspended solids concentration in urban runoff are fairly high in comparison with treatment plant discharges. Urban runoff control is strongly indicated where water quality problems associated with total suspended solids, including build-up of contaminated sediments, exist.
7. A summary characterization of urban runoff has been developed and is believed to be appropriate for use in estimating urban runoff pollutant discharges from sites where monitoring data are scant or lacking, at least for planning level purposes.

With respect to the impact of urban stormwater runoff on receiving waters—rivers and streams, lakes, estuaries and embayments, and groundwater—NURP's principal conclusions (quoted from Athayde et al., 1983) are:

Rivers and Streams

1. Frequent exceedances of heavy metals ambient water quality criteria for freshwater aquatic life are produced by urban runoff.
2. Although a significant number of problem situations could result from heavy metals in urban runoff, levels of freshwater aquatic life use impairment suggested by the magnitude and frequency of ambient criteria exceedances were not observed.
3. Copper, lead and zinc appear to pose a significant threat to aquatic life uses in some areas of the country. Copper is suggested to be the most significant of the three.
4. Organic priority pollutants in urban runoff do not appear to pose a general threat to freshwater aquatic life.
5. The physical aspects of urban runoff, such as erosion and scour, can be a significant cause of habitat disruption and can affect the type of fishery pres-

ent. However, this area was studied only incidentally by several of the projects under the NURP program and more concentrated study is necessary.
6. Several projects identified possible problems in the sediments because of the build-up of priority pollutants contributed wholly or in part by urban runoff. However, the NURP studies in this area were few in number and limited in scope, and the findings must be considered only indicative of the need for further study, particularly as to long-term impacts.
7. Coliform bacteria are present at high levels in urban runoff and can be expected to exceed EPA water quality criteria during and immediately after storm events in most rivers and streams.
8. Domestic water supply systems with intakes located on streams in close proximity to urban runoff discharges are encouraged to check for priority pollutants which have been detected in urban runoff, particularly those in the organic category.

Lakes

1. Nutrients in urban runoff may accelerate eutrophication problems and severely limit recreational uses, especially in lakes. However, NURP's lake projects indicate that the degree of beneficial use impairment varies widely, as does the significance of the urban runoff component.
2. Coliform bacteria discharges in urban runoff have a significant negative impact on the recreational uses of lakes.

Estuaries and Embayments

1. Adverse effects of urban runoff in marine waters will be highly specific local situations. Though estuaries and embayments were studied to a very limited extent in NURP, they are not believed to be generally threatened by urban runoff, though specific instances where use is impaired or denied can be of significant local and even regional importance. Coliform bacteria present in urban runoff is the primary pollutant of concern, causing direct impacts on shellfish harvesting and beach closures.

Groundwater Aquifers

1. Groundwater aquifers that receive deliberate recharge of urban runoff do not appear to be imminently threatened by this practice at the two locations where it was investigated.

In summary, the massive NURP indicates that stormwater runoff from urban areas generally contains significant quantities of potential pollutants, including heavy metals, coliform bacteria, nutrients, oxygen-demanding substances, and total suspended solids. Furthermore, the constituents in stormwater runoff can have an adverse effect on aquatic life in and on uses of surface water resources.

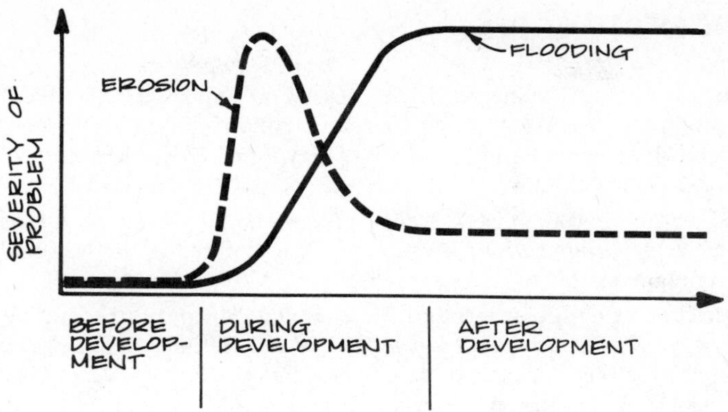

FIGURE 2.12 Sequential nature of erosion–sedimentation and flooding problems in the absence of a surface water management program

2.4 THE SEQUENTIAL NATURE OF EROSION/SEDIMENTATION AND FLOODING PROBLEMS

Preceding sections strongly suggest that in the absence of mitigating measures, urbanization will usually result in a significant increase in flooding problems, erosion/sedimentation problems, and related nonpoint-source pollution problems. As suggested by Figure 2.12, the flooding problems and erosion–sedimentation and other nonpoint-source pollution problems will tend to occur sequentially—not simultaneously.

Erosion–sedimentation problems will appear during development, because that is the time of greatest earth-disturbing activity. In contrast, flooding problems tend to appear in the latter stages of development or after development because they are largely controlled by random meteorologic events interacting with the newly constructed impervious surfaces.

Given the preceding premise, it follows that successful control of erosion–sedimentation and related nonpoint-source pollution problems does not necessarily mean that there will not be subsequent flooding problems. Similarly, provision for flood control measures does not necessarily preclude the occurrence of erosion–sedimentation and related nonpoint-source pollution problems during land development.

Both quantity and quality problems can be traced to the same cause—adverse interaction with the hydrologic cycle—and at least ideally, both problems should be controlled. It follows that remedial and preventive measures should be sought to serve the twofold purpose of flood control and control of erosion–sedimentation and other nonpoint-source pollution. A few such measures are available, such as sedimentation basins in series with detention/retention facilities. Available management measures are discussed in Chapter 11.

REFERENCES

American Society of Civil Engineers (ASCE), Task Force on Effect of Urban Development on Flood Discharges, "Effect of Urban Development on Flood Discharges—Current Knowledge and Future Needs." *J. Hydraul. Div., Am. Soc. Civ. Eng.* **95**(HY1), 287–309 (1969a).

American Society of Civil Engineers, *Report: International Workshop on the Hydrological Effects of Urbanization, Warsaw, 1973.* Submitted to the National Science Foundation, January 1974.

American Society of Civil Engineers and Water Pollution Control Federation, *Design and Construction of Sanitary and Storm Sewers,* ASCE Man. Rep. Eng. Pract. No. 37/WPCF Man. Pract. No. 9, pp. 42–55. ASCE/WPCF, New York, 1969b.

Athayde, D. N., P. E. Shelley, E. D. Driscoll, D. Gaboury, and G. Boyd, *Results of the Nationwide Urban Runoff Program,* Exec. Summ. U.S. Environ. Prot. Agency, Washington, DC, December 1983.

Barrett, B. R., "Monitors Solve Fish-Kill Mystery." *Civ. Eng. (N.Y.).* **41**(1), January, pp. 40–42 (1971).

Bedient, P. B. and W. C. Huber, *Hydrology and Floodplain Analysis,* Chapter 1, Addison Wesley, New York, 1988.

Bowlby, J., and S. L. Ponce, *The Effect of Lake-Based Recreation and Mountain Home Development on Surface Water Quality.* Presented at the American Water Resources Association Conference, October 31-November 3, 1977.

Changnon, S. A., Jr., "Inadvertent Weather Modification." *Water Resour. Bull.* **12**(4), 695–718 (1976).

Changnon, S. A., Jr., "Flooding on the Increase in Illinois: A Call for Awareness and Action." *Ill. Munic. Rev.* January, pp. 7–8 (1984).

Cherkauer, D. S., "Urbanization Impact on Water Quality During a Flood in Small Watersheds." *Water Resour. Bull.* **11**(5), 987–998 (1975).

Cherkauer, D. S., "Effects of Urban Lakes on Surface Runoff and Water Quality." *Water Resour. Bull.* **13**(5), 1057–1067 (1977).

Chow, V. T., D. R. Maidment, and L. W. Mays, *Applied Hydrology,* Chapters 1–6, McGraw-Hill, New York, 1988.

Detwyler, T. R., and M. G. Marcus, *Urbanization and Environment,* Chapter 3. Wadsworth (Duxbury Press), Belmont, CA, 1972.

Donigian, A. S., *Nonpoint Source Pollution From Land Use Activities.* Prepared by Hydrocomp, Inc. for the Northeastern Illinois Planning Commission, August 1977.

Donohue & Associates, Inc., *Smith Ditch Lagoon No. 1 and Hotter Lagoon Investigation, Valparaiso, IN,* June 1984.

Eisen, C., *The Groundwater/Surface Water Interaction in the Menomonee River Watershed—Southeastern Wisconsin.* M.S. Thesis in Geology, University of Wisconsin-Madison, 1977.

Hollis, G. E., "The Effect of Urbanization on Floods of Different Recurrence Interval." *Water Resour. Res.* **11**(3), 431–435 (1975).

James, L. D., "Using a Digital Computer to Estimate the Effects of Urban Development on Flood Peaks." *Water Resour. Res.* **1**(2), 223–234 (1965).

James, L. D., and A. M. Lumb, "Flood Hydrograph Simulation for Urban Frequency Analysis: Application to a Watershed." *Proc. Natl. Symp. Urban Hydrol., Sediment Control, 1975* pp. 181–192 (1975).

Kibler, D. F. (Ed.), *Urban Stormwater Hydrology,* Water Resour. Monogr. No. 7. Am. Geophys. Union, Washington, DC, 1982.

Klein, R. D., "Urbanization and Stream Quality Impairment." *Water Resour. Bull.* **15**(4), 948–963 (1979).

Leopold, L. B., "Hydrology for Urban Land Planning—A Guidebook on the Hydrologic Effects of Urban Land Use." *Geol. Surv. Circ. (U.S.)* **544** (1968).

Linsley, R. K., Jr., M. A. Kohler, and J. L. H. Paulhus, *Hydrology for Engineers,* 3rd ed. McGraw-Hill, New York, 1982.

Malcom, R. H., *Culverts, Flooding, and Erosion.* Presented at the Engineering Foundation Conference on Water Problems in Urban Areas, Henniker, NH, July 16-21, 1978.

Sheaffer, J. R., K. R. Wright, W. C. Taggart, and R. M. Wright, *Urban Storm Drainage Management,* Chapter 11. Dekker, New York, 1982.

Southeastern Wisconsin Regional Planning Commission, *A Comprehensive Plan for the Menomonee River Watershed,* Vols. 1 and 2, Plann. Rep. No. 26. SEWRPC, Waukesha, WI, October 1976.

Torno, H. C., J. Marsalek, and M. Desbordes (Eds.), *Urban Runoff Pollution,* NATO Adv. Sci. Inst. Ser., Ser. G, Vol. 10, Chapter 1. Springer-Verlag, New York, 1986.

Viessman, W., Jr., J. W. Knapp, G. L. Lewis, and T. E. Harbaugh, *Introduction to Hydrology,* 2nd ed. IEP-A Dun-Donnelley, New York, 1977.

Walesh, S. G., and R. M. Videkovich, "Urbanization: Hydrologic-Hydraulic-Damage Effects." *J. Hydraul. Div., Am. Soc. Civ. Eng.* **104**(HY2), 141–155 (1978).

Whipple, W., N. S. Grigg, T. Grizzard, C. W. Randall, R. P. Shubinski, and L. Scott Tucker, *Stormwater Management in Urbanizing Areas,* Chapter 4. Prentice-Hall, Englewood Cliffs, NJ, 1983.

Wilber, W. G., and J. V. Hunter, *The Impact of Urbanization on the Distribution of Heavy Metals and Bottom Sediments on the Sattle River.* Presented at the American Water Resources Association Conference, October 31–November 3, 1977.

3

TECHNIQUES FOR HYDROLOGIC ANALYSES

From the perspective of the practicing engineer, the pragmatic purpose of urban hydrology is calculation of hydrologic loads—surface water runoff rates and volumes—under existing and future land development conditions. One use of existing and future condition hydrologic loads is quantifying the impact of urbanization. As discussed in Chapter 2, hydrologic loads can increase markedly as urban development occurs.

Hydrologic loads are also used to define floodplains and floodways and to size surface water control works such as storm sewers, bridges and culverts, channels, pumping facilities, detention/retention facilities, and dikes and floodwalls. Hydrologic techniques enable the engineer to plan and develop a surface water system that can accommodate existing and future runoff rates and volumes in a safe, economic, and environmentally compatible manner.

This chapter begins with a discussion of hydrologic data types, sources, and uses. Delineation of watersheds, subwatersheds, and subbasins is discussed, followed by an introduction to hydrologic techniques, including a categorization of methods. Most of the remainder of the chapter is devoted to discussion of hydrologic methods in the following categories: statistical analysis of streamflow records, regional methods, transfer methods, and rainfall–runoff methods.

Hydrologic analysis is just one step, and often the first, in a series of analytic steps leading to the sizing of a surface water control facility, such as a storm sewer, culvert, or detention area. A thorough hydrologic analysis is necessary, but not sufficient, in the planning and design process. Although a careful hydrologic analysis may minimize the uncertainty in the size of stormwater control facilities, other areas of potential uncertainty must be addressed, such as planning and design criteria, hydraulics, cost estimating, material selection, construction monitoring, and service life (Yen, 1978a). The challenge in planning and designing surface water control facilities is to achieve an optimum distribution of effort in all areas of uncertainty and to avoid overemphasis in one or a few areas—such as hydrologic analysis.

3.1 HYDROLOGIC DATA TYPES, SOURCES, AND USES

Various natural and cultural features determine the hydrologic–hydraulic response of an urban or urbanizing watershed. Collecting and organizing data usually requires time and effort equal to or in excess of that required to subsequently perform hydrologic–hydraulic analyses. Because hydrologic–hydraulic data are important and because a major effort is required to obtain the data, hydrologic–hydraulic analyses should be preceded by a selective, carefully planned inventory of watershed data. Although this chapter deals with hydrologic techniques, hydraulic, as well as hydrologic, data types, sources, and uses are discussed in this section for ease and completeness of presentation.

Data Types

Hydrologic and hydraulic data may be placed in three categories to guide the inventory process: (1) completed or ongoing studies, (2) natural resources data, and (3) infrastructure data. The three data types, typical sources, and examples of use are presented in Table 3.1. Data drawn from each of the three data types will be needed in any hydrologic–hydraulic investigation. Specific data requirements for a given investigation, however, depend on the hydrologic–hydraulic methods selected.

Data Sources

Identifying the numerous potential sources of hydrologic–hydraulic data can be difficult and the subsequent contacting process can be time consuming. To assist with this problem, typical data sources are presented in Table 3.1. Listed sources range from federal agencies such as the Federal Emergency Management Agency, through state-level agencies such as departments of natural resources, through regional agencies such as regional planning commissions, down to local governmental units. The local units, that is, counties, cities, villages, or towns, are usually the best sources of data or the most logical starting point.

Uses of Data

Examples of uses of data obtained during the inventory are presented in Table 3.1. The principal use is quantification of the hydrologic–hydraulic character-

TABLE 3.1 Hydrologic and Hydraulic Data Types, Sources, and Uses

Type			
Category	Example	Typical Sources[a]	Examples of Use
Completed or ongoing studies	Stormwater master plan	Drainage, sewerage, flood control, or other special district	Establish type and configuration of stormwater control facilities
	208 Plan	U.S. Environmental Protection Agency Regional planning agency Council of governments	Delineate watersheds and subbasins
	SCS PL 566 plan	U.S. Soil Conservation Service	Establish flood flows, stages, and area of inundation on principal streams
	Floodplain information report	U.S. Army Corps of Engineers U.S. Geological Survey Regional planning agency	Establish flood flows, stages, and area of inundation on principal streams
	Flood insurance study	U.S. Federal Emergency Management Agency State	Establish flood flows, stages, and area of inundation on principal streams
Natural resource	Topographic map	Regional planning agency Drainage, sewerage, flood control, or other special districts Field survey	Delineate watershed and subbasins Identify potential detention sites Determine land slopes
	Soils	U.S. Soil Conservation Service Construction logs Soil borings	Determine runoff coefficients, runoff curve numbers, and other runoff factors Evaluate erosion potential Project construction conditions
	Vegetation	State Department of Natural Resources	Identify potential environmentally sensitive areas

(continued)

TABLE 3.1 (Continued)

Type			
Category	Example	Typical Sources[a]	Examples of Use
		Regional planning agency Field survey	Determine runoff coefficients, runoff curve numbers, and other runoff factors
	Historic inundation areas and high water	Drainage, sewerage, flood control, or other special district Regional planning agency News media—newspapers, radio, TV Museums, historical societies Residents Field survey	Document location and severity of historic inundation and other problem areas
	Precipitation intensity–duration–frequency curves	National weather service Drainage, sewerage flood control, or other special district	Develop design storms
	Historic precipitation	National Weather Service Drainage, sewerage, flood control, or other special district	Assess severity of historic floods
	Stream stage and discharge	U.S. Geological Survey Drainage, sewerage, flood control, or other special district	Develop discharge-probability relationships Assess severity of historic floods
Infrastructure	Existing land use	Regional planning agency Field survey	Determine runoff coefficients, runoff curve numbers, and other runoff factors

TABLE 3.1 (Continued)

Type: Category	Type: Example	Typical Sources[a]	Examples of Use
	Land use plan	Regional planning agency	Determine runoff coefficients, runoff curve numbers, and other runoff factors
	Zoning map and ordinance		Project future land use
	Subdivision plats		Project future land use Establish type and configuration of future stormwater control facilities
	Agricultural and other land management measures	U.S. Soil Conservation Service Regional planning agency Field survey	Determine runoff coefficients, runoff curve numbers, and other runoff factors
	Transportation, sewerage, and other public facility-utility systems and plans	Regional planning agency State/Federal Department of Transportation	Establish future watershed and subbasin divides Project future land use
	Stormwater system maps, plans, profiles, as-builts	Drainage, sewerage, flood control, or other special district	Delineate existing/future watershed and subbasin divides Develop hydraulic characteristics
	Bridge, culvert, channel, and other hydraulic structure as-builts or plans	Drainage, sewerage, flood control, or other special district State/Federal Department of Transportation Field surveys	Delineate existing/future watershed and subbasin divides Develop hydraulic characteristics
	Land ownership—public vs. private		Identify potential sites for detention and other facilities

Source: Adapted from Walesh (1981a).

[a]The local community—county, city, village, town—is usually the most logical starting point and may be the best single source of data.

istics of the watershed to permit computation of surface water runoff volumes and rates. Quantification of hydrologic–hydraulic characteristics usually must be done for both existing and future conditions because, as discussed elsewhere in this book, land use markedly influences the rainfall–runoff process.

3.2 WATERSHED DELINEATION

A subbasin is the smallest unit into which the land surface is subdivided for hydrologic studies. All the area within the subbasin discharges at one point. Subbasin boundaries are determined by surface topography and by the configuration of swales, storm sewers, channels, and other drainage elements. A subwatershed is a drainage area composed of two or more subbasins. A watershed is a drainage area composed of two or more subbasins or two or more subwatersheds. A divide is the imaginary line indicating the limits of a subbasin, subwatershed, or watershed. A watershed with its subbasins and subwatersheds is illustrated in Figure 3.1. Kinori and Mevorach (1984) discuss watershed characteristics, including the effect of watershed shape on the form of the hydrograph.

The Importance of Delineation

Experience, and some systematic studies, suggest that delineations of subbasins and the subwatersheds and watersheds defined by those subbasins are sometimes inadequate, particularly in urban areas. Poor delineations may have a major, adverse effect on the accuracy of calculated surface water runoff rates and volumes because area enters directly into most hydrologic calculations.

The importance of careful delineation of subbasins is illustrated by a study

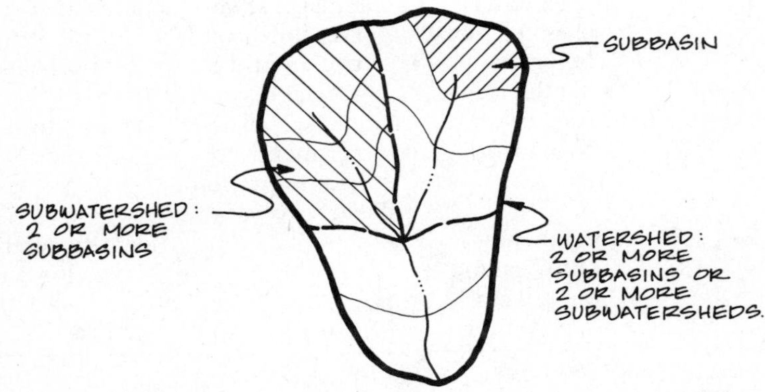

FIGURE 3.1 Subbasin, subwatershed, and watershed

(Alley and Veenhuis, 1979) conducted in the Denver, Colorado, area. U.S. Geological Survey 7-minute quadrangle maps (1 in. = 2000 ft, 10-ft contour interval) were used to delineate 12 subbasins. A second, field-verified delineation was then performed. The two delineations were compared by superposition and calculating the ratio of the map-based area to the field-verified area.

The superimposed maps for the 12 subbasins, along with the area ratios, are shown in Figure 3.2. Extreme differences in area occurred as indicated by comparing map-based subbasin sizes to field-verified subbasin sizes. The ratios of map-based areas to field-verified areas range from 0.13 to 4.90. Differences between map-based areas and field-verified areas were attributed to the effects of the street grading, storm sewers, irrigation and other ditches, and resolution of the maps.

The Denver area study is extreme in that only small-scale maps were used for the map-based delineation. However, the study illustrates the types and sizes of errors that may occur in delineating subbasins—especially in urban areas—if supplemental data sources and field verification are not used. The resulting large subbasin, subwatershed, and watershed delineation errors may produce gross errors in calculated surface water runoff rates and volumes, which, in turn, may cause serious undersizing or oversizing of stormwater facilities.

Delineation Guidelines

The following factors should be considered when delineating subbasins within subwatersheds or watersheds:

1. Utilize the best available topographic maps and other topographic information. Possible sources of topographic information include large-scale (e.g., 1 in. = 200 ft, 2-ft contour) topographic maps, small-scale topographic maps such as U.S. Geological Survey quadrangle sheets; aerial photographs; storm and combined sewer plans, profiles, and system maps; and street elevation, slope, and grade data.
2. Select a point of interest on a stream or drainageway and begin delineating the divide assuming that surface water runoff moves perpendicular to topographic contours. Examples of points of interest are:
   (a) Corporate limits—village, city, and county limits and boundaries of special districts
   (b) Streamflow or stage gauging stations or sites of high water marks
   (c) Prominent structures or features such as high, long railroad or roadway embankments
   (d) Outlets of detention/retention facilities or the downstream end of stream reaches that will be analyzed as impoundments or may have the potential for development of storage facilities

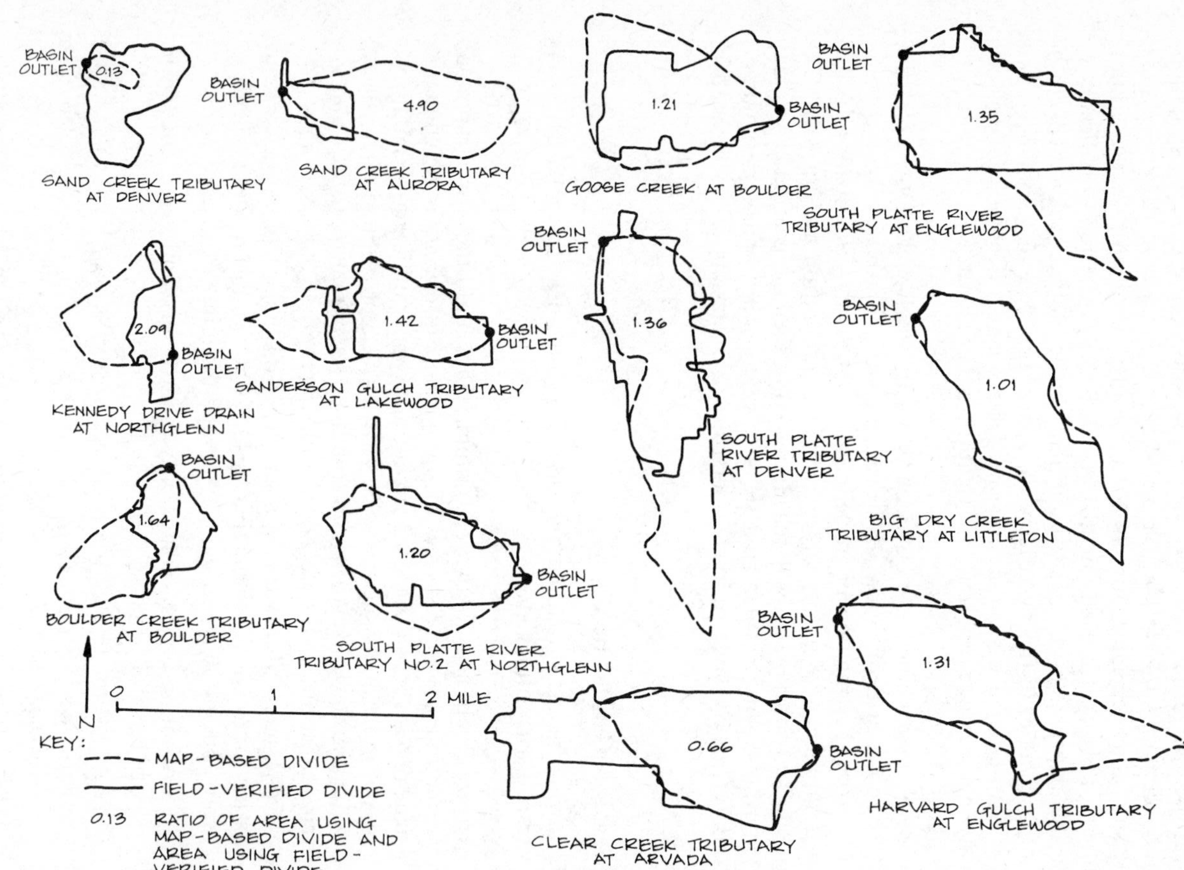

FIGURE 3.2 Map-based and field-verified subbasin delineations (*Source:* Adapted from Alley and Veenhuis, 1979)

(e) Existing or proposed hydraulic transitions, such as points at which swales enter sewers, at inlets or catch basins, and at junctions of channels or sewers
(f) Changes in hydrologic parameters, such as soil type, slope, land use, and imperviousness, to encourage the hydrologic homogeneity of subbasins

3. Scan the area for topographic highs as indicated by circular or oval contour line patterns. Such high points are likely to be near the divide.
4. Examine the legend of maps to determine if the map distinguishes between intermittent and perennial streams. The former are more likely to be near the divide.
5. In the vicinity of a swale or stream, topographic contours tend to point upstream, that is, toward the divide.
6. Evaluate storm or combined sewer flow direction data because these drainage works can significantly alter the size and shape of subbasins as would be defined solely by surface topography. For example, surface water runoff may be carried across topographic divides by gravity sewers or by pumping.
7. Be aware of internally drained areas which often occur in glacial drift and in karst topography. Depending on the area, depth, and other features of these special geomorphologic or geologic features, give consideration to subtracting their areas from the total tributary area.
8. If data are available on the location of groundwater divides, recognize that groundwater divides are not necessarily coincident with surface water divides.
9. Consider the likely effects of future land development on the associated surface water system. The surface water pattern for future development may be oriented to planned or platted streets, which may have the effect of squaring off or otherwise altering subbasin divides. That is, future development may not only change the runoff characteristics of a subbasin, but may also change its size and shape.
10. If surface water quality is of concern, give particular attention to the location of known or potential pollution sources, particularly point sources. That is, determine if such sources are within or outside the subbasin.
11. Conduct a field check to resolve uncertainties which inevitably remain even after a thorough office analysis of available data.

A common concern in subbasin delineation is the desired size or, more precisely, the allowable maximum size recognizing that subbasin delineation can be tedious and time consuming and that delineation time increases as subbasin size decreases. If the preceding delineation guidelines are followed, the size question will be largely implicitly resolved. That is, the resulting size range or average size will be appropriate because all the applicable delineation guidelines will be satisfied.

However, one additional factor should be considered in checking the adequacy of subbasin size. Average subbasin size, more specifically average subbasin time of concentration (t_c) or other runoff time parameters, should be consistent with the computational interval to be used in the subsequent hydrologic–hydraulic analysis. If t_c is very large compared to the time interval, that is, the subbasins are too large, the detail inherent in the input and in the computations will be "averaged out" and not properly reflected in the output.

In contrast, if t_c is very small compared to the time interval, that is, if subbasins are too small, no physically meaningful simulation or computation can occur on each subbasin. A rule of thumb is to size subbasins such that the t_c is approximately twice the time interval.

3.3 HYDROLOGIC TECHNIQUES: AN OVERVIEW

Many hydrologic tools and techniques are available to calculate surface water runoff rates and volumes. Knowing about them, selecting among them, and correctly using them is a challenge for practitioners.

Selection Factors

Some of the ways in which hydrologic techniques differ and, therefore, some factors that warrant consideration in identifying and selecting a hydrologic technique for a particular application, are:

- The ease with which the method can be applied as measured in terms of staff time and cost
- Computational requirements in terms of manual efforts or computer needs
- Data requirements
- Applicability to rural, urban, and other special land uses or covers
- Maximum or minimum size limitations on tributary area
- Types of causative meteorological events that can be accommodated such as rainfall and snowmelt
- Ability to incorporate the effect of detention/retention facilities and other forms of storage
- Kind of output provided, such as peak discharge, runoff volume, and complete hydrograph
- Expected accuracy

Categories of Hydrologic Techniques

For the purposes of this and subsequent chapters, hydrologic techniques are organized into the following five categories:

1. Statistical analysis of streamflow records
2. Regional methods
3. Transfer methods
4. Manual rainfall–runoff methods
5. Computerized rainfall–runoff methods

At least one example from each of the first four categories is included in this chapter. In each example, the method is described, important constraints and assumptions are identified, and positive and negative features are noted. Literature references are provided for many methods in each of the first four categories. The fifth category, computer modeling, is discussed separately in Chapter 10.

A common error in hydrologic investigations is to move too quickly into selection and application of one or more hydrologic techniques and, as a result, fail to search for and consider hydrologic–hydraulic results of earlier studies. Experience suggests that the results of other investigations will eventually be discovered and considerable effort may be required to reconcile results of previous investigations with the current investigation.

As suggested by Table 3.1, surface water flow data may be available from a variety of sources. Examples include U.S. Geological Survey analyses of gauging station data; Federal Emergency Management Agency flood insurance studies; and special investigations, plans, and designs carried out by state, regional, and local governmental units and agencies.

Use of Two or More Techniques

Careful practice suggests the use of more than one hydrologic method for each particular application. A detailed and complete application of a primary method followed by, or parallel with, use of a second method helps to guard against blunders in hydrologic analysis.

Envelope Curves

One means of testing the reasonableness of calculated flood flows is the use of envelope curves. In their simplest form, envelope curves present largest recorded discharges as a function of tributary area for hydrologically homogeneous regions. Typically, the discharges are the largest that have been observed

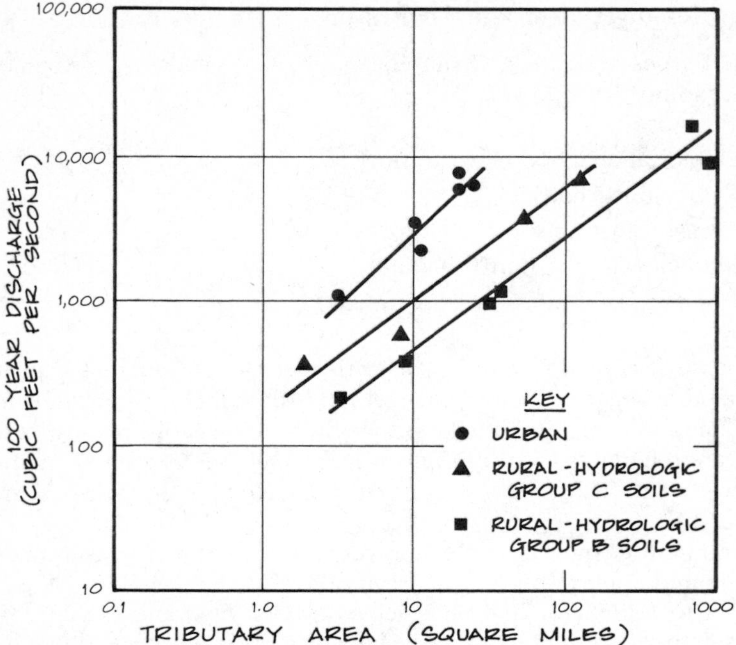

FIGURE 3.3 Regional 100-year-recurrence-interval experience curves for southeastern Wisconsin (*Source:* Adapted from Southeastern Wisconsin Regional Planning Commission, 1977)

and are not necessarily associated with a recurrence interval. The curves envelope the largest known flood flows. Crippen (1982) presents discharge versus area envelope curve equations for 17 regions in the United States. Similar envelope curves could be developed for one or more urban areas in a hydrologic region.

Regional Experience Curves

Although more difficult to assemble, regional flood flow experience curves usually prove to be more useful than evelope curves. Regional flood flow experience curves present flood flows of a specified probability as a function of tributary area parameters, such as area, land slope, soil type, and land use or cover.

Data for these curves can be assembled over a period of time by one or more public or private organizations that collect hydrologic data or perform hydrologic analyses in a particular area. Flood flow experience curves for southeastern Wisconsin are shown in Figure 3.3.

Such relationships should not be used as the primary hydrologic technique unless they are based on a large amount of carefully analyzed data, in which case they would become regional methods, as discussed later in this chapter. Regional flood flow experience curves permit comparison of the results of a current analysis with flows previously measured or calculated in the same region. If a calculated discharge differs markedly from the experience curve value, an error has not necessarily occurred. However, the variance should be examined to see if watershed physical conditions, such as an unusually large amount of storage or an extremely efficient conveyance system, warrant the variance.

Accuracy and Reproducibility of Hydrologic Techniques

Little documentation is available to indicate the accuracy likely to be achieved with different hydrologic techniques. For example, a literature evaluation of methods for determining flows at ungauged sites (McCuen et al., 1977) stated: "The most important conclusion of this literature evaluation is that the literature, while voluminous, does not contain sufficient information to provide a general statement about either specific procedures or general techniques as to their accuracy, reproducibility, or practicality."

A subsequent investigation (U.S. Water Resources Council, 1981a) evaluated hydrologic techniques on the basis of accuracy, reproducibility, and practicality. The last factor was measured by the time required to apply the method.

Statistical methods were found to be the most accurate, followed by manual and computerized rainfall–runoff techniques. Similarly, statistical methods were the most reproducible, followed by manual or computerized rainfall–runoff techniques. With respect to practicality, the investigation concluded that computer models required much more time to implement than did manual methods.

Although not explicitly discussed in the report on the investigation, the initial relatively large effort required to establish a hydrologic–hydraulic computer model of a surface water system may be more than offset by the value of the resulting more organized data base. Furthermore, the greater detail with which the system is represented in the computer model provides the ability to simulate responses to various hypothetical land development conditions and numerous alternative surface water control systems.

Rounding of Calculated Flows

Calculated flows should be rounded off or shown with a limited number of significant digits to avoid implying an accuracy greater than that which can normally be expected in hydrologic analyses. Examples of rounding guidelines based on accuracies of up to 1 and 10 percent of the discharge magnitude are presented in Table 3.2. These are only examples and are meant solely to illustrate the importance of rounding computed flows to indicate the expected accuracy. The concept also applies to computed volumes, velocities, and stages.

TABLE 3.2 Guidelines for Rounding of Calculated Flows to Be Consistent with Expected Accuracy

Range of Flows (ft³/sec)		Round to Nearest (ft³/sec)	
Equal to or Greater Than	Less Than	For Accuracy of Up to 1 Percent	For Accuracy of Up to 10 Percent
0	500	5	50
500	1,000	10	100
1,000	5,000	50	500
5,000	10,000	100	1,000
10,000	20,000	200	2,000
20,000	50,000	500	5,000

Urban Hydrology: An Art and a Science

Urban hydrology is a science in that many documented techniques are available, most are based on fundamental principles, and each consists of a series of logical, mathematical steps culminating in numerical results. Furthermore, many of the methodologies are so well defined and organized that they can readily be computerized.

Urban hydrology is also an art. There are few, if any, clearly defined criteria for selection of methods and accuracy limits are lacking for most methodologies. In addition, although many methods appear to be largely systematic and mathematical, they require many assumptions the validity of which is based on the user's experience and judgment.

In summary, the application of hydrologic techniques to urban and urbanizing areas is an engineering function, not a technical exercise. Urban hydrology should be carried out under the careful guidance of an experienced engineer.

3.4 STATISTICAL ANALYSES FOR DETERMINING PEAK DISCHARGE AND VOLUME

Measured streamflow or other hydrologic events can be analyzed by statistical methods. In addition, statistical methods can be applied to streamflow and other data produced by continuous computer models (e.g., Walesh and Raasch, 1978; Walesh and Videkovich, 1978) and quasi-continuous computer models (e.g., Walesh and Snyder, 1979; Walesh et al., 1979).

Statistical methods produce a probabilistic statement about the future occurrence of a streamflow or other hydrologic event of specific magnitude. A series of measured annual maximum peak flow rates, for example, could be analyzed statistically to estimate the discharge that will be reached or exceeded with a 25 percent probability in a given year. Underlying the application of statistical methods is the concept that the available streamflow or other records

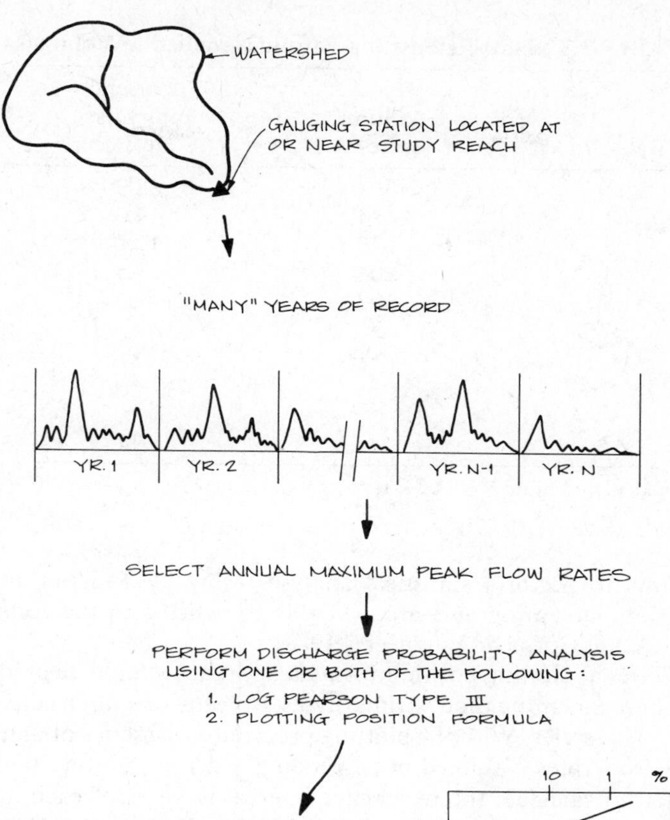

FIGURE 3.4 Statistical analysis of streamflow records

constitute a representative sample of the universe or population of streamflow or other records.

Methodologies

The statistical analysis procedure, using streamflow as an example, is shown schematically in Figure 3.4. The log-Pearson type III method is recommended by the U.S. Water Resources Council (1981b). The Council's report provides a detailed description of this statistical procedure, including presentation of underlying assumptions, example applications, and discussion of special circumstances. The log-Pearson type III method is also described and illustrated in hydrology textbooks (e.g., Bedient and Huber, 1988; Chow et al., 1988; Linsley et al., 1982; Viessman et al., 1977). Voluminous calculations are typi-

TABLE 3.3 Plotting Position Formula Applied to Volumes

Rank, m	Volume (acre-ft)	Recurrence Intervals[a] (years)
1	351	38.0
2	274	19.0
3	243	12.7
4	204	9.5
5	193	7.6
•	•	•
•	•	•
•	•	•
•	•	•
37	•	1.03

[a]Calculated from $T = (N + 1)/m$, where $N = 37$.

cally required to perform statistical analyses using log-Pearson and similar methods. Computer programs are available to assist with the computations (e.g., U.S. Army Corps of Engineers, 1983).

The traditional plotting position formula approach should be used in parallel to the log-Pearson method, primarily as a means of graphically depicting the historic data series. With the plotting procedure, the series of annual maximum peak flow rates is ranked in descending order. A plotting position formula is used to calculate the recurrence interval in years of each item in the series. A typical plotting position formula is $(N + 1)/m$, where N is the number of years of available record and m is the rank of the peak flows arranged from largest to smallest. This and other available plotting position formulas are discussed and illustrated in hydrology texts (e.g., Linsley et al., 1982; Viessman et al., 1977). Table 3.3 is an illustration of the application of a plotting position formula to the analysis of storage volumes for a detention/retention facility.

Urban hydrologic investigations typically require the determination of flows having recurrence intervals up to 100 years. Unfortunately, available streamflow records rarely exceed 50 years. The usual disparity between the desired recurrence interval and the period of record raises the question of the effect of record length on reliability of the resulting discharge–probability analysis. A few studies have been conducted to determine the effect of length of historic record on discharge–probability analysis (Benson, 1960; Hydrocomp, 1977; Linsley et al., 1982; Ott, 1971; Victorov, 1971). Based on these investigations, one may tentatively conclude that there is a tendency to overestimate flows. As a rule of thumb, statistical methods should not be used to estimate recurrence intervals in years that are more than twice the number of years of available homogeneous data.

Positive Features

The strength of statistical analysis of streamflow records is that the procedure is usually based on flows actually measured at or near the study area. In contrast, no other hydrologic method presented in this chapter incorporates measured streamflow.

Negative Features

A significant negative characteristic of statistical analysis of streamflow records is that it is not likely to be applicable in urban and urbanizing areas. First, gauging stations are usually not located in the small watersheds typically analyzed in urban surface water management. Even if the gauging station exists, the length of record is likely to be much shorter than gauges found on larger rural basins, mainly because streamflow monitoring in urban areas is a relatively recent development. Finally, the urban watershed may have been significantly altered as a result of various aspects of urbanization during the period of streamflow records, and therefore the record would not be suitable for statistical analysis because it would be heterogeneous.

Another negative feature of statistical analysis of streamflow records is lack of a procedure for forecasting the effect of hypothetical or planned watershed changes on flood flows. That is, statistical analyses cannot be used to determine how urbanization or alternative surface water management measures will alter discharge–probability relationships. Most other hydrologic methods described in this chapter can be used to determine the impact of watershed changes.

3.5 REGIONAL METHODS FOR DETERMINING PEAK DISCHARGE

A regional method is a correlation of a dependent variable (e.g., 50-year-recurrence-interval discharge) with one or more causative or physically related and readily determined watershed and stream system factors (e.g., watershed area or stream slope) for a defined geographic area. This category of hydrologic methods is specified as being regional because any given method is applicable only within the region that provided the streamflow and watershed data used to develop the method.

In the United States, the most common region selected for development of a regional method is the state. This arrangement reflects the administrative structure of the U.S. Geological Survey, which usually takes the lead in preparing regional methods, and its state agency cooperators. A list of regional method documents, most of which apply to the rural portions of states, is included in a recent report by Sauer et al. (1983).

An extensive data base and a major analytic effort are required to develop a regional method. For example, the development of the regional method applicable to most of Wisconsin required statistical analysis of flow records from 184 gauging stations (Conger, 1981). A statistical analysis, such as the log-Pearson type III method, is applied to the streamflow data series for each gauging station to produce a discharge–probability relationship for that station. Flows of specified recurrence intervals (e.g., 2-, 5-, 10-, 25-, 50-, and 100-year) are selected from the many discharge–probability relationships and correlated, by a largely trial-and-error procedure, with readily measurable watershed characteristics.

Although a major effort is required to develop a regional method, regional methods are among the easiest to use. Regional methods typically require substitution of readily obtainable watershed parameters into an equation.

In applying a regional method, it is important to be aware of the methodology used to develop the method and therefore the restrictions inherent in the method. Regional methods must be applied to watersheds and under conditions consistent with those used to develop the method. For example, most regional methods are applicable only to large rural watersheds.

Methodology

Area, stream slope, percent imperviousness, and other watershed parameters are quantified. Watershed parameters are then substituted into one or more regional method equations to calculate flows of specified recurrence interval.

For example, a regional method was developed for use in urban and urbanizing areas of northeastern Illinois (U.S. Geological Survey, 1979). The equations used and the independent variables in the equations are typical of regional methods. The 100-year recurrence interval equation taken from this method is

$$Q_{100} = 48.0\, A^{0.660} S^{0.349} I^{0.172}$$

where Q_{100} = 100-year-recurrence-interval discharge in cubic feet per second
A = total drainage area in square miles
S = main channel slope in feet per mile using points located 10 percent and 85 percent of the distance along the channel from the point of interest to the basin divide
I = impervious area expressed as percent of the area of the watershed

Although uncommon, regional methods have been developed for a few urban areas, such as Washington, D.C. (Anderson, 1970) and New Jersey (Stankowski, 1974). Sauer et al. (1983) present a generalized regional method which is applicable to urban areas throughout the United States provided that the user has an independent estimate of equivalent rural discharges.

Regional methods are not common in urban areas, but more urban regional methods are likely to be developed, for two reasons. First, increasing interest in urban surface water management in general and nonpoint-source pollution in particular has resulted in an increase in runoff monitoring on small urban watersheds. Therefore, more streamflow data are becoming available. Second, the existing and future streamflow data base for urban areas can be extended using the typical long rainfall and other meteorological data bases. The extension may be accomplished with computer modeling techniques. The resulting extended flow series may then be used to develop a regional method applicable throughout the geographic area.

Positive Features

Modest data requirements and simple computations are the advantages of regional methods. Furthermore, these methods can readily be used to estimate the impact of additional urbanization. For example, the northeastern Illinois method could be used to predict urbanization impacts simply by increasing the percentage of impervious area.

Negative Feature

As already suggested, the principal weakness of regional techniques in urban surface water management is the lack of urban methods. That is, regional methods have been developed for relatively few urban areas, mainly because of the lack of sufficient stream gauging data.

3.6 TRANSFER METHOD FOR DETERMINING PEAK DISCHARGE

Using the transfer method, a flood flow of specified recurrence interval for a tributary area of given size and runoff characteristics is used to estimate a flood flow of the same recurrence interval for a larger or smaller portion of the watershed having similar runoff characteristics. The discharge transfer is made partly in proportion to the ratio of tributary area (Mertes, 1968; Wiitala et al., 1961). This simple method may be applicable in certain situations in urban hydrology. For example, the transfer method may be used to interpolate flood flows between locations on a stream system at which flows were determined by some other hydrologic method.

Underlying the transfer method is the assumption that the tributary area to which it is being applied has runoff characteristics similar to the tributary area for which a flow of specified recurrence interval is known. The only significant difference between the two watersheds, or two points in a given watershed, should be the size of the tributary areas.

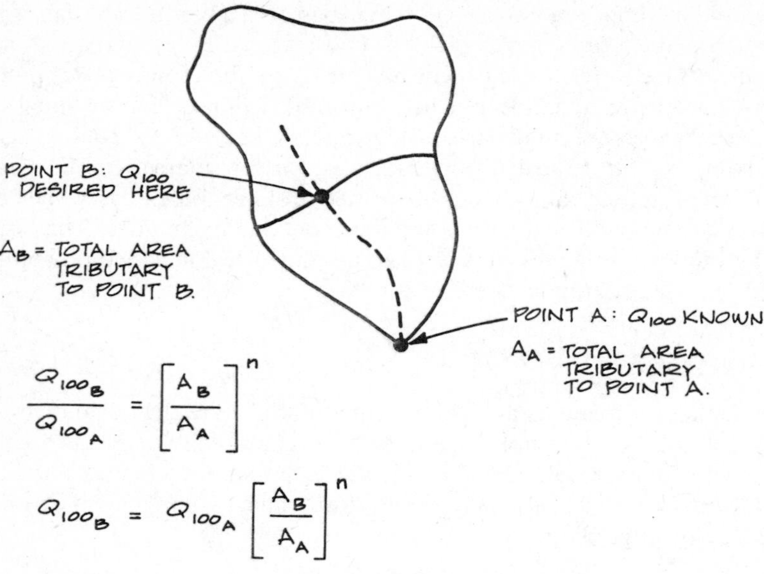

$$\frac{Q_{100_B}}{Q_{100_A}} = \left[\frac{A_B}{A_A}\right]^n$$

$$Q_{100_B} = Q_{100_A}\left[\frac{A_B}{A_A}\right]^n$$

FIGURE 3.5 Transfer method

Methodology

The transfer method is shown schematically in Figure 3.5. The 100-year-recurrence-interval flood flow is known for point *A* in the watershed and desired for point *B*, another location in the watershed. The required flood flow at point *B* is calculated as a function of the known discharge at point *A* and ratio of the tributary areas raised to the exponent *n*. Points *A* and *B* could also be in different watersheds provided that the total tributary areas have similar characteristics, including being subject to similar meteorologic conditions.

Two means are available for establishing a value for the exponent *n*. First, if a regional method exists, *n* is the exponent on tributary area in the regional method equation. For example, in northeastern Illinois *n* is 0.660, based on the applicable regional method (U.S. Geological Survey, 1979). Another means of quantifying *n* is to establish the mean annual flood for gauging stations in the geographic area. The exponent *n* is the slope of a log-log graph of mean annual flood versus tributary area (Wiitala et al., 1961).

If flood flows of a specified recurrence interval are available for two or more points along the stream system in a watershed and flows are required for one or more additional points in a watershed, a graphical interpolation or extrapolation might be used. The known flows are plotted versus tributary area on log-log graph paper and a line is fitted to the points. Flow estimates for additional points are obtained from the graph by entering it with the tributary area corresponding to each point and reading the estimated flow.

Assume, for example, that a watershed-wide hydrologic analysis has determined 100-year flood flows for relatively widely spaced locations along the watershed stream system. Because of the wide spacing, significant flow differences occur from point to point. Assume further that backwater computations are to be conducted for purposes of calculating flood stages and plotting the stages and mapping the floodplains. Prior to performing the backwater computations, the transfer method could be used to interpolate flood flows.

Positive Feature

The key advantage of the transfer method is the minimal data requirements, assuming that flood flows are available for one or more points having drainage areas similar to the location at which flood flows are desired.

Negative Features

A disadvantage of the transfer method is that tributary areas for points of known and unknown flows are not likely to have similar runoff characteristics. Dissimilarity in runoff characteristics is likely to be the case in urban areas because they usually exhibit a heterogeneous land use and land cover pattern. In addition, urban areas have varying and complex stormwater storage and conveyance facilities, including detention/retention facilities, channels, and storm sewers.

Determination of the exponent n may also pose a difficulty when applying the transfer method, particularly to urban areas. It should be based on long-term gauging records for stations in the area, which are not likely to be available in urban areas. As noted, another approach is to use the exponent on tributary area in a regional method. However, if an applicable regional method is available, it may be preferable to use that technique rather than the transfer method.

3.7 RAINFALL–RUNOFF METHODS: INTRODUCTION

Hydrologic loads, that is, surface water runoff rate and volume, are caused by rainfall and snowmelt. However, none of the preceding methods—statistical analysis of streamflow records, regional methods, or transfer methods—explicitly incorporate precipitation. Rainfall-runoff methods share a common trait; they all explicitly utilize rainfall either as an intensity or in the form of a hyetograph.

Design Storms

Most rainfall–runoff methods require selection of a design storm having a specified recurrence interval. These methods assume that the probability or recurrence interval of the resulting surface water runoff discharge or volume is equal to the probability or recurrence interval of the design storm.

The assumption that the recurrence interval of the storm equals the recurrence interval of the resulting discharge or volume should be questioned on a case-by-case basis because factors in addition to the design storm may influence peak flow or runoff volume. Examples of such factors are antecedent moisture conditions, presence and initial state of lakes and retention/detention facilities in the system, and spatial distribution of rainfall on the tributary area.

Marsalek (1978) compared discharge and volume probability relationships obtained with a design storm commonly used in Illinois to discharge and volume probability relationships obtained by inputting a series of historic storms to a computer model. The study concluded that the two approaches produced significantly different results and recommended use of historic storms rather than the design storm approach.

In two modeling studies, Wenzel and Voorhees (1978, 1979) compared the design storm approach to the use of series of historic storms. They concluded that the form of the design storm hyetograph and the assumed antecedent moisture conditions significantly influenced the resulting discharge–probability relationship.

Urbonas (1979) conducted a similar investigation and concluded that design storms generally significantly overestimated discharge–probability relationships. Additional studies and other information on the use of the design storm and alternatives to it are found in the literature [e.g. Adams and Howard, 1986; American Society of Civil Engineers (ASCE), 1983; Chow et al., 1988; Patry and McPherson, 1979; Walesh, 1979; Walesh and Snyder, 1979; Walesh et al., 1979].

The temporal distribution of the design storm hyetograph may be based on rainfall investigations in a particular geographic area. For example, extensive rainfall monitoring studies in Illinois (Huff, 1967, 1979) resulted in the characterization of rainfall events. Similar distributions were developed for the Lafayette region of Indiana (Burke, 1981). Figure 3.6 illustrates the kind of data available in Illinois for selecting the temporal distribution of a design rainfall event. Unfortunately, this kind of geographically specific summary data is rare, although, as discussed later in this chapter, the Soil Conservation Service (U.S. Department of Agriculture, 1986) has developed four 24-hour dimensionless hyetographs that cover the United States.

Development of a design storm hyetograph for a particular area usually requires rainfall amounts for various durations and recurrence intervals. If intensity–duration–frequency relationships are not available, such relationships can be developed using data available for almost any location in the

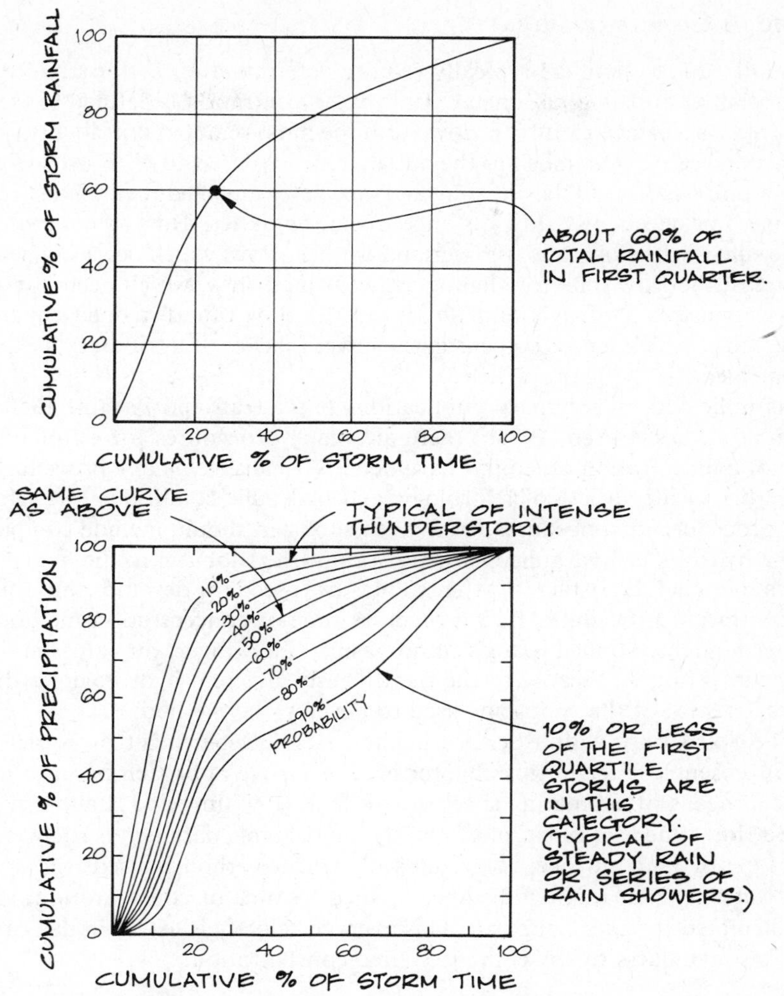

FIGURE 3.6 Cumulative dimensionless hyetographs for Illinois (*Source:* Adapted from Huff, 1967)

United States [National Oceanic and Atmospheric Administration (NOAA), 1977; U.S. Department of Commerce, 1961]. These reports provide rainfall volumes for durations of 5, 10, and 15 minutes and of 0.5, 1, 2, 3, 6, 12, and 24 hours; and for recurrence intervals of 1, 2, 5, 10, 25, 50, and 100 years. Refer to Sheaffer et al. (1982) for additional discussion of the availability of precipitation data, interpretation of the data, and development of design rainfalls.

Time of Concentration

Rainfall–runoff methods typically require determination of the time of concentration or a related parameter. Time of concentration is defined as the time required for surface runoff to flow from the most remote point in a subbasin, subwatershed, or watershed to the outlet. The word ''remote'' is used to designate a point on the subbasin, subwatershed, or watershed divide most remote in time, not necessarily flow distance, from the outlet. Time of concentration is the sum of the flow time for overland or sheet flow, which occurs in headwater areas; the flow time for shallow concentrated flow, which occurs immediately downstream of overland flow; and the flow time for open channel or sewer flow, which tends to occur in the lower reaches of a tributary area (U.S. Department of Agriculture, 1986).

As indicated by summary publications (e.g., Hall and Austin, 1980; McCuen et al., 1984; Yen, 1978b) there are many procedures for estimating the time of concentration or similar parameters. Because available procedures are based on a wide variety of hydrologic and hydraulic conditions, the selection of a procedure or procedures for a given subbasin should include comparison of the hydrologic–hydraulic characteristics of the subbasin to the hydrologic–hydraulic characteristics of the subbasins used to develop the time-of-concentration procedure. If two or more time-of-concentration methods are indiscriminately applied to a given subbasin, widely divergent values are likely to occur, probably because of the significantly different hydrologic–hydraulic characteristics of the subbasins used to develop the method.

The Soil Conservation Service method (U.S. Department of Agriculture, 1986) presented later in this chapter is an effective approach because it provides a means of estimating sheet or overland flow time and shallow concentrated flow time as a function of readily quantifiable parameters such as slope and type of land surface. Regardless of which method is used, the average flow velocity, that is, flow distance divided by time of concentration, should be calculated for each subbasin. This average velocity is useful in determining the reasonableness of the computed time concentration.

3.8 RATIONAL METHOD FOR DETERMINING PEAK DISCHARGE

The rational method is one of the most frequently used urban hydrology methods in the United States. The formula used in the rational method is

$$Q_T = C i_T A$$

where Q_T = peak discharge for the recurrence interval T in cubic feet per second

C = dimensionless runoff coefficient

i_T = average rainfall intensity during a period of time equal to the time of concentration for the recurrence interval T in inches per hour

A = tributary area in acres

Background

Emil Kuichling, City Engineer of Rochester, New York, developed the rational method a century ago (Kuichling, 1889; McPherson, 1969). Kuichling was concerned with commonly used methods based on rainfall periods of 1 hour or more. He hypothesized that shorter periods of more intense rainfall may cause larger flows and, accordingly, perhaps storm sewers were being undersized. Kuichling performed rainfall–runoff measurements on five subbasins in Rochester, ranging in size from 25 to 357 acres. Based on his measurements, he concluded the following:

1. Runoff volume is proportional to imperviousness. This effect is accounted for by the runoff coefficient in the rational method formula.
2. Maximum discharge occurs when the rainfall lasts long enough for the entire tributary area to contribute flow. This is the basis for defining t_c, the time of concentration, and using it to determine the average rainfall intensity.
3. Peak discharge is proportional to rainfall intensity. This is the basis for including intensity in the rational method equation.
4. Antecedent moisture levels are likely to have a significant effect on peak flows. However, Kuichling was not able to provide a means of accounting for antecedent moisture conditions.

Kuichling published the results of his Rochester research to establish a base for additional investigations that might be done by other engineers. He states in his paper: ''and hence it is obvious that a more rational method of sewer computation is urgently demanded . . . it is sincerely hoped that the efforts of the writer will be amply supplemented by many valuable suggestions and experiments each other member of the Society [American Society of Civil Engineers] may generously contribute'' (Kuichling, 1889). Now known as the rational method, the technique developed by Kuichling is used extensively in the United States. However, and despite the request for additional investigations as stated in his 1889 paper, few studies have been conducted (e.g., Schaake et al., 1967) and the rational method has changed little since Kuichling introduced it into U.S. engineering practice a century ago.

Methodology

Explanations of the rational method and procedures, including tabular formats for performing the computations, are readily available. Procedural de-

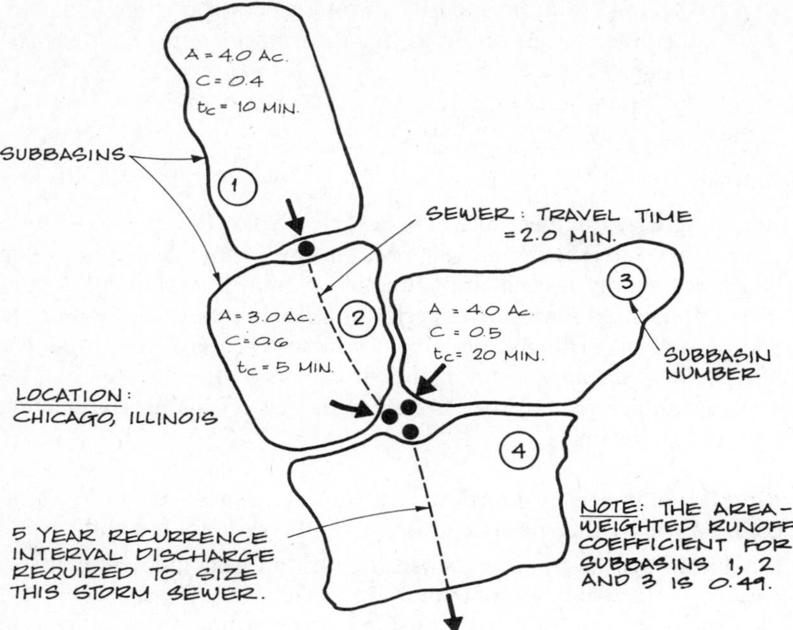

FIGURE 3.7 Subbasins and physical parameters for rational method example

scriptions are presented in design manuals (e.g., ASCE, 1969), hydrology textbooks (e.g., Viessman et al., 1977), and special publications and reports (Bauer, 1965; Rossmiller, 1980; Yen, 1978c). Rainfall intensity–duration–frequency curves which are used in the rational method are discussed by Wenzel (1978) with emphasis on the fact that the curves are based on parts or portions of storms.

A simple rational method example is presented to illustrate the basic procedure and to serve as a basis for subsequent discussion of potential misuse of the method. Figure 3.7 shows a four-subbasin watershed and presents area, runoff coefficient, and time-of-concentration data for three of the subbasins. Assume that the rational method is to be used to determine the 5-year recurrence interval design flow for a storm sewer that will receive stormwater from subbasins 1, 2, and 3 and convey it through subbasin 4, the downstream most subbasin. The watershed is located in Chicago and the corresponding intensity–duration–frequency relationships are shown in Figure 3.8.

Given the subbasin and rainfall data, the first step is to determine the time of concentration to the upstream end and of the storm sewer in subbasin 4. Three possible flow paths involving subbasins 1, 2, and 3 are examined to determine that path having the longest flow time and therefore defining the time of concentration for the combined areas of subbasins 1, 2, and 3. The

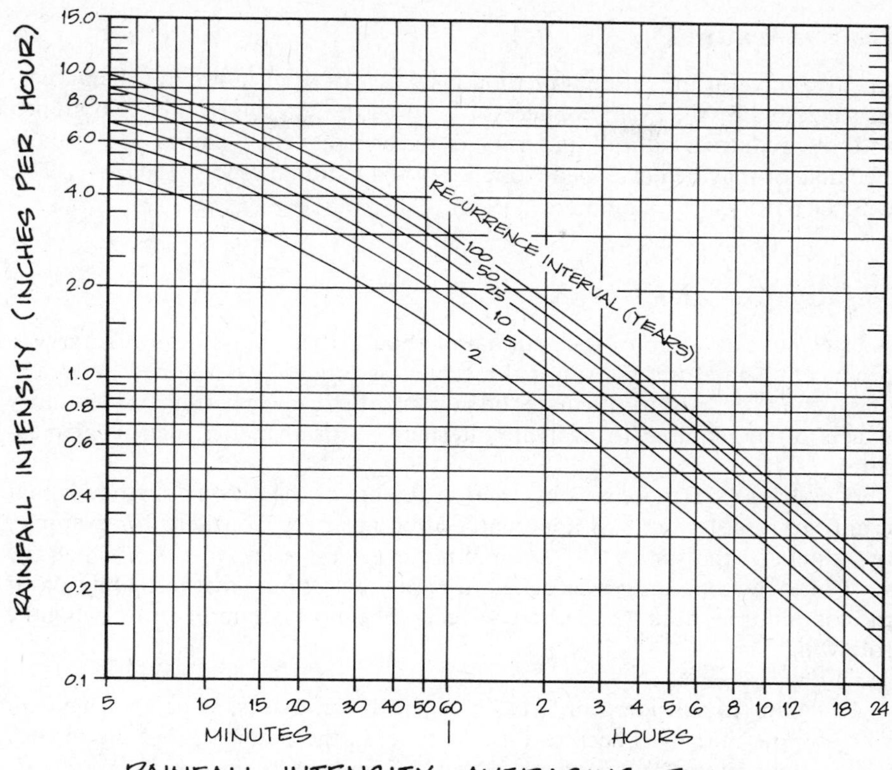

FIGURE 3.8 Rainfall intensity–duration–frequency relationships for Chicago, Illinois (*Source:* Adapted from Burke, 1981)

first flow path consists of overland or sheet flow across subbasin 1 followed by sewer flow through subbasin 2 for a total travel time of 12 minutes. The second flow path involves only sheet or overland flow on subbasin 2 and represents a travel time of 5 minutes. The third flow path consists only of sheet or overland flow on subbasin 3 and has a travel time of 20 minutes. The longest travel time and therefore the time of concentration for the area comprised of subbasins 1, 2, and 3 is 20 minutes.

For a 20-minute time of concentration or rainfall intensity averaging time, Figure 3.8 indicates that the rainfall intensity is 3.5 in./hr. As shown in Figure 3.7, the area-weighted runoff coefficient for the total area represented by subbasins 1, 2, and 3 is 0.49. Therefore, the desired 5-year recurrence interval design flow for the storm sewer in subbasin 4 is

$$\begin{aligned} Q_5 &= C i_5 A \\ &= (0.49)(3.5 \text{ in./hr})(11 \text{ acres}) \\ &= 18.9 \text{ acre-in./hr or } 18.9 \text{ ft}^3\text{/sec} \end{aligned}$$

Positive Features

An advantage of the rational method is its ease of application in urban areas. As suggested by the example, necessary subbasin data can readily be obtained and the required rainfall intensity–duration–frequency curves are usually available or may be developed from published national reports (NOAA, 1977; U.S. Department of Commerce, 1961).

Negative Features

A principal limitation of the rational method is that only a peak discharge is produced. Therefore, the rational method, as originally developed by Kuichling, cannot be used to calculate the volume of stormwater runoff, a quantity that is often required to evaluate alternative stormwater management measures.

A perhaps even more serious negative feature of the rational method is that it appears simple, but it is apparently often misused. Consider, for example, the results of the survey of 32 communities (Ardis et al., 1969) in which the staff in each community was asked to apply the rational method. Only six of the communities used the method correctly, the most common errors being the following:

1. Failing to consider rainfall as being variable, that is, as being dependent on the time of concentration. Assume, in the example, that the time of concentration is incorrectly set at 10 minutes, the time of concentration of the subbasin farthest upstream. For a rainfall-intensity averaging period of 10 minutes, Figure 3.8 indicates that the rainfall intensity would be 4.8 in./hr. The 5-year recurrence interval design flow for the total area represented by subbasins 1, 2, and 3 would be

$$\begin{aligned} Q_5 = Ci_5 A &= (0.49)(4.8 \text{ in./hr})(11 \text{ acres}) \\ &= 25.9 \text{ ft}^3/\text{sec} \end{aligned}$$

   This incorrectly calculated discharge is 37 percent larger than the correct value.

2. Calculating flows for individual subbasins and simply summing them to get the total flow for the watershed rather than determining a rainfall intensity and weighted runoff coefficient for each successive downstream point in the drainage system. Assume, in the example, that 5-year recurrence discharges are calculated for each of three subbasins and then incorrectly summed. The 5-year discharge for subbasin 1 is

$$\begin{aligned} Q_5 = Ci_5 A &= (0.4)(4.8 \text{ in./hr})(4.0 \text{ acres}) \\ &= 7.7 \text{ ft}^3/\text{sec} \end{aligned}$$

where 4.8 in./hr is the rainfall intensity corresponding to the time of concentration for subbasin 1. Similarly, the 5-year recurrence interval discharges for subbasins 2 and 3 are calculated, respectively, to be 10.8 and 7.0 ft^3/sec. The sum of the 5-year recurrence discharges of each of the three subbasins is 25.5 ft^3/sec, which is 35 percent larger than the correct value.

3. Failing to account for flow time in storm sewers in computing the time of concentration. This error would have no effect in the example because of the relatively long time of concentration of subbasin 3.

Therefore, at least one documented investigation combined with experience suggests that the apparently simple rational method is occasionally misunderstood and applied incorrectly. The method should always be used under the close supervision of an experienced engineer.

Another apparently widely held misconception about the rational method is that the time of concentration corresponds to the duration of a storm. The time of concentration is simply the critical time period used to determine average rainfall intensity from intensity–duration–frequency curves. Inasmuch as rainfall is a random event, it is possible that a rainfall event could have a duration equal to the time of concentration of a subbasin. However, it is much more likely that the total duration of a storm will be longer than the time of concentration used in the rational method. Furthermore, the critical duration of interest could occur anywhere within the storm.

The misunderstanding that time of concentration equals duration of the rainfall rather than the rainfall intensity averaging time has resulted in the development of detention/retention sizing methodologies that are likely to significantly underestimate storage volumes. These procedures fail to account for the rainfall and the resulting runoff volume that is generated before and after the rainfall intensity averaging time.

3.9 MODIFIED RATIONAL METHOD FOR DETERMINING DISCHARGE AND VOLUME

This procedure extends the rational method to the development of runoff hydrographs. The runoff hydrographs are used to compute the size of a detention/retention facility for a specified recurrence interval and concurrent release rate (Kao, 1975; Poertner, 1974; Rao, 1975; Walesh, 1981a).

Theory

The modified rational method (MRM) relies on the assumptions of the rational method. Therefore, the MRM is subject to the same misunderstandings and

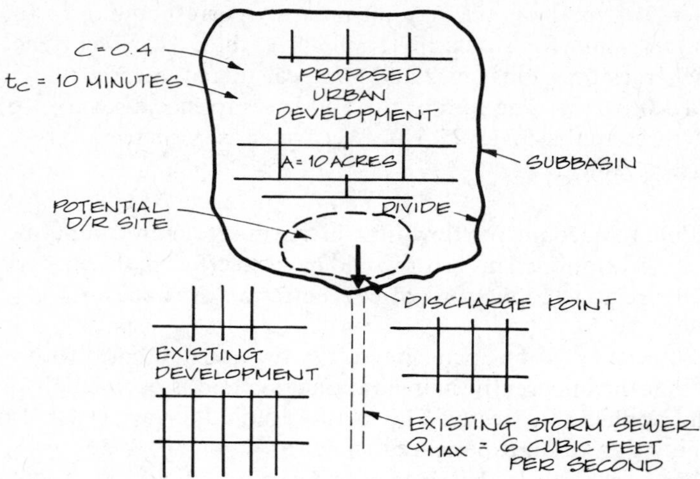

FIGURE 3.9 Subbasin and physical parameters needed to apply modified rational method

misapplications as the rational method. An important additional assumption is that the rainfall intensity averaging time used in the MRM equals the storm duration. As introduced in the preceding discussion of the rational method, this assumption means that the rainfall, and the runoff generated by the rainfall that occurs before or after the rainfall averaging period are not accounted for. Accordingly, the MRM may seriously underestimate required storage volume.

The MRM also is based on the assumption that the runoff hydrograph for an urban area can be approximated as being either triangular or trapezoidal in shape. This is equivalent to assuming that the runoff from the tributary area increases linearly with time or that there is a linear area–time relationship for the subbasin.

MRM hydrograph shape assumptions may be explained using the subbasin illustrated in Figure 3.9. The subbasin has an area of 10 acres, a runoff coefficient of 0.4, and a time of concentration of 10 minutes. As illustrated in Figure 3.10, three types of hydrographs may be developed for the subbasin using the MRM procedure. Hydrograph type is a function of the length of the rainfall averaging time, *d,* with respect to the subbasin time of concentration. The following three types are possible:

Type 1 d is greater than t_c. The resulting trapezoidal hydrograph has a uniform maximum discharge of Ci_TA, as determined from the rational method. The linear rising and falling limbs each has a duration of t_c.

Type 2 d is equal to t_c. The resulting triangular hydrograph has a peak

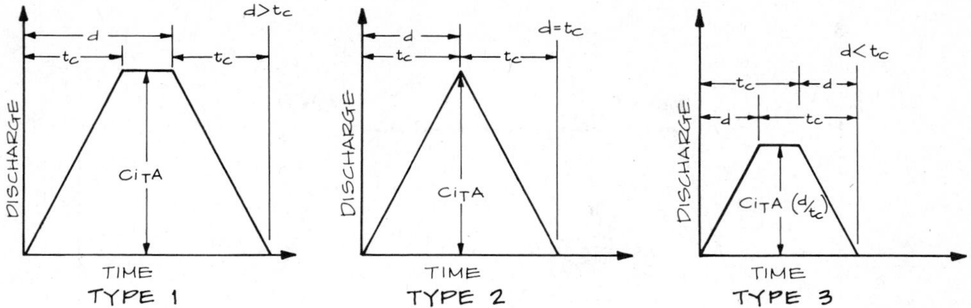

FIGURE 3.10 Hydrograph types for the modified rational method

discharge of Ci_TA. The linear rising and falling limbs each has a duration of t_c.

Type 3 d is less than t_c. The resulting trapezoidal hydrograph has a uniform maximum discharge of Ci_TA times (d/t_c). The linear rising and falling portions of the hydrograph each has a duration of d.

Assume, for example, that the rainfall averaging time for the subbasin shown in Figure 3.9 is 30 minutes, which exceeds the 10-minute time of concentration. Therefore, a type 1 hydrograph will occur. Assume further that a 10-year-recurrence-interval intensity duration curve is available and that the rainfall intensity for a rainfall averaging period of 30 minutes is 3.5 in./hr. The uniform maximum flow calculated using Ci_TA is 14 ft^3/sec, as shown in Figure 3.11. The linear rising and falling portions of the hydrograph each covers a time period of t_c, or 10 minutes, as also shown in Figure 3.11.

In summary, hydrograph type in the MRM is determined by the relationship between rainfall averaging time and the time of concentration of the subbasin. Given the hydrograph type, the peak discharge is determined using the rational method equation or a variation on the equation.

Methodology

Step 1: Obtain and Display Subbasin Data. Use of the MRM can be viewed as a four-step process. The first step is to obtain and display the following data for the subbasin for which a detention/retention site is being considered: tributary area, runoff coefficient, time of concentration, and allowable release rate from the detention/retention site.

These data for an example subbasin are shown in Figure 3.9. A potential detention/retention facility site exists at the downstream end of the subbasin. The maximum allowable release rate from the detention/retention facility, 6 ft^3/sec, reflects the available capacity in a downstream storm sewer.

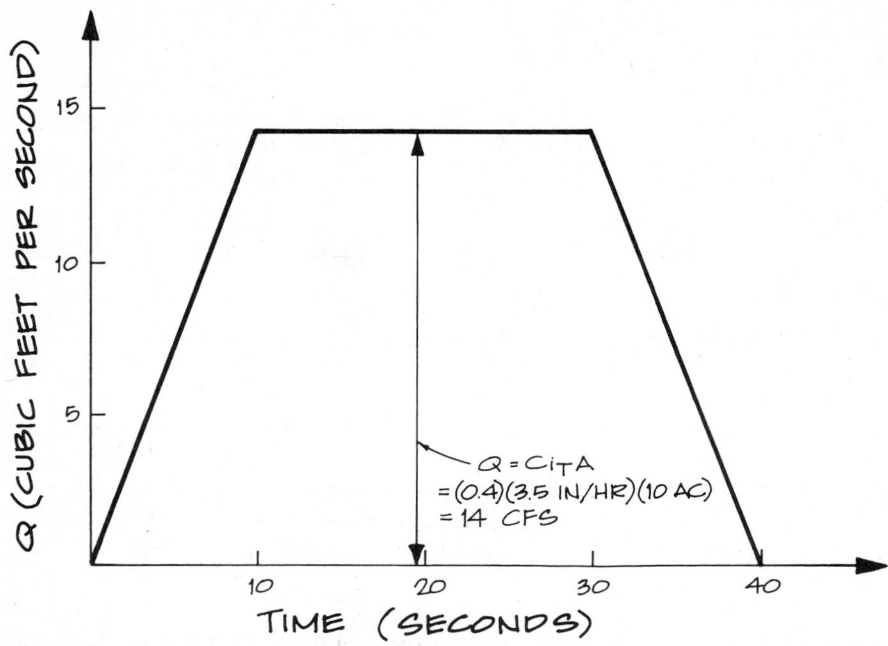

FIGURE 3.11 Example of a hydrograph developed with the modified rational method

Step 2: Develop a Family of Hydrographs. A family of hydrographs are developed under the second step. Assume that the detention/retention facility is to be sized for a 10-year recurrence interval event and that the corresponding intensity–duration relationship, although not shown here, is available. Rainfall intensity averaging periods are selected arbitrarily. In the case of the example, periods of 10, 20, 30, 40, 50, 60, and 70 minutes were used. The shortest period selected normally approximates the time of concentration of the subbasin, which in this case is 10 minutes. After completing the first pass through the sizing procedure, it may be necessary to add additional rainfall averaging periods.

MRM hydrographs are then developed for each of the rainfall intensity averaging periods. For the example, the resulting hydrographs are shown in Figure 3.12. The hydrograph corresponding to the 10-minute rainfall averaging period is a type 2, or triangle, because the rainfall averaging period of 10 minutes equals the time of concentration. All other hydrographs are type 1 trapezoids because the rainfall intensity averaging period exceeds the time of concentration.

Step 3: Determine Set of Runoff Volumes. To begin this step, the maximum allowable release rate of 6 ft^3/sec is superimposed on the MRM hydrographs

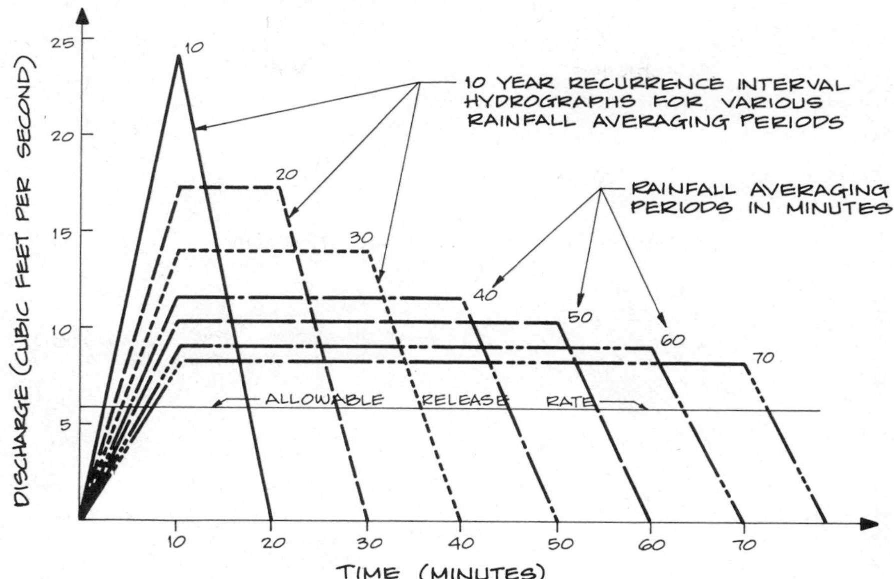

FIGURE 3.12 Hydrographs developed for use in the modified rational method

as shown in Figure 3.12. The required detention/retention facility storage for each hydrograph is approximated as the area beneath the hydrograph and above the maximum allowable release rate.

The basis for this approximation is illustrated in Figure 3.13, where the uniform release rate is assumed to approximate the curvilinear outlet hydrograph typical of a detention/retention facility. Because of the approximation, the resulting required storage slightly underestimates the actual required storage, which would be defined as the area above the curvilinear outflow hydrograph and below the inflow hydrograph. However, because the release rate is usually small compared to the peak discharge of the inflow hydrograph, the resulting volume error will be small compared to the volume of storage required.

Step 4: Select Critical Storage. The final step in the MRM procedure is shown in Table 3.4 and consists of a tabulation of the required storage values for each of the seven rainfall averaging periods. The critical value is 12,200 ft^3 because it is the largest one in the series. If the tabulated values do not contain a maximum, additional rainfall averaging periods would be used.

Positive Features

A strength of the MRM is the relative ease of application and its similarity to the familiar rational method.

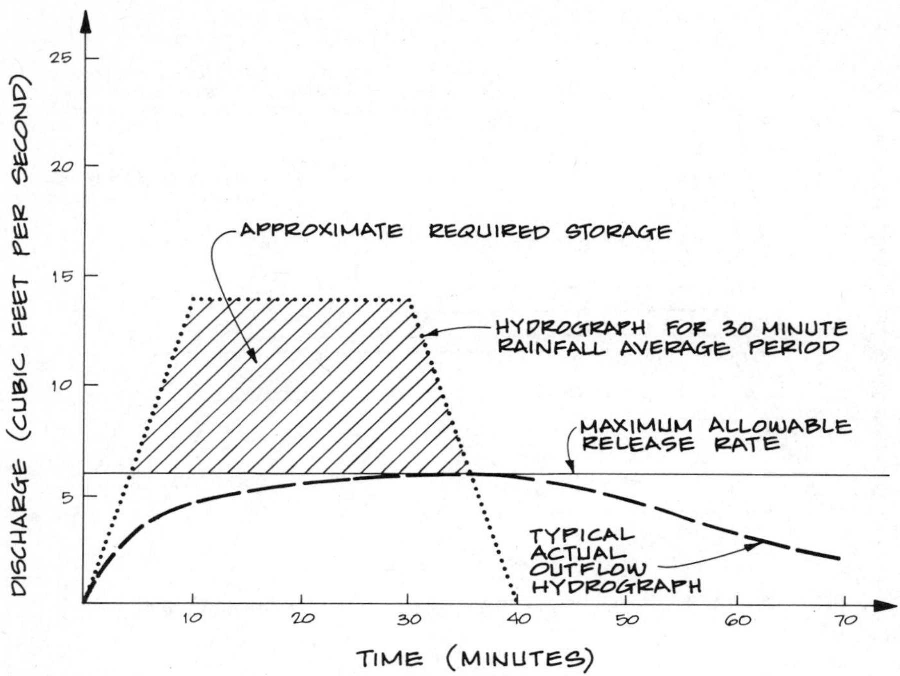

FIGURE 3.13 Method used to approximate required storage in the modified rational method

Negative Features

The most serious deficiency of the MRM, particularly if it is not recognized by the user, is the implicit assumption that the rainfall intensity averaging time incorporated in the method equals the duration of a storm. That is, the MRM assumes that the rainfall occurring under design conditions continues only as long as the rainfall intensity averaging time. Although possible, the much more likely situation is one in which the rainfall intensity averaging period will be preceded and followed by other rainfall. The runoff volume generated by the rainfall prior to and after the rainfall intensity averaging period is not included in the storage volume calculated with the method. Thus the MRM method will generally underestimate the required volume of storage.

Experience with the MRM used in parallel with other sizing methods indicates that significant underestimation of required storage may occur. Users of the MRM should carefully evaluate the results obtained with this approach and should probably restrict use of the MRM to preliminary determination of detention/retention facility volumes.

Another negative aspect of MRM is use of a uniform release rate, which,

TABLE 3.4 Ten-Year-Recurrence-Interval Storage Values Obtained with the Modified Rational Method

Rainfall Averaging Period (min)	Storage (ft^3)
10	8,100
20	11,100
30	12,200[a]
40	11,800
50	11,700
60	10,300
70	9,000

[a]Largest value = required storage.

as noted earlier, causes a small underestimation of required storage volume. The disadvantage of representing the release rate as a uniform discharge can be eliminated by routing the triangular or trapezoidal hydrograph using reservoir routing techniques (Baker, 1977). Refinement of the routing, however, has no effect on the inherent weakness of neglecting rainfall and runoff prior to and subsequent to the rainfall averaging period.

Other versions of the MRM are available in addition to the graphically oriented version described in this chapter. For example, a similar method has been specified for use in the Chicago metropolitan area since the early 1970s (Metropolitan Sanitary District of Greater Chicago, 1972). Calculations are performed in a specified tabular format, but the underlying concepts and assumptions are very similar to the MRM described in this chaper.

An analytic or equation version of the method is also available (Burton, 1980). An equation was developed in which the required storage volume can be directly calculated as a function of subbasin area, allowable release rate, and intensity–duration curve parameters. Once again, the concepts and assumptions are similar to the MRM presented in this chapter (Walesh, 1981b).

Another graphical means of forming the MRM is to plot cumulative required storage versus rainfall intensity averaging time. Superimpose on this graph the linear relationship between cumulative allowable discharge and rainfall intensity averaging time. The required storage is determined graphically as the maximum vertical difference between the plotted linear and curvilinear relationships. Because of the shape of the resulting pair of graphs, this technique is sometimes called the "bow" method.

Again, whether the technique is performed graphically, in tabular fashion, or in equation form, the underlying concepts and assumptions are the same. All versions are subject to the same negative features—the most important being underestimation of required storage.

3.10 SOIL CONSERVATION SERVICE TR55 METHOD FOR DETERMINING DISCHARGE AND VOLUME

An urban hydrology method developed by the Soil Conservation Service, released in 1975 (U.S. Department of Agriculture, 1975) and subsequently significantly revised (U.S. Department of Agriculture, 1986), is often referred to as TR55. The method calculates peak discharge and volume for one or more recurrence intervals. Given a specified recurrence interval, the TR55 method can calculate corresponding peak discharge and with the addition of an allowable peak outflow discharge can be used to calculate the required volume of a detention/retention facility. The TR55 method can also be used to develop runoff hydrographs for subbasins in a watershed and to combine them as they move through the stream system.

In addition to the primary technical report (U.S. Department of Agriculture, 1986), other reports and user aids are available from the federal government (e.g., U.S. Department of Agriculture, 1970) and from other sources (e.g., McCuen, 1982; Wisner et al., 1980). Available user aids include a personal computer program (U.S. Department of Agriculture, 1986).

Origin

The TR55 method was developed by performing many computer simulations using the SCS computer program Project Formulation-Hydrology (U.S. Department of Agriculture, 1965; 1983). Computer modeling results were summarized as a set of graphs and tables which are presented in the technical report. These graphs and tables, used in conjunction with other technical data, all of which is included in the report, facilitate the quick manual application of TR55.

In performing the computer simulations needed to develop TR55, a range of physical conditions was established. Accordingly, TR55 is applicable to rural and urban watersheds only under the following conditions:

1. Use of one of four 24-hour rainfall hyetographs, as shown in Figure 3.14, the selection of which depends on the location in the United States, as shown on a map included in the TR55 manual.
2. Average antecedent moisture conditions are defined as a total of 1.4 to 2.1 in. of rainfall during the 5-day period immediately preceding the design rainfall. Adjustments may be made to simulate dry or wet antecedent moisture conditions.
3. All flow is assumed to be sheet flow, shallow concentrated flow, or open channel flow. Conditions such as surcharged pipe flow and flow on street surfaces are not accounted for explicitly.

Numerous other conditions and limitations may apply, depending on which part of the procedure is being used. TR55 is introduced in this chapter by using

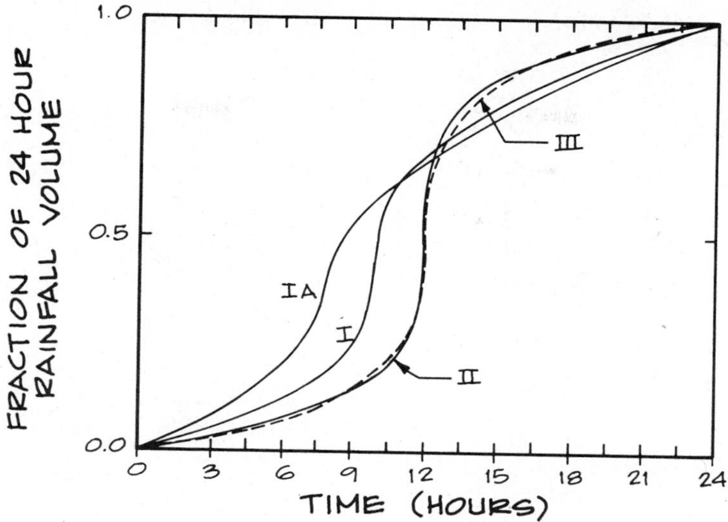

FIGURE 3.14 Twenty-four-hour rainfall distributions as developed by the Soil Conservation Service (*Source:* Adapted from U.S. Department of Agriculture, 1986)

an example to illustrate some of its many features. The method should be applied to actual situations only after careful study of the procedures and limitations presented in the manual (U.S. Department of Agriculture, 1986).

Methodology

Application of TR55 to determine peak discharge and the volume of detention storage may, in keeping with the presentation format used in the manual, be viewed as the following four-step process: (1) determine runoff volume, (2) estimate time of concentration, (3) compute peak discharge, and (4) compute storage volume. The homogeneous urban subbasin shown in Figure 3.15 is used to illustrate the procedure.

The indicated data required for application of the TR55 method include recurrence interval(s) of interest; subbasin location, because this determines meteorological conditions; subbasin area; hydrologic soil group or groups; land use (more explicitly referred to as cover type and hydrologic condition); the amount of impervious surface and the nature of its connection to the subbasin drainage system; type, orientation, length, slope, and roughness of flow paths; and allowable peak outflow discharge from the subbasin.

For this example, assume that the 100-year recurrence interval discharge from the subbasin is to be calculated and that the detention storage required to reduce that peak flow to an acceptable downstream discharge is to be determined.

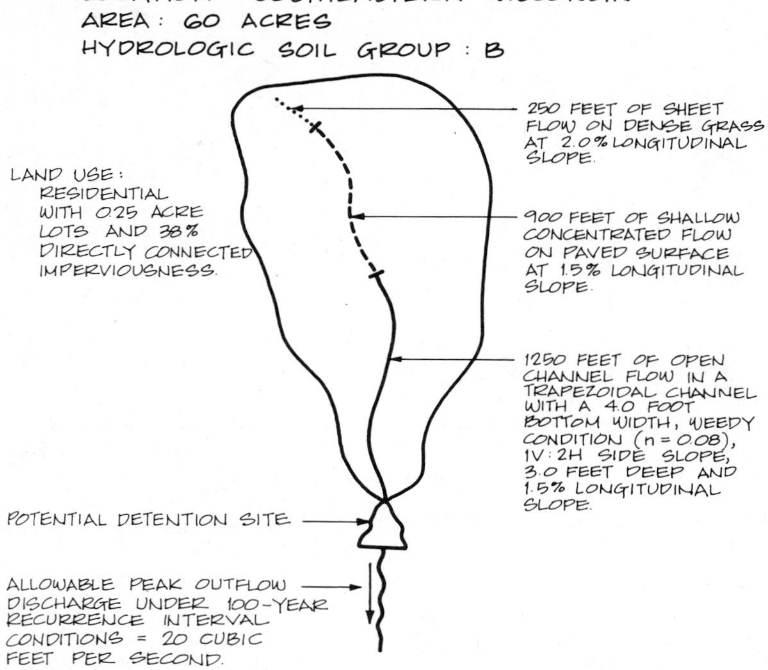

FIGURE 3.15 Subbasin used for illustration of the TR55 method

Step 1: Determine Runoff Volume. The runoff curve number (RCN) is a runoff coefficient because it provides a way to convert rainfall volume to surface runoff volume. Table 3.5, which is adapted from the TR55 manual (U.S. Department of Agriculture, 1986), provides for the estimation of the RCN as a function of land use and hydrologic soil group. Entering the table with residential land use at an imperviousness of 38 percent and hydrologic soil group B, the RCN is determined to be 75.

TR55 accommodates various possible Step 1 complexities that are not relevant to this example. These include determining the RCN for impervious percentages not listed in Table 3.5, setting the RCN when all of the impervious surface is not directly connected to the drainage system, and determining the RCN when the drainage area contains more than one land use–hydrologic soil group combination.

As specified for the example, flow and volume are to be determined for a 100-year-recurrence-interval condition and, as also noted, TR55 utilizes a 24-hour rainfall. For southeastern Wisconsin, the location of the example subbasin, the 100-year-recurrence-interval 24-hour rainfall volume is 5.5 in. This quantity was taken from maps included in the TR55 manual (U.S. Department of Agriculture, 1986), which are reproductions of rainfall maps available elsewhere (e.g., U.S. Department of Commerce, 1961).

Entering Table 3.6, which is adapted from the TR55 manual, with a RCN of 75 and the rainfall volume of 5.5 in., the associated runoff volume is determined by interpolation to be 2.9 in. More detailed tables and charts which virtually eliminate the need for interpolation are available (e.g., U.S. Department of Agriculture, 1970).

Step 1 indicates that 5.5 in. of rain will fall on the subbasin under the specified design storm conditions. Given the characteristics of the subbasin, the RCN is 75 and 2.9 in. of surface runoff will occur. This runoff volume is equivalent to 14.5 acre-feet when related to the 60-acre subbasin.

Step 2: Estimate Time of Concentration. Time of concentration is the sum of flow time for overland or sheet flow, flow time for shallow concentrated flow, and flow time for open channel flow. TR55 recommends a systematic approach in which each of the three components of the time of concentration is analyzed and determined separately and then summed.

For sheet flow, the following equation is recommended in TR55:

$$t = \frac{(0.007) \times (n \times L)^{0.8}}{P_2^{0.5} S^{0.4}}$$

where t = sheet flow time, in hours
n = Manning roughness coefficient for sheet flow
L = length of sheet flow path in feet, not to exceed 300 ft
P_2 = 2-year-recurrence-interval 24-hour rainfall in inches
S = slope along sheet flow path in feet per foot

Table 3.7, which is taken from the TR55 manual (U.S. Department of Agriculture, 1986), provides special values of Manning roughness coefficients. For the dense grass specified in the example, $n = 0.24$. The 2-year-recurrence-interval 24-hour rainfall volume for southeastern Wisconsin is 2.75 in. Substituting $n = 0.24$, $L = 250$ ft, $P_2 = 2.75$ in., and $S = 0.02$, the sheet flow time is 0.53 hour.

Flow time for shallow concentrated flow is obtained from Figure 3.16, which is taken from the TR55 manual. Entering Figure 3.16 for a paved surface at $S = 0.015$ yields an average velocity of 2.5 ft/sec. The flow time is 360 seconds or 0.10 hour, which is obtained by dividing the length of the flow path (900 ft) by the average velocity.

Average velocity for open channel flow in the lower reaches in the subbasin can be obtained using the Manning equation for open channel flow, assuming bankfull or less than bankfull conditions. For the given trapezoidal channel, the average velocity at bankfull conditions is 3.3 ft/sec. Incidentally, the average velocity at half depth is about 60 percent of the bankfull condition veloc-

TABLE 3.5 Runoff Curve Number as a Function of Land Use and Hydrologic Soil Group[a]

Land Use		Curve Numbers (CN) for Indicated Hydrologic Soil Groups			
Cover type and Hydrologic Condition	Average Percent Impervious[b]	A	B	C	D
Fully Developed Urban Areas (Vegetation Established)					
Open space (lawns, parks, golf courses, cemeteries, etc.)[c]					
Poor condition (grass cover less than 50%)		68	79	86	89
Fair condition (grass cover 50% to 75%)		49	69	79	84
Good condition (grass cover greater than 75%)		39	61	74	80
Impervious areas:					
Paved parking lots, roofs, driveways, etc. (excluding right-of-way)		98	98	98	98
Streets and roads:					
Paved: curbs and storm sewers (excluding right-of-way)		98	98	98	98
Paved: open ditches (including right-of-way)		83	89	92	93
Gravel (including right-of-way)		76	85	89	91
Dirt (including right-of-way)		72	82	87	89

Western desert urban areas:					
Natural desert landscaping (pervious areas only)		63	77	85	88
Artificial desert landscaping (impervious weed barrier, desert shrub with 1- to 2-inch sand or gravel mulch and basin borders)		96	96	96	96
Urban districts:					
Commercial and business	85	89	92	94	95
Industrial	72	81	88	91	93
Residential districts by average lot size:					
$\frac{1}{8}$ acre or less (town houses)	65	77	85	90	92
$\frac{1}{4}$ acre	38	61	75	83	87
$\frac{1}{3}$ acre	30	57	72	81	86
$\frac{1}{2}$ acre	25	54	70	80	85
1 acre	20	51	68	79	84
2 acres	12	46	65	77	82
Developing Urban Areas					
Newly graded areas (pervious areas only, no vegetation)		77	86	91	94

Source: Adapted from U.S. Department of Agriculture (1986).

[a]For average runoff condition and $I_a = 0.2S$, that is, initial abstraction equals 0.2 times potential maximum retention after runoff begins.

[b]The average percent impervious area shown was used to develop the composite CNs. Other assumptions are as follows: impervious areas are directly connected to the drainage system, impervious areas have a CN of 98, and pervious areas are considered equivalent to open space in good hydrologic condition.

[c]CNs shown are equivalent to those of pasture. Composite CNs may be computed for other combinations of open space cover type.

TABLE 3.6 Runoff in Inches as a Function of Rainfall and Runoff Control Number

Rainfall	Runoff Curve Number												
(in.)	40	45	50	55	60	65	70	75	80	85	90	95	98
1.0	0.00	0.00	0.00	0.00	0.00	0.00	0.00	0.03	0.08	0.17	0.32	0.56	0.79
1.2	0.00	0.00	0.00	0.00	0.00	0.00	0.03	0.07	0.15	0.27	0.46	0.74	0.99
1.4	0.00	0.00	0.00	0.00	0.00	0.02	0.06	0.13	0.24	0.39	0.61	0.92	1.18
1.6	0.00	0.00	0.00	0.00	0.01	0.05	0.11	0.20	0.34	0.52	0.76	1.11	1.38
1.8	0.00	0.00	0.00	0.00	0.03	0.09	0.17	0.29	0.44	0.65	0.93	1.29	1.58
2.0	0.00	0.00	0.00	0.02	0.06	0.14	0.24	0.38	0.56	0.80	1.09	1.48	1.77
2.5	0.00	0.00	0.02	0.08	0.17	0.30	0.46	0.65	0.89	1.18	1.53	1.96	2.27
3.0	0.00	0.02	0.09	0.19	0.33	0.51	0.71	0.96	1.25	1.59	1.98	2.45	2.77
3.5	0.02	0.08	0.20	0.35	0.53	0.75	1.01	1.30	1.64	2.02	2.45	2.94	3.27
4.0	0.06	0.18	0.33	0.53	0.76	1.03	1.33	1.67	2.04	2.46	2.92	3.43	3.77
4.5	0.14	0.30	0.50	0.74	1.02	1.33	1.67	2.05	2.46	2.91	3.40	3.92	4.26
5.0	0.24	0.44	0.69	0.98	1.30	1.65	2.04	2.45	2.89	3.37	3.88	4.42	4.76
6.0	0.50	0.80	1.14	1.52	1.92	2.35	2.81	3.28	3.78	4.30	4.85	5.41	5.76
7.0	0.84	1.24	1.68	2.12	2.60	3.10	3.62	4.15	4.69	5.25	5.82	6.41	6.76
8.0	1.25	1.74	2.25	2.78	3.33	3.89	4.46	5.04	5.63	6.21	6.81	7.40	7.76
9.0	1.71	2.29	2.88	3.49	4.10	4.72	5.33	5.95	6.57	7.18	7.79	8.40	8.76
10.0	2.23	2.89	3.56	4.23	4.90	5.56	6.22	6.88	7.52	8.16	8.78	9.40	9.76
11.0	2.78	3.52	4.26	5.00	5.72	6.43	7.13	7.81	8.48	9.13	9.77	10.39	10.76
12.0	3.38	4.19	5.00	5.79	6.56	7.32	8.05	8.76	9.45	10.11	10.76	11.39	11.76
13.0	4.00	4.89	5.76	6.61	7.42	8.21	8.98	9.71	10.42	11.10	11.76	12.39	12.76
14.0	4.65	5.62	6.55	7.44	8.30	9.12	9.91	10.67	11.39	12.08	12.75	13.39	13.76
15.0	5.33	6.36	7.35	8.29	9.19	10.04	10.85	11.63	12.37	13.07	13.74	14.39	14.76

Source: U.S. Department of Agriculture (1986).

TABLE 3.7 Manning's Roughness Coefficients for Sheet Flow

Surface Description	Roughness Coefficient n^a
Smooth surfaces (concrete, asphalt, gravel, or bare soil)	0.011
Fallow (no residue)	0.05
Cultivated soils	
Residue cover (less than 20%)	0.06
Residue cover (greater than 20%)	0.17
Grass	
Short grass prairie	0.15
Dense grasses[b]	0.24
Bermuda grass	0.41
Range (natural)	0.13
Woods[c]	
Light underbrush	0.40
Dense underbrush	0.80

Source: Adapted from U.S. Department of Agriculture (1986).

[a]Intended for use with the time-of-flow equation for sheet flow as presented in this chapter.

[b]Includes species such as weeping lovegrass, bluegrass, buffalo grass, blue grama grass, and native grass mixtures.

[c]When selecting n, consider cover to a height of about 0.1 ft. This is the only part of the plant cover that will obstruct sheet flow.

ity, indicating that velocity is relatively insensitive to flow depth. Therefore, the flow time for the open channel flow segment is 379 seconds or 0.11 hour, obtained as a quotient of the flow path length (1250 ft) and the average velocity under bankfull conditions.

Based on step 2, the flow times for sheet flow, shallow concentrated flow, and open channel flow are, respectively, 0.53, 0.10, and 0.11 hour. Therefore, the time of concentration for the example subbasin is the sum of these, 0.74 hour.

The TR55 manual briefly discusses various possible Step 2 complications not pertinent to the simplified example. Included are estimating the time of concentration for a lake or pond, accounting for surcharging of storm sewers, and being aware of storage effects in the system.

Step 3: Compute Peak Discharge. The design rainfall (5.5 in.) and the RCN (75) obtained from step 1 and the time of concentration (0.74 hour) obtained from step 2 are combined in step 3 with the rainfall distribution and the subbasin area to compute the peak discharge. The step begins with the determination of initial abstraction (I_a), which is the initial portion of a rainfall event in inches that occurs before surface runoff begins. The quantity is "abstracted

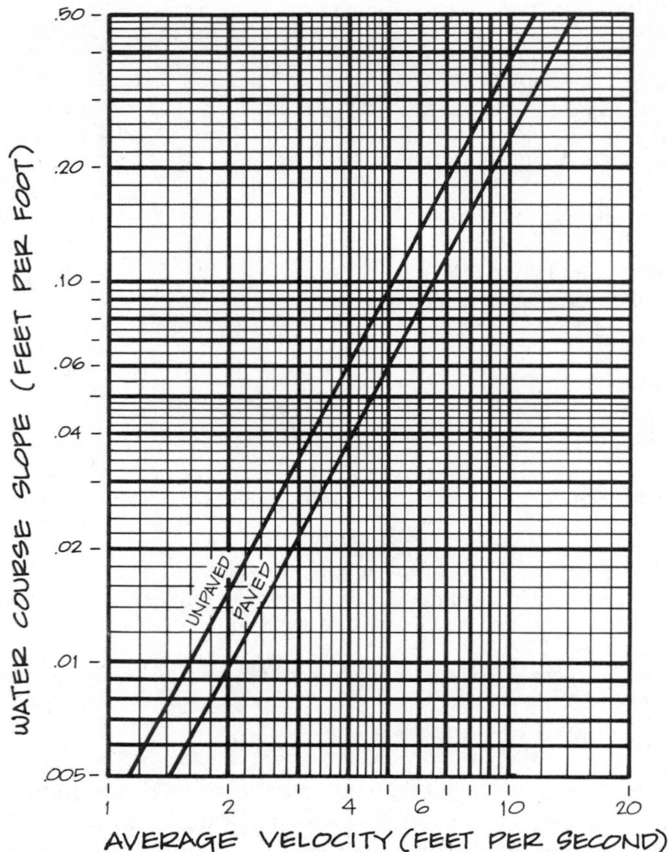

FIGURE 3.16 Average velocity for shallow concentrated flow as a function of slope and type of surface (*Source:* Adapted from U.S. Department of Agriculture, 1986)

from'' or ''lost to'' runoff by processes such as depression storage, infiltration, interception, and evaporation, as discussed in Chapter 2. The initial abstraction is only a portion of the total amount of rainfall that does not appear as surface runoff. It is that portion which is abstracted before surface runoff begins.

The initial abstraction is obtained as a function of the RCN using Table 3.8, which is taken from the TR55 manual. Entering Table 3.8 with an RCN of 75 yields an initial abstraction of 0.667 inch. The ratio of initial abstraction to the 100-year 24-hour rainfall amount of 5.5 in. is 0.121.

Time of concentration and the ratio of initial abstraction to rainfall are used to determine the unit peak discharge by means of one of four graphs, each corresponding to one of the four temporal rainfall distributions incorporated in the TR55 and shown in Figure 3.14. Figure 3.17, taken from the TR55

TABLE 3.8 Initial Abstraction as a Function of Runoff Curve Number

Runoff Curve Number	Initial Abstraction, I_a (in.)	Runoff Curve Number	Initial Abstraction, I_a (in.)
40	3.000	70	0.857
41	2.878	71	0.817
42	2.762	72	0.778
43	2.651	73	0.740
44	2.545	74	0.703
45	2.444	75	0.667
46	2.348	76	0.632
47	2.255	77	0.597
48	2.167	78	0.564
49	2.082	79	0.532
50	2.000	80	0.500
51	1.922	81	0.469
52	1.846	82	0.439
53	1.774	83	0.410
54	1.704	84	0.381
55	1.636	85	0.353
56	1.571	86	0.326
57	1.509	87	0.299
58	1.448	88	0.273
59	1.390	89	0.247
60	1.333	90	0.222
61	1.279	91	0.198
62	1.226	92	0.174
63	1.175	93	0.151
64	1.125	94	0.128
65	1.077	95	0.105
66	1.030	96	0.083
67	0.985	97	0.062
68	0.941	98	0.041
69	0.899		

Source: U.S. Department of Agriculture (1986).

manual, is for the SCS type II 24-hour rainfall distribution, which applies to southeastern Wisconsin, the site of the example subbasin.

Entering Figure 3.17 with the time of concentration of 0.74 hour and the initial abstraction to rainfall ratio of 0.12 yields a unit peak discharge for a type II rainfall distribution of 420 ft^3/sec per square mile per inch of runoff. This value is a "unit" or normalized peak discharge in two ways: (1) it relates to the surface runoff unit—inches—and (2) it relates to the area unit—square miles.

The peak discharge is the product of the peak unit discharge (420 ft^3/sec

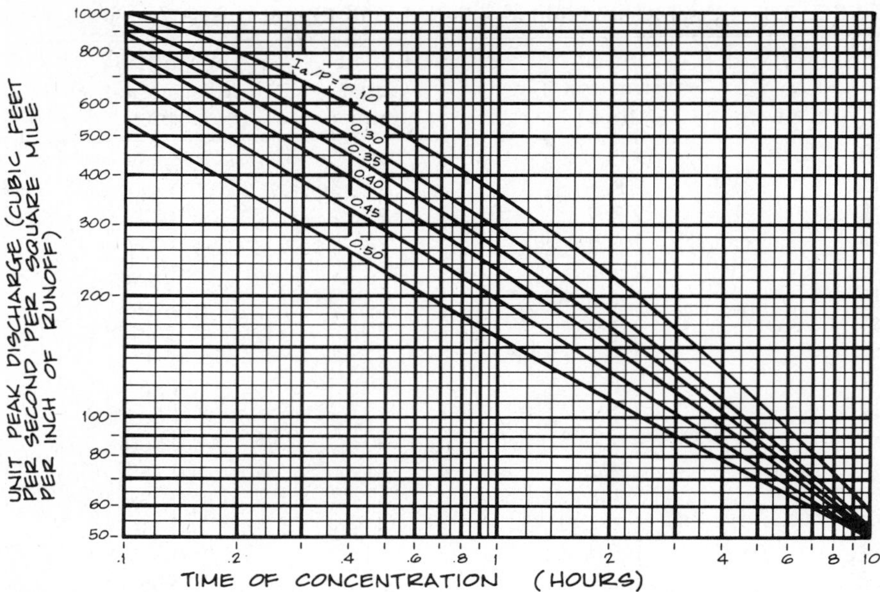

FIGURE 3.17 Unit peak discharge as a function of time of concentration, initial abstraction, and precipitation amount for a type II rainfall distribution (*Source:* Adapted from U.S. Department of Agriculture, 1986)

per square mile), the surface runoff (2.9 in.), and the subbasin area (60 acres or 0.094 square mile). The product of these three factors is 114 ft^3/sec.

TR55 provides various cautions and offers guidance on possible step 3 complexities not relevant to the simplified example. Complexities include accommodating more than one main stream or drainageway, and adjusting for ponds and wetlands scattered around the watershed.

Step 4: Compute Storage Volume. Four given or calculated factors are used to determine required storage. The first factor is peak discharge from the subbasin, and therefore the peak flow rate into a potential detention/retention facility as obtained from step 3 (114 ft^3/sec). The second factor is the volume of subbasin surface runoff as developed in step 1 (14.5 acre-ft). The third factor is the allowable peak outflow discharge from the potential detention/retention facility (20 ft^3/sec). The fourth and last factor is the type of rainfall distribution applicable to the subbasin (type II).

The ratio of peak outflow discharge to peak inflow discharge is 0.175. Entering Figure 3.18, which is taken directly from the TR55 manual, with the discharge ratio and reading the figure for a type II rainfall distribution yields a volume ratio of 0.48. This is the ratio of required storage volume to the previously determined runoff volume. Therefore, the required storage volume is 0.48 times 14.5 acre-ft, or 7.0 acre-ft.

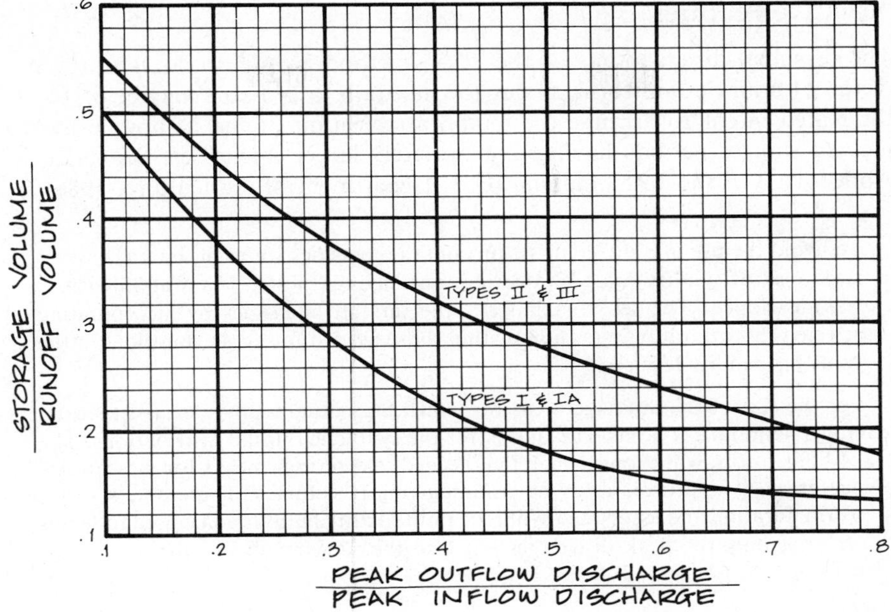

FIGURE 3.18 Volume ratio–discharge ratio relationships for different rainfall types (*Source:* Adapted from U.S. Department of Agriculture, 1986)

The TR55 manual discusses various limitations to the volume determination process, such as diminished accuracy for very high and very low discharge or volume ratios and the conservancy—tendency to oversize storage—built into the method. The manual also presents uses of step 4 different than those for the example presented here. One use is in evaluating the flow control effectiveness of any existing detention/retention facility, and another is as an aid in designing a single facility to control design rainfall events of different recurrence intervals. The latter topic is also discussed in Chapter 8.

Positive Features

TR55 has the advantage of being easily applied in urban areas. Because the method includes input parameters such as land use and imperviousness, it can be used to demonstrate the impact of existing or proposed development on peak discharges and runoff volumes. This method, unlike many manual methods, also provides estimates of both peak discharge and volume. Although not presented here, TR55 also contains other potentially useful features, such as development and combination of hydrographs as they move through the stream system. Finally, TR55 is very well documented by virtue of the manual and various published supporting papers and reports.

Negative Features

The principal disadvantage of the TR55 method is its restriction to the use of the 24-hour design storm. Fixing the duration to 24 hours may be offset by the stated intent to arrange the temporal variation of the 24-hour event to include shorter periods having appropriately larger intensities. As stated in Appendix B of the TR55 manual (U.S. Department of Agriculture, 1986):

> To avoid the use of a different set of rainfall intensities for each drainage area's size, a set of synthetic rainfall distributions having ''nested'' rainfall intensities was developed. The set ''maximizes'' the rainfall intensities by incorporating selected short duration intensities within those needed for larger durations at the same probability level.
>
> For the size of the drainage areas for which SCS usually provides assistance, a storm period of 24 hours was chosen for the synthetic rainfall distributions. The 24-hour storm, while longer than that needed to determine peaks for these drainage areas, is appropriate for determining runoff volumes. Therefore, a single storm duration and associated synthetic rainfall distribution can be used to represent not only the peak discharges but also the runoff volumes for a range of drainage areas sizes.

In some situations, other design storm durations and temporal distributions may be more appropriate. Refer to Chapter 10 for an example of the use of computer modeling to determine critical storm duration.

3.11 BRITISH ROAD RESEARCH LABORATORY METHOD FOR DETERMINING DISCHARGE AND VOLUME

The British Road Research Laboratory method (BRRL) was developed in Great Britain after World War II using rainfall–runoff data from urban subbasins (Terstriep and Stall, 1969). Subsequently evaluated in the United States, the BRRL proved to be accurate, subject to certain conditions presented in this section (Stall and Terstriep, 1972).

The BRRL method requires a little more effort than the rational method but has the advantage of providing a complete hydrograph. Therefore, the BRRL method is presented in this chapter as a potential alternative to the rational method. In addition, certain features of the BRRL method, such as development of a hydrograph by combining a hyetograph with a time–area curve, are occasionally used in hydrologic analysis and incorporated in some computer models (Kinori and Mevorach, 1984; Linsley et al., 1982). Therefore, an understanding of the BRRL method aids in the understanding of these computer models.

Methodology

The BRRL method is based on the assumption that only the directly connected impervious area in a subbasin contributes direct runoff during a major rainfall event. This fundamental aspect of the BRRL method has been tested under field conditions in the United States (Stall and Terstriep, 1972). These tests indicate that the method is valid subject to the following conditions:

1. The directly connected impervious area must cover at least 15 percent of the subbasin.
2. The design storm must have a recurrence interval of less than 20 years.
3. The total tributary area should not exceed 5 square miles.
4. Exercise caution if grassed areas are very steep, if tight soils are present, or if high antecedent moisture conditions exist.

The BRRL method is unusual in that, unlike most manual hydrologic techniques, this method has been rigorously field tested. Field testing results may seem to impose unduly restrictive conditions. However, most of these restrictions are not applicable to the computer model Illinois Urban Drainage Area Simulator (ILLUDAS) (Terstriep and Stall, 1974) and a proprietary version ILUDRAIN (Terstriep and Noel, 1985) which were developed after field testing of the manual method.

An overview of the four-step procedure used in the BRRL method is presented in Figure 3.19. The first step is to obtain and display physical data for the subbasin, focusing on directly connected impervious area. Directly connected impervious area is all impervious surface that drains, in a continuous path, to the watershed outlet. This contrasts with indirectly connected impervious area, which is impervious surface that drains directly to pervious surfaces. In the second step, subbasin physical data are used to develop an area–time relationship. A design storm hyetograph is selected in the third step. The fourth and final step consists of combining the area–time curve from the second step with the design storm hyetograph from the third step to obtain a runoff hydrograph for the subbasin. An optional fifth step, not presented here but described elsewhere (Terstriep and Stall, 1969), modifies the hydrograph to account for storage within the conveyance system such as might be provided by storm sewers or detention facilities.

Having introduced the four-step procedure in general terms, a detailed example is presented. The basis for the example is the urban subbasin illustrated in Figure 3.20. Assume that the objective is to develop a direct runoff hydrograph at the subbasin discharge point for a historic storm hyetograph.

Step 1: Obtain, Quantify, and Display Subbasin Physical Data. Under this step, the subbasin divide is delineated. Pervious and impervious areas are iden-

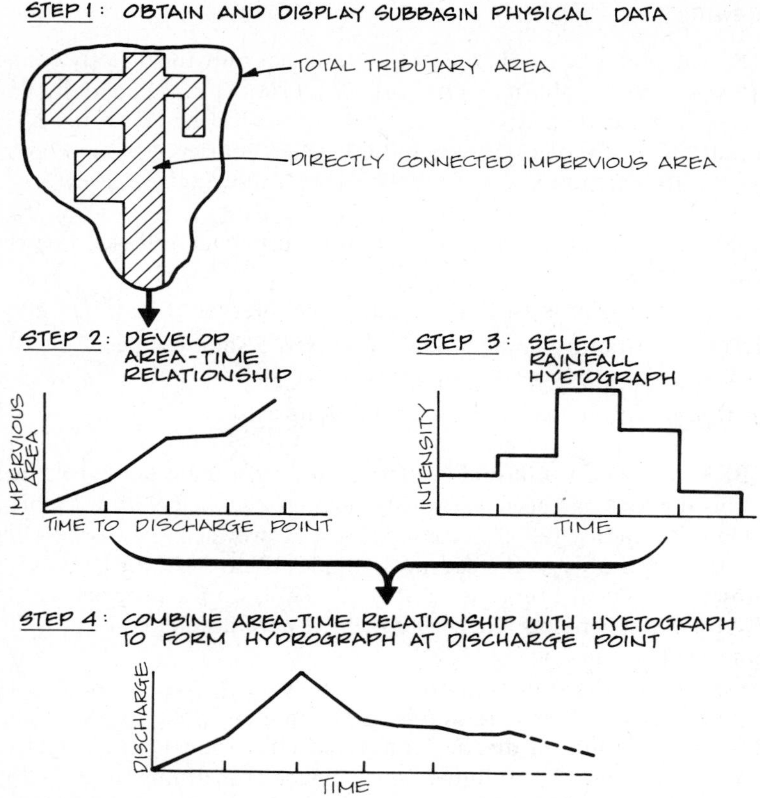

FIGURE 3.19 Procedure for the British Road Research Laboratory method (*Source:* Adapted from Walesh, 1981a)

tified and mapped carefully, distinguishing between directly connected impervious area and indirectly connected impervious area. Finally, the surface and subsurface drainage system are determined.

The results of step 1 for the example subbasin are shown partially in Figure 3.20, which indicates types of land use and land cover and the basic sewer system. Other results of step 1 are illustrated in Figure 3.21 in the form of a map showing, by means of crosshatching, the directly connected impervious surface.

Runoff from all areas indicated by crosshatching flows to the outlet over a continuous impervious surface. In contrast, the roof of the single-family residential structure in the upper right-hand portion of the subbasin is presumed to have drainage directed to the lawn areas, that is, to the pervious surface. Therefore, although the rooftop is impervious, drainage from the

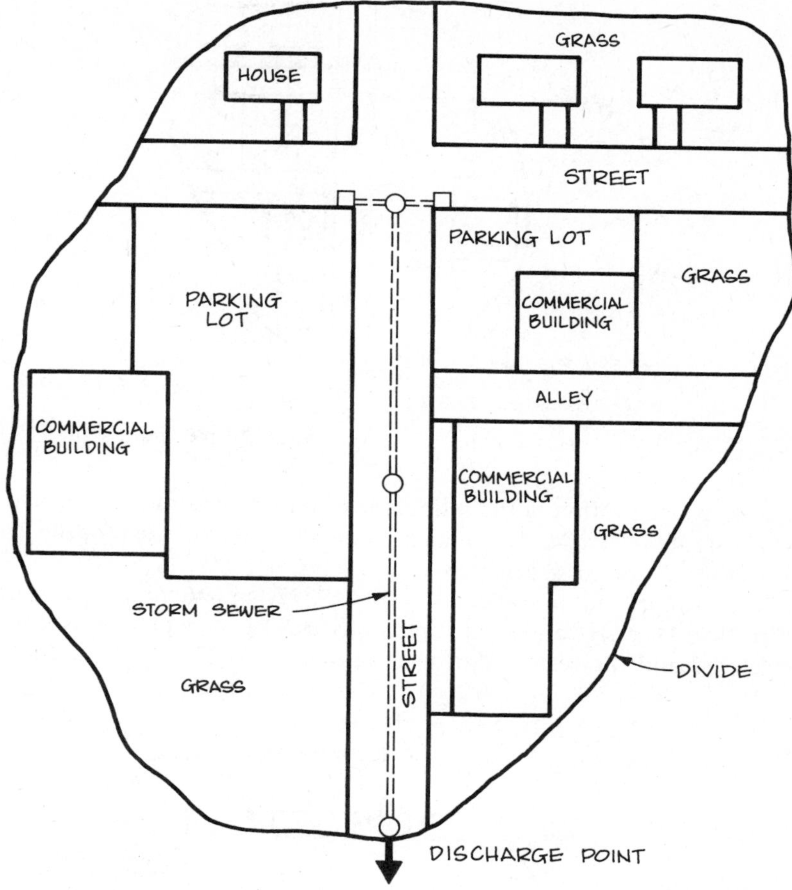

FIGURE 3.20 Subbasin used for example application of the Road Research Laboratory method (*Source:* Adapted from Walesh, 1981a)

rooftop is directed to a pervious surface and the rooftop is not part of the directly connected impervious area of the subbasin.

Step 2: Develop Time–Area Relationship. Isochrones—lines showing equal travel time to the subbasin outlet—are placed on the directly connected impervious area as illustrated in Figure 3.22. One way to do this is to select points scattered around the directly connected impervious area and calculate the travel time from those points to the subbasin outlet using overland flow, shallow concentrated flow, channel flow, and pipe flow methods. For any point, the travel time computation procedure is the same as the process used to calcu-

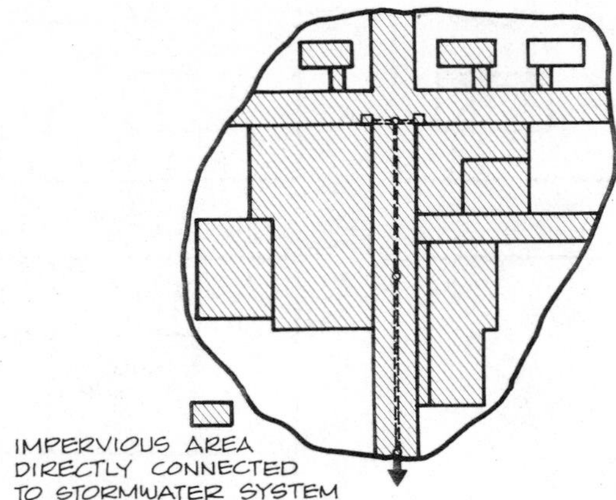

FIGURE 3.21 Directly connected impervious area (*Source:* Adapted from Walesh, 1981a)

late time of concentration in the rational method or TR55 method. Isochrone time intervals are selected to be identical to the time interval used for the design storm hyetograph discussed in step 3.

Isochrones representing travel times of 4, 8, 12, and 16 minutes are shown in Figure 3.22. Directly connected impervious area between pairs of isochrones are measured and the results shown as illustrated in Figure 3.22. This step

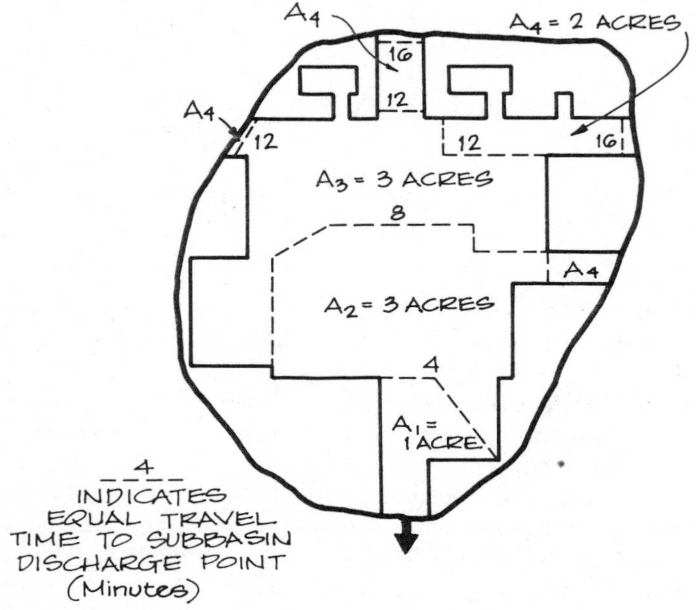

FIGURE 3.22 Isochrones for subbasin (*Source:* Adapted from Walesh, 1981a)

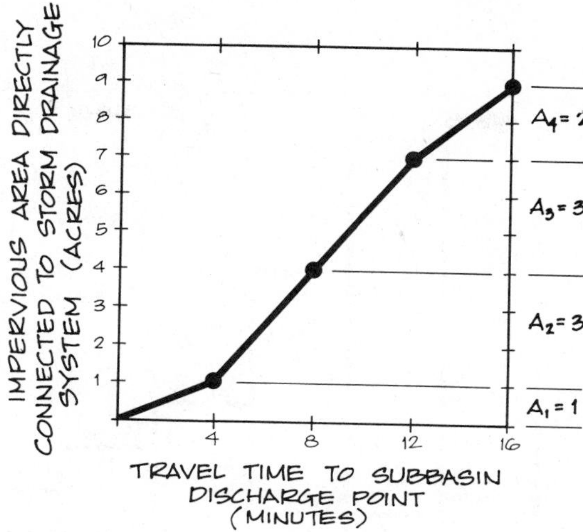

FIGURE 3.23 Area–time relationship for subbasin (*Source:* Adapted from Walesh, 1981a)

concludes with the development of a graph of directly connected impervious area in acres versus travel time to the subbasin discharge point in minutes, as shown in Figure 3.23.

Step 3: Develop the Rainfall Hyetograph. The rainfall hyetograph may be developed in several ways. For example, perhaps the time distribution of rainfall has been determined for the study area. Such is the case in Illinois (Huff, 1967), where a most probable or representative hyetograph shape has been established in each part of the state for a specified rainfall duration and recurrence interval, that is, rainfall volume. Another approach is to utilize one or more major historic storms that have occurred in or near the study area. This second approach will not provide probability or recurrence interval information, but may be effective because the analysis will be based on an actual, readily remembered rainfall event.

The rainfall hyetograph used for the example is presented in Figure 3.24. The effect of depression storage on impervious surfaces can, as illustrated by the crosshatching in Figure 3.24, be represented in the rainfall hyetograph. The crosshatched area represents 0.1 in. of depression storage. In general, the depression storage refinement is not necessary because the volume of depression storage will be very small compared to the volume of the rainfall event.

Step 4: Synthesize the Runoff Hydrograph. The runoff hydrograph is constructed by combining the area–time curve developed in step 2 with the rainfall hyetograph developed in step 3. The discharge rate at the subbasin outlet is calculated at the end of each time increment by determining the combinations

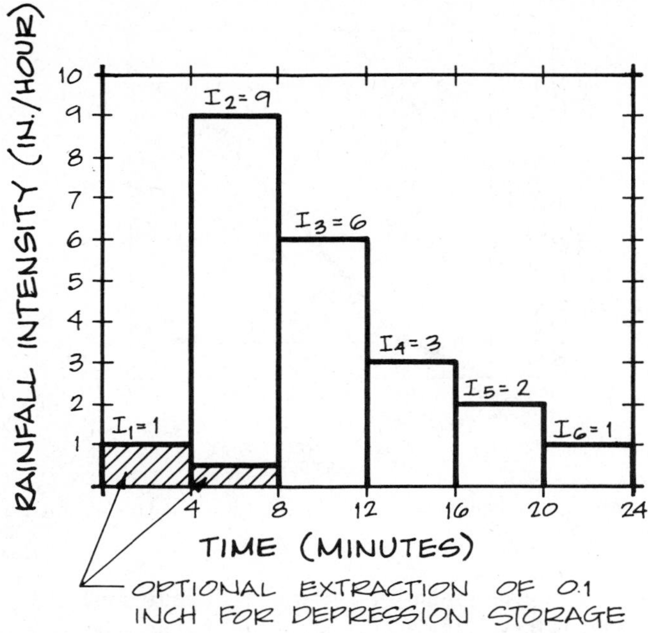

FIGURE 3.24 Rainfall hyetograph (*Source:* Adapted from Walesh, 1981a)

of rainfall increment and area increment contributing to the basin discharge.

For example, at the end of the first time increment, that is, 4 minutes after the beginning of the rainfall event, only that portion of the directly connected impervious area downstream of the 4-minute isochrone will be contributing to discharge at the subbasin outlet. Therefore, the discharge at the end of the first time increment is

$$\begin{aligned} Q_1 &= A_1 I_1 = (1 \text{ acre}) \times (1 \text{ in./hr}) \\ &= 1 \text{ acre-in./hr} = 1 \text{ ft}^3/\text{sec} \end{aligned}$$

That is, all the directly connected impervious area within 4 minutes of the subbasin discharge point is contributing to flow at the discharge point. Flow from the directly connected impervious area more than 4 minutes from the subbasin discharge point is not contributing to subbasin discharge because it has not yet reached the discharge point.

At the end of the second time increment, that is, 8 minutes after the start of the rainfall event, only that directly connected impervious area downstream of the 8-minute isochrone will be contributing to subbasin discharge. That portion of the directly connected impervious area upstream of the 8-minute isochrone will not yet have arrived at the discharge point.

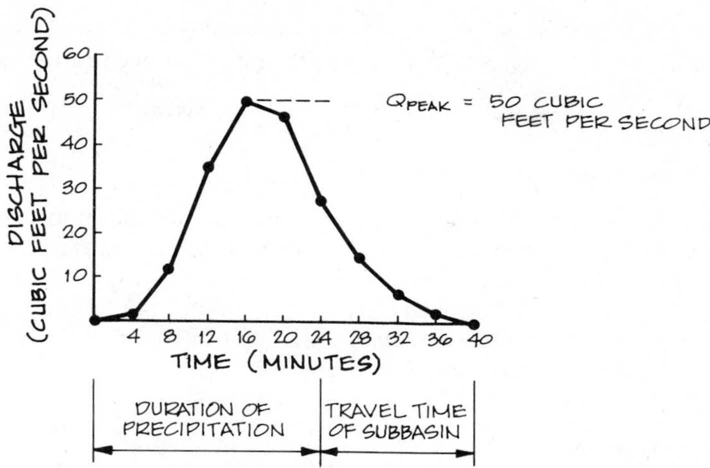

FIGURE 3.25 Hydrograph developed with the British Road Research Laboratory method (*Source:* Adapted from Walesh, 1981a)

Rainfall that occurred during the second time increment but on the first area increment will now be passing the subbasin discharge point. In addition, rainfall that occurred during the first time increment but on the second area increment will also now be passing the subbasin discharge point. Therefore, the discharge at the end of the second time increment is

$$Q_2 = A_1I_2 + A_2I_1 = (1 \times 9) + (3 \times 1) = 12 \text{ ft}^3/\text{sec}$$

Similarly, the discharge from the subbasin at the end of third through ninth time increments can be calculated from the apparent pattern established by the expressions for the first and second time increments and is given by

$$\begin{aligned}
Q_3 &= A_1I_3 + A_2I_2 + A_3I_1 &&= 36 \text{ ft}^3/\text{sec}\\
Q_4 &= A_1I_4 + A_2I_3 + A_3I_2 + A_4I_1 &&= 50 \text{ ft}^3/\text{sec}\\
Q_5 &= A_1I_5 + A_2I_4 + A_3I_3 + A_4I_2 &&= 47 \text{ ft}^3/\text{sec}\\
Q_6 &= A_1I_6 + A_2I_5 + A_3I_4 + A_4I_3 &&= 29 \text{ ft}^3/\text{sec}\\
Q_7 &= A_2I_6 + A_3I_5 + A_4I_4 &&= 13 \text{ ft}^3/\text{sec}\\
Q_8 &= A_3I_6 + A_4I_5 &&= 7 \text{ ft}^3/\text{sec}\\
Q_9 &= + A_4I_6 &&= 2 \text{ ft}^3/\text{sec}
\end{aligned}$$

The resulting discharges were used to plot the subbasin hydrograph shown in Figure 3.25. The peak discharge is 50 ft^3/sec and occurs at the end of the fourth time increment. Runoff continues for four time increments or 16 minutes after the termination of the 24-minute rainfall event. The 16 minutes of postrainfall runoff represents the 16-minute time of concentration of the directly connected impervious area.

Positive Features

Significant advantages of the BRRL method are applicability to urban areas and the ability to produce a complete hydrograph rather than just a peak discharge, as is the case with the rational method. Furthermore, and as noted earlier, the BRRL method is relatively unique among rainfall–runoff hydrologic methods in that it has been field tested and the limits of its application have been established (Stall and Terstriep, 1972).

Negative Features

Development of isochrones is a cumbersome feature of the manual application of the BRRL methods. However, experience with the method indicates that the area–time curve may usually be approximated as a linear relationship. For example, this is the approach used in the computer model versions of BRRL (Terstriep and Noel, 1985; Terstriep and Stall, 1974). When the linear assumption is used, the linear area–time curve is constructed by measuring the area of the directly connected impervious surface and estimating the flow time or time of concentration for the point on the directly connected impervious area hydraulically most remote from the subbasin outlet. This effort is identical to that required for an application of the rational method.

As with most manual rainfall–runoff techniques when they are used to determine a discharge or volume of specified recurrence interval, the BRRL method has the disadvantage of requiring the user to select a design rainfall hyetograph for a particular recurrence interval and assume that the resulting hydrograph has the same recurrence interval. In addition, the BRRL method must be used within certain limitations, including limiting it to a maximum recurrence interval of about 20 years and to subbasins having at least 15 percent directly connected impervious area. However, as noted, these limitations do not apply to the computer models ILLUDAS and ILUDRAIN, which incorporate the basic BRRL method.

3.12 OTHER HYDROLOGIC METHODS

Discussed in detail in this chapter are commonly used and representative manual hydrologic methods. Other methods are available for application in urban and urbanizing areas. Examples include variations of the unit hydrograph method as described by Sherman (1932); the inlet method presented by Kaltenbach (1963); Chow's technique for determining the size of drainage structure waterway openings (1962); and highway drainage methods (U.S. Department of Transportation, 1979).

REFERENCES

Adams, B. J. and C. D. D. Howard, "Design Storm Pathology", *Can. Water Resources J.,* Canadian Water Resources Association, **11**(3), pp. 49–55, 1986.

Alley, W. M., and J. E. Veenhuis, "Determination of Basin Characteristics of an Urban Distributed Routing Rainfall-Runoff Model." *Proc. Stormwater Manage. Model (SWMM) Users Group Meet., 1979* EPA-600/9-79-026, pp. 1–27 (1979).

American Society of Civil Engineers and Water Pollution Control Federation, *Design and Construction of Sanitary and Storm Sewers,* ASCE Man. Rep. Eng. Pract. No. 37 WPCF Man. Pract. No. 9. ASCE/WPCF, New York, 1969.

American Society of Civil Engineers, Urban Water Resources Research Council, *Annotated Bibliography on Urban Design Storms.* ASCE, New York, 1983.

Anderson, D. G., "Effects of Urban Development on Floods in Northern Virginia." *Geol. Surv. Water-Supply Pap. (U.S.)* **2001-C** (1970).

Ardis, C. V., K. J. Dueker, and A. T. Lenz, "Storm Drainage Practices of Thirty-Two Cities." *J. Hydraul. Div., Am. Soc. Civ. Eng.* **95**(HY1), 383–408 (1969).

Baker, W. R., "Stormwater Detention Basin Design for Small Drainage Areas." *Public Works* March, pp. 75–79 (1977).

Bauer, K. W., "Determination of Runoff for Urban Stormwater Drainage System Design." *Tech. Rec., Southeast. Wis. Reg. Plann. Comm.* **2**(4), April-May (1965).

Bedient, P. B., and W. C. Huber, *Hydrology and Floodplain Analysis,* Chapter 3, Addison Wesley, New York, 1988.

Benson, J., "Characteristics of Frequency Curves Based on a Theoretical 1,000-Year Record." *Geol. Surv. Water-Supply Pap. (U.S.)* **1543-A,** 51–71 (1960).

Burke, C. B., *County Storm Drainage Manual,* Chapter 2. Highway Extension and Research Project for Indiana Counties, Eng. Exp. Stn., Purdue University, West Lafayette, May 1981.

Burton, K. R., "Stormwater Detention Basin Sizing," Tech. Notes. *J. Hydraul. Div., Am. Soc. Civ. Eng.* **106**(HY3), 437–439 (1980).

Chow, V. T., D. R. Maidment, and L. W. Mays, *Applied Hydrology,* Chapters 11 and 14, McGraw-Hill, New York, 1988.

Chow, V. T., "Hydrologic Determination of Waterway Areas for Design of Drainage Structures in Small Drainage Basins." *Bull.—Ill. Univ., Eng. Exp. Stn.* **462** (1962).

Conger, D. H., "Techniques of Estimating Magnitude and Frequency of Floods for Wisconsin Streams." *Water Resour. Invest. Open File Rep. (U.S. Geol. Surv.)* **80-1214** (1981).

Crippen, J. R., "Envelope Curves for Extreme Flood Events." *J. Hydraul. Div., Am. Soc. Civ. Eng.* **108**(HY10), 1208–1212 (1982).

Hall, C. A., Jr., and T. A. Austin, *A Review of Methods of Estimating Time of Concentration and Watershed Lag Time.* Iowa State Water Resour. Res. Inst., Iowa State University, Ames, 1980.

Huff, F. A., "Time Distribution of Rainfall in Heavy Storms." *Water Resour. Res.* **3**(4), 1007–1019 (1967).

Huff, F. A., "Hydrometerological Characteristics of Severe Rain Storms in Illinois." *Rep. Invest.—Ill. State Water Surv.* **90** (1979).

Hydrocomp, Inc., *Flood Frequency Estimation,* Hydrocomp Tech. Note No. 3. Hydrocomp, Inc., Palo Alto, CA, 1977.

Kaltenbach, A. B., "Storm Sewer Design by the Inlet Method." *Public Works* January, pp. 86–89 (1963).

Kao, T. Y., "Hydraulic Design of Stormwater Detention Basin." *Proc. Natl. Symp. Urban Hydrol., Hydraul., Sediment Control, 1975* Mini-Course No. 3 (1975).

Kinori, B. Z., and J. Mevorach, *Manual of Surface Drainage Engineering,* Vol. II. Am. Elsevier, New York, 1984.

Kuichling, E., "The Relation Between Rainfall and the Discharge of Sewers in Populous Districts." *Trans. Am. Soc. Civ. Eng.* **20,** 1–56 (1889).

Linsley, R. K., Jr., M. A. Kohler, and J. L. H. Paulhus, *Hydrology for Engineers,* 3rd ed., McGraw-Hill, New York, 1982.

Marsalek, J., *Research on the Design Storm Concept,* Tech. Memo. No. 33. ASCE Urban Water Resources Research Program, New York, September 1978.

McCuen, R. H., *Guide to Hydrologic Analysis Using SCS Methods.* Prentice-Hall, Englewood Cliffs, NJ, 1982.

McCuen, R. H., H. J. Rawls, G. T. Fisher, and R. L. Powell, "Flood Flow Frequency for Ungauged Watersheds: A Literature Evaluation." *U.S., Agric. Res. Serv., Northeast. Reg. [Rep.] ARS-NE* **ARS-NE-86,** November (1977).

McCuen, R. H., S. L. Wong, and W. J. Rawls, "Estimating Urban Time of Concentration. *J. Hydraul. Eng., Am. Soc. Civ. Eng.* **110**(HY7), 887–904 (1984).

McPherson, M. B., *Some Notes on the Rational Method of Storm Drain Design,* Tech. Memo. No. 6. ASCE Urban Water Resources Research Program, New York, January 1969.

Mertes, F., "Hydrology of Regional Flood Magnitude Determination." In *Proceedings of the Workshop on Hydrologic Aspects for Floodplain Construction.* Wisconsin Department of Natural Resources, Madison, 1968.

Metropolitan Sanitary District of Greater Chicago, *Detailed Steps for Determining Allowable Release Rate and Required Flood Storage,* Inf. Pam. MSDGC, IL, 1972.

National Oceanic and Atmospheric Administration (NOAA), *5-to-60 Minute Precipitation Frequency for the Eastern and Central United States,* NOAA Tech. Memo. NWS Hydro-35. NOAA, Silver Spring, MD, June 1977.

Ott, R. F., *Streamflow Frequency Using Stochastically Generated Hourly Rainfall,* Tech. Rep. No. 151. Dep. Civ. Eng., Stanford University, Stanford, CA, December 1971.

Patry, G., and M. B. McPherson, *The Design Storm Concept.* Urban Water Resour. Res. Group, École Polytechnique de Montréal, December 1979.

Poertner, H. G., *Practices in Detention of Urban Stormwater Runoff,* Chapter 3, Spec. Rep. No. 43. Am. Public Works Assoc., Chicago, IL, 1974.

Rao, G. V. V., "Methods for Sizing Stormwater Detention Basins—A Designer's Evaluation." *Proc. Natl. Symp. Urban Hydrol., Hydraul., Sediment Control, 1975.*

Rossmiller, R. L., "The Rational Formula Revisited." *Proc. Int. Symp. Urban Storm Runoff, 1980.*

Sauer, V. B., W. O. Thomas, Jr., V. A. Stricker, and K. V. Wilson, "Flood Characteristics of Urban Watersheds in the United States." *Geol. Surv. Water-SupplyPap. (U.S.)* **2207** (1983).

Schaake, J. C., Jr., J. C. Geyer, and J. W. Knapp, "Experimental Examination of the Rational Method." *J. Hydraul. Div., Am. Soc. Civ. Eng.* **93**(HY6), 353–370 (1967).

Sheaffer, J. R., K. R. Wright, W. C. Taggart, and R. M. Wright, *Urban Storm Drainage Management,* Chapter 4. Dekker, New York, 1982.

Sherman, L. K., "Stream-Flow From Rainfall by the Unit-Graph Method." *Eng. News Rec.* April, pp. 501–505 (1932).

Southeastern Wisconsin Regional Planning Commission, *Validation of Flood Discharges and Stages,* Staff Memo. SEWRPC, Waukesha, WI, April 18, 1977.

Stall, J. B., and M. L. Terstriep, *Storm Sewer Design—An Evaluation of the BRRL Method.* Prepared for the Office of Research and Monitoring of the U.S. EPA, October 1972.

Stankowski, S. J., *Magnitude and Frequency of Floods in New Jersey with Effects of Urbanization,* Spec. Rep. No. 38. U.S. Geological Survey and State of New Jersey, 1974.

Terstriep, M. L., and D. Noel, *ILUDRAIN Version 2.1—Users Manual,* Champaign, IL, May 1985.

Terstriep, M. L., and J. B. Stall, "Urban Runoff by Road Research Laboratory Method." *J. Hydraul. Am. Soc. Civ. Eng.* **95**(HY6), 1809–1834 (1969). (See also the following discussions in the *Journal of the Hydraulics Division:* R. K. Linsley, April 1970; H. W. Duff and G. C. C. Hsieh, July 1970; D. E. Jones, September 1970; M. L. Terstriep and J. B. Stall, April 1971.)

Terstriep, M. L., and J. B. Stall, "The Illinois Urban Drainage Area Simulator, ILLUDAS." *Bull.—Ill. State Water Surv.* **58** (1974).

Urbonas, B., "Reliability of Design Storms in Modeling." *Proc. Int. Symp. Urban Storm Runoff, 1979,* pp. 27–35.

U.S. Army Corps of Engineers, Hydrologic Engineering Center, *HECWRC, Flood Flow Frequency Analysis* (Computer Program). Davis, CA, December, 1983.

U.S. Department of Agriculture, Soil Conservation Service, *Computer Program for Project Formulation-Hydrology,* Tech. Release No. 20. USDA, Washington, DC, May 1965.

U.S. Department of Agriculture, Soil Conservation Service, *Rainfall-Runoff Depths for Selected Runoff Curve Numbers—Charts and Tables from SCS TR-16.* USDA, Washington, DC, 1970.

U.S. Department of Agriculture, Soil Conservation Service, *Urban Hydrology for Small Watersheds,* Tech. Release No. 55. USDA, Washington, DC, January 1975.

U.S. Department of Agriculture, Soil Conservation Service, *Computer Program for Project Formulation-Hydrology,* Tech. Release No. 20 (Draft). USDA, Washington, DC, 1983.

U.S. Department of Agriculture, Soil Conservation Service, *Urban Hydrology for Small Watersheds,* 2nd ed., Tech. Release No. 55. USDA, Washington, DC, June 1986.

U.S. Department of Commerce, *Rainfall Frequency Atlas of the United States,* Tech. Pap. No. 40. USDC, Washington, DC, May 1961.

U.S. Department of Transportation, *Design of Urban Highway Drainage—The State of the Art,* FHWA-TS-79-225. USDT, Washington, DC, August 1979.

U.S. Geological Survey, "Effects of Urbanization on the Magnitude and Frequency of Floods in Northeastern Illinois." *Water Resour. Invest. (U.S. Geol. Surv.)* **79-36,** May (1979).

U.S. Water Resources Council, Hydrology Committee, *Estimating Peak Flow Frequencies for Natural Ungauged Watersheds—A Proposed Nationwide Test.* USWRC, Washington, DC, 1981a.

U.S. Water Resources Council, Hydrology Committee, *Guidelines for Determining Flood Flow Frequency,* Bull. No. 17B. USWRC, Washington, DC, September 1981b (revised).

Victorov, P., "Effect of Period of Record on Flood Prediction." *J. Hydraul. Div., Am. Soc. Civ. Eng.* **97**(HY1), 1853–1866 (1971).

Viessman, W., Jr., J. W. Knapp, G. L. Lewis, and T. E. Harbaugh, *Introduction to Hydrology,* 2nd ed. IEP-A Dun-Donnelly, New York, 1977.

Walesh, S. G., "Summary—Seminar on the Design Storm Concept." *Proc. Stormwater Manage. Model (SWMM) Users Group Meet., 1979* EPA 600/9-79-026, June (1979).

Walesh, S. G., *Urban Hydrology,* Chapter 5, Spec. Rep. No. 49. Am. Public Works Assoc., Chicago, IL, 1981a.

Walesh, S. G., Discussion of Burton (1980). *J. Hydraul. Div., Am. Soc. Civ. Eng.* **107**(HY2), 241–242 (1981b).

Walesh, S. G., and G. E. Raasch, *Calibration: Key to Credibility in Modeling.* Presented at the American Society of Civil Engineers Hydraulics Division Conference, University of Maryland, College Park, August 1978.

Walesh, S. G., and D. F. Snyder, "Reducing the Cost of Continuous Hydrologic-Hydraulic Simulation." *Water Resour. Bull.* **15**(3), 644–659 (1979).

Walesh, S. G., and R. M. Videkovich, "Urbanization: Hydrologic-Hydraulic-Damage Effects." *J. Hydraul. Div., Am. Soc. Civ. Eng.* **104**(HY2), 141–155 (1978).

Walesh, S. G., D. H. Lau, and M. D. Liebman, "Statistically-Based Use of Event Models." *Proc. Int. Symp. Urban Storm Runoff, 1979* pp. 75–81 (1979).

Wenzel, H. G., *Rainfall Data for Sewer Design,* Sect. III. Dep. Civ. Eng., University of Illinois, Urbana-Champaign, 1978.

Wenzel, H. G., and M. L. Voorhees, *Evaluation of the Design Storm Concept.* Presented at the Fall Meeting of the American Geophysical Union, San Francisco, CA, December 1978.

Wenzel, H. G., and M. L. Voorhees, *Sensitivity of Design Storm Frequency.* Presented at the Spring Meeting of the American Geophysical Union, Washington, DC, May 28–June 1, 1979.

Wiitala, S. W., K. R. Jetter, and A. J. Sommerville, "Hydraulic and Hydrologic Aspects of Floodplain Planning." *Geol. Surv. Water Supply Pap. (U.S.)* **1526** (1961).

Wisner, P. E., S. Gupta, and A. M. Kassem, *Considerations Regarding the Application of SCS TR55 Procedures for Runoff Computations.* Presented at SWMM Users Group Meeting, Toronto, Ontario, June 19-20, 1980.

Yen, B. C. *Design Philosophy Constraints and Assumptions,* Sec. II. Dep. Civ. Eng., University of Illinois, Urbana-Champaign, 1978a.

Yen, B. C., "Inlet Hydrographs," Sect. V. Dep. Civ. Eng., University of Illinois, Urbana-Champaign, 1978b.

Yen, B. C., "The Rational Method," Sect. VI. Dep. Civ. Eng., University of Illinois, Urbana-Champaign, 1978c.

4

FLOODPLAIN HYDRAULICS

Given one or more river or creek discharges of specified recurrence interval, the desired result is often the corresponding water surface profile and a delineation of continuous land area that would be inundated. Floodplain hydraulics, as described in this unit, consist of the methodologies used to determine flood stage profiles and areas of inundation corresponding to specific flood flows.

Floodplain maps and flood stage and streambed profiles are illustrated and briefly discussed. The theory of open channel hydraulics is reviewed, followed by discussions of data needs and a data acquisition. Use of computer programs to perform floodplain hydraulics calculations is introduced. Published sensitivity studies are reviewed because of their usefulness in guiding the data acquisition phase of floodplain hydraulics. The chapter concludes with a description of the floodway determination process. For a complement to the narrow focus of this chapter and to the other river engineering topics discussed in this book, refer to Petersen's (1986) comprehensive applications oriented and highly descriptive treatment of river engineering.

4.1 FLOODPLAIN MAP AND FLOOD STAGE AND STREAMBED PROFILES

Floodplain Map

A floodplain map, as illustrated in Figure 4.1, shows the area inundated by one or more flood events of specified probability or, less often, by a historic flood event. A floodplain map is usually the principal desired product in floodplain hydraulic studies.

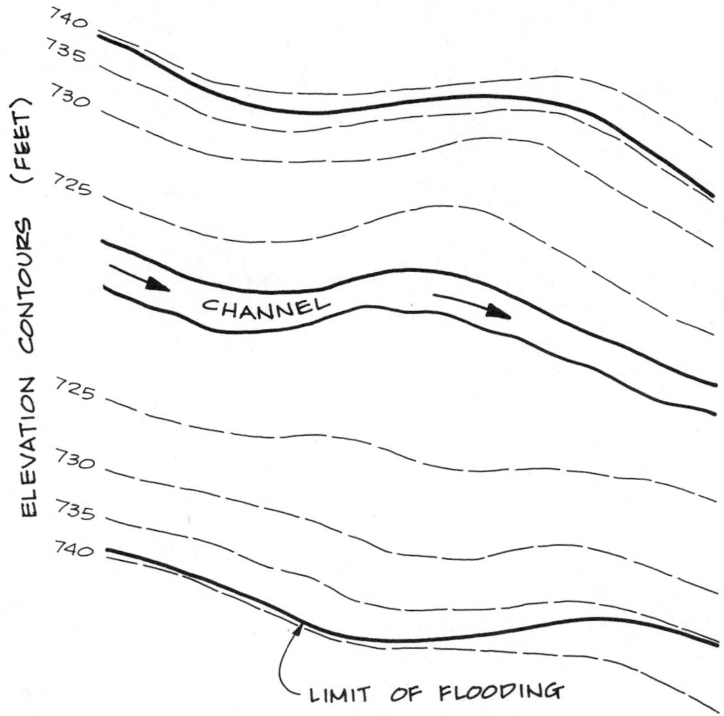

FIGURE 4.1 Floodplain map

Floodplain maps, which usually include the 100-year-recurrence-interval floodplain and floodway, are often used as the basis for development of floodplain land use and other regulations. Flood insurance studies require delineation of 100-year and 500-year-recurrence-interval floodplains. Flood insurance premiums for residential and other structures are based, in part, on location relative to floodplain limits.

Floodplain maps are also used to identify riverine property suitable for public park and open space use which can be secured through mechanisms such as public acquisition or dedication to the public by developers. Floodplain mapping is also fundamental to an examination of alternative structural surface water management measures, such as dikes and floodwalls. Finally, floodplain maps define areas of concern to be addressed in community emergency action programs.

Streambed and Flood Stage Profiles

A graph of streambed and water surface elevation versus distance along a river or stream is referred to as a streambed and flood stage profile. Each flood stage profile may correspond to an event of specified probability or to a historic flood event. Bridges, culverts, dams, and other hydraulic structures are usually shown on the profile, as illustrated in Figure 4.2.

Profile development precedes the floodplain map; that is, the flood stage profile is required to develop the floodplain map. Flood stage elevations are transferred from the profile to a topographic map and the resulting points are connected to construct the floodplain map.

In addition to delineating floodplains, streambed and flood stage profiles have other uses. Culverts, bridges, and other hydraulic structures having significant backwater effects are identified by examination of streambed and flood stage profiles. Profiles can also be used very effectively to display the results of eliminating or reducing restrictions caused by hydraulic structures. Similarly, before and after profiles offer an effective way to graphically illustrate potential adverse effects of introducing hydraulic restrictions, such as culverts, bridges, floodplain fill, levees and floodwalls, and floodways.

Flood stage profiles are also needed to establish flood insurance premiums. For example, premiums are based, in part, on the vertical distance between the 10-year- and 100-year-recurrence-interval flood stages. Finally, as described in detail in Chapter 6, flood stage profiles are an integral part of the process used to estimate average annual monetary flood damages.

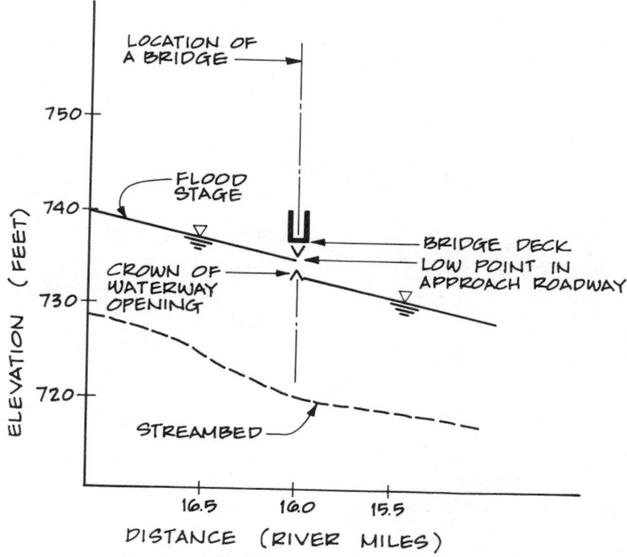

FIGURE 4.2 Streambed and flood stage profile

4.2 FUNDAMENTALS OF OPEN CHANNEL HYDRAULICS

Computation of Water Surface Profiles

Figure 4.3 is a schematic representation of a typical river–floodplain reach. Computation of water surface profiles for a series of reaches is usually subject to the following assumptions (Chow, 1959; French, 1985; U.S. Department of the Army, 1975):

(1) Steady flow, that is, no change in flow characteristics with time at any point
(2) Nonuniform flow, that is, the cross-section shape generally changes in the direction of flow, in contrast with a cross section of uniform shape
(3) Gradually varied flow, that is, no abrupt spatial changes in flow characteristics, such as the occurrence of a hydraulic jump
(4) One-dimensional flow
(5) Mild longitudinal slope, that is, less than about 10 percent

These restrictions, although numerous, are not overly confining. They are satisfied in most floodplain hydraulics situations.

Given physical characteristics such as shape and roughness for a channel–floodplain reach and given the water surface elevation and other hydraulic characteristics at the downstream end of a reach, the computation of water surface profiles focuses on determining the water surface elevation at the upstream end of the reach. Having determined the elevation at the upstream end,

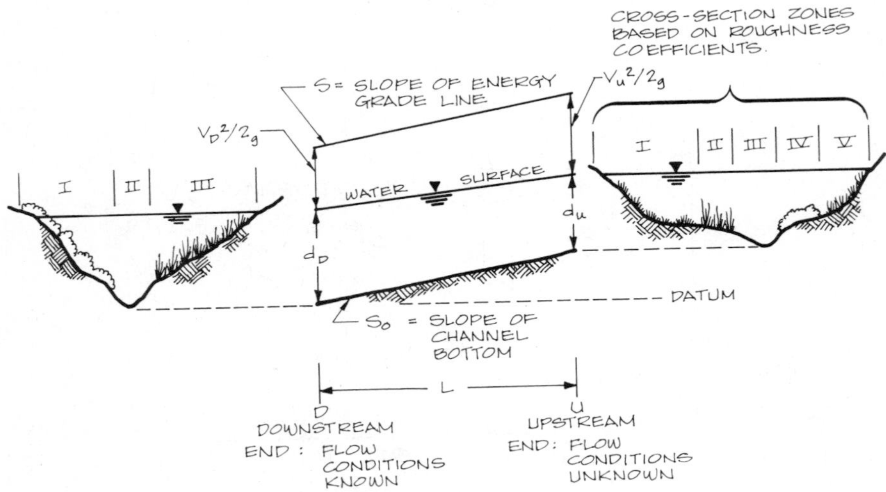

FIGURE 4.3 Generalized river–floodplain reach

that elevation becomes the water surface elevation at the downstream end of the next upstream reach and the process is repeated. Computations proceed in an upstream direction on a reach-by-reach basis.

The conservation of mechanical energy is the principle used to compute, by a trial-and-error procedure, the water surface elevation at the upstream end of a reach. The conservation of mechanical energy, when applied to a river–floodplain reach, states that the mechanical energy at the downstream end of the reach plus the loss in mechanical energy through the reach equals the mechanical energy at the upstream end of the reach. Using the generalized river–floodplain reach presented in Figure 4.3, the conservation of mechanical energy for a river reach is

$$d_D + \frac{V_D^2}{2g} + SL = d_U + \frac{V_U^2}{2g} + S_0 L$$

where d_D = maximum water depth at the downstream end

V_D = average velocity at downstream end
= Q/A_{DTOT}, where Q is discharge and A_{DTOT} is the total cross-section area of flow at the downstream end

S = slope of the energy gradeline (dimensionless)

L = reach length

V_U = average velocity at upstream end
= Q/A_{UTOT}, where A_{UTOT} is the total cross-sectional area of flow at the upstream end

S_0 = slope of the channel bottom (dimensionless)

In addition to the conservation of mechanical energy, water surface profile computations utilize the conservation of mass, or the continuity principle, and the Manning equation.

One trial-and-error technique, sometimes referred to as the standard step method, is as follows:

1. Assume a water surface elevation at the upstream end of the channel–floodplain reach shown in Figure 4.3.
2. Calculate the total flow area at the upstream end of the reach:

$$A_{UTOT} = A_{UI} + \cdots + A_{UIV} + A_{UV}$$

3. Calculate the hydraulic radius (R) for each zone at the upper end of the reach:

$$R_{UI} = \frac{A_{UI}}{P_{UI}}; \quad \ldots; \quad R_{UV} = \frac{A_{UV}}{P_{UV}}$$

where P is the wetted perimeter. "Wet–wet" boundaries are not included in the calculation of P.

4. Calculate the total conveyance at the upstream end. Recall the Manning equation, $Q = (1.49/n) \times (AR^{2/3})(S^{1/2})$, where $(1.49/n) \times (AR^{2/3}) = K =$ conveyance.

$$K_{UTOT} = (1.49)\,\frac{A_{U1}R_{U1}^{2/3}}{n_{U1}} + \cdots + \frac{A_{U1}R_{UV}^{2/3}}{n_{UV}}$$

5. Calculate the average conveyance for the reach:

$$K_{AVE} = \frac{K_{DTOT} + K_{UTOT}}{2}$$

6. Calculate the slope of the energy grade line (the friction slope):

$$S = \left(\frac{Q}{K_{AVE}}\right)^2$$

7. Calculate the mechanical energy loss caused by friction:

$$h_f = SL$$

8. Calculate the average velocity at the upstream end of the reach:

$$V_U = \frac{Q}{A_{UTOT}}$$

9. Apply the conservation of energy equation:

$$d_D + \frac{V_D^2}{2g} + h_f = d_U + \frac{V_U^2}{2g} + S_0 L$$

10. If the conservation of energy equation is not satisfied, repeat steps 1 through 9. When the equation is satisfied, let the upstream end of the reach be the downstream end of the next upstream reach, and apply the trial-and-error process to the next upstream reach.

Expansion and contraction losses and horizontal distribution of velocity are not included.

The preceding basic 10-step process must typically be repeated many times for a given reach in order to balance the conservation of mechanical energy equation. Therefore, water surface profile computations are time consuming and error prone if performed manually. Accordingly, water surface profile computations are typically performed with one of several available computer programs, as introduced later in this chapter and discussed further in Chapter 10.

Normal Depth Analysis

Normal depth open channel flow is steady flow in a uniform cross section. Although a uniform cross section is unlikely to occur over a significant reach in natural channel–floodplain conditions, situations may arise where water surface profiles may be approximated with a normal depth analysis.

Given discharge and channel–floodplain geometry and roughness, the Manning equation can be used to calculate flow depth. As an alternative, graphical aids are available to expedite normal depth analysis (e.g., Al-Khafaji and Orth, 1968; Chow, 1959; French, 1985; Gulliver, 1973; Jeppson, 1965; Zanker, 1980).

Effect of Hydraulic Structures

Culverts, bridges, dams, channel bottom drops and other hydraulic structures are typically located along channel–floodplain reaches. These structures often result in discontinuities in the water surface profile. The differential in water surface profiles across these structures is typically accounted for by use of weir and orifice formulas and coefficients (e.g., Linsley and Franzini, 1979; U.S. Department of the Army, 1975). Nomographs and other aids are available (e.g., Portland Cement Association, 1964; U.S. Department of Transportation, 1965, 1973) for the calculation of water surface profiles at hydraulic structures.

Various combinations of open channel, orifice, and weir flow may occur at a given hydraulic structure, depending on the physical characteristics of the structure and the discharge. Allocation of flow through and over a hydraulic structure often requires a trial-and-error solution. Therefore, and as is the case with water surface profile calculation through a river–floodplain reach, the analysis of hydraulic structures, if performed manually, is time consuming and error prone. Computer programs used to perform water surface profile computations typically include algorithms for computation of the effect of hydraulic structures. Refer to Chapter 5 for information on the hydraulics of culverts, weirs, and orifices and to Chapter 10 for discussions of computer models that simulate flow through or over hydraulic structures.

4.3 DATA NEEDS AND USES: OVERVIEW

As illustrated in Figure 4.4, computation of flood stage profiles requires the following data: one or more discharges, channel–floodplain geometry and roughness, hydraulic structure geometry, and historic flood stage or areas of

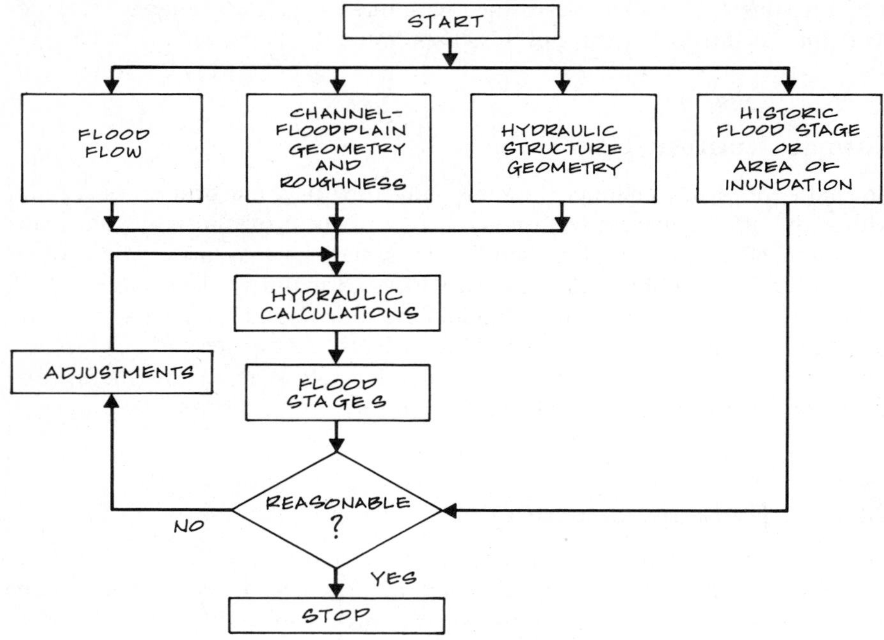

FIGURE 4.4 Data needs and uses

inundation. Although it is mathematically possible to compute flood stage profiles without historic stage or area of inundation data, the calculations are likely to be much more credible if they have been calibrated against actual data. Refer to Chapter 10 for a discussion of the calibration of computer models.

Flood flows in such computations are usually of a specified probability and obtained from a hydrologic analysis. Occasionally, measured or estimated historic flows may be used. For example, if backwater computations are conducted for a river reach in the vicinity of a gauging station, the gauging station stage–discharge relationship may be used to calibrate the hydraulic computations.

An example of a channel–floodplain cross section is presented in Figure 4.5. As a matter of convention, channel–floodplain cross sections are stationed from left to right, with a cross section viewed looking in the downstream direction. In addition to the channel–floodplain geometry, a complete description of the channel–floodplain cross section includes the Manning roughness coefficients for the entire cross section.

A bridge cross section is presented in Figure 4.6 to serve as an example of data typically required to represent a hydraulic structure in floodplain hydraulic computations. Similar data are required for culverts, dams, and drop struc-

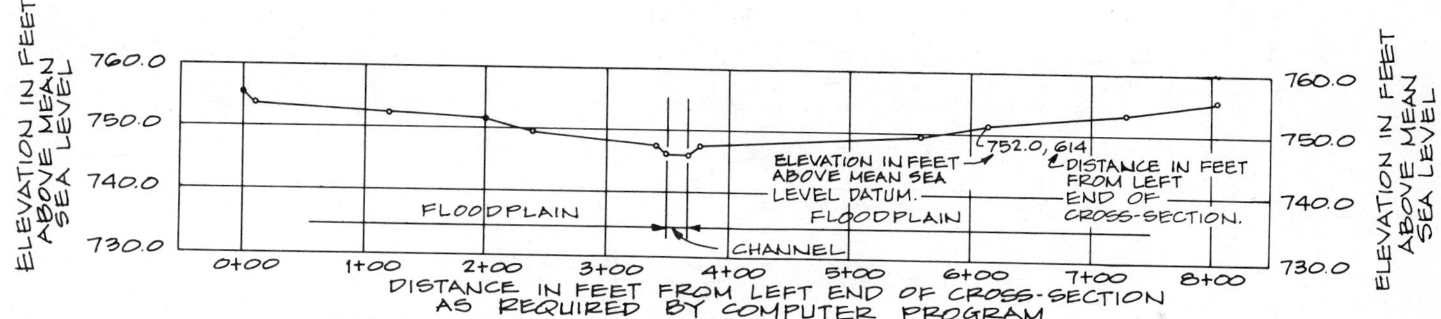

FIGURE 4.5 Channel–floodplain cross section (*Source:* Adapted from Southeastern Wisconsin Regional Planning Commission, 1976)

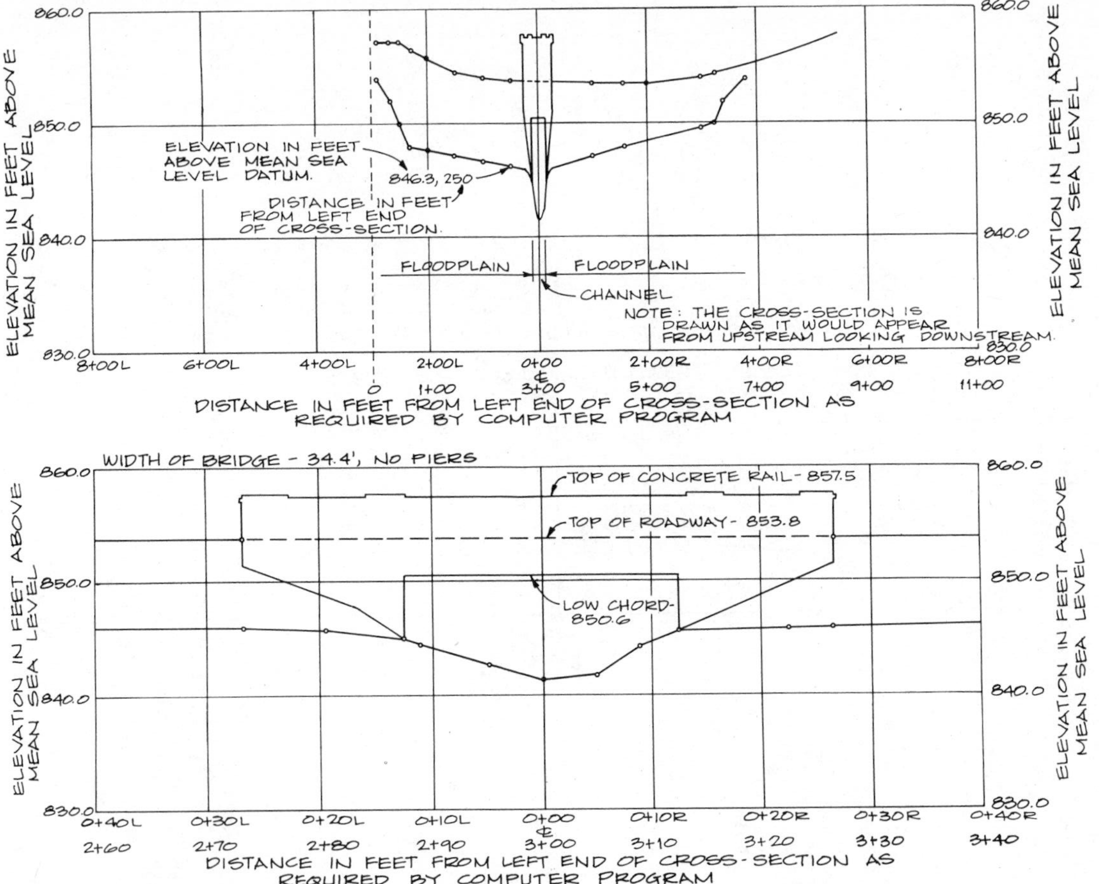

FIGURE 4.6 Bridge cross section (*Source:* Adapted from Southeastern Wisconsin Regional Planning Commission, 1976)

tures. As is the case with the channel–floodplain cross sections, stationing is from left to right, with the left side being defined when looking in the downstream direction.

Examples of historic flood stage and area of inundation data include continuous and discrete stage records, ground photos, aerial photos, and scattered discrete ground observations. As suggested by Table 4.1, historic data may be available from a variety of sources, although considerable research effort may be required to locate the sources and obtain the information.

A simple but effective means of managing historic flood stage data is to plot the data—with appropriate annotations—on computer-generated or reproducible streambed and flood stage profiles. The results of floodplain hydraulics calculations are then plotted on prints of these profiles for easy comparison with the historic data. A similar procedure can be used with data for historic area of inundation. Stage and area of inundation data can be converted to a common base of either stage or area of inundation through the use of topographic maps or special field surveys.

4.4 ACQUISITION OF CHANNEL–FLOODPLAIN GEOMETRY AND ROUGHNESS DATA

The acquisition of channel–floodplain geometry and roughness data should be supervised in the field by an engineer. The engineer should be experienced in floodplain hydraulics, particularly the computer program or other analytic method that will be used to calculate flood stage profiles. Establishing channel–floodplain geometry and roughness begins with determining the location of cross sections, which requires consideration of hydraulic and nonhydraulic factors. The orientation of cross sections and the termination points for each end of the cross section must also be established. Next, Manning roughness coefficients are assigned for the entire length of a cross section, and finally, cross-section elevation and distance coordinates are obtained. Various aspects of acquiring channel–floodplain geometry and roughness data are discussed in subsequent sections to supplement published guidelines (e.g., Lee, 1968, 1971).

Location of Cross Sections: Hydraulic Factors

Many factors influence the location of channel–floodplain cross sections. Cross sections should be located so that they are representative of hydraulic characteristics of a reach, not of a specific location along a reach. For example, cross sections should be located so that they do not extend into lateral backwater areas which would not convey flood flows. Cross sections 2 and 3 on Figure 4.7 are properly located because they do not include the intervening right-bank backwater area.

Cross sections should be positioned to exclude localized high areas or depressions in the floodplain or channel such as shown on Figure 4.7 between

TABLE 4.1 Types and Sources of Historic Data

Type		Source							
Category	Subcategory	USGS and Other Federal Agencies	State Agency	Regional Districts and Agencies	Cities, Villages, and Towns	News Media	Museums and Historical Societies	Consulting Firms	Residents
Stage	Continuous record	×							
	Partial record	×		×					
	Special	×	×	×	×			×	×
	Ground photos			×	×	×	×	×	×
Area of inundation	Aerial photos		×			×	×		
	Ground photos		×	×	×	×	×	×	×
	Discrete ground observations				×			×	

Source: Walesh and Raasch (1978).

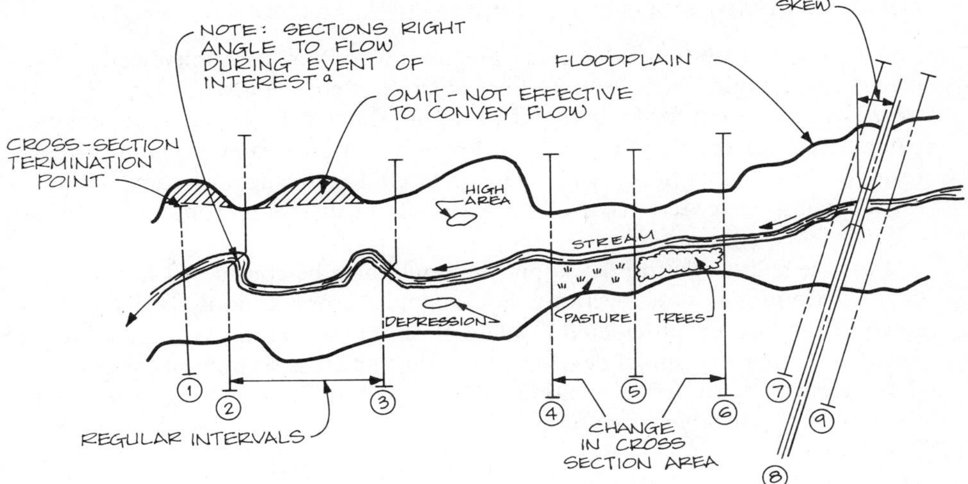

FIGURE 4.7 Channel–floodplain reaches (*Source:* Adapted from Lee, 1968)

cross sections 3 and 4. If a cross section must intersect a localized high or low area, the coordinates representing the localized area should be omitted from the cross section.

Cross sections should be placed at abrupt changes in cross-sectional area of the channel or the floodplain, as illustrated by cross sections 4 and 6 in Figure 4.7. Similarly, cross sections should be located at abrupt changes in the roughness of the channel or the floodplain, as illustrated by the vegetation change on the left floodplain at cross section 5 in Figure 4.7. Abrupt changes in longitudinal slope of the channel or the floodplain also determine cross-section locations.

Channel–floodplain cross sections should be positioned at or adjacent to dams, culverts, bridges, drops, and other hydraulic structures, as suggested by cross sections 7, 8, and 9 on Figure 4.7. The precise location for hydraulic structure cross sections is determined by the input requirements of the computer program or other analytic method used to perform the hydraulic computations through and over the structure.

Cross sections should be placed at positions coincident with the location and orientation of cross sections used in previous floodplain hydraulic computations. This facilitates comparison of computed stages at identical locations.

Finally, cross sections should be located at sites of historic high water marks, near areas of documented historic flood inundation, or at points where discharge measurements have been made. Channel–floodplain cross sections coincident with locations at which historic flood data are available facilitates verification of the computed flood stage profiles.

Location of Cross Sections: Nonhydraulic Factors

Cross-section location may also be influenced by important nonhydraulic factors. For example, cross sections should be located at county, city, village, town, and other corporate boundaries. Many states and jurisdictions require computation of the stage increase occurring in an upstream community as a result of an actual or proposed floodplain alteration in a downstream community. Placing cross sections at corporate limits facilitates responding to these requirements.

Possible sites of future development should also be considered in locating channel–floodplain cross sections. For example, cross sections should be located at planned or anticipated residential or commercial areas, parks, highways, and other floodplain developments. Placing cross sections at these locations facilitates future evaluation of the hydraulic impact of proposed floodplain fill or other alterations.

Orientation of Cross Sections

The orientation of the channel–floodplain cross section is determined, in part, by some of the preceding hydraulic and nonhydraulic location factors. For example, cross sections used to represent hydraulic structures are typically oriented approximately parallel to the structures.

The principal criterion, however, for orientation of cross sections is to place them approximately perpendicular to the expected direction of the flow during the flow event of interest, which will usually be a major flood. Therefore, the channel-floodplain cross section is usually linear and generally oriented perpendicular to the longitudinal orientation of the floodplain. If low flows approximating the capacity of the channel are to be included in floodplain hydraulics, the cross section in the vicinity of the channel should be oriented perpendicular to the localized longitudinal axis of the channel, as illustrated by the channel portions of cross sections 2 and 3 in Figure 4.7.

Termination of Cross Sections

The principal criterion for cross-section termination is that the cross section must represent the entire portion of the channel–floodplain available for conveying flood flow. For example, cross section 2 in Figure 4.7 extends to and beyond the limits of the flow conveying area. In contrast, the right side of cross section 1 is terminated to exclude a backwater area that would not be expected to convey flow. Cross-section termination points should be clearly indicated, perhaps with short hash marks as used at the end of all cross sections shown on Figure 4.7.

A special situation in which cross-section termination may be used to represent available conveyance on a floodplain is depicted in Figure 4.8, where a cluster of relatively dense urban development or a stand of tall, dense vegeta-

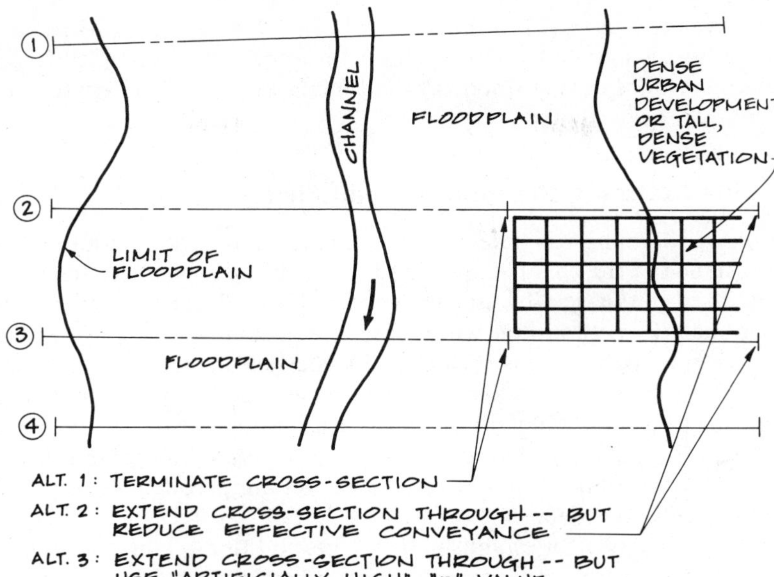

FIGURE 4.8 Use of cross sections to represent urban development or tall, dense vegetation

tion projects, in peninsula fashion, into the floodplain. Assume that the development or vegetation is sufficiently dense by virtue of many major and auxiliary buildings, thick vegetation, and other obstacles as to obstruct flow significantly or completely. In this case, cross sections should be terminated at the riverward side of the protruding development, as shown on the left side of cross sections 2 and 3 in Figure 4.8.

In contrast, assume less dense development or vegetation such that some flow can move on the left floodplain through the development or vegetation. In this case, a sawtoothed cross-section representation might be used to approximate flow through the development or vegetation. Another approach, which also involves extending the cross section completely across the left floodplain, is to represent reduced conveyance by means of an artificially high Manning roughness coefficient. This approach is less desirable than the preceding approach because of the difficulty of rationally establishing the value of roughness coefficients.

If a liberal approach is taken in representing conveyance associated with protruding or encroaching obstructions into the floodplain, that is, if an excessive amount of conveyance is assumed, hydraulic and other problems are likely to occur in subsequent floodway determinations. This problem is particularly acute if those subsequent delineations attempt to exclude existing development from the floodway under the constraint of small or no stage increases. If excessive conveyance is assigned to the portion of this development protruding into the floodplain as part of the floodplain delineation effort, complete exclusion of that conveyance during floodway determination effort is likely to cause an unacceptable stage increase.

Assigning Manning Roughness Coefficients

Selection of Manning roughness coefficients is equivalent to establishing the flow resistance of the channel and floodplains. Roughness coefficients should be established in the field by an engineer familiar with the computer program or other analytic method that will be used to calculate flood stage profiles.

Because spatial variations in roughness coefficients are important factors in determining cross-section locations, roughness coefficients should be determined prior to or in conjunction with the establishment of cross-section location, orientation, and termination. If large-scale aerial photographs are available, they can be taken in the field and Manning roughness coefficients recorded directly on them.

Table 4.2 is a guide for establishing channel and floodplain roughness coefficients. A component approach is used for determining the channel roughness in that five independent factors are evaluated and combined to form a composite Manning roughness coefficient.

The five conditions that affect channel roughness are material, degree of irregularity, relative effect of obstructions, amount of vegetation, and degree of meandering (Chow, 1959). The five conditions are self-explanatory, with the possible exception of degree of irregularity, which refers to the relative changes in channel cross section along the longitudinal axis of the channel. Roughness coefficients will not be significantly affected by gradual changes in cross section, but abrupt changes over short distances can influence the roughness coefficient. Photographs of actual natural and altered channels for which roughness coefficients have been determined (e.g., Barnes, 1967; Chow, 1959) offer an alternative to or supplement to the roughness determination procedure set forth in Table 4.2. Floodplain photographs and other channel and floodplain roughness determination aids are provided by the U.S. Department of Transportation (1984).

For floodplains, the normal condition roughness coefficients presented in Table 4.2 should be used in most cases. In some situations, such as when seasonal effects are important, the maximum or minimum roughness coefficient for a particular type of floodplain cover should be utilized.

Obtaining Cross-Section Coordinates

Three fundamentally different approaches are available for obtaining cross-section coordinate data similar to that illustrated on Figure 4.5. The first approach is use of large-scale topographic maps. If such maps are available, this method is usually least expensive. Based on a site visit, an engineer knowledgeable about the flood stage computation procedures to be used shows the location, orientation, and termination of cross sections on the large-scale topographic maps. An office technician can then obtain distance and elevation coordinates for the entire cross section, with the exception of the channel bottom.

The channel bottom configuration can be obtained by supplemental field survey; by interpolation between known channel bottom elevations such as at bridges, culverts, and other hydraulic structures; or by judgment based on approximate depth of flow. Depending on the relative conveyance of the channel in contrast with the floodplains, the position of the channel bottom may have no significant effect on the calculated stage of a major flood. That is, during flood stage, the channel may carry an insignificant fraction of the total flow. For example, refer to Figure 4.5 and compare the channel cross-sectional area beneath elevation 748 ft mean sea level to the total channel–floodplain cross-sectional area beneath elevation 756 ft mean sea level. The channel cross-sectional area is very small relative to the total cross-sectional area of the channel and floodplain.

Sensitivity studies can be conducted on actual cross sections representative of a project area to determine the importance of accurate representation of the channel bottom. Several actual channel bottom cross sections can be obtained by field survey for a sample reach. Flood stage profiles can then be calculated for the reach with and without the inclusion of the channel and compared to determine the importance of representing the channel.

A second method for obtaining cross-section coordinates is to rely totally on field survey. Vertical control is established by surveyors. An engineer, based on field reconnaissance and using a suitable map or aerial photograph, designates cross-section location, orientation, and termination. Cross-section lines are then surveyed to provide distance and elevation coordinates. Although channel bottom elevations are relatively easy to obtain when the field survey method is used, particularly on shallow narrow streams, such coordinates may not be necessary.

The third and final means of obtaining cross-section coordinates is use of photogrammetry. With this approach, surveyors establish horizontal and vertical control points along the stream system. Control points are typically placed on bridges, culverts and other hydraulic structures. Aerial photographs suitable for subsequent preparation of maps by stereoscopic methods are obtained. Prints of the aerial photos are taken into the field by an engineer, who uses them to show the desired location, orientation, and termination of cross sections. Office technicians then use stereoscopic methods to obtain coordinates along each cross-sectional line. If the channel bottom configuration is required, it may be obtained by supplemental field survey, by interpolation between known channel bottoms elevations, or by judgment based on approximate depth of flow on the day when the aerial photographs were obtained.

TABLE 4.2 **Manning Roughness Coefficients for Channels and Floodplains**

Channel				Floodplain				
						Roughness Coefficient		
Condition			Roughness Coefficient Component[a]		Condition	Minimum	Normal	Maximum
Material involved	Earth		0.020	Pasture	Short grass	0.025	0.030	0.035
	Rock cut	n_1	0.025		High grass	0.030	0.035	0.050
	Fine gravel		0.024	Cultivated Areas	No crop	0.020	0.030	0.040
	Coarse gravel		0.028		Mature row crops	0.025	0.035	0.045
Degree of irregularity	Smooth		0.000		Mature field crops	0.030	0.040	0.050
	Minor	n_2	0.005	Brush	Scattered brush, heavy weeds	0.035	0.050	0.070
	Moderate		0.010		Light brush and trees, in winter	0.035	0.050	0.060
	Severe		0.020		Light brush and trees, in summer	0.040	0.060	0.080
Relative effect of obstructions	Negligible		0.000		Medium to dense brush, in winter	0.045	0.070	0.110
	Minor		0.010–0.015		Medium to dense brush, in summer	0.070	0.100	0.160

	Appreciable	n_3	0.020–0.030	Trees	Dense willows, summer, straight	0.110	0.150	0.200
	Severe		0.040–0.060		Cleared land with tree stumps, no sprouts	0.030	0.040	0.050
Vegetation	Low		0.005–0.010		Same as above, but with heavy growth of sprouts	0.050	0.060	0.080
	Medium		0.010–0.025					
	High	n_4	0.025–0.050		Heavy stand of timber a few down trees, little undergrowth, flood stage below branches	0.080	0.100	0.120
	Very high		0.050–0.100					
Degree of meandering	Minor		1.000		Same as above, but with flood stage reaching branches	0.100	0.120	0.160
	Appreciable	k	1.150					
	Severe		1.300					

Source: Adapted from Chow (1959).

[a]The composite Manning roughness coefficient for a channel reach = $k \times (n_1 + n_2 + n_3 + n_4)$.

4.5 ACQUISITION OF HYDRAULIC STRUCTURE GEOMETRY DATA

Figure 4.6 illustrates the type of hydraulic structure data typically required for computation of flood stage profiles. Although exact data requirements are determined by the computer program or other computational procedure used, required hydraulic structure data typically include a profile along the roadway or crest of the structure, the location and shape of the waterway opening, and all or part of the channel–floodplain cross section immediately downstream of the structure.

Depending on the existence and size of a waterway opening and the characteristics of the approach roads, hydraulic structures such as bridges and culverts can be important elements in the hydraulics of a watershed, particularly with respect to localized upstream effects. The hydraulic constriction caused by a bridge or culvert may, under flood conditions, result in a large backwater effect immediately upstream of the structure. The large backwater may create upstream flood stage profiles that are significantly higher and an upstream floodplain that is significantly wider than would exist in the absence of the structure.

However, hydraulic structure geometry may not be required for every bridge and culvert located along a stream reach for which flood stage profiles are to be calculated. The hydraulic significance of each structure should be determined by an engineer who examines the structure in the field prior to any surveying work or other data acquisition in order to reduce the number of structures for which detailed data are required. In examining each bridge, culvert, or other hydraulic structure to evaluate its hydraulic significance, the structure is considered to consist of the approaches as well as structural components such as abutments, piers, and the bridge deck.

A hydraulic structure is hydraulically significant if field inspection by the engineer suggests that the structure would have a backwater for the flood flow of interest in excess of a standard, such as 0.5 ft. Hydraulically significant bridges and culverts generally are characterized by relatively small waterway openings in combination with approaches that are constructed well above the elevation of the floodplain. Most bridges, culverts, and other hydraulic structures will be hydraulically significant.

One category of hydraulically insignificant bridges and culverts are those having a relatively small superstructure relative to the combined width of the channel and floodplain. Such structures typically have approaches that do not rise significantly above the floodplain, while the portion of the structure in the immediate vicinity of the channel simply spans the channel. Pedestrian crossings, such as typically found in golf courses and parks, and private roadway bridges and culverts are examples of hydraulically insignificant structures.

The second category of hydraulically insignificant structures are those bridges that are elevated on piers well above channel and floodplain, utilize little or no floodplain fill for the approaches, and therefore offer little impedance to flow even during major flood events. Some bridges on the interstate highway system are examples of this type of hydraulically insignificant structure.

Two methods are available for obtaining hydraulic structure geometry data. The first is use of design or as-built plans. If such information is available, particularly as-built plans, this approach can be very cost-effective even if some supplemental field survey is needed. However, particularly for older bridges, culverts, and other hydraulic structures, design or as-built plans are often not readily available, or the structure has been modified so as to require significant field survey.

The second and more costly approach for obtaining hydraulic structure geometry data is to survey and measure the structure in the field. Because of the very specific and limited data requirements, surveys should be carefully planned to minimize the amount of field survey time while obtaining the necessary information.

4.6 PERFORMING FLOOD STAGE PROFILE CALCULATIONS WITH COMPUTER PROGRAMS

The iterative nature of flood stage profile computation combined with the voluminous data requirements means that most flood stage profile calculations are usually performed with computer programs rather than by manual methods. Many programs are available in the public domain, such as the HEC-2 program, which is supported by the Corps of Engineers Hydrologic Engineering Center. Proprietary programs are also available from consultants and software vendors. Chapter 10 includes an introduction to HEC-2 with a discussion of data requirements and output. Literature citations for other flood stage computer programs are also presented in Chapter 10. A major effort is usually required to learn how to use flood stage profile computer programs properly and fully, because of size of the programs, the voluminous data requirements, and the many computational options.

Contingency Checks

Because of the voluminous input data requirements and the voluminous output generated by flood stage profile computer programs, a disciplined contingency check is required to guard against errors and blunders. A formalized procedure should be developed and implemented on all computer runs. Examples of contingency checks that should be performed on computer program output are:

1. Do changes in Manning roughness coefficients occur at the proper locations?
2. Is the relative amount of flow in the left floodplain, in the channel, and on the right floodplain reasonable?
3. Are channel and floodplain discharges consistent between consecutive cross sections?
4. Do channel velocities exceed floodplain velocities, and if not, is there a physically meaningful explanation?
5. Do plotted profiles look reasonable, particularly with respect to the historic high water data?
6. Are profile discontinuities in the vicinity of hydraulic structures consistent with the type and size of the structures?
7. Does critical depth occur in consecutive cross sections?
8. Are there error or warning messages?

4.7 SENSITIVITY STUDIES

If the flood stage profile computation procedure using a computer program is viewed as a two-step process of field data collection followed by office engineering analysis, the cost of the first step could easily be three-fourths of the total cost of the process. Accordingly, field activities should be carefully planned to be cost-effective, that is, to strike a satisfactory balance between minimizing the cost of data collection and the apparent accuracy of the resulting computed floodstage profiles.

Cost influencing aspects of field data collection include spacing and accuracy of channel–floodplain cross sections and the extent to which the spatial variations in Manning roughness coefficients are recorded. The literature contains relatively little guidance with respect to practical considerations, such as the impact of cross-section spacing and the impact of roughness coefficients on profile accuracy. A few investigations providing at least qualitative guidance are summarized here.

The Corps of Engineers investigated a 6-mi-long reach of Line Creek, a tributary to Tibbee River, in eastern Mississippi (U.S. Department of the Army, 1970). The creek reach had a longitudinal slope of 3 ft/mi, the valley was approximately 1.5 miles wide, and there were apparently no significant hydraulic structures within the reach. The 100-year-recurrence-interval discharge was 30,000 ft^3/sec.

HEC-2 was used to compute a base profile for the 100-year discharge using cross sections based on U.S. Geological Survey quadrangle maps supplemented with field survey. Combinations of the following variations on the base condition were then simulated: longitudinal slopes of 1 and 10 ft/mi, doubled Manning roughness coefficients, cross sections based solely on quadrangle maps and cross sections based on quadrangle maps supplemented with field survey, and plus and minus 25 percent changes in discharge.

The sensitivity investigation indicated that the influence of Manning roughness coefficients increased as the longitudinal slope of the stream reach decreased. That is, Manning roughness coefficients were found to be more important for mild slopes than for steep slopes. Furthermore, the influence of cross-section inaccuracy increased as longitudinal slope increased. That is, cross-section accuracy was more important for steeper slopes. Finally, the influence of discharge increased as longitudinal slope decreased. That is, discharge accuracy was found to be more important for milder slopes.

A similar sensitivity investigation was conducted on Minnesota streams (Barr Engineering Company, 1972). After establishing a base profile, channel–floodplain cross sections were progressively removed from the computer model to observe the effect on floodstage profiles. In addition, alternate cross sections were raised and lowered by 1 ft to observe the effect on flood stage profiles.

The Minnesota study concluded that the influence of cross-section spacing increased as longitudinal slope increased. That is, closely spaced cross sections are more important on steeper slopes. In addition, the influence of cross-section inaccuracy increased as longitudinal slope increased. That is, cross-section accuracy is more important for steep slopes.

The two sensitivity studies do not provide quantitative guidance on cross-section accuracy and spacing. However, the findings of the sensitivity investigations are consistent and provide qualitative guidance. If the range of possible flood stage computation situations is viewed as a spectrum from very mild to very steep longitudinal slopes, Manning roughness coefficient accuracy and discharge accuracy are more important near the low-slope end of the spectrum. Cross-section accuracy and reduced spacing are more important near the steep slope end of the spectrum.

By simulating variations on 98 preexisting natural stream data sets, Burnham and Davis (1986) studied the accuracy of computed water surface profiles as a function of survey technology and Manning roughness coefficients. The research was prompted in part by concern with the high cost of field data collection and the related desire to optimize data collection vis-à-vis flood stage profile accuracy. Project constraints included simulating only the 100-year discharge, use of the HEC-2 computer program, consideration of only subcritical flow, and excluding the effects of hydraulic structures and other localized effects.

One conclusion is that for a given contour interval, photogrammetrically determined cross-section coordinates are more accurate than coordinates obtained from topographic maps developed by photogrammetric methods. A second key conclusion, which reinforces the conclusions of the two previously cited sensitivity studies, is that the accuracy of flood stage profiles is very sensitive to the Manning roughness coefficient.

Motayed and Dawdy (1979) computed flood stage profiles for a 4-mi river reach using the same input data and current versions of the Corps of Engineers, Soil Conservation Service, and U.S. Geological Survey computer programs. There were no hydraulic structures in the river reach, the longitudinal slope was about 4 ft/mi, and the river was large, having a 100-year discharge of 294,000 ft^3/sec. For identical starting elevations, and using the profile developed with the USGS program as a point of reference, comparison of the resulting profiles at all cross sections indicated that the profiles obtained with the SCS and Corps programs average about 0.70 ft below the reference profile. This study suggests that although the various computer programs are based on the same theory, algorithms in the computer programs can result in significant differences in computed flood stage profiles.

4.8 FLOODWAY DETERMINATION

As illustrated in Figure 1.13, the floodway is that portion of the 100-year floodplain required to convey the 100-year discharge ''safely''. More precisely, the floodway is a designated portion of the 100-year-recurrence-interval floodplain that will convey the 100-year discharge with small upstream and downstream stage increases. The floodway typically includes the channel and defines that portion of the floodplain not suited for human habitation because of the hazard to life and danger to property. The floodway sometimes includes areas having depths and velocities in excess of established critical values.

Figure 1.13 suggests that use of a large (e.g., 0.5 ft or more) allowable stage increase may subject a significant amount of additional land to floodplain regulations. Therefore, use of a floodway under conditions of a large allowable stage increase to gain developable land in a river reach may result in an increase in regulated land in the reach immediately upstream.

The flood fringe, as illustrated in Figure 1.13, is the area outside the floodway but within the 100-year floodplain. The premise of the hydraulic analyses and other considerations used to establish the floodway limits is that the flood fringe will be completely filled and otherwise developed so as to eliminate all floodwater conveyance outside the floodway. Although filling the entire flood fringe is unlikely, filling, development, and obstruction of the area could proceed, particularly in urbanized areas, to the point where the flood fringe has no significant remaining conveyance.

In contrast with the 100-year-recurrence-interval floodplain, which is a natural and unique landscape feature, the floodway is neither natural nor unique. Whereas there is in concept only one 100-year-recurrence-interval floodplain along a given reach, there may be an infinite number of floodways. Both hydraulic and nonhydraulic factors must be considered in determining a floodway (Evenson, 1979; Jones and Jones, 1987; Lee, 1971; Minnesota Department of Natural Resources, 1977; Walesh, 1973, 1976; Wisconsin Department of Natural Resources, 1978).

Hydraulic Factors

The most important hydraulic factor is the specified allowable stage increase, which, depending on the state or jurisdiction, can range from essentially zero up to 1 ft. The hydraulic analysis assumes that there is no conveyance in the portion of the floodplain outside the floodway. Considering only allowable stage increase, the floodway limits may be positioned anywhere within the floodplain limits subject to the condition that the 100-year recurrence interval flood stage does not exceed the prescribed allowable stage increase at any point along the stream.

Another important hydraulic factor is to provide smooth hydraulic transitions for flow into and out of the waterway opening of hydraulic structures as suggested by the bridge farthest upstream in Figure 1.13. Assuming the analysis indicates that all the flow will pass through the bridge waterway opening, floodway limits immediately upstream and downstream of the bridge should gradually flare out to approximate the area of moving, as opposed to stagnant, water. For example, the Wisconsin Department of Natural Resources (1978) suggests that the floodway limits converge on the upstream side of the waterway openings at a rate no greater than 1 unit laterally to 1 unit in the longitudinal direction. Similarly, the standard suggests that the floodway limits expand on the downstream side of the waterway opening at a rate no greater than 1 unit laterally to 4 units longitudinally.

Velocity and depth combinations are another hydraulic factor influencing floodway determination. In the interest of safety, and recognizing that development will be discouraged within the floodway but permitted in the flood fringe, those portions of the overall floodplain having hazardous velocity and depth combinations should be included within the floodway. For example, the Minnesota Department of Natural Resources (1977) suggests that areas having a depth of greater than 4 ft or a velocity of greater than 2 ft/sec should be included within the floodway.

The final important factor in floodway determination in some jurisdictions is the concept of equal degree of encroachment. Equal degree of encroachment is intended to provide equity to riverine landowners. The principle is that if a landowner on one side of a stream is permitted to encroach into the floodplain by virtue of establishing a riverward floodway line, the landowner on the opposite side should be provided with the same option. That is, property owners on opposite sides of a stream should have the right of "equal degree of encroachment," whether or not they act on that right by filling or developing up to the floodway limit. One approach is to allow the same reduction in conveyance, as defined by the Manning equation, on both sides of the floodplain, subject to satisfying other hydraulic factors. Another, less hydraulically correct approach is to permit equal reduction in floodplain cross-sectional areas.

Nonhydraulic Factors

Assuming that the applicable hydraulic factors are satisfied, numerous nonhydraulic factors are critical because typically at the time the floodplain and floodway are being determined in a community, some of the floodplain is already developed, probably in incompatible uses, and other areas are targeted for development.

To the extent feasible, the number of existing residential, commercial, industrial, and other structures that will be included in a floodway should be minimized. If such structures are placed in the floodway, they may be considered nonconforming under floodplain and other regulations. Similarly, areas committed to development by virtue of actions such as approved plans or constructed utilities and facilities, should be excluded from the floodway because zoning of the floodway district will generally prohibit development.

Conversely, existing or planned park and open space should be included in the floodway because such use is consistent with floodway function. Considerations should be given to making the alignment of planned dikes and floodwalls coincident with the floodway limits because the expected function of dikes and floodwalls, when they are constructed, is similar to the theoretical function of the floodway as determined by the hydraulic analysis.

Adequate provision must be made to maintain lateral drainageways, which of necessity must pass through the flood fringe. That is, although the flood fringe may generally be suitable for fill and development, lateral drainageways must be kept open across the floodways through mechanisms such as zoning, easements, or public acquisition.

Although flood fringe areas in or near communities may not be required to convey or store floodwater, they should not necessarily be filled or otherwise developed. Some areas may have significant aesthetic, ecological, cultural, or other values. Consideration should be given to protecting these areas by acquisition or by floodplain, conservancy, or other special-purpose zoning (Jones and Jones, 1987; Walesh, 1973, 1976).

Hydrologic Considerations

A word of caution is in order. Whereas hydraulic considerations almost always enter into a floodway determination, hydrologic considerations rarely do. There is a danger here. The gradual filling and development of floodplain fringes in most of the communities in a given watershed may significantly reduce the floodwater storage capacity of the floodplain system. Although the storage loss within any given community in a watershed may be small, the cumulative effect of storage loss through many communities along the watershed stream system may be significant, particularly in downstream communities. As a result, flood flows of a specified recurrence interval may increase, particularly in the lower reaches of the watershed. The use of computer modeling to illustrate the impact of various forms of urbanization, including floodplain fill and development, is discussed in Chapters 2 and 10.

Stated differently, floodway determination regulations and guidelines typically address only hydraulic factors, probably because these can be considered by a community unilaterally. Floodway determination regulations and guidelines typically exclude consideration of hydrologic factors, probably because they require intercommunity, watershed-wide analysis.

In summary, partitioning of the floodplain into a floodway and floodplain fringe is a land use regulatory tool based only in part on hydrologic–hydraulic analyses. That is, if the floodway and the floodplain fringe are to be used in an effective, meaningful manner, their determination must embody, but be much more than, a hydraulic analysis. As discussed in more detail in Chapter 1, there is strong judicial support for a comprehensive approach to floodplain regulations, which presumably includes the right to consider much more than passage of floodwaters when determining floodway boundaries.

REFERENCES

Al-Khafaji, A., and R. J. Orth, "The Use of Normal Depth Analysis for Encroachment Studies." In *Proceedings of the Workshop on Hydrologic and Hydraulic Aspects of Flood Plain Construction.* Wisconsin Department of Natural Resources, Madison, August 28-29, 1968.

Barnes, H. H., Jr., "Roughness Characteristics of Natural Channels." *Geol. Surv. Water-Supply Pap. (U.S.)* **1849** (1967).

Barr Engineering Company, *The Effect of Cross Section Spacing and Cross Section Data Errors on Water Surface Profile Determination.* Prepared for the Minnesota Department of Natural Resources, Minneapolis, 1972.

Burnham, M. W., and D. W. Davis, *Accuracy of Computed Water Surface Profiles,* Tech. Pap. No. 114. U.S. Army Corps of Engineers, Hydrologic Engineering Center, Davis, CA, December 1986.

Chow, V. T., *Open Channel Hydraulics.* McGraw-Hill, New York, 1959.

Evenson, P. E., *Tailoring Floodland Regulations to Meet Community Needs.* Presented at the University of Wisconsin-Extension Institute on Floodplain Management, Madison, May 17-18, 1979.

French, R. H., *Open Channel Hydraulics.* McGraw-Hill, New York, 1985.

Gulliver, E. A., "Preliminary Flow Profiles for Highway Engineers," Engineer's Notebook. *Civ. Eng. (N.Y.)* **43**(12), pp. 49–51 (1973).

Jeppson, R. W., *Graphical Solutions to Frequently Encountered Fluid Flow Problems.* Utah Water Res. Lab., Utah State University, Logan, June 1965.

Jones, D. E., Jr., and J. E. Jones, "Floodway Delineation and Management." *J. Water Resour. Plann. Manage. Div., Am. Soc. Civ. Eng.* **113** (2), 228–242 (1987).

Lee, T. M., "Engineering Judgement and Field Surveys for Water Surveys Profile Calculation." In *Proceedings of the Workshop on Hydrologic and Hydraulic Aspects of Floodplain Construction.* Wisconsin Department of Natural Resources, Madison, August 28-29, 1968.

Lee, T. M., *Factors in Floodway Selection.* Presented at the American Society of Civil Engineers, Hydraulics Division Conference, Iowa City, IA, August 18-20, 1971.

Linsley, R. F., and J. B. Franzini, *Water Resources Engineering,* 3rd ed., McGraw-Hill, New York, 1979.

Minnesota Department of Natural Resources, *The Regulatory Floodway in Floodplain Management,* Tech. Rep. No. 6. MDNR, St. Paul, MN, September 1977.

Motayed, A., and D. R. Dawdy, "Uncertainty in Step Backwater Analysis," Tech. Notes. *J. Hydraul. Div., Am. Soc. Civ. Eng.* **105**(HY5), 617–622 (1979).

Petersen, M. S., *River Engineering.* Prentice-Hall, Englewood Cliffs, NJ, 1986.

Portland Cement Association, *Handbook of Concrete Pipe Hydraulics.* PCA, Chicago, IL, 1964.

Southeastern Wisconsin Regional Planning Commission, *A Comprehensive Plan for the Menomonee River Watershed,* Plann. Rep. No. 26. SEWRPC, Waukesha, WI, October 1976.

U.S. Department of the Army, Hydrologic Engineering Center, *Sensitivity Analysis of Factors that Influence Water Surface Profiles.* USDA, Washington, DC, October 1970.

U.S. Department of the Army, Hydrologic Engineering Center, *Hydrologic Engineering Methods for Water Resources Development,* Vol. 6. Davis, CA, July 1975.

U.S. Department of Transportation, Federal Highway Administration, *Hydraulic Charts for the Selection of Highway Culverts,* Hydraul. Eng. Circ. No. 5. USDT, Washington, DC, December 1965.

U.S. Department of Transportation, Federal Highway Administration, *Hydraulics of Bridge Waterways.* USDT, Washington, DC, September 1973.

U.S. Department of Transportation, Federal Highway Administration, *Guide for Selecting Manning's Roughness Coefficients for Natural Channels and Flood Plains,* Rep. No. FHWA-TS-84-204. USDT, Washington, DC, April 1984.

Walesh, S. G., "Floodland Management: The Environmental Corridor Concept." *Proc. 21st Annu. Hydraul. Div. Spec. Conf., Am. Soc. Civ. Eng., 1973.* 105–111.

Walesh, S. G. "Floodplain Management: The Environmental Corridor Concept." *Tech. Rec., Southeast. Wis. Reg. Plann. Comm.,* **3**(6), 1–13 (1976).

Walesh, S. G., and G. E. Raasch, "Calibration: Key to Credibility in Modeling." Presented at the American Society of Civil Engineers, Hydraulics Division Conference, College Park, MD, August 9-11, 1978.

Wisconsin Department of Natural Resources, *Floodplain Regulation Administration Manual.* WDNR, Madison, WI, January 1978.

Zanker, A., "Calculating Channel Discharge." *Water Sewage Works* **127**(6), June 30 (1980).

5

STORMWATER FACILITY HYDRAULICS

Stormwater facility hydraulics, as described in this chapter, consists of the application of basic hydraulic concepts, equations, and relationships to the analysis and design of stormwater facilities. Examples of basic hydraulic concepts used in this chapter are velocity head, hydraulic grade line, energy grade line, and critical depth. Hydraulic equations and computation procedures used in this chapter include open channel flow calculations with the Manning equation, pipe flow computations with the Darcy–Weisbach equation, and orifice and weir flow calculations.

The emphasis in this chapter is on applications not typically presented in reference books or textbooks. Analysis and design guidelines for the hydraulic aspects of stormwater facilities are presented to establish the framework for the examples that comprise the remainder of the chapter and to provide support for other chapters. The drag equation is used to calculate the forces exerted by floodwaters. Culvert design to satisfy specified flow and headwater–tailwater conditions is described, as is the use of headwater and tailwater conditions to determine flows through the culvert. Approaches to determining the flow capacity and the storage capacity of urban streets are presented. The unit concludes with an example of the development of a storage versus discharge relationship for a detention/retention (D/R) outlet control structure without submergence. Examples presented in this chapter assume that design flows are given, that is, hydrologic analysis and design are completed or are being done concurrently with hydraulic analysis and design.

5.1 HYDRAULIC ANALYSIS AND DESIGN GUIDELINES

Guidelines provide the basis for, and give direction to, analysis and design. Guidelines presented in this section are examples of the kinds of factors that should be considered in the analysis and design of surface water facilities, particularly the hydraulic aspects of those facilities [American Society of Civil Engineers (ASCE), 1985; Donohue & Associates, 1985a,b; Metropolitan Sanitary District of Greater Chicago, 1978].

Gravity Driven

Gravity operation of stormwater system components is preferred over the use of pumping and other facilities, particularly by directors of public works, city engineers, and other public works officials. To the extent feasible, electrical or mechanical controls and equipment should be avoided. Simple gravity-operated facilities reduce the likelihood of failure during rainfall–runoff events, when their operation is critical, and minimize future operation and maintenance costs.

Subcritical Flow

Flow in swales, channels, and other stormwater facilities should, to the extent feasible, be maintained in the subcritical range. Supercritical flow conditions are characterized by potentially dangerous and damaging high velocities. Furthermore, at or near critical depth, flow conditions can become very unstable, and widely fluctuating velocity and depth conditions may occur.

Backwater Effects

Where a conveyance facility, such as a swale, sewer, or channel, enters another conveyance facility, the effect of the receiving facility on stages in the tributary facility should be determined.

Safety Measures

Storm sewers, swales, channels, and D/R facilities can pose hazards to residents of urban areas, particularly children, who are often attracted to the facilities when they are conveying or storing stormwater. Following are examples of safety provisions and devices applicable to sewers, channels, and swales:

1. Safety cages or grates mounted on the entrance to open sewers
2. Fences or guardrails on the top edge of vertical or steep channel sidewalls
3. Foot or hand rungs installed intermittently along the length of concrete-

lined or other armored channel walls to provide for escape or rescue from flowing waters

4. Drop structures installed intermittently along channels to reduce flow velocity to safer levels
5. Warning signs mounted near steep-walled channels or along channels that may carry high-velocity flow
6. Freeboard provided above the design stage

Safety provisions potentially applicable to culverts and bridges include:

1. Cages or grates installed on entrances to long culverts
2. Fences or guardrails placed near the top edge of headwalls and wingwalls

On-site safety provisions and devices that may be applicable to D/R facilities include the following:

1. Removable safety cages or grates mounted on the entrance to otherwise open storm sewers which flow either into or out of the D/R facility. Installation of safety cages or grates is critical where inlet and outlet pipes are connected directly to a long or extensive underground pipe system, that is, where such pipes are not simply short culverts beneath roadways or through berms. Cages or grates installed on the entrance to outlet pipes should be sloped so that water moving through the grate will tend to exert an upward force component on a person or object trapped against the grate. The total grate area should be large enough to reduce to safe levels drag forces at the face of the grate.
2. Guardrails or fences installed near the top edge of vertical or steep walls or slopes, especially along the top of headwalls and wingwalls at inlet and outlet structures.
3. Steps, including hand rails, strategically located on the periphery of a D/R facility if there are no or few mildly sloped areas to provide access to and exit from the lower areas of a D/R facility.
4. Signs placed around the perimeter of a D/R facility to indicate its occasional use for storage of water.
5. Use of mild side slopes (e.g., 7 horizontal to 1 vertical or flatter) under water around the periphery of a retention facility.
6. Maximum lateral and longitudinal slopes on concrete cunnettes or trickle channels of 4 percent (about 0.5 in./ft) to minimize the possibility of falling on wet, slippery surfaces.
7. Positioning of active recreation areas such as ballfields and playgrounds away from busy streets, and locating these facilities so that they are easily visible from areas outside, but close to, the D/R facility.

8. Provision of rescue equipment, such as lifesaving rings and small boats, near retention facilities.
9. Freeboard above design stages.

D/R facilities may also pose potential off-site hazards, particularly in downstream areas. Off-site-oriented safety provisions, primarily structural, that may be applicable to D/R facility outlet works include:

1. An emergency spillway sized and adequately armored to pass at least the 100-year-recurrence-interval flood flow.
2. Provision for an emergency downstream flow path to safely carry discharge passing over the emergency spillway.
3. Seepage collars on outlet pipes through earthen dams to mitigate piping supplemented with use of carefully selected and adequately compacted fill material.
4. Installation of a cutoff trench beneath the outlet control works to mitigate piping.
5. Riprap, an energy dissipater, or other protection at the downstream end of the D/R facility outlet pipe to mitigate erosion, particularly erosion that might occur at the toe of, and damage, an earthen embankment.
6. Mildly sloped embankment faces to provide structural stability.
7. Grates, cages, hoods, and other devices to resist movement of objects and debris into and the resulting blockage of D/R outlet works.
8. Adoption of downstream floodplain or similar regulations, or modification of existing regulations, to control the extent and type of flood-prone activity or land use.

Aesthetics

Stormwater facilities such as swales, channels, culverts, and D/R facility inlet and outlet control works may require aesthetic features because of the facility's location in or proximity to residential or other aesthetically sensitive areas. Potential aesthetic treatments include vegetation and other landscaping; use of riprap, gabions, or other visually pleasing material for armoring swales and channels and for forming transition sections; use of gabions, natural stone, or other aesthetically pleasing materials for bridge faces and wingwalls; and installation of decorative guardrails on bridges and along steep-walled channels.

Aesthetic treatments specifically applicable to D/R facilities are:

1. Curvilinear patterns, as opposed to rectilinear forms, in plan and cross section to approximate or suggest the topography of the natural landscape

2. Preservation of trees and other vegetation, topographic relief, rock outcrops, and other natural features
3. Placement of trees, shrubs, grass, and other landscaping
4. Shielding or treatment of inlet and outlet structures by means such as planting shrubs and vines; placing natural rock against and near the structures; use of irregular and rough wooden forms to provide a textured surface on reinforced concrete; and installation of gabions
5. Locating all or part of an inlet or outlet control structure in an underground vault
6. Excavating organic soil from shallow areas around the periphery of a retention facility and placing a clay layer or other relatively impermeable liner on the pond bottom to retard the growth of emergent vegetation

Low Flow Control

A D/R facility may be subject to continuous or frequent intermittent flow. If continuous or frequent flow is expected, a means of accommodating the flow is necessary to facilitate maintenance, such as grass cutting, and to discourage the creation of nuisances such as standing water, weed growth, and odors. Low-flow-control provisions sometimes used in D/R facilities are:

1. A cunette or trickle channel constructed of reinforced concrete or lined with gabion or riprap. Armoring with riprap is not generally recommended, unless firmly held in place with grout, because vandals are likely to move individual rocks. In addition, flat longitudinal grades characteristic of cunettes or trickle channels are difficult to construct and to maintain, especially when riprap is used.
2. An out-of-sight underflow storm sewer connected to the overlying D/R facility by manholes and area drains.
3. Inclusion of drop structures along the cunette or trickle channel to limit the longitudinal slope to about 4 percent or less for safety purposes.

Antiplugging Measures

Stormwater typically picks up and carries buoyant objects and other debris. If these materials block the entrance of or lodge within a component of the stormwater system, the resulting flow constriction could cause serious disruption, damage, or danger. Accordingly, measures such as the following should be used.

1. Circular storm sewers should have a minimum diameter of at least 12 in. The minimum inside dimension for elliptical or arch pipe should also be 12 in.

2. D/R outlet works components such as orifices, pipes, and fully opened sluice gates and other hydraulic controls should have a minimum clear space or diameter of 4 in. to minimize the possibility of full or partial blockage. In addition, outlet works components should be protected by cages or grates with a maximum clear space less than the minimum diameter of clear space of the components positioned immediately downstream. That is, objects and debris that may pass through the cage or grate should readily pass through the downstream orifice, pipe, gate, or other hydraulic device.
3. To minimize the likelihood of complete blockage of a protective cage or grate by waterborne debris or objects, the total open area of the cage or grate should be at least 10 times the cross-sectional area of the protected orifice, pipe, gate, or valve.
4. If an outlet works or other stormwater control facility is particularly vulnerable to debris blockage, an alternative discharge conduit controlled by a manually operated sluice or other gate should be provided. In the event the primary outlet works is inoperable because of debris blockage, the gate on the alternate outlet could be opened to permit discharge from the detention facility and then removal of the debris.

Vortex Protection

Outlet control works, especially vertical risers located in a D/R facility and surrounded by temporarily impounded water, may cause vortexing, which is a swirling or circular motion of water in the horizontal plane as the water moves toward and into the riser. Vortexing is undesirable because the associated air entrainment may reduce the capacity of the outlet works. Furthermore, vortex-induced currents may cause damaging erosion to the side or bottom of the D/R facility.

The possibility of vortexing should be investigated. Available preventive measures include vertical vanes or other baffles installed on the vertical riser, and modification to the size or configuration of the outlet, particularly the vertical riser and the horizontal pipe that receives flow from the riser.

Antivandalism Provisions

Some components of D/R facility outlet works are vital to the effective functioning of the D/R facility. Examples are cages and grates, orifices, pipes, sluice gates, and gate or other valves. These components should be planned and designed to resist vandalism. Examples of ways in which robust materials, mechanical parts, and connections can be used to protect critical components from vandalism are:

1. Using heavy-gauge steel or thick reinforced concrete for orifices and weirs.
2. Fastening protective cages and grates so that they cannot readily be moved or removed.
3. Use of heavy-gauge brackets, bolts, nuts, and other connecting devices.
4. Locking, bolting, and otherwise securing manholes and other access points.
5. Securing sluice gates, stop logs, and other adjustable components of inlet and outlet works to prevent unauthorized, potentially damaging, or disruptive modifications to the intended positions or settings.
6. Perimeter asphalt curbing is subject to unintentional damage or unauthorized removal and therefore should not be used to provide stormwater storage on a parking lot.

Drainage of Recreational Areas

Grassed areas in D/R facilities normally intended for recreational use should have a minimum slope of 2 percent to encourage rapid post-flood-event drainage. In situations where adequate slope cannot be achieved, alternative measures, such as underdrains or peripheral swales, should be provided. Tennis or basketball courts and other carefully constructed paved areas located in D/R facilities may have a slope of less than 2 percent.

Maximum Slopes for Grassed Areas

If slopes in D/R facilities or along channels have grass and are to be maintained by power mowing, the maximum slope should be 1 vertical to 4 or 5 horizontal. In addition to mowing considerations, determination of the maximum allowable slope should consider the stability of the slope, which will depend, in part, on soil properties and the position of the water table. Typically, slopes must exceed the recommended maximum in proximity to inlet and outlet works. Riprap, gabions, or other suitable armoring and protective material should be used in these areas.

Emptying Times

Except for parking lots and other special facilities, D/R facilities should be designed so that detained water discharges in within a few days. Active recreation portions of such facilities may require a shorter emptying time. In contrast, parking lots should be designed so that detained water discharges in a matter of hours.

Erosion and Sedimentation Control

Both erosion and sedimentation can interfere with the functioning of a stormwater system and result in high operation and maintenance costs. Examples of hydraulic erosion and sedimentation control measures are:

1. Storm sewers and paved or armored channels should be designed to flow frequently with velocities in excess of 2 ft/sec to minimize the deposition of silt, sand, and debris.
2. Where culverts or storm sewers enter unprotected channels or swales, an adequate transition must be provided to prevent erosion in the receiving channel or swale.
3. If erosive velocities are expected in swales and channels under design flow conditions, drop structures should be considered to reduce velocities. An alternative is to armor the swale or channel with riprap, gabions, concrete, or other suitable material.
4. If swales or channels are lined with concrete or armored with other impermeable material, weep holes, drain pipes, or other means should be utilized to control hydrostatic uplift.

Inspection and Maintenance

The type and frequency of required inspection and maintenance of stormwater facilities should be identified and documented during the design process. A fundamental requirement is safe and easy access and adequate working space for personnel and equipment. Examples of inspection and maintenance procedures which may be required for stormwater facilities are:

1. Removal of sediment, debris, and other obstructions
2. Clearing of safety grates and grills
3. Control of brush and weeds by cutting and application of herbicides
4. Restoration of eroded areas and sloughed banks and slopes
5. Repair of armoring, such as cracked concrete lining or deteriorating riprap
6. Repair or replacement of fences, rails, and other safety features
7. Repair of transition sections
8. Revegetation by means such as seeding, sodding, fertilization, and mulching
9. Inspection and repair of sluice gates and other control devices
10. Inspection, painting, lubrication, and repair of recreational facilities located in D/R facilities
11. Control of algae, mosquitos, and other nuisances

12. Periodic investigation of earthen dams and berms for erosion, slumping, excessive seepage, woody growth, and rodent damage
13. Examination of D/R spillways for stability and integrity
14. Periodic removal of sediment and debris from sedimentation basins
15. Provision for and servicing of waste receptacles in D/R facilities
16. Cleaning of sediment, debris, and obstructions from inlets and catch basins
17. Examination and cleaning or repair of flow regulating devices installed in catch basins, inlets, or other control works

Some facilities, particularly D/R facilities, require inspection and maintenance during and immediately after a runoff event. For example, the outlet control works should be visited to determine if it is working properly and to remove accumulated debris. After the D/R facility empties, unsightly waterborne debris should be removed and other minor cleaning and maintenance performed.

5.2 FORCE EXERTED ON A PERSON BY MOVING FLOODWATER

The public often fails to appreciate the danger of moving floodwaters, particularly the dynamic forces exerted when a person enters or tries to cross a flooded stream or roadway either on foot or in a vehicle. Basic hydraulics can be used to calculate the drag force on people or objects, thus helping to illustrate and quantify the danger of moving floodwaters.

Hydraulic Analysis Procedure

Figure 5.1 illustrates a hypothetical situation of a person standing on a floodplain and facing toward the current. The dynamic force exerted by the floodwaters on the person can be calculated for a variety of floodplain velocity and depth conditions using the drag equation (Streeter and Wylie, 1985):

$$F_D = C_D A m \frac{V^2}{2}$$

where F_D = drag force

C_D = drag coefficient (dimensionless)

A = submerged area perpendicular to the direction of flow = wd

m = mass density of water

V = average velocity of flow in the vicinity of the object

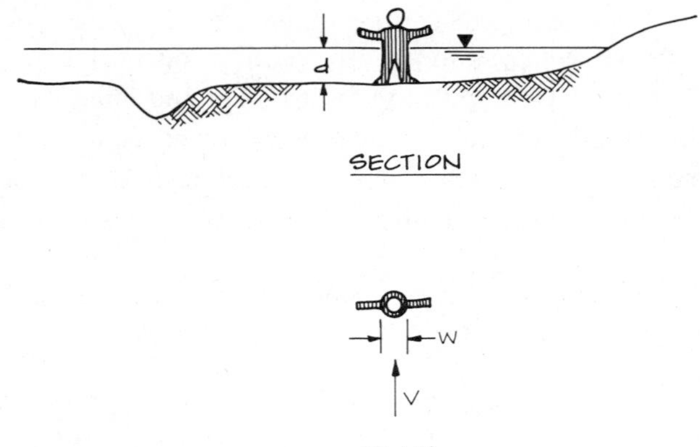

FIGURE 5.1 Person standing on a floodplain

Drag coefficients for various shapes and for specified Reynolds numbers are presented in Figure 5.2.

Assume water at 60° Fahrenheit which has a mass density of 1.94 (lb-sec^2)/ft^4. Approximate the person in Figure 5.1 as a cylinder, and based on Figure 5.2, use a drag coefficient of 1.2. The width of the person, *w*, perpendicular to the flow is assumed to be 1.5 ft, to represent an adult. The drag force exerted on the person shown in Figure 5.1 was calculated for depths of 1 and 3 ft and for velocities of 1 to 10 ft/sec. The resulting forces are summarized in Table 5.1.

Results

The drag force is linearly proportional to depth and proportional to the square of velocity. Therefore, a doubling of depth doubles the drag force, whereas a doubling of velocity quadruples the drag force.

For a typical floodplain situation such as water flowing 3 ft deep at 6 ft/sec, Table 5.1 indicates that the drag force is approximately 190 lb. Not only is this drag force large, but it is likely to be applied abruptly and unexpectedly, thereby knocking a person down, causing further danger.

Even seemingly safe situations can lead to accidents. For example, water flowing at a velocity of 4 ft/sec and a depth of 1 ft over a roadway or across a parking lot exerts a force of about 30 lb on a person, which, if not expected, could easily cause a pedestrian to fall, be injured, and possibly be swept downstream.

Similar dangerous situations arise when people try to drive vehicles across roadways and bridges being overtopped by moving floodwaters. Large drag

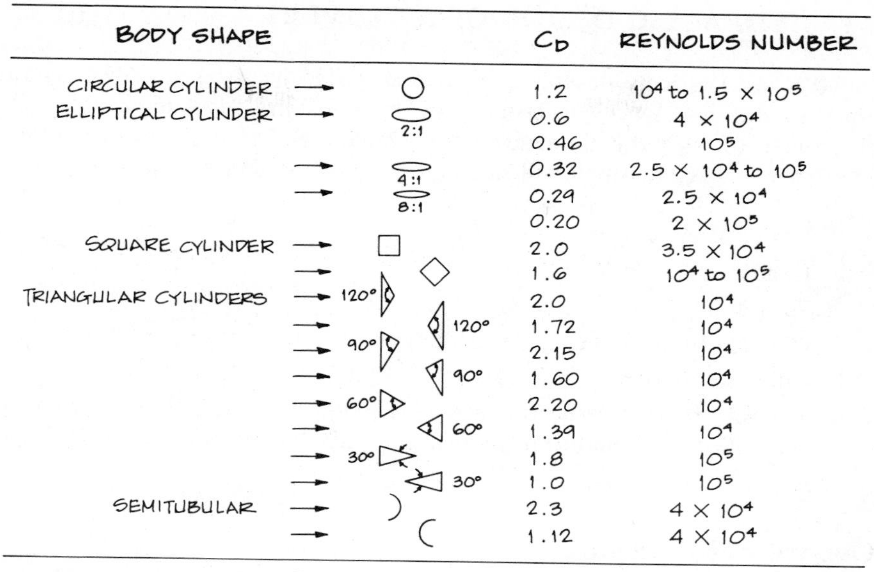

BODY SHAPE		C_D	REYNOLDS NUMBER
CIRCULAR CYLINDER		1.2	10^4 to 1.5×10^5
ELLIPTICAL CYLINDER	2:1	0.6	4×10^4
		0.46	10^5
	4:1	0.32	2.5×10^4 to 10^5
	8:1	0.29	2.5×10^4
		0.20	2×10^5
SQUARE CYLINDER		2.0	3.5×10^4
		1.6	10^4 to 10^5
TRIANGULAR CYLINDERS	120°	2.0	10^4
	120°	1.72	10^4
	90°	2.15	10^4
	90°	1.60	10^4
	60°	2.20	10^4
	60°	1.39	10^4
	30°	1.8	10^5
	30°	1.0	10^5
SEMITUBULAR		2.3	4×10^4
		1.12	4×10^4

FIGURE 5.2 Drag coefficients for cylinders (*Source:* Adapted from Streeter and Wylie, 1985)

forces can be exerted on the side of the vehicle, tending to push the vehicle off the roadway or bridge and into the swollen stream. The risk of being pushed off the roadway or bridge is increased with depth because of the increased upward buoyant force and the resulting decrease in resisting lateral friction force on the vehicle.

TABLE 5.1 Forces Exerted on a Person by Moving Floodwater[a]

Velocity (ft/sec)	Depth (ft)	Drag Force (lb)
1	1	1.7
	3	5.2
2	1	7.0
	3	21.0
4	1	27.9
	3	83.8
6	1	62.9
	3	188.6
8	1	111.7
	3	335.2
10	1	174.6
	3	523.8

[a]For the hypothetical situation illustrated in Figure 5.1.

5.3 HYDRAULIC DESIGN OF A CULVERT

Culverts are routinely designed to meet specified headwater, tailwater, and other conditions. A typical design problem is illustrated in Figure 5.3. The objective is to select the diameter of corrugated metal pipe, having a Manning coefficient of 0.024, which will satisfy the following design conditions:

1. Design flow = Q = 50 ft/sec.
2. Headwater = HW = 6.0. ft
3. Tailwater = TW = 1.0 ft.
4. Channel bottom and culvert slope = S_o = 0.1 percent or 0.001 ft/ft.
5. Culvert length = L = 75 ft.
6. The culvert has a projecting entrance condition; that is, there is no headwall. With this condition, known as type 3, the entrance loss coefficient is 0.9.

Overview of Procedure

Culvert design is a four-step procedure using design aids prepared by the U.S. Department of Transportation (1965). Under the first step, inlet control is assumed and the required pipe diameter is determined. As illustrated in Figure 5.4, inlet control means that discharge is a function of headwater depth and entrance geometry.

The second step is to determine the required pipe capacity assuming outlet control. As illustrated in Figure 5.5, under outlet control, discharge is a function of the headwater depth, entrance conditions, the Manning roughness coefficient of the culvert, culvert length, culvert slope, and outlet conditions.

Pipe diameters obtained under steps 1 and 2 are compared in step 3 to determine if the culvert is operating under inlet control or outlet control. The

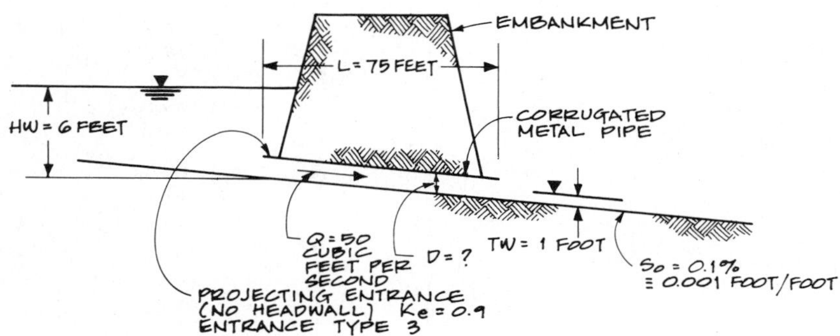

FIGURE 5.3 Design conditions for a culvert

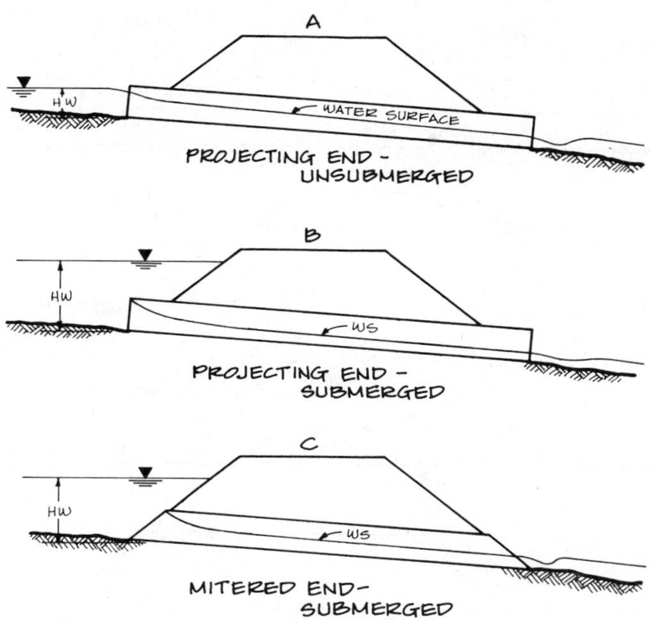

FIGURE 5.4 Inlet control conditions (*Source:* Adapted from U.S. Department of Transportation, 1965)

type of control is determined by which of the two conditions requires the largest pipe size. Using the selected type of control, the culvert diameter is refined under the fourth and last step.

Example

Step 1: Determine Culvert Diameter Assuming Inlet Control. Assume that the culvert diameter, D, is 36 in. Enter the appropriate nomograph from the U.S. Department of Transportation report (1965), which is included for illustration purposes as Figure 5.6, with D = 36 in., Q = 50 ft^3/sec, and entrance type 3. Read HW/D = 1.5, and given that D = 3.0 ft, calculate HW = 4.5 ft. Because the calculated HW is less than the allowable HW of 6.0 ft, a smaller culvert can be assumed.

Assume that D = 30 in. Enter Figure 5.6 with D = 30 in., Q = 50 ft^3/sec, and entrance type 3, and read HW/D = 3.0. Given that D = 2.5 ft, calculate HW = 7.5 ft. The calculated HW exceeds the allowable HW of 6.0 ft. Therefore, a larger pipe diameter is required.

Assume that D = 33 in. Enter Figure 5.6 with D = 33 in, Q = 50 ft^3/sec,

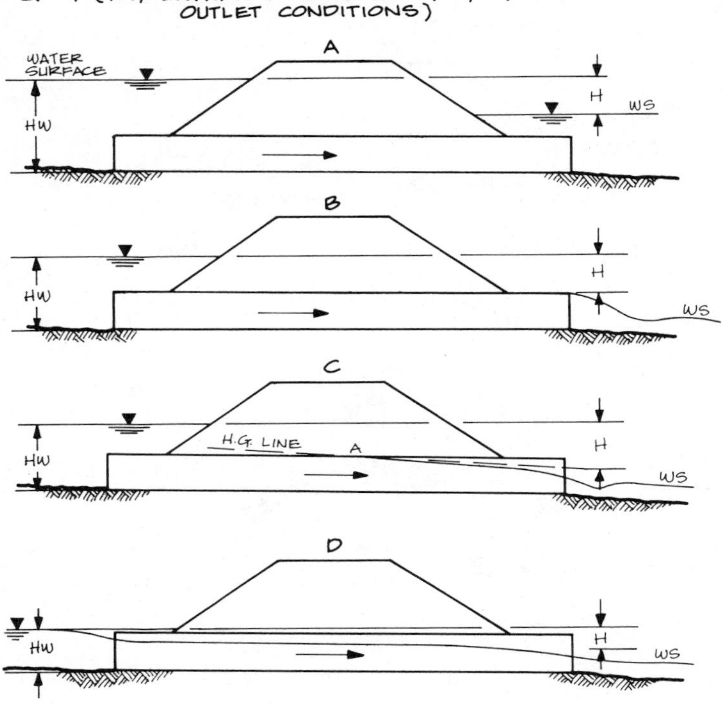

FIGURE 5.5 Outlet control conditions (*Source:* Adapted from U.S. Department of Transportation, 1965)

an entrance type 3, and read HW/D = 2.0. Given that D = 2.75 ft, calculate HW = 5.5 ft. The calculated HW is slightly less than the maximum allowable HW of 6.0 ft. Therefore, assuming inlet control, the required culvert diameter is 33 in.

Step 2: Determine Culvert Diameter Assuming Outlet Control. Assume that D = 33 in. or 2.75 ft. Enter the appropriate graph from the U.S. Department of Transportation report (1965), which is provided as Figure 5.7, with D = 2.75 ft and Q = 50 ft^3/sec. Read critical depth = d_c = 2.3 ft. The critical depth, that is, the depth of 2.3 ft at the discharge end of the circular culvert, is greater than the downstream flow depth. Therefore, the depth of flow in the culvert at its downstream end will be 2.3 ft.

The vertical distance from the outlet invert to the outlet hydraulic grade line is approximated by the following, based on backwater computations:

$$h_o = \frac{d_c + D}{2} = \frac{2.3 + 2.75}{2} = 2.5 \text{ ft}$$

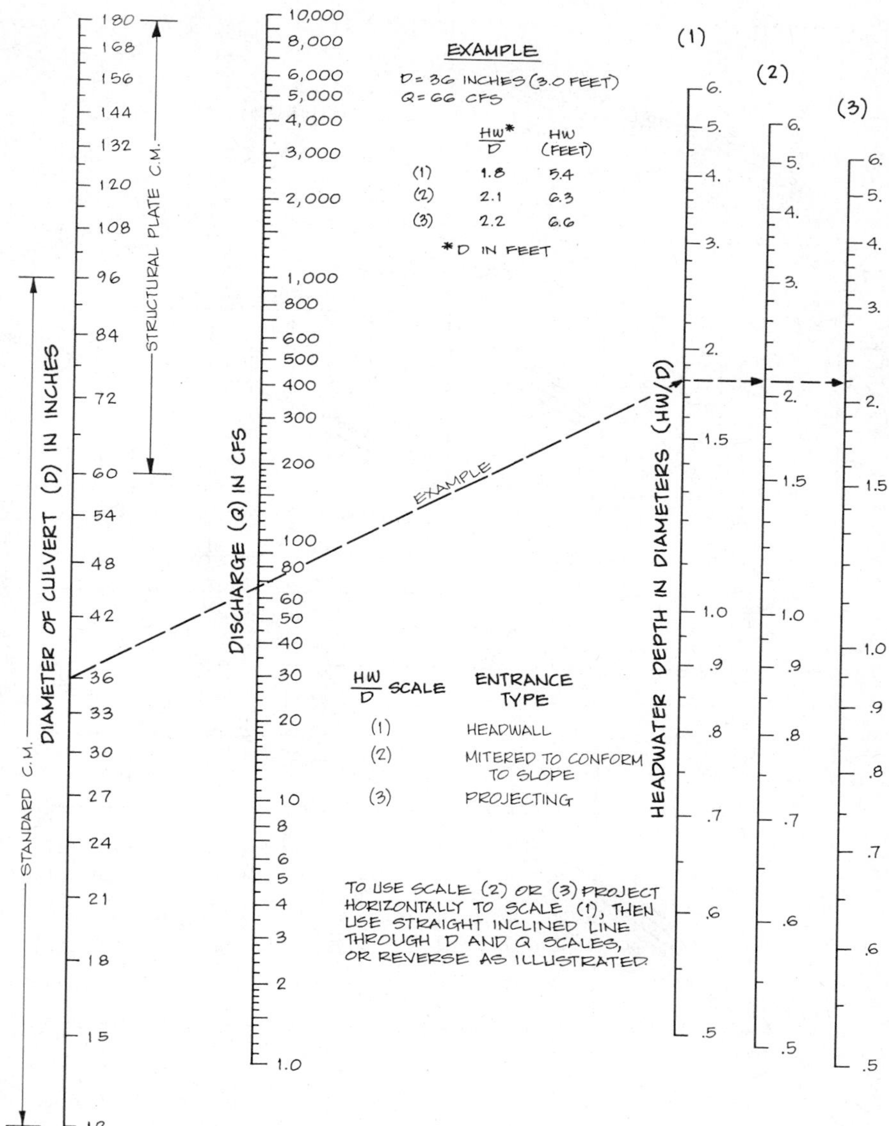

FIGURE 5.6 Nomograph for corrugated metal pipe culverts under inlet control (*Source:* Adapted from U.S. Department of Transportation, 1965)

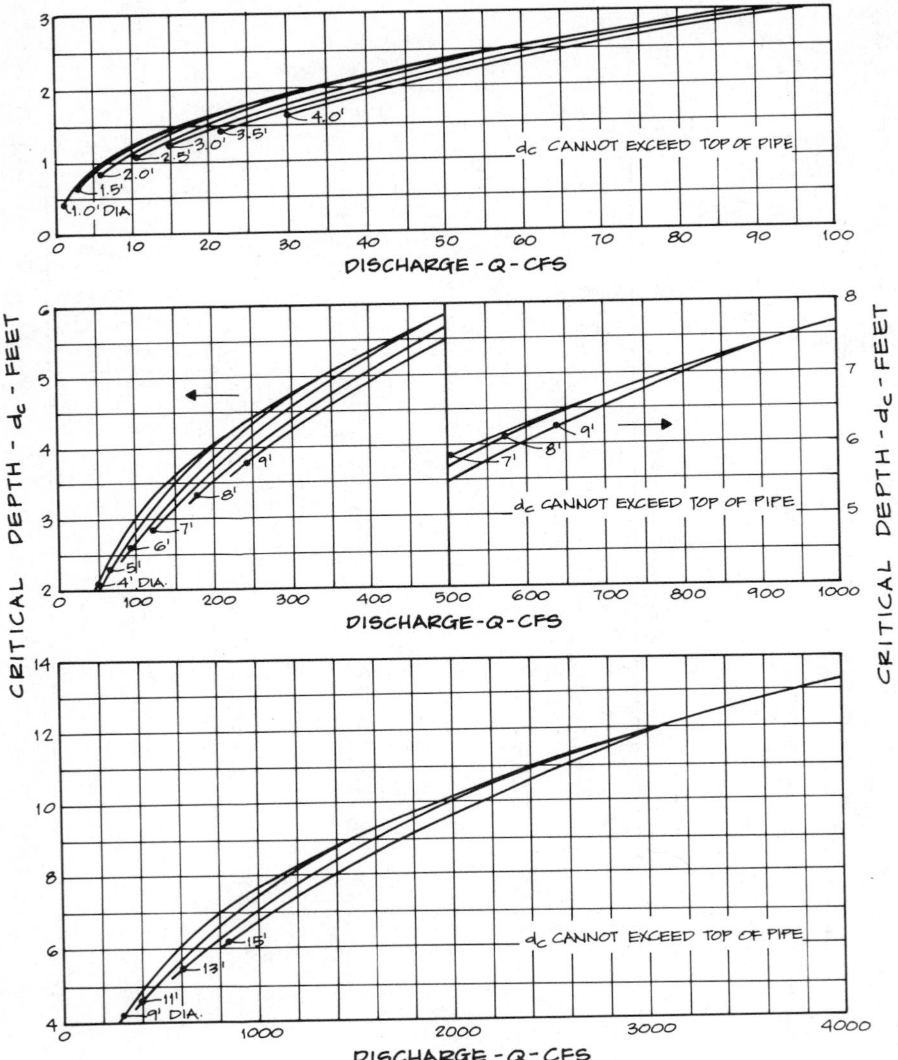

FIGURE 5.7 Critical depth in circular pipes (*Source:* Adapted from U.S. Department of Transportation, 1965)

The headwater depth is given by

$$HW = H + h_o - LS_o$$

As illustrated by the appropriate U.S. Department of Transportation monograph (1965), which is included as Figure 5.8, H is the head differential

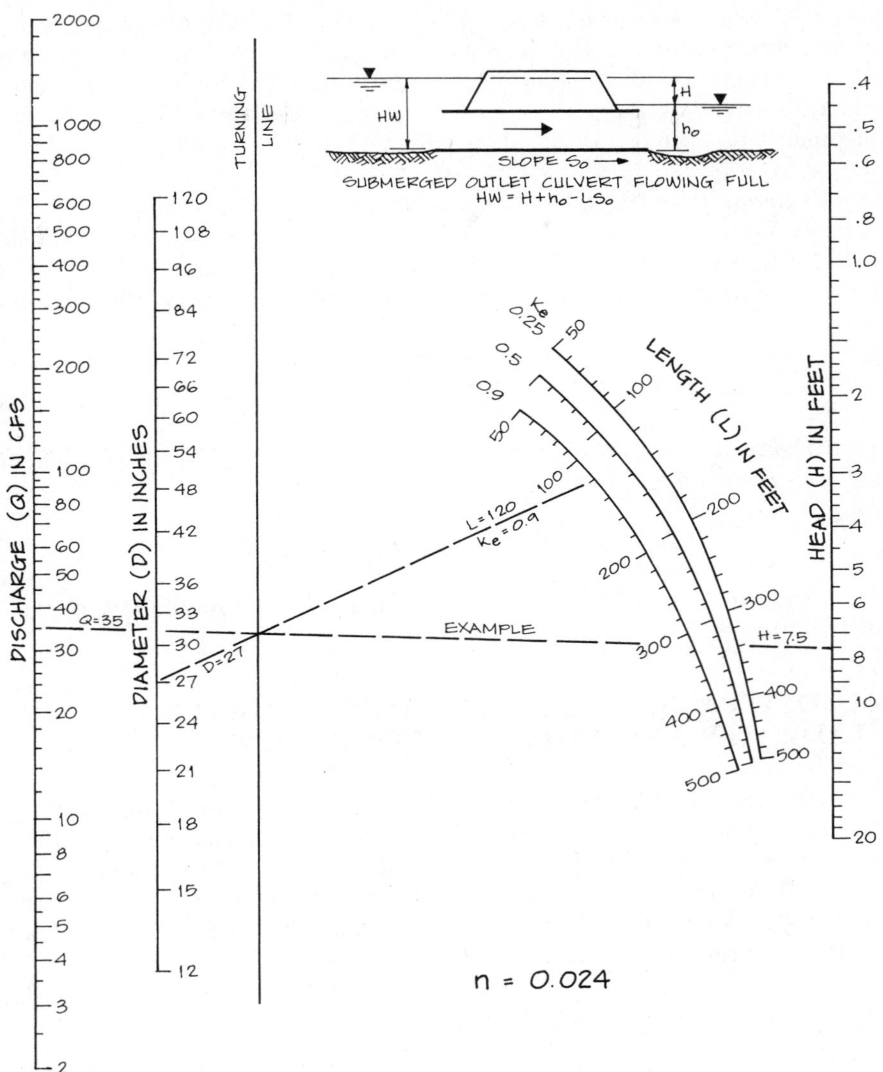

FIGURE 5.8 Nomograph for corrugated metal pipe culvert under outlet control (*Source:* Adapted from U.S. Department of Transportation, 1965)

across the culvert, that is, the drop in the hydraulic grade line. Enter Figure 5.8 with D = 33 in., Q = 50 ft^3/sec, k_e = 0.9, and L = 75 ft, and read H = 4.0 ft. Then, using the preceding equation, calculate:

$$\text{HW} = 4.0 + 2.5 - 75 \times 0.001 = 6.4 \text{ ft}$$

Step 3: Compare Results of Steps 1 and 2. Assuming a 33-in.-diameter pipe, outlet control requires a HW of 6.4 ft, whereas inlet control requires a HW of 5.5 ft. Therefore, outlet control is more stringent and governs. Because the outlet control HW for D = 33 in. is 6.4 ft, which is greater than the allowable maximum HW of 6.0 ft, a pipe larger than 33 in. is required.

Step 4: Refine Pipe Diameter Assuming Outlet Control. Assume that D = 36 in. or 3.0 ft. Enter Figure 5.7 with D = 3.0 ft and Q = 50 ft³/sec and read d_c = 2.3 ft, which is greater than the downstream flow depth. Calculate the vertical distance from the outlet invert to the outlet HGL using the equation

$$h_o = \frac{2.3 + 3.0}{2} = 2.6 \text{ ft}$$

Enter Figure 5.8 with D = 36 in., Q = 50 ft³/sec, k_e = 0.9, and L = 75 ft and read H = 3.0 ft. Then

$$\text{HW} = 3.0 + 2.6 - 75 \times 0.001 = 5.53 \text{ ft}$$

The calculated HW is slightly less than the maximum allowable HW of 6.0 ft. Therefore, select a 36-in.-diameter culvert.

5.4 DETERMINING CULVERT FLOW FOR GIVEN HEADWATER AND TAILWATER CONDITIONS

Peak coincident headwater and tailwater elevations are sometimes available for historic flood events. These data can be used to calculate the peak culvert flow, which, in turn, can be used to determine the severity of a flood, that is, its peak flow and recurrence interval, or be used to calibrate or verify a computer model. An example situation is illustrated in Figure 5.9, where the objective is to determine the discharge through the culvert.

Hydraulic Analysis Using Nomographs

Select the appropriate nomograph from the U.S. Department of Transportation report (1965), which is included as Figure 5.10. This nomograph was selected because it is for circular reinforced-concrete pipe under outlet control conditions.

The upstream end of the culvert projects beyond the embankment; that is, there is no headwall. Table 5.2 presents pipe entrance loss coefficients for a variety of pipe types and conditions. The sample problem is defined as concrete pipe projecting from fill with a square cut and, according to Table 5.2, has an entrance loss coefficient k_e of 0.5.

Enter Figure 5.10 with L = 110 ft and D = 60 in. to establish a pivot point on the turning line. Then enter Figure 5.10 with H = 4.5 ft, extend through

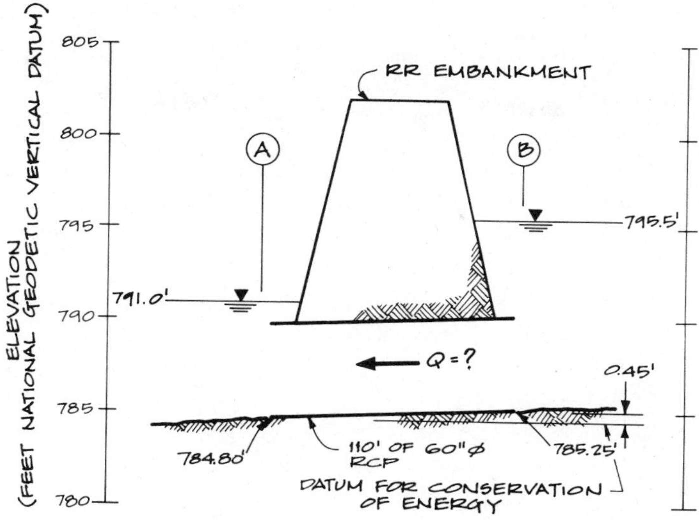

FIGURE 5.9 Analysis conditions for a culvert

the turning line, and read $Q = 240$ ft³/sec. Therefore, for the given conditions, the discharge is estimated to be 240 ft³/sec using the USDOT monographs.

Hydraulic Analysis Using the Conservation of Mechanical Energy

As a check on the preceding calculation, and in order to illustrate an alternative, more generic approach to solving the flow determination problem, the conservation of mechanical energy, can be used. The conservation of mechanical energy equation for the culvert configuration presented in Figure 5.9 and written from section A to section B is

$$d_A + \frac{V_A^2}{2g} = d_B + \frac{V_B^2}{2g} + Z_B - \frac{k_e V^2}{2g} - \frac{k_{ex} V^2}{2g} - f(L/D)(V^2/2g)$$

where d_A = water depth at section A, the downstream end

V_A = average velocity at section A

d_B = water depth at section B, the upstream end

V_B = average velocity at section B

Z_B = rise in culvert invert from section A to section B

k_e = entrance loss coefficient

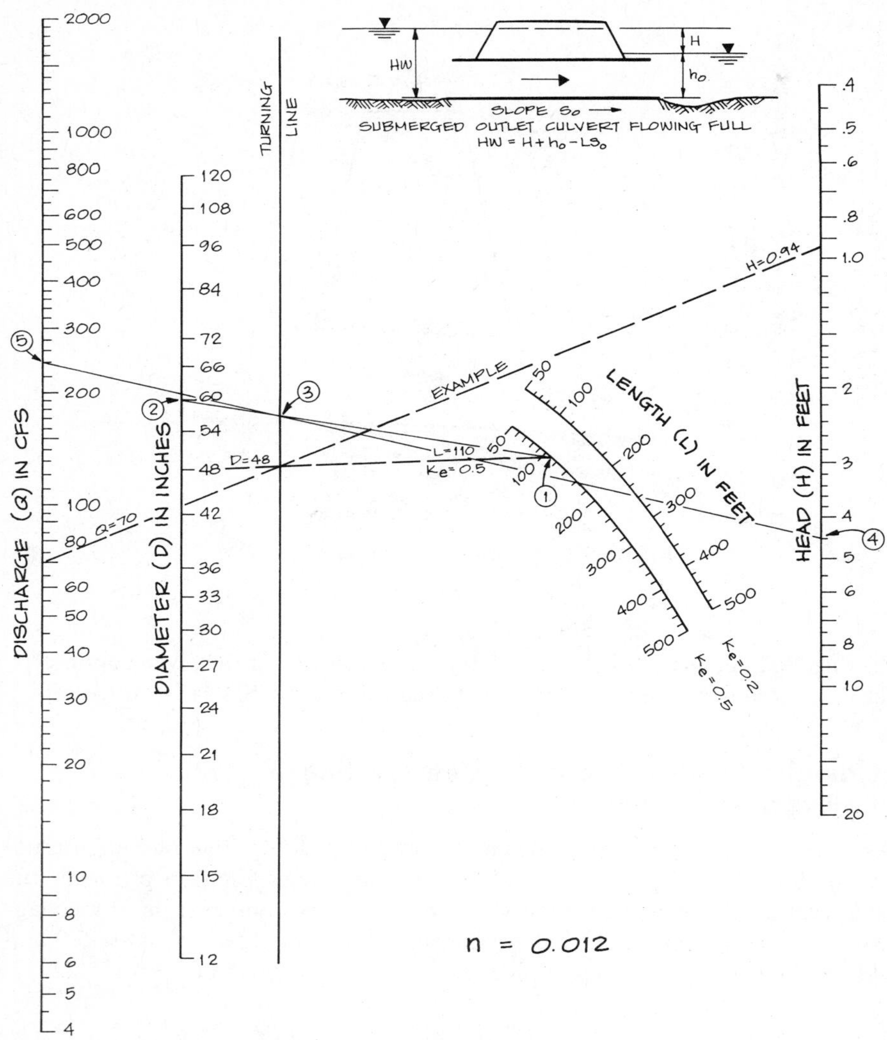

FIGURE 5.10 Nomograph for concrete pipe culvert under outlet control (*Source:* Adapted from U.S. Department of Transportation, 1965)

V = average velocity in culvert
k_{ex} = exit loss coefficient
f = friction factor used in Darcy–Weisbach equation
L = culvert length
D = culvert diameter

TABLE 5.2 Entrance Loss Coefficients[a]

Type of Structure and Design of Entrance	Coefficient, k_e
Pipe, concrete	
Projecting from fill, socket end (groove end)	0.2
Projecting from fill, square-cut end	0.5
Headwall or headwall and wingwalls	
Socket end of pipe (groove end)	0.2
Square edge	0.5
Rounded (radius = 1.12D)	0.2
Mitered to conform to fill slope	0.7
End section conforming to fill slope[b]	0.5
Beveled edges, 33.7° or 45° bevels	0.2
Side or slope-tapered inlet	0.2
Pipe, or pipe arch, corrugated metal	
Projecting from fill (no headwall)	0.9
Headwall or headwall and wingwalls square edge	0.5
Mitered to conform to fill slope, paved or unpaved slope	0.7
End section conforming to fill slope[b]	0.5
Beveled edges, 33.7° or 45° bevels	0.2
Side or slope-tapered inlet	0.2
Box, reinforced concrete	
Headwall parallel to embankment (no wingwalls)	
Square edged on three edges	0.5
Rounded on three edges to radius of 1/12 barrel dimension, or beveled edges on three sides	0.2
Wingwalls at 30° to 75° to barrel	
Square-edged at crown	0.4
Crown edge rounded to radius of 1/12 barrel dimension, or beveled top edge	0.2
Wingwall at 10° to 25° to barrel, square-edged at crown	0.5
Wingwalls parallel (extension of sides), square-edged at crown	0.7
Side or slope-tapered inlet	0.2

Source: Adapted from U.S. Department of Transportation, 1965

[a]Outlet control flowing, full or partly full.

[b]Sections commonly available from manufacturers and determined, by hydraulic tests, to be equivalent in operation to headwall installations.

The equation can be rearranged to yield

$$(d_B + Z_B) - d_A = V^2/2g\ [k_e + k_{ex} + f(L/D)]$$

The expression $(d_B + Z_B) - d_A$ is the difference in water surface elevation from section A to section B, which is 795.5 − 791.0 = 4.5 ft.

Based on Table 5.2, the entrance loss coefficient, k_e, is 0.5. At the downstream end of the culvert, a sudden expansion occurs, and therefore the exit loss coefficient, k_{ex}, is 1.0.

Referring to the friction factor diagram included as Figure 5.11, the absolute roughness for concrete is assumed to be 0.005 ft. Then the relative roughness = 0.005/5.0 = 0.001. Entering the Moody diagram with relative roughness = 0.001 and, assuming complete turbulence, the friction factor is f = 0.0195. Substituting the preceding values for $(d_B + Z_B) - d_A$, k_e, k_{ex}, and f along with L = 110 ft and D = 5 ft into the equation yields an average velocity through the pipe of V = 12.3 ft/sec. The earlier assumption of complete turbulence should be verified. Therefore, the computed discharge using the conservation of mechanical energy approach is given by

$$Q = VA = \frac{V\pi D^2}{4} = 12.3 \times \pi \times \frac{5^2}{4} = 241 \text{ ft}^3/\text{sec}$$

Therefore, the two solutions provide essentially the same discharge, approximately 240 ft³/sec. The first approach, using the USDOT nomographs, gives the quickest solution and is therefore preferable in most applications. However, the second approach, using the conservation of mechanical energy, is useful because it is more general; that is, it can be used in unusual situations, such as for different sizes and types of culverts in series and for culverts with bends and other unusual configurations.

5.5 FLOW CAPACITY OF AN URBAN STREET

Urban streets can be a vital element in the emergency stormwater system by conveying stormwater to a safe discharge point during a major rainfall–runoff event. The Manning open channel flow equation can be used to calculate depth versus discharge relationships for urban streets.

Some actual street cross sections, including the adjacent parkway, sidewalk, and lawn up to the street side of residences, are shown in Figure 5.12. A typical half cross section of a street, based in part on the configurations of the actual street cross sections shown in Figure 5.12, is presented in Figure 5.13. Longitudinal slopes, S_0, of 0.1, 1.0, and 3.0 percent are assumed for the subsequent analysis.

Analysis Procedure

The objective is to determine, assuming normal depth, the flow capacity of the street cross section for a range of depths and a range of longitudinal slopes. The total flow in the half section can be determined as the sum of the flow in subsection A, the street portion of the cross section, and subsection B, the

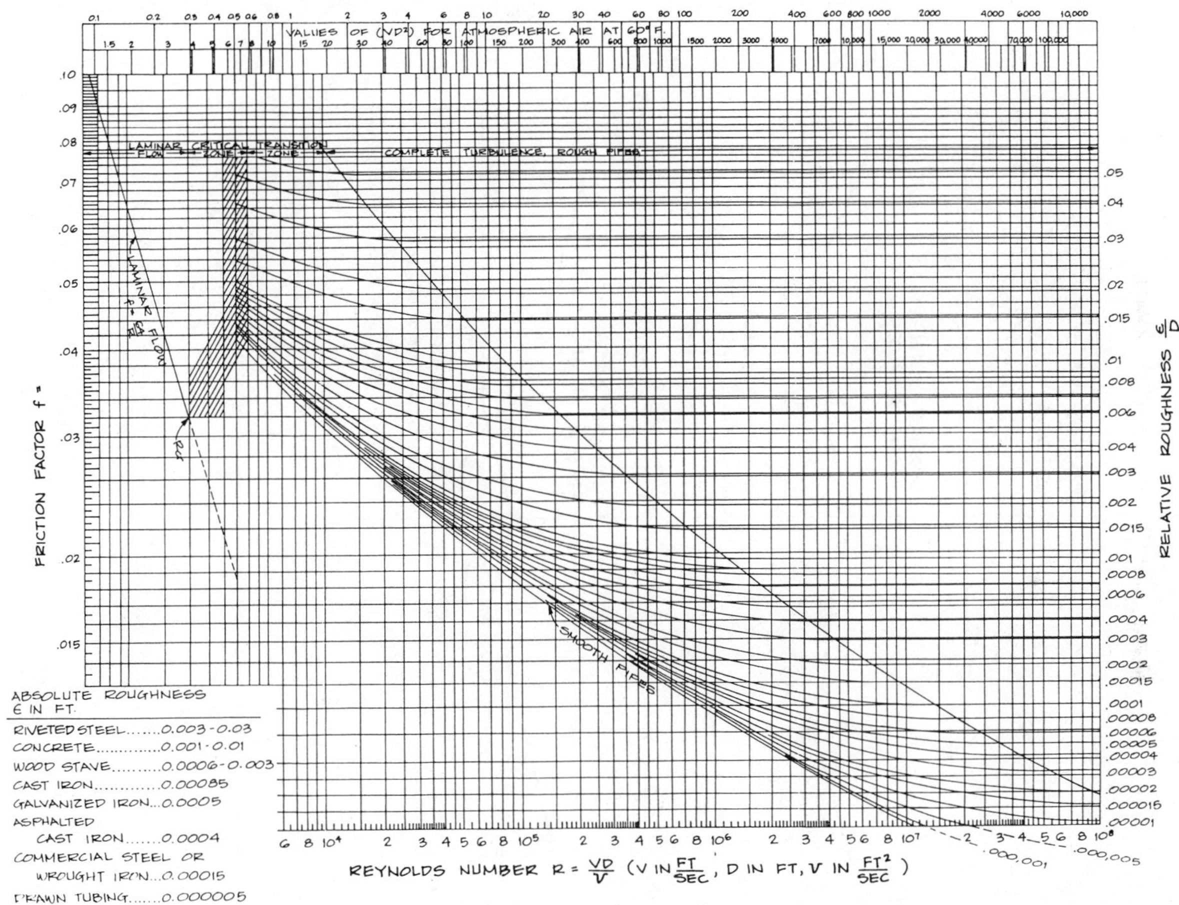

FIGURE 5.11 Friction factor diagram for pipes (*Source:* Moody, 1944)

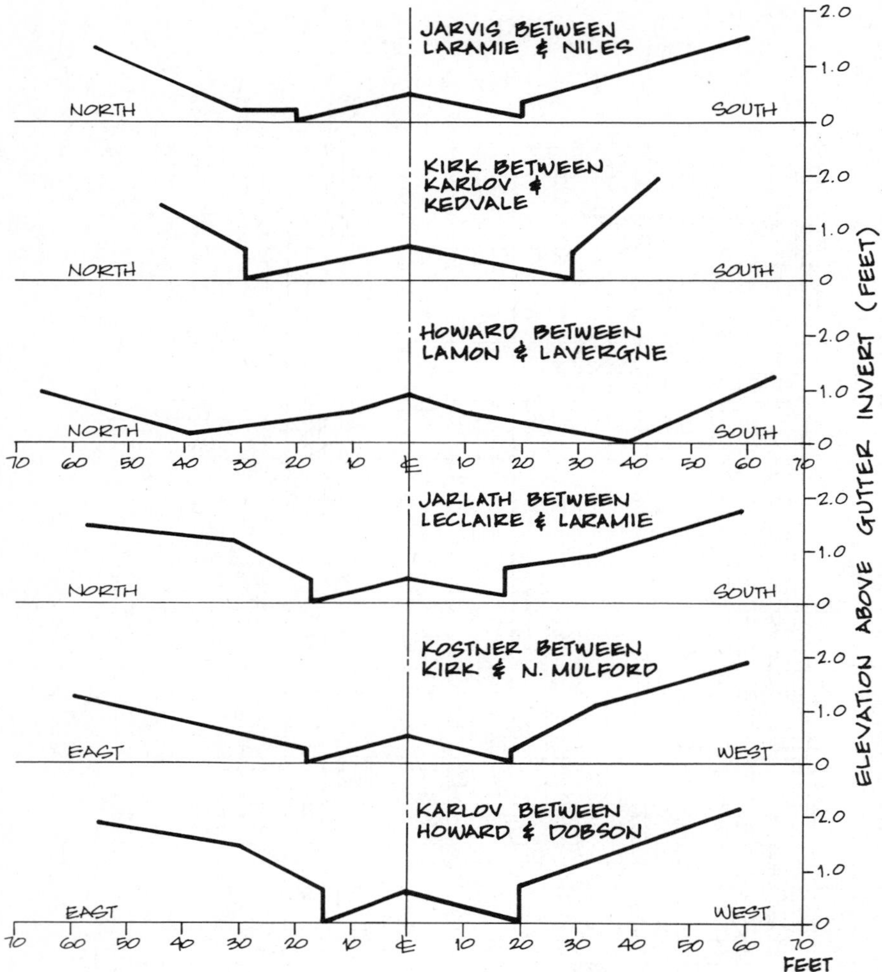

FIGURE 5.12 Selected street cross sections from Skokie, Illinois (*Source:* Adapted from Donohue & Associates, 1982)

lawn portion of the cross section. With this approach; the Manning equation becomes

$$Q = Q_A + Q_B = 1.49 S_0^{0.5} \left(\frac{A_A R_A^{2/3}}{n_A} + \frac{A_B R_B^{2/3}}{n_B} \right)$$

where Q_A = discharge in subsection A

Q_B = discharge in subsection B

S_0 = longitudinal slope for both subsections (dimensionless)

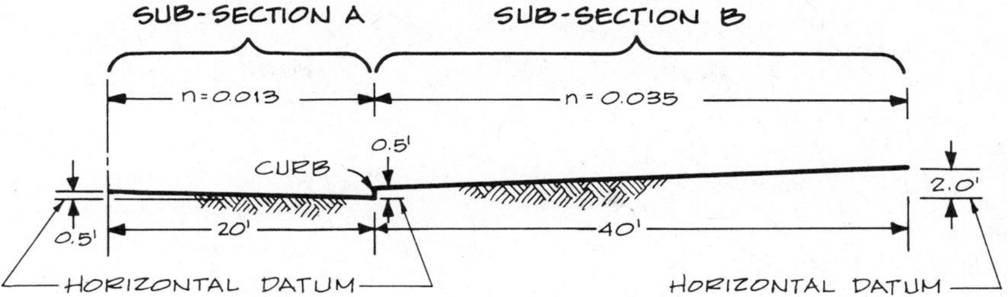

FIGURE 5.13 Typical street and lawn cross section

A_A = flow cross-sectional area in subsection A
R_A = hydraulic radius for subsection A = A_A/P_A, where P_A = the wetted perimeter
n_A = Manning roughness coefficient for subsection A
A_B = flow cross-sectional area in subsection B
R_B = hydraulic radius for subsection B = A_B/P_B, where P_B = the wetted perimeter
n_B = Manning roughness coefficient for subsection B

Assume a depth of flow at the gutter of 0.5 ft. Then

$$A_A = (0.5)\,(0.5)\,(20) = 5 \text{ ft}^2$$
$$P_A = 0.5 + (20^2 + 0.5^2)^{0.5} = 20.5 \text{ ft}$$
$$R_A = \frac{5}{20.5} = 0.244 \text{ ft}$$
$$A_B = 0$$
$$Q = Q_A + Q_B = 1.49 S_0^{0.5}\,\frac{(5 \times 0.244^{2/3})}{0.013} + 0 = 224 S_0^{1/2}$$

Substituting S = 0.1, 1.0, and 3.0 percent yields half-street discharges of 7.1, 22.4, and 38.8 ft^3/sec, respectively. The corresponding average velocities are, 1.42, 4.48, and 7.76 ft/sec, respectively. The preceding process is repeated for depths at the gutter of 1.0 and 2.0 ft.

Results

Depth versus discharge relationships for the complete street cross section, including adjacent lawns, are summarized in graphic form in Figure 5.14. A separate curve is presented for each of the three longitudinal slopes.

The analysis indicates that streets can carry very large flows relative to typical storm sewers at similar slopes. For example, consider the case with a depth at the gutter of 1.0 ft. For the three longitudinal slopes, flows on the full width

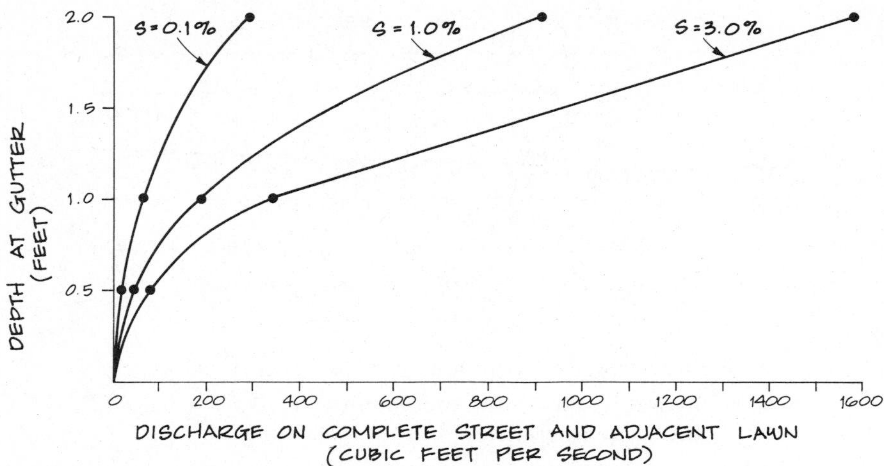

FIGURE 5.14 Depth versus discharge relationships for typical street and lawn cross sections

of street and adjacent lawn areas are approximately 14 to 18 times greater than those that would be carried in a 24-in. reinforced-concrete pipe laid at the same slope and flowing full. The analysis suggests that engineers could make better use of streets by designing some of them to be channels that function as part of the emergency stormwater system.

5.6 STORAGE CAPACITY OF AN URBAN STREET

Urban streets can also constitute a vital element in the emergency stormwater system by temporarily storing stormwater until it can be safely discharged to storm or combined sewers. Actual street cross sections shown in Figure 5.12 suggest the volume of storage available.

Consider again the typical street cross section presented in Figure 5.13 and cross-sectional areas calculated and presented in Section 5.5. A plan view of a typical single-family residential area with paved streets and curb and gutter is shown in Figure 5.15.

Consider the east half of the 600-ft-long section of Easy Street and the directly tributary area of 67,500 ft^2. The runoff coefficient for the area is 0.5; that is, half of the rainfall on the total tributary area is directed toward the east half of Easy Street.

Analysis Procedure

Assuming that the street has a zero longitudinal slope, the cross-sectional area of the east side of the street and cumulative storage on the east side of the

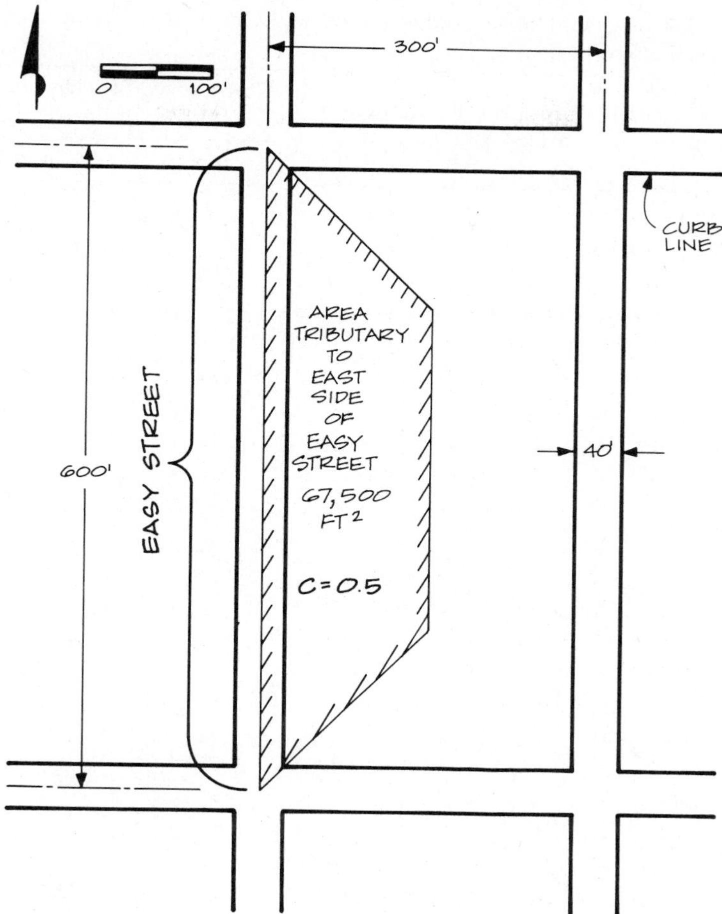

FIGURE 5.15 Typical urban street plan

street may be calculated as a function of depth of water relative to the gutter. Results are presented in Table 5.3 and Figure 5.16. The depth versus volume relationship for the east side of the 600-ft-long street has a shape similar to the depth versus volume relationship for a natural river valley. That is, as depth increases, the relative volume of incremental storage per unit of depth increases at least over the first 1 ft of depth.

Assume rainfall amounts of 0.5, 1.0, 2.0, and 4.0 in., which may be typical of moderate to very severe rainfall events. Assuming that half of the rainfall is directed to and remains in the street, the depth versus storage relationship presented in Figure 5.16 can be used to determine the depth of ponded water for each rainfall amount. The results are presented in Table 5.4.

TABLE 5.3 Depth, Cross-Sectional Area, and Cumulative Volume Data for Half of Hypothetical Street

Depth at Gutter (ft)	Cross-Sectional Area on East Side of Street (ft²)	Cumulative Storage on East Side of Street[a] (ft³)	Cumulative Storage on East Side of Street[a] (acre-ft)
0.0	0.0	0	0
0.5	5.0	3,000	0.07
1.0	18.33	11,000	0.25
2.0	65.00	39,000	0.90

[a]For 600-ft street segment and assuming zero longitudinal grade.

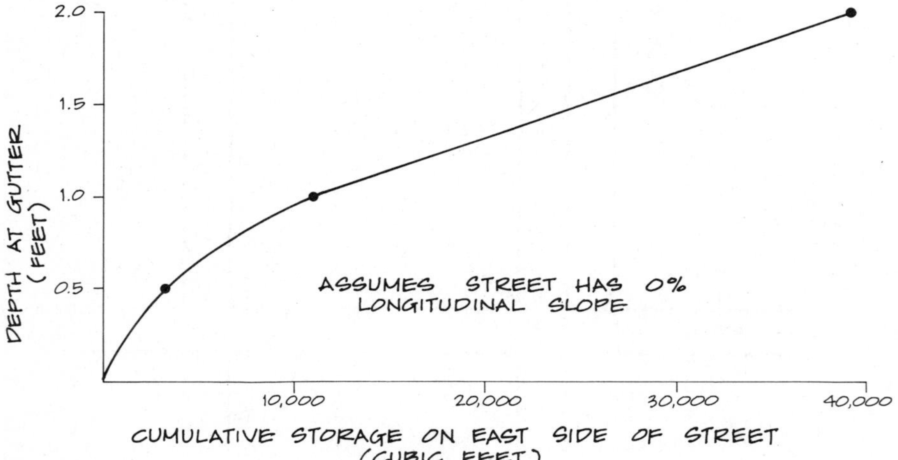

FIGURE 5.16 Depth versus volume relationship for typical street and lawn cross section

TABLE 5.4 Rainfall and Depth of Ponding for Typical Street[a]

Rainfall (in.)	Runoff (in.)	Runoff (ft³)	Depth of Ponding in Street Relative to Gutter (ft)
0.50	0.25	1,410	0.30
1.00	0.50	2,810	0.45
2.00	1.00	5,625	0.75
4.00	2.00	11,250	1.00

[a]Assumes zero longitudinal grade.

Results

As indicated, even with 4 in. of rainfall, and assuming that 2 in. of runoff is stored in the street, the peak depth of street ponding relative to the gutter would be 1.0 ft. The simple analysis suggests that streets have the capacity to store large volumes of runoff. Situations may arise where new streets can be designed to store stormwater, or existing streets can be retrofitted to serve a storage function as part of the emergency system.

5.7 DEVELOPMENT OF STAGE–DISCHARGE RELATIONSHIP FOR A DETENTION/RETENTION OUTLET CONTROL STRUCTURE

Numerous D/R outlet control configurations are possible consisting of combinations of culverts, orifices, and weirs. The cumulative or total stage versus discharge relationship for any given outlet control structure configuration may be constructed by first developing the stage versus discharge relationship for individual components and then summing the component relationships.

Figure 5.17 is a D/R outlet structure consisting of three culverts, a concrete spillway, and an earthen berm. Assume no downstream submergence; that is, for all discharges considered, downstream stages are below the spillway crest.

The objective of the analysis is to determine the cumulative stage versus discharge relationship for the outlet control works assuming no submergence. The overall approach is to develop the stage versus discharge relationship for the weirs and for the culverts and then to determine the total relationship as the sum of the individual relationships.

Development of the stage versus discharge relationship for the weir segments is presented in Table 5.5. Upstream stages, beginning at the spillway crest elevation of 791.0 ft, are increased progressively in 1-ft increments. For each upstream stage, the total length of weir is segmented and for each segment the average head over the weir is determined.

The weir is assumed to be a broad-crested weir of shape D, as illustrated in Figure 5.18. Table 5.6 presents coefficients for broad-crested weirs as a function of weir shape and head on the weir. For shape D and the head range in the example, a weir coefficient of 3.6 is appropriate.

Although not needed for the example, Table 5.7 presents coefficients for sharp-crested weirs. Coefficients for sharp and broad-crested weirs are also provided by Hulsing (1967). Finally, French (1985) discusses various sharp- and broad-crested weir shapes and includes equations and coefficients.

Weir discharge is calculated with the equation

$$Q = C_w Lh^{1.5}$$

where Q = discharge

C_w = weir coefficient

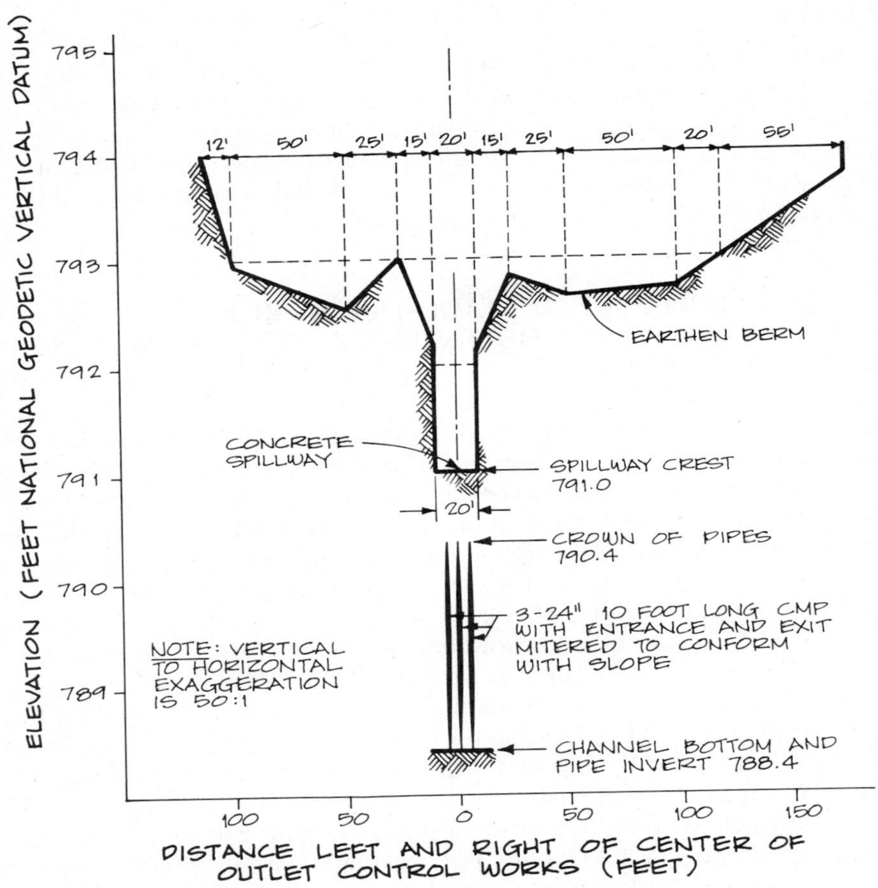

FIGURE 5.17 Outlet control works for a detention facility

L = length of weir segment

h = average head on weir segment

The stage versus discharge relationship for the broad-crested weirs comprising the outlet control structure is presented in Table 5.8.

Because they are relatively short, the three 24-in.-diameter corrugated metal pipes were assumed to function under inlet control conditions. Accordingly, Figure 5.6 was used to determine stage versus discharge relationships for the three culverts using the procedure described in Section 5.3. The resulting stage versus discharge relationship for the three culverts is presented in Table 5.8. Also presented in Table 5.8 is the cumulative stage versus discharge relationship, which is also shown graphically in Figure 5.19.

TABLE 5.5 Stage versus Discharge Calculation for Weir Segments

Upstream Stage (ft NGVD)	Weir Lengths (ft)	Head, h (ft)	Weir Coefficient, C_w	Discharge (ft^3/sec)
791.0	20	0	3.6	0
792.0	20	1.0	3.6	72
793.0	50	0.30	3.6	30[a]
	25	0.25	3.6	11[a]
	15	0.40	3.6	14[a]
	20	2.0	3.6	204[a]
	15	0.50	3.6	19[a]
	25	0.30	3.6	15[a]
	50	0.35	3.6	37[a]
	20	0.15	3.6	4[a]
794.0	12	0.5	3.6	15[b]
	50	1.30	3.6	267[b]
	25	1.25	3.6	126[b]
	15	1.40	3.6	89[b]
	20	3.0	3.6	374[b]
	15	1.50	3.6	99[b]
	25	1.30	3.6	133[b]
	50	1.35	3.6	282[b]
	20	1.15	3.6	89[b]
	55	0.6	3.6	92[b]

[a]Total weir discharge for stage of 793.0 = 334 ft^3/sec.
[b]Total weir discharge for stage of 794.0 = 1566 ft^3/sec.

If a D/R facility outlet is submerged, that is, the downstream stage is above the weir crest, the free condition discharge must be reduced to account for the retarding effect of submergence. Figure 5.20 provides a means of accounting for submergence.

If submergence occurred for the specific example presented in this section,

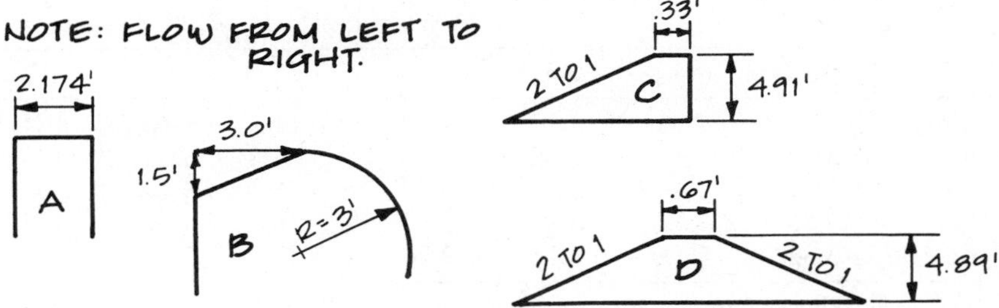

FIGURE 5.18 Types of broad-crested weirs (*Source:* Adapted from Horton, 1907)

TABLE 5.6 Coefficients for Broad-Crested Weirs

	Head on Weir, *h* (ft)						
Shape	0.5	1.0	1.5	2.0	3.0	4.0	5.0
A	2.70	2.64	2.64	2.70	2.89	—	—
B	3.27	3.38	3.46	3.51	3.58	3.67	3.83
C	—	3.85	3.82	3.79	3.75	3.70	3.64
D	—	—	3.58	3.56	3.58	3.62	3.68

Source: Horton (1907).

a trial-and-error procedure would be required to determine the upstream stage and weir flow plus culvert flow distribution corresponding to each downstream stage–discharge combination. The trial-and-error procedure would proceed as follows:

1. Assume an upstream stage.
2. Calculate the corresponding weir flow, accounting for possible submergence, and the corresponding culvert flow, accounting for probable outlet control conditions, and determine the total discharge.
3. Compare the total flow to the total downstream flow.
4. If the discharges are not equal, assume a new upstream stage and repeat steps 1, 2, and 3 until there is concurrence between the upstream stage, the discharge over the weirs and through the culverts, and the downstream discharge and stage.

If a D/R facility outlet control includes orifices, the relationship between stage and discharge is given by

$$Q = C_d A(2gh)^{1/2}$$

TABLE 5.7 Coefficients for Rectangular Sharp-Crested Weirs

	Head on Weir, *h* (ft)						
H/h[a]	0.2	0.4	0.6	0.8	1.0	2.0	5.0
0.5	4.18	4.13	4.12	4.11	4.11	4.10	4.10
1.0	3.75	3.71	3.69	3.68	3.68	3.67	3.67
2.0	3.53	3.49	3.48	3.47	3.46	3.46	3.45
10.0	3.36	3.32	3.30	3.30	3.29	3.29	3.28
Above 10	3.32	3.28	3.26	3.26	3.25	3.25	3.24

Source: Linsley and Franzini (1979).

[a] *H*, height of weir; *h*, head on weir.

TABLE 5.8 Stage versus Discharge Relationship for Weir and Culverts

Upstream Stage (ft NGVD)	Weir Flow (ft^3/sec)	Culvert Flow (ft^3/sec)	Total Flow (ft^3/sec)
791	0	51	51
792	72	66	138
793	334	78	412
794	1566	90	1656

where Q = discharge
C_d = discharge coefficient (dimensionless)
A = orifice cross-sectional area
h = head

Various types of orifices and the corresponding orifice coefficients are presented in Figure 5.21.

5.8 OTHER HYDRAULIC METHODS

Many other areas of hydraulics are of interest in urban SWM. The literature is rich with papers and reports providing analysis and design information. For example, useful analysis and design information on manholes is provided by Marsalek (1987) and U.S. Department of Transportation (1979), and inlet hydraulic analysis and design are covered by ASCE (1969) and U.S. Department of Transportation (1979).

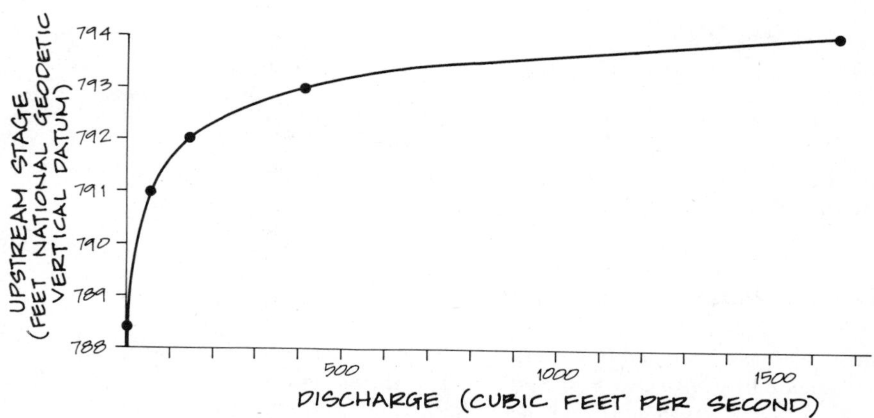

FIGURE 5.19 Stage versus discharge relationship

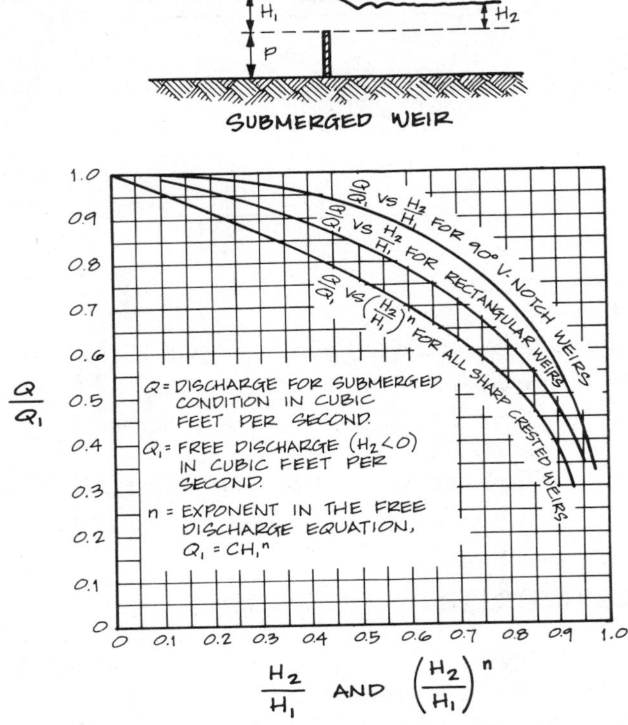

FIGURE 5.20 Submerged weir relationships (*Source:* Adapted from King and Brater, 1963)

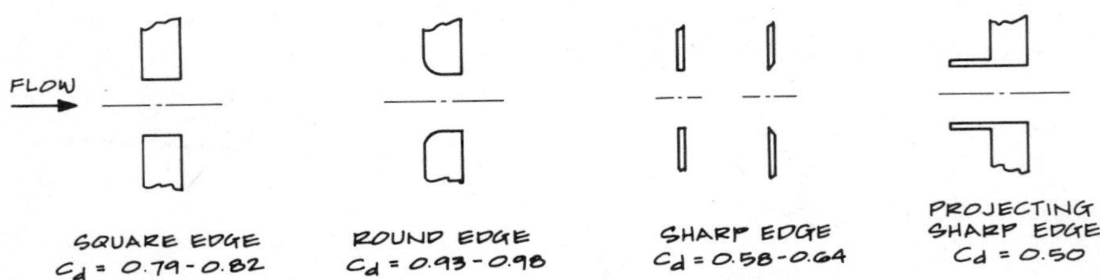

FIGURE 5.21 Discharge coefficients for orifices (*Source:* Adapted from Metropolitan Sanitary District of Greater Chicago, 1978)

REFERENCES

American Society of Civil Engineers (ASCE) and Water Pollution Control Federation, *Design and Construction of Sanitary and Storm Sewers,* ASCE Man. Rep. Eng. Pract. No. 37/WPCF Man. Pract. No. 9. ASCE/WPCF, New York, 1969.

American Society of Civil Engineers, *Stormwater Detention Outlet Control Structures.*

Report of the Task Committee on the Design of Outlet Control Structures of the Committee on Hydraulic Structures of the Hydraulics Division, 1985.

Donohue & Associates, Inc., *Preliminary Engineering—Runoff Control Program, Howard Street Sewer District,—Skokie, IL,* July 1982.

Donohue & Associates, Inc., *Stormwater Management Guidelines for Bettendorf, IA,* 1985a.

Donohue & Associates, Inc., *Stormwater Management Guidelines for DeKalb, IL,* 1985b.

French, R. H., *Open Channel Hydraulics.* McGraw-Hill, New York, 1985.

Horton, R. E., "Weir Experiments, Coefficients and Formulas." *Geol. Surv. Water-Supply Pap. (U.S.)* **200,** 59–134 (1907).

Hulsing, H., "Measurement of Peak Discharge at Dams by Indirect Method." In *Techniques of Water Resource Investigations of the United States Geological Survey,* Book 3, Chapter A5. USGS, Washington, DC, 1967.

King, H., and E. F. Brater, *Handbook of Hydraulics,* 5th ed. McGraw-Hill, New York, 1963.

Linsley, R. K., and J. F. Franzini, *Water Resources Engineering,* 3rd ed. McGraw-Hill, New York, 1979.

Marsalek, J., "Manhole Junction Flow." *Civ. Eng. (N.Y.)* January, pp. 68–69 (1987).

Metropolitan Sanitary District of Greater Chicago, *Recommendations for Design of Outflow Control Devices for Stormwater Detention Facilities,* Inter-office Memo. MSDGC, IL, April 6, 1978.

Moody, L. F., "Friction Factors for Pipe Flow," *Trans. ASME,* **66**,8, November 1944, pp. 671–684.

Streeter, V. L., and E. B. Wylie, *Fluid Mechanics,* 8th ed., McGraw-Hill, New York, 1985.

U.S. Department of Transportation, Federal Highway Administration, "Hydraulic Charts for Selection of Highway Culverts." *Hydraul. Eng. Circ.* No. 5, December (1965).

U.S. Department of Transportation, Federal Highway Administration, *Design of Urban Highway Drainage—The State of the Art,* FHWA-TS-79-225. USDT, Washington, DC, August 1979.

6

COMPUTATION OF AVERAGE ANNUAL MONETARY FLOOD DAMAGE

Calculation of average annual monetary damage (AAMD) may be desirable and feasible in certain urban surface water management situations. An example of an often desirable and usually technically feasible application is development, comparison, and selection from among various structural alternatives for mitigating flood damage along a river reach.

This chapter begins with a discussion of types of monetary flood losses. The potential use of AAMD to determine average annual monetary benefits (AAMB) for comparison to average annual costs is explained. A seven-step process used to calculate AAMD is presented. The chapter concludes with a discussion of how various alternative measures, such as channel enlargement or removal of flood-prone structures, are reflected in the determination of before-and-after condition AAMD.

6.1 TYPES OF FLOOD LOSSES

Monetary and other losses incurred as a result of flooding can be categorized by means of the matrix presented as Table 6.1. The vertical axis of the matrix includes two forms of flood damage or loss: direct and indirect. Direct damage is caused by floodwater being in physical contact with flood-prone objects and materials. These objects or materials subsequently require cleaning, repair, or replacement. Indirect damage is that which occurs as a consequence of the direct damage.

The horizontal axis of the matrix includes the two ways in which flood damage or loss is measured: tangible and intangible. Tangible damage is that

TABLE 6.1 Typology of Flood Loss

Form of Damage or Loss	Tangible Loss (May Be Measured in Monetary Terms)		Intangible Loss (Not Currently Possible to Measure in Monetary Terms)	
	Private Sector	Public Sector	Private Sector	Public Sector
Direct	Cost of cleaning, repairing or replacing, residential, commercial, and industrial buildings; contents and land Cost of cleaning, repairing, or replacing contents and cost of lost crops and livestock	Cost of repairing or replacing roads, bridges, culverts, and dams. Cost of repairing damage to stormwater systems, sanitary sewerage systems, and other utilities. Cost of restoring parks and other public lands.	Loss of life Health hazards Psychological stress caused by current flood	Disruption of normal community activities Loss of archaelogical site
Indirect	Cost of temporary evacuation and relocation Lost wages Lost production and sales Incremental cost of transportation Cost of postflood floodproofing Cost of purchasing and storing flood-fighting equipment and materials	Incremental costs to governmental units as a result of flood fighting measures Cost of post-flood engineering and planning studies and of implementing structural and nonstructural management measures Cost of purchasing and storing flood-fighting equipment and materials	Reluctance by individuals to inhabit flood-prone areas, thereby depreciating riverine area property values an unknown amount Psychological stress caused by possibility of future floods	Reluctance by business interests to continue development of flood-prone commercial–industrial areas, thereby adversely affecting the community tax base

Source: Adapted from Southeastern Wisconsin Regional Planning Commission (1976) and Green et al. (1983).

which can be measured or quantified in monetary terms. Intangible damage is that which cannot now be expressed in monetary terms. Further research may lead to ways to convert some intangible effects to the tangible category (Green et al., 1983). The tangible and intangible categories are each further subdivided into private sector and public sector.

The ideal engineering approach would be to define the severity of a flooding problem by estimating the AAMD associated with tangible direct and indirect damage and then identify and somehow quantify the nonmonetary, intangible damage. In the best realistic situations, the methodology presented in this chapter can be used to calculate direct monetary damage to some areas within the private and public sector and to grossly quantify some indirect monetary damage. Intangible direct and indirect damage can be identified subjectively, but, as noted earlier, cannot yet be readily quantified in a monetary or other sense.

6.2 THE USES OF AVERAGE ANNUAL MONETARY DAMAGE

The two principal uses of AAMD are:

1. *Problem Quantification and Prioritization.* Average annual monetary damage on a reach-by-reach basis can be used to identify the most serious existing or future floodprone areas throughout a river system. Based on relative AAMD, reaches can be prioritized according to the severity of the flooding problem, so that remedial or preventive measures can be focused where they are needed most.
2. *Selection of Most Economic Solution.* AAMD can be used to identify the most economic structural or nonstructural solution for a particular flood-prone reach or area.

More specifically, and with respect to the second of the two principal uses of AAMD, monetary benefits and costs can be calculated for each technically feasible and otherwise feasible alternative (Grigg, 1985; Helweg, 1985; James and Lee, 1971; Linsley and Franzini, 1979). The concept is illustrated in Figure 6.1 using a channel modification as a potential solution to a flooding problem along a river reach. The average annual cost of the channel modification alternative is the sum of the annual amortization of the capital costs plus the annual operation and maintenance costs. The AAMB that would be realized as a result of the channel modification alternative is the reduction in AAMD attributable to the channel modification. The AAMB is the difference between the AAMD without the channel modification, that is, the existing AAMD, minus the AAMD with the channel modification. Given the benefit and cost of each alternative, as determined with the preceding approach, the most cost-effective solution can be selected.

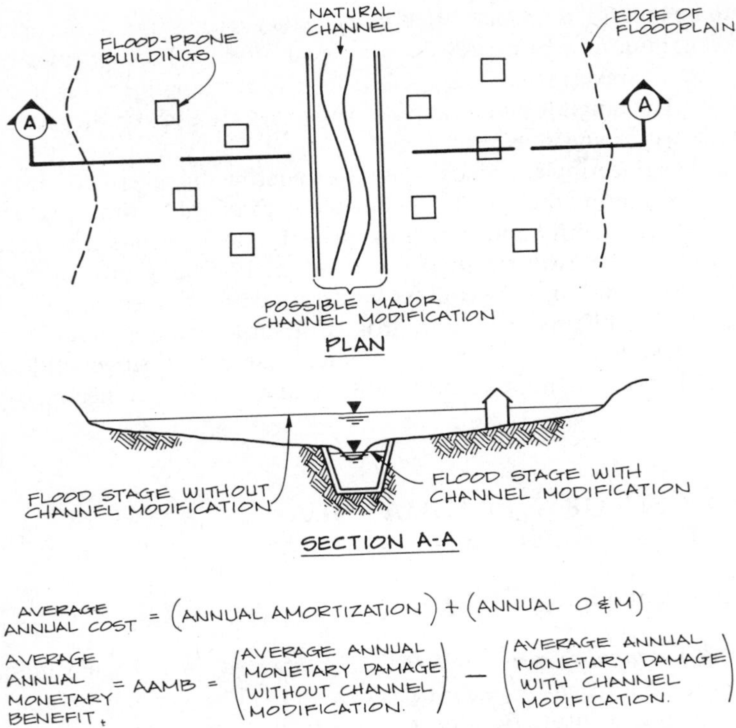

FIGURE 6.1 Benefit-cost analysis for a channel modification

Calculation of monetary benefits and costs is not limited to consideration of traditional structural solutions, such as channel modifications, dikes and floodwalls, detention/retention facilities, and reservoirs. James (1972) presents a simplified example of benefit-cost analysis applied to floodplain zoning and to floodproofing of individual new buildings by specifying a minimum first-flow elevation. Johnson (1976) reports on a benefit-cost analysis of the alternative of purchasing flood-prone structures and removing them from the floodplain. Changes in AAMD were used by Walesh and Videkovich (1978) as one indicator of the impact of various urbanization scenarios for a watershed. Debo (1982) used the AAMD computation technique to develop regional flood damage estimating curves.

The preceding discussion of principal potential uses of AAMD seems logical, but computation of AAMD is rare relative to the number of urban SWM projects undertaken. In most urban SWM investigations, monetary damage is not calculated. Instead, the approach is to select a design condition, such as the 100-year-recurrence-interval event, and search for the least costly means of controlling or accommodating the stormwater discharges and volumes associated with that event. This "least cost" approach is used primarily because it simplifies the engineering investigation and reduces engineering costs.

The least cost procedure usually used in urban SWM is similar to the least cost approach commonly used in water quality management. For example, wastewater treatment plants are typically designed to provide the least costly facility needed to achieve a specified effluent quality without specifically determining the benefits that will accrue as a result of the investment.

The problem associated with the least cost approach in urban SWM is the concern that solutions may be implemented which, although they are technically, financially, and administratively feasible, are uneconomical. That is, a community or agency may be choosing a course of action that, in effect, has a benefit/cost ratio of less than 1 or has a benefit-cost relationship that is less favorable than that of an alternative.

6.3 CALCULATION PROCEDURE

Calculation of AAMD is described here as a seven-step process. As suggested by the steps in the process, the computations are voluminous, tedious, and therefore, not generally amenable to manual computation. Computer programs are available to compute AAMD and to calculate the costs of structural and nonstructural flood mitigation measures. Flood damage calculation and related programs are available from government agencies such as the Corps of Engineers. Proprietary programs are also available from consultants and software vendors. Even with the help of a computer program, determination of AAMD and the annualized cost of alternative structural and nonstructural management measures requires considerable data and analysis. Furthermore, and as suggested by Plazak (1986), various errors can enter into the AAMD computation process.

Step 1: Obtain General Stage–Damage Relationships for all Types of Structures

Figure 6.2 is a generalized stage–damage relationship for a residential structure with a basement. The relationship is generalized or normalized in the sense that monetary damage to the structure and its contents is expressed in "percent of total value of structure and contents" as a function of depth of inundation relative to the first floor of the building. Stage–damage relationships for other structures, including residential structures without basements and small business structures with and without basements, are presented in Figure 6.3.

Use of generalized stage–damage curves requires an estimate of the monetary value of the structure and its contents. The market value of a structure plus the land on which it is located (excluding the value of the contents) can usually be obtained from a community assessor or other similar sources. Un-

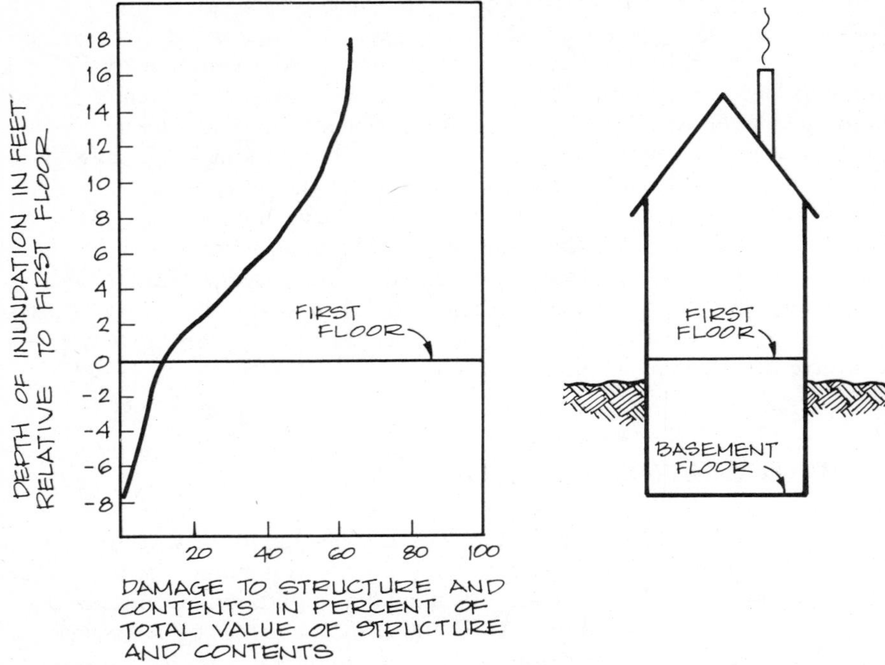

FIGURE 6.2 Generalized stage–damage relationship for a residential structure with a basement (*Source:* Adapted from Southeastern Wisconsin Regional Planning Commission, 1976)

fortunately, structure value and land value are usually not assessed separately. However, the market value of the structure alone can be estimated from the market value of the structure plus land. For example, one investigation (Southeastern Wisconsin Regional Planning Commission, 1976) concluded that the value of the typical single-family residential structure was approximately 80% of the value of the structure plus land. This conclusion was based on two communities which made separate appraisals of structure and land value.

Having estimated the value of a structure, the value of the structure plus contents must then be estimated as a factor times the value of the structure. One study (Grigg, 1975) suggested that a factor of 1.3 and values as high as 1.5 have been used. Local insurance agents may be a good source of information on the value of contents as a percentage of the value of the structure.

Step 2: Develop the Specific Stage–Damage Relationship for Each Structure

Reach lengths must be selected so that any given flood stage is approximately equal throughout the length of the reach. Therefore, on streams with relatively

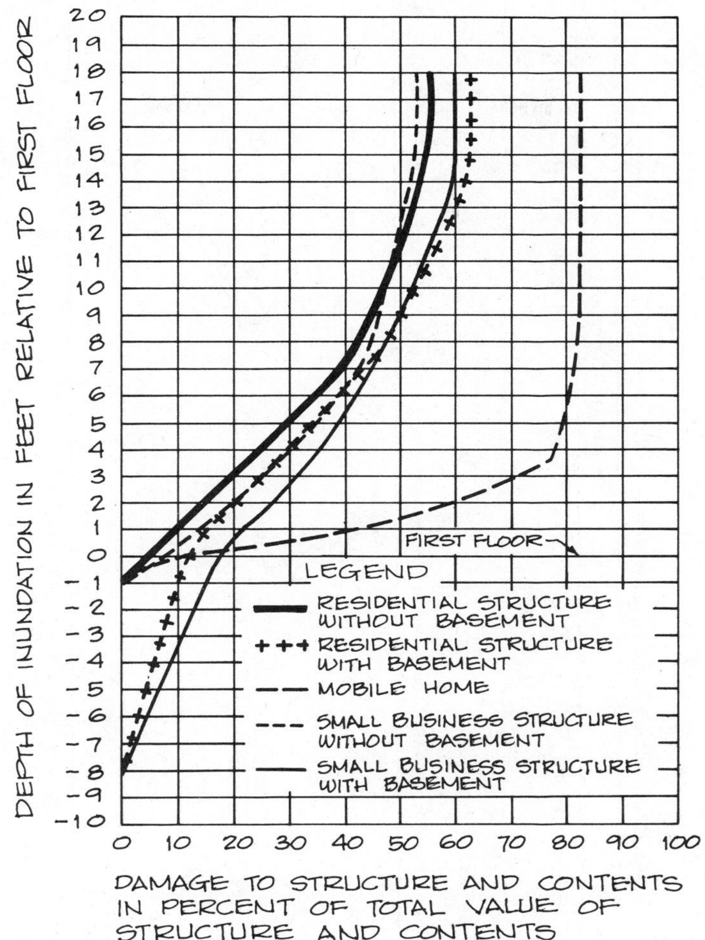

FIGURE 6.3 Generalized stage–damage relationships for selected structures (*Source:* Adapted from Southeastern Wisconsin Regional Planning Commission, 1976)

flat longitudinal slopes, reaches defined for flood damage computation purposes can be long relative to reaches used for flood damage calculations on streams with steep longitudinal slopes.

Table 6.2 presents a format for calculating flood damage versus stage for various types of structures in a reach. The table is intended to illustrate the computational procedure and not suggest that computations be done manually.

Consider, for example, structure 1, a single-family residence with a basement. As indicated in column 3 of Table 6.2, the value of structure and con-

TABLE 6.2 Calculation of the Stage–Damage Relationship for a River Reach

Structure Information				Stage–Damage Information								
(1) Identification Number	(2) Type	(3) Value of Structure and Contents (thousands of dollars)	(4) First-Floor Elevation (ft. NGVD)	(5) Flood Stage (ft. NGVD)	(6) Damage (thousands of dollars)	(7) Flood Stage	(8) Damage	(9) Flood Stage	(10) Damage	(11) Flood Stage	(12) Damage	...
1	Single-family residential with basement	90	646.0	650.0 (+4.0)[a]	25.2	648.0 (+2.0)	17.1	646.0 (0.0)	10.8	644.0 (−2.0)	7.2	...
2	Business without basement	250	644.0	650.0 (+6.0)	107.5	648.0 (+4.0)	87.5	646.0 (+2.0)	67.5	644.0 (0.0)	42.5	...
⋮	⋮	⋮	⋮	⋮	⋮	⋮	⋮	⋮	⋮	⋮	⋮	...
			Totals:		—		—		—		—	...

[a]Flood stage relative to first floor, in feet.

tents is estimated to be $90,000 and the first floor of the residence is at elevation 646.0 National Geodetic Vertical Datum (NGVD).

Columns 5, 7, 9, 11, . . . include a wide range of flood stages—from very high to very low—that might be experienced in the study reach. For a flood stage of 650.0 NGVD, the flood stage will be 4.0 ft above the first-floor elevation of structure 1. Referring to Figure 6.3, a 4.0-ft inundation on the first floor corresponds to a flood damage equal to 28 percent of the total value of the structure and contents. Multiplying 0.28 times $90,000 yields $25,200, which is entered into column 6. This damage computation process is repeated in columns 8, 10, 12,. . . . In summary, columns 5, 6, 7, 8, 9, 10, 11, 12, . . . are the specific stage–damage relationship for structure 1. The calculation is repeated for other structures in the study reach.

Step 3: Develop the Reach Stage–Damage Relationship

Refer again to Table 6.2, which represents all structures in the study reach. Damage totals for columns 6, 8, 10, 12, . . . represent study reach damages for, respectively, the stages in columns 5, 7, 9, 11,. . . . A typical stage–damage curve for a reach is presented in Figure 6.4. The stage–damage relationship usually exhibits an exponential, concave-down shape.

Step 4: Develop the Reach Stage–Probability Relationship

The stage–probability curve is based solely on the reach hydrology and hydraulics. Development of the reach stage–probability relationship presumes that a range of flood discharges (e.g., 2-, 5-, 10-, and 100-year recurrence interval) have been determined and that backwater calculations or other means have been used to convert the range of discharges to a range of flood stage profiles through the study reach. Resulting flood stages are then graphed versus probability to produce the stage–probability relationship. An example stage–probability relationship is presented in Figure 6.4.

Step 5: Develop the Reach Damage–Probability Relationship

The damage–probability relationship is developed by graphically or otherwise eliminating "stage" between the stage–probability relationship and stage–damage relationship for a reach, thus combining them into a damage–probability relationship. Figure 6.4 presents a reach damage–probability relationship. This relationship was developed by eliminating "stage" between the stage–probability and stage–damage relationships shown in Figure 6.4.

Step 6: Compute Average Annual Direct Monetary Damage Using the Damage–Probability Relationship

AAMD is the sum of the damages of floods of all probabilities each weighted according to its probability of occurrence. AAMD for a reach is equivalent to

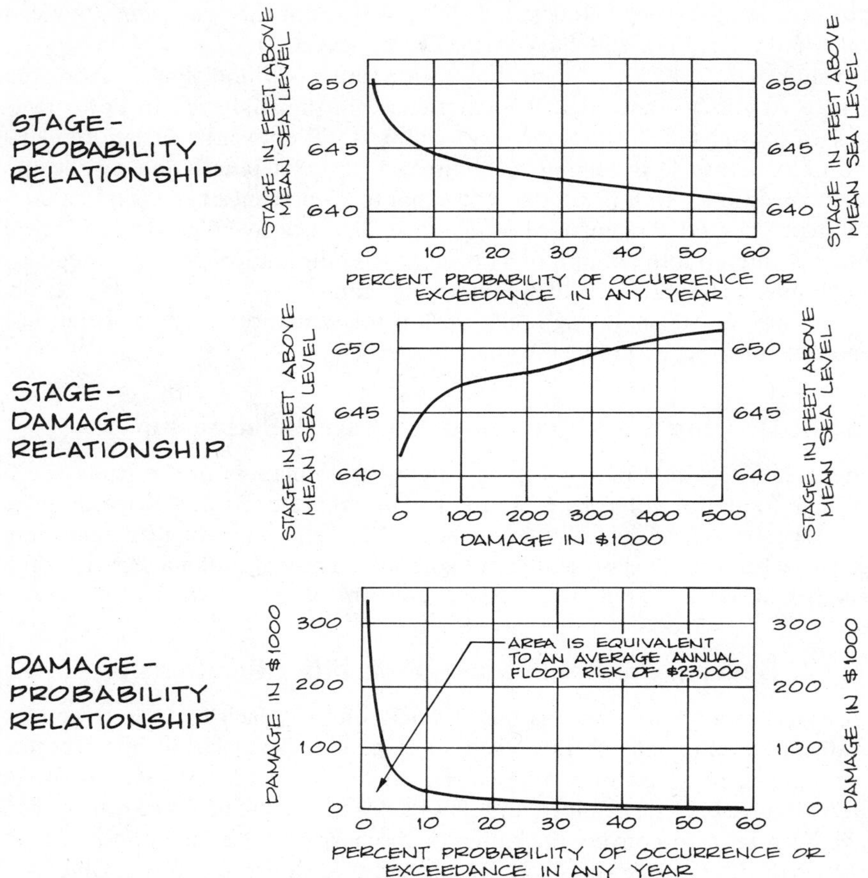

FIGURE 6.4 Examples of reach stage–damage–probability relationships and their use in determining average annual monetary damage (*Source:* Adapted from Southeastern Wisconsin Regional Planning Commission, 1976)

the area beneath the damage probability curve (Franzini, 1961; Grigg, 1985; Helweg, 1985; Hydrocomp, 1977; James and Lee, 1971).

For example, the area beneath the reach damage–probability curve in Figure 6.4 represents an AAMD of $23,000. Whereas the damage incurred in this reach is likely to vary widely from year to year, over a long period the average annual damage will be $23,000.

Step 7: Adjust Upward to Account for Indirect Monetary Damages

As explained earlier in this chapter, the two forms of monetary damage are direct and indirect. Steps 1 through 6 result in the estimation of direct AAMD.

Unfortunately, few investigations have been conducted to estimate indirect monetary damage as an absolute quantity or as a fraction of direct monetary damage.

One source (Kates, 1965) estimated that indirect damage incurred as a result of flooding of residential structures was 15 percent of the direct damage. This same investigation added 40 percent to the average annual direct damage incurred by commercial–industrial structures to account for indirect damage.

6.4 DETERMINING THE AVERAGE ANNUAL MONETARY BENEFIT OF ALTERNATIVE FLOOD DAMAGE MITIGATION MANAGEMENT MEASURES

A hypothetical watershed with a flood-prone community near the watershed outlet is illustrated in Figure 6.5. Potential solutions, either individually or in combination, to the flooding problem are also shown on Figure 6.5. The potential solutions are an upstream reservoir, an upstream diversion from the watershed, floodproofing structures within the flood-prone reach, enlarging the channel through the reach, and constructing a dike–floodwall system along the channel through the reach.

As noted earlier in this chapter, one of the two principal uses of the flood damage computation procedure is determination of the monetary benefits of each flood control alternative. In general, any given flood control alterna-

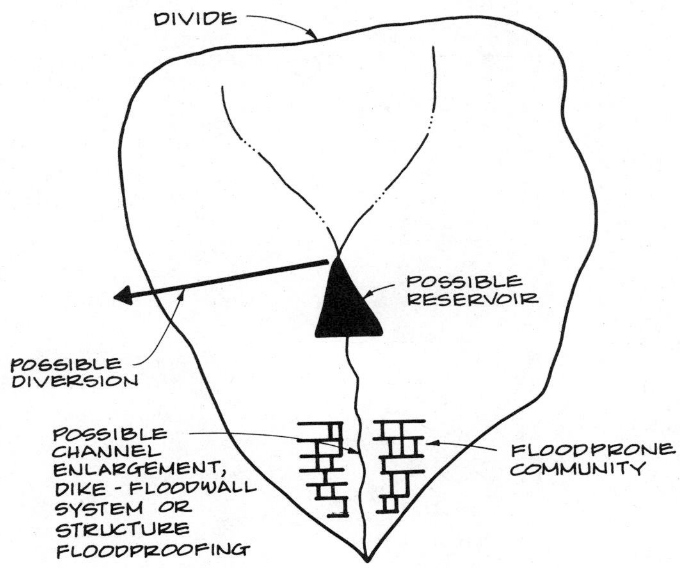

FIGURE 6.5 Alternative flood damage mitigation measures

tive, such as those illustrated in Figure 6.5, will affect either the reach stage–probability relationship or the reach stage–damage relationship, which will in turn influence the damage–probability relationship. The incremental change in the damage–probability relationship represents the resulting average annual benefit of the flood control alternative.

For example, consider the alternative of constructing an upstream flood control or detention/retention facility. Assuming that the facility is of sufficient size to affect the hydrology and hydraulics of the flood-prone reach, the stage–probability relationship for the reach would be affected. More specifically, construction of the storage facility would have the effect of shifting the stage–probability relationship down and to the left as illustrated in Figure 6.6. The reservoir would have no effect on the stage–damage relationship. The shifted stage–probability relationship, when combined with the unchanged stage–damage relationship, would shift the damage–probability relationship down and to the left, as also illustrated in Figure 6.6. The reduction in area

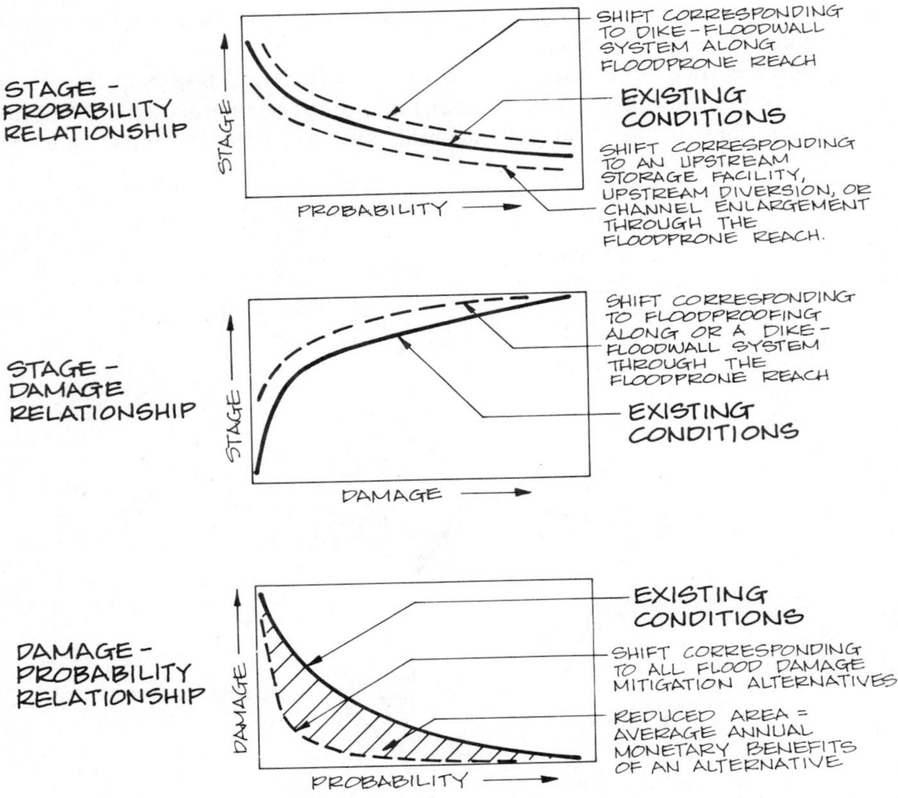

FIGURE 6.6 Effect of alternative flood damage mitigation measures on the procedure used to calculate average annual monetary damage

beneath the damage–probability relationship corresponds to the reduction in AAMD and therefore is the AAMB of the reservoir.

A diversion upstream of the flood-prone reach or a channel enlargement through the flood-prone reach will, as illustrated in Figure 6.6, have an effect very similar to that of the upstream reservoir. That is, the reach stage–probability relationship would shift down and to the left and the reach–damage relationship would be unaffected. Again, the reach–damage probability relationship would shift down and to the left, and the resulting reduced area would represent the AAMB.

As another example, assume that a dike–floodwall system is evaluated for the flood-prone reach. Because the dikes or floodwalls will constrict flood flows, the reach stage–probability relationship would tend to shift up and to the right, as illustrated in Figure 6.6. The reach stage–damage relationship would also be affected, in that it will shift up and to the left. The likely net affect of the shifts in the preceding two relationships would be to move the reach damage–probability relationship down and to the left, with the reduced area again representing the AAMB.

The last alternative presented in Figure 6.5 is floodproofing of residential and other structures in the flood-prone reach. Structure floodproofing has no effect on the stage–probability relationship but, as illustrated in Figure 6.6, would shift the stage–damage relationship up and to the left. Combining these two relationships, the resulting reach damage–probability relationship would shift down and to the left, with the reduced area representing the AAMB.

In summary, the AAMB of essentially any flood damage mitigation alternative can be evaluated by reflecting its impact on either the reach stage–probability relationship, the reach stage–damage relationship, or on both of these relationships. The adjusted relationship or relationships are combined to form the damage–probability relationship and the corresponding reduction in area beneath that curve is the AAMB of the alternative.

REFERENCES

Debo, T. N., "Urban Flood Damage Estimating Curves." *J. Hydraul. Div., Am. Soc. Civ. Eng.* **108**(HY10), 1059–1069 (1982).

Franzini, J. B., "Flood Control—Average Annual Benefits." *Consult. Eng.* May, pp. 107–109 (1961).

Green, C. H., D. J. Parker, P. Thompson, and E. C. Penning-Rowsell, *Indirect Losses from Urban Flooding: An Analytical Framework,* Geogr. Plann. Pap. No. 6. Flood Hazard Research Centre, Middlesex Polytechnic, Great Britain, March 1983.

Grigg, N. S., "State-of-the-Art of Estimating Flood Damage in Urban Areas." *Water Resour. Bull.* **11**(2), 379–390 (1975).

Grigg, N. S., *Water Resources Planning.* McGraw-Hill, New York, 1985.

Helweg, O. J., *Water Resources Planning and Management.* Wiley, New York, 1985.

Hydrocomp, Inc., *Flood Damage Mitigation,* Hydrocomp Tech. Note No. 2., Hydrocomp, Inc., Mountain View, CA, 1977.

James, L. D., and R. R. Lee, *Economics of Water Resources Planning,* Chapter 10. McGraw-Hill, New York, 1971.

James, L. D., "The Role of Economics in Planning Flood Plain Land Use." *J. Hydraul. Div., Am. Soc. Civ. Eng.* **98**(HY6), 981–992 (1972).

Johnson, N. L., Jr., "Economics of Permanent Flood-Plain Evacuation." *J. Irrig. Drain. Div., Am. Soc. Civ. Eng.* **102**(IR3), 272–283 (1976).

Kates, R. W., *Industrial Flood Losses: Damage Estimation in the Lehigh Valley,* Res. Pap. No. 98. Department of Geography, University of Chicago, Chicago, IL, 1965.

Linsley, R. K., and J. B. Franzini, *Water-Resources Engineering,* 3rd ed. McGraw-Hill, New York, 1979.

Plazak, D. J., "Flood Control Benefits Revisited." *J. Hydraul. Div., Am. Soc. Civ. Eng.* **112**(2), 265–276 (1986).

Southeastern Wisconsin Regional Planning Commission, *A Comprehensive Plan for the Menomonee River Watershed,* Plann. Rep. No. 26. SEWRPC, Waukesha, WI, October 1976.

Walesh, S. G., and R. M. Videkovich, "Urbanization: Hydrologic-Hydraulic-Damage Effects. *J. Hydraul. Div., Am. Soc. Civ. Eng.* **104**(HY2), 141–155 (1978).

7

NONPOINT-SOURCE POLLUTION LOAD TECHNIQUES

Concern continues to rise over the adverse impacts of nonpoint-source (NPS) pollution, based in part on the findings of the Nationwide Urban Runoff Program, which are summarized in Chapter 2. Engineers and other professionals will increasingly be required to calculate NPS yields under existing and future land use and management conditions. NPS yield is the average annual, or other period, quantity of the NPS substance transported from a subbasin, subwatershed, or watershed by its surface water system.

For example, the design of a detention/retention facility should include, at minimum, an estimate of the average annual load of suspended solids and perhaps other potential NPS pollutants to determine if the quantities are significant and, accordingly, if mitigation measures such as land treatment, sedimentation basins, or periodic cleaning and maintenance will be required. Diagnostic studies of eutrophic lakes in urban areas usually include estimates of average annual or seasonal loading of suspended solids, nutrients, and other potential NPS pollutants. Typically, these studies also try to identify the major sources within the tributary watershed of each apparent NPS pollutant. Estimates of NPS loads are required in the design of facilities such as sedimentation basins, artificial wetlands, and infiltration basins intended to remedy or prevent surface and groundwater pollution problems attributable to NPS pollution.

Relative to urban hydrology and hydraulics, knowledge of urban NPS pollution—how much there is, where it comes from, its adverse effects, and what can be done about it—is limited. The field of NPS pollution is in its infancy relative to urban hydrology and hydraulics. NPS phenomena are less well understood, NPS analysis and design tools and techniques are more primitive, fewer structural and nonstructural NPS management measures are available, and there is more uncertainty regarding their effectiveness.

Accordingly, this chapter focuses on one fundamental, pragmatic aspect of NPS pollution: simplified, manual methods for estimating average pollutant yield for a period such as a year or a season. More sophisticated approaches are available, as suggested by the water quality modeling concepts and water quality modeling examples presented in Chapter 10. Nevertheless, urban NPS pollution phenomena, in contrast with urban hydrologic and hydraulic phenomena, will often be analyzed by manual methods in the near future. These simplified, sometimes called "desktop," methods are emphasized in this chapter.

The chapter begins with the definition of NPS pollution, emphasizing the unique features of NPS pollution in contrast to point-source pollution. An overview of NPS pollution load calculation techniques is presented, followed by presentations of five different techniques, each illustrated with at least one example and supported with references to the literature.

7.1 DEFINITION AND ESSENTIAL FEATURES OF NONPOINT-SOURCE POLLUTION

The infancy of the NPS pollution field is evident in disagreement over the definition of NPS pollution. For example, some professionals argue that a NPS pollutant is one that enters the surface water system in a diffuse fashion rather than at a point, that is, as sheet flow rather than via a storm sewer. Other professionals would maintain that physical conditions at the point of entry are secondary, the prime consideration being the source of the NPS pollutants, that is, numerous and varied sources dispersed in a continuum over the urban landscape. Still others argue that one must go even further and specify that NPS pollutants are those that become operational and are transported because of and during and immediately after rainfall and snowmelt events.

In the interest of consistency and clarity, and recognizing that there are diverse opinions on the subject, NPS pollution is defined in this book as that which is transported from the land surface by means of direct runoff (surface runoff and interflow) during and immediately after rainfall or rainfall and snowmelt events. For completeness, NPS pollution also includes potentially troublesome substances carried to the surface water system via groundwater discharge, which constitutes the source of streamflow between rainfall and rainfall–snowmelt events. NPS pollution, including to some extent the groundwater component, differs from point-source pollution in that the former enters a surface water system at highly irregular rates relative to the latter, with most of the transport tending to occur during and immediately after rainfall or snowmelt events.

The preceding definition, which emphasizes the diffuse source of the pollutants and the meteorologically driven means by which they are transported to the surface water system, deemphasizes physical conditions at the point of entry of the pollutants into the surface water system. For example, the NPS pollutant sediment moving toward the surface water system during a runoff event might be transported to and enter the surface water system in open channels or through storm sewers, that is, at a point, or might enter the surface water system in a nondiffuse fashion, that is, in sheet flow.

In the context of NPS as defined in this book, NPS and point-source pollution are similar in the sense that each can cause toxic, organic, nutrient, pathogenic, sediment, radiological, and aesthetic pollution problems. The Nationwide Urban Runoff Program reviewed in Chapter 2 indicates that the types and quantities of NPS pollutants approximates the types and quantities of point-source pollutants and that the adverse impacts of NPS pollutants are similar to those of point-source pollutants. Furthermore, the seriousness of NPS pollutants is likely to increase because there is a longer history of engineering, legislative, and legal actions intended to mitigate point-source pollution, and therefore the relative impact of point-source pollution is on the decline.

7.2 NONPOINT-SOURCE POLLUTION LOAD TECHNIQUES: AN OVERVIEW

Unlike hydrologic techniques, which are numerous, NPS techniques are limited in number and capability. Five techniques are presented for determination of the average annual or seasonal yield of NPS pollutants from watersheds. Although the methods are generally applicable to urban areas, any particular method may not be usable in a given urban area because data and information needs are not satisfied, or there are other complexities.

The order of presentation of NPS techniques is generally from simple to complex, that is, from those methods that are usually easy to apply and require relatively few site-specific data to those that are typically more difficult to apply because they require significantly more site-specific data. For each example, the method is described, important constraints and assumptions are identified, and positive and negative features are noted. Literature references are provided for many methods in each category. Another category, computer modeling, is discussed separately in Chapter 10.

Analogous to the Chapter 3 recommendation for hydrologic analysis, careful practice suggests use of more than one NPS method for each application. In addition to guarding against blunders, which is the primary reason for using two or more techniques in hydrologic analysis, the use of two or more techniques in NPS analysis is also motivated by the general uncertainty of this area of analysis.

7.3 UNIT LOAD METHOD

The unit load method of estimating NPS pollutant quantities is probably the easiest to use of the methods presented in this chapter and the least reliable. With this method, the total NPS pollutant contribution of a subbasin, subwatershed, or watershed is simply the sum of the products of the area of each land use times the unit load of the potential pollutant for each land use.

Methodology

Unit loads are expressed in dimensions of weight or mass per unit area per unit of time. Typical units are pounds per acre per year. For any NPS constituent, the product of area and unit load is typically mass or weight contributed per year.

Unit loads may be obtained from sampling programs established at discharge points of representative relatively homogeneous land areas in the watershed or other areas under study. However, because of the high cost of monitoring and the duration of months to years typically required to capture a variety of runoff conditions, unit loads are usually obtained from the literature.

Numerous references are found in the literature. For example, unit loads are reported by Beaulac and Reckhow (1982), Bryan (1972), Dendy and Bolton (1976), Donigian (1977), Huber et al. (1982), Loehr (1972), Marsalek (1978), Novotny and Chesters (1981), Roesner (1982), Sheaffer et al. (1982), Torno et al. (1986), Uttormark et al. (1974), and Whipple et al. (1983). When using literature values, close attention should be given to the conditions under which the data were selected. Conditions to consider include size of the contributing areas; when monitoring was conducted, such as only during or immediately after runoff events or continually so as to include low-flow periods; whether or not bedload is included; and the extent to which hydrologic conditions during the sampling period were representative of long-term conditions.

Example

The unit load method was used to estimate the production or movement of NPS pollutants from various land uses in the 25-mi^2 Kinnickinnic River watershed in the Milwaukee area [Southeastern Wisconsin Regional Planning Commission (SEWRPC), 1978]. NPS pollutants included in the analysis as indicated in Table 7.1 are suspended sediment, 5-day biochemical oxygen demand (BOD5), total phosphorus, and total nitrogen. A total of 10 urban and rural land uses were included in the analysis. Unit loads were generalized from data collected within the planning region and from the literature.

This example, as is the case with most applications of the unit load method, emphasizes how unit loads of a given NPS pollutant typically vary markedly with land use. For example, assuming that the unit loads in Table 7.1 are reasonably representative, areas under development are expected to generate ap-

TABLE 7.1 Use of the Unit Load Method to Estimate the Average Annual Production off Non-Point-Source Pollutants in the Kinnickinnic River Watershed

	Area		Average Annual Unit and Total Loads for Selected Pollutants											
			Suspended Sediment			5-Day Biochemical Oxygen Demand			Total Phophorus			Total Nitrogen		
	Square Miles	Percent of Watershed	lbs/Acre/Yr	Tons	% Water-shed Total	lbs/Acre/Yr	Tons	% Water-shed Total	lbs/Acre/Yr	Tons	% of Water-shed Total	lbs/Acre/Yr	Tons	% of Water-shed Total
Urban land use														
Residential	12.61	50.74	545	2200	5.4	24	97	31.6	0.32	1.3	11.5	4.0	16	27.9
Commercial	2.07	8.33	745	490	1.2	98	65	21.2	0.75	0.5	4.4	9.0	6	10.4
Industrial	2.31	9.30	975	720	1.8	37	27	8.8	0.70	0.5	4.4	8.4	6	10.4
Extractive	0.02	0.08	150,000	960	2.4	120	0.8	0.3	45	0.3	2.6	60	0.4	0.7
Transportation	2.86	11.51	2,900–43,000	13,000	32.0	18–159	86	28.1	1.40–2.70	1.9	16.8	12–23	14	24.3
Recreation	1.45	5.84	420	195	0.5	1.3	0.6	0.2	.06–0.2	0.03	0.3	2.30–4.40	1	1.7
Construction	0.45	1.81	150,000	21,600	53.2	120	17	5.6	45	6.5	57.4	60	9	15.7
Subtotal	21.77	87.61	—	39,765	97.9	—	293	95.8	—	11.03	97.4	—	524	91.1
Rural land use														
Cropland, pasture and Unused Rural land	3.00	12.07	420–10,000	830	2.0	2.1–30	10	3.3	0.09–0.64	0.3	2.6	0.9–23	5.0	8.7
Silviculture	0.03	0.13	250	2.4	0.0	4.6	—	—	0.14	—	—	2.3	—	—
Surface water	0.05	0.20	665	11	0.0	162	2.6	0.8	0.50	—	—	8.9	0.1	0.2
Subtotal	3.08	12.39	—	843	2.1	—	12.6	4.1	—	0.3	2.6	—	5.1	8.9
Total	24.85	100.0	—	40,608	100.0	—	306	100.0	—	11.33	100.0	—	57.5	100.0

Source: Adapted from Southeastern Wisconsin Regional Planning Commission (1978).

proximately 275 times as much suspended solids per unit area as are established residential areas.

Another feature of most unit load analyses is that they help to identify those land uses likely to be generating a disproportionate amount of the total amount of a given NPS pollutant being produced in a watershed. Referring to Table 7.1, although areas under construction comprise only about 2 percent of the watershed, these areas account for over half of the suspended solids and total phosphorus generated within the watershed.

Reported unit loads are typically for small areas with relatively homogeneous land use. NPS materials carried from such areas will not necessarily move through the watershed hydrologic–hydraulic system at the same rate. NPS materials, particularly suspended solids and substances absorbed into or adsorbed onto suspended solids, will tend to settle out and remain in low areas and other discontinuities in the landscape. To reiterate, the annual mass or weight of most NPS pollutants transported past the outlet of a typical urban watershed containing many different land uses will tend to be less than the sum of the contributions of each of the individual land uses in the watershed when those contributions are calculated using the unit load method. For example, Table 7.1 indicates that the total production or movement of suspended solids from the 25-mi^2 Kinnickinnic River watershed is 40,000 tons per year. As indicated later in this chapter in the discussion of the constituent rating curve–flow duration curve method, the actual sediment yield of the Kinnickinnic River watershed is probably about half the value obtained with the unit load method.

Positive Features

The unit load method offers simplicity, assuming the availability of unit load values applicable to the study area. Another positive feature of the unit load method is that it helps to identify those land uses which, on a unit area basis, contribute the largest amounts of any given NPS substance. The method also reveals those land uses in a watershed which contribute a disproportionately large amount of a given potential pollutant.

Negative Features

The major weakness of the unit load method is uncertainty in the selection of unit loads, particularly if measurements are not available within the watershed to which the method is being applied. For a given land use and a given NPS substance, unit loads reported in the literature vary widely. Selecting the most representative value within the available range is difficult.

Another weakness of the unit load method is that for a given NPS pollutant, the sum of average annual contributions of that pollutant from all land uses in a watershed is not likely to be the total average annual yield of that pollutant from the watershed. Unit loads are usually based on measurements from small study areas, and therefore unit loads do not account for settling or accumulation of NPS pollutants in low points and other discontinuities in the land surface.

7.4 PRELIMINARY SCREENING PROCEDURE

The preliminary screening procedure consists of equations that may be used, with minimum input requirements, to estimate average annual NPS contributions of the following five potential pollutants: 5-day biochemical oxygen demand (BOD_5), suspended solids, volatile solids, total phosphate (PO_4) phosphorus as total phosphorus (P), and total nitrogen. The procedure was developed with the support of the U.S. Environmental Protection Agency by reviewing and using, as appropriate, nationwide NPS pollution and other data (Heaney et al., 1976). Based on the techniques used to develop and apply the procedure, the preliminary screening method is similar to regional methods used in hydrology as described in Chapter 3. Regional hydrologic methods and the preliminary screening method both are developed by analyzing a large data base for a given region, identifying the most important independent variables, and summarizing the results in the form of simple, easy-to-use equations.

The preliminary screening procedure may also be viewed as a sophisticated unit load method because the equations used in the preliminary screening procedure provide for the calculation of unit loads as a function of land use as well as other significant factors. The typical unit load method usually involves selection of the unit load as a function only of land use.

As part of a Connecticut project, the preliminary screening procedure was used to estimate average annual pollutant loads (Binder et al., 1979). Annual loads obtained by monitoring were compared to annual pollutant loads obtained with the preliminary screening procedure and the results were favorable.

Methodology

The equations that constitute the preliminary screening procedure provide estimates of unit loadings of each of the five potential pollutants as a function of land use, type of sewer system, precipitation, population density, and street-sweeping frequency. Each of these independent variables, which are relatively easy to obtain for most urban areas, are determined for each land use segment in a particular watershed, subwatershed, or subbasin. The resulting unit NPS pollutant loads for each land use are multiplied by the area of that land use and the products are then summed to obtain the total load of each potential pollutant from the study area.

The basic equation is

$$L = u(i, j) \times P \times \text{PDF} \times \text{SWF}$$

where L = average annual amount of pollutant j generated per unit of land use i in pounds per acre per year

$u(i, j)$ = load of pollutant j generated per unit of runoff from land use i, as shown in Table 7.2, in pounds per acre-inch

P = average annual precipitation in inches

PDF = population density factor, a dimensionless parameter with a value for residential areas of $0.142 + (0.218)(\text{PD})^{0.54}$, where PD is population density in persons per acre, a value of 1.0 for commercial and industrial areas, and a value of 0.142 for other developed areas (e.g., parks, cemeteries, and schools)

SWF = street-sweeping factor, a dimensionless parameter; SWF = 1.0 if the average time between street sweeping is greater than 20 days, that is, streets are swept infrequently; for frequent street sweeping, that is, when the average time between sweeping is 20 days or less, SWF is less than 1.0 and calculated as the average number of days between sweeping divided by 20

Example

Assume that the preliminary screening procedure is to be used to estimate the average annual contribution of suspended solids from the 7.5-mi^2 urban watershed shown in Figure 7.1. Note that the watershed contains four land uses and has separate sewers.

Referring to Table 7.2, the suspended solids unit load factor for residential, commercial, industrial, and other land uses are respectively, 16.3, 22.2, 29.1, and 2.7 lb/acre-in. As shown in Figure 7.1, the annual precipitation, P, is 30.1 in.

The population density factor for the residential area is calculated as a function of population density. Substituting the given population density of 30.5 people per acre into the equation yields a population function of 1.52 for the residential land use. The population functions for the commercial and industrial areas are each 1.0 and for the other area is 0.142. The street-sweeping frequency is given in Figure 7.1 as being 30 days. Therefore, the street-sweeping effectiveness factor is 1.0.

Substituting the preceding independent variables into the unit load equation for each of the four land uses yields the following unit loads:

Residential: (16.3 lb/acre-in.) $\times$ (30.1 in. per year) $\times$ 1.52 $\times$ 1.0 = 746 lb/acre per year

Commercial: 22.2 $\times$ 30.1 $\times$ 1.0 $\times$ 1.0 = 668 lb/acre per year

TABLE 7.2 Loads of Nonpoint-Source Pollutants as a Function of Land Use and Type of Sewer System for Use with the Preliminary Screening Procedure

Type of Sewer System	Land Use, i	Nonpoint-Source Pollutant, j (pounds per acre-inch of runoff)				
		5-Day Biochemical Oxygen Demand	Suspended Solids	Volatile Solids	Total Phosphorus as Phosphate	Nitrogen
Separate	Residential	0.799	16.3	9.45	0.0336	0.131
	Commercial	3.20	22.2	14.0	0.0757	0.296
	Industrial	1.21	29.1	14.3	0.0705	0.277
	Other developed	0.113	2.70	2.6	0.00994	0.0605
Combined	Residential	3.29	67.2	38.9	0.139	0.540
	Commercial	13.2	91.8	57.9	0.312	1.22
	Industrial	5.00	120.0	59.2	0.291	1.14
	Other developed	0.467	11.1	10.8	0.0411	0.250

Source: Adapted from Heaney et al. (1976).

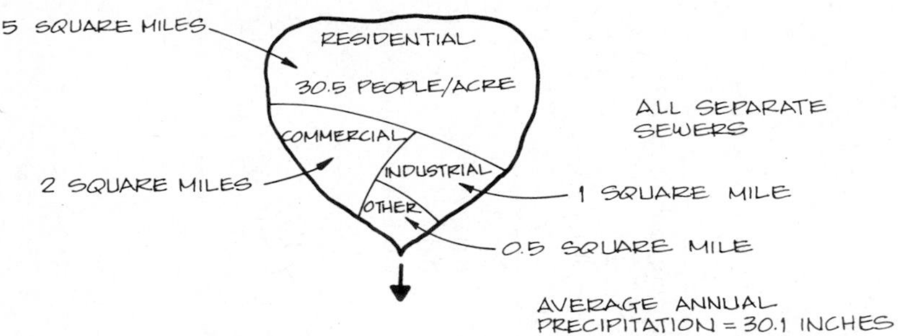

FIGURE 7.1 Urban watershed used for example application of the preliminary screening procedure

Industrial: $29.1 \times 30.1 \times 1.0 \times 1.0 = 876$ lb/acre per year
Other: $2.7 \times 30.1 \times 0.142 \times 1.0 = 11.5$ lb/acre per year

The total contribution of suspended solids from all four land uses in the 8.5-mi^2 watershed is then calculated as the sum of the individual contributions, yielding

$$(746 \text{ lb/acre per year}) \times (5.0 \text{ mi}^2) \times (640 \text{ acres/mi}^2) + 668 \times 2.0 \times 640 + 876 \times 1.0 \times 640 + 11.5 \times 0.5 \times 640 = 3{,}800{,}000 \text{ lb per year}$$
or 1900 tons per year of suspended solids

As a further illustration of the use of this method, assume that a retention facility is to be located at the outlet of the watershed. Assume further that all of the suspended solids reaching the retention facility will settle out at a submerged dry weight of 40 lb/ft^3. Then the annual loss of storage in the retention facility would be 3520 cubic yards or 2.2 acre-ft. This is approximately equivalent to about 2 ft of depth over a standard football field.

Conversion of sediment yield on a mass or weight basis to volume of settled sediment, as done in this and subsequent examples, requires an estimate of submerged dry unit weight of sediment. Guidance on selecting unit weights is provided by Chen (1975), Heinemann (1962), Linsley and Franzini (1979), Linsley et al. (1982), U.S. Department of Agriculture (1975), and Wylie (1979).

Positive Features

Ease of application is one strength of the preliminary screening procedure inasmuch as the input data are relatively easy to obtain and the calculations are simple. A second advantage is that the method focuses on different land uses within a watershed. This permits at least an approximation of the relative contribution of each land use to the total NPS pollution generated by the watershed. A third advantage of the method, compared to other methods, is that the preliminary screening procedure may be used for five different potential pollutants.

Negative Features

The principal weakness of the preliminary screening procedure is that the equations and parameters are based on a generalization of data and information for the entire United States. Therefore, neither the equations nor most of the input parameters can be tailored to a particular situation. As clearly suggested by the title of the method, results obtained with the method should be used only for preliminary screening purposes.

7.5 UNIVERSAL SOIL LOSS EQUATION

The universal soil loss equation (USLE) was originally developed in the 1950s for use in agricultural areas to estimate the annual sediment yield from small, relatively homogeneous parcels of land as a function of various natural features of the agricultural practices being used on the land. The method is limited to sheet erosion caused by water. Erosion caused by snowmelt waters and by wind are excluded, as is gully erosion.

Furthermore, the USLE does not necessarily predict the amount of material moving from the land surface to the receiving waters. In a watershed the latter is generally significantly less than the amount predicted by the equation. It is important to note that the USLE is intended primarily to predict long-term average annual soil movement from relatively homogeneous parcels of land (U.S. Department of Agriculture, 1977; Wischmeier, 1976; Wischmeier and Smith, 1978; Wischmeier et al., 1971).

In recent years, the USLE has been increasingly used to estimate sediment yield from small relatively homogeneous subbasins in urban and urbanizing areas (e.g., Wylie, 1979). In that context, the method also is used to quantify, in at least a very approximate fashion, the relative effects of various erosion control measures typically used in urban areas (Meyer and Ports, 1976; Wylie, 1979).

Methodology

The USLE is

$$E = A \times R \times K \times \text{LS} \times C \times P$$

where E = soil loss by water erosion in rill and inter-rill areas in tons per year

A = area in acres

R = rainfall factor, which accounts for the erosive forces of rainfall and runoff, as shown in Figure 7.2, in erosion index units per year

K = soil erodibility factor reflecting the physical and chemical properties of a particular soil, with typical values as shown in Table 7.3, in tons per acre per erosion index unit

LS = slope length or topographic factor reflecting the effects of the length and steepness of the land surface, a dimensionless parameter obtained from Figure 7.3

C = cover and management factor reflecting the influence of vegetation and mulch, a dimensionless factor obtained from Table 7.4

P = erosion control practice factor that is similar to the cover-management factor but accounts for practices superimposed on the land surface, such as contouring, terracing, compacting, sedimentation basins, and control structures; a dimensionless factor obtained from Table 7.5

Example

A hypothetical parcel of land undergoing development is shown in Figure 7.4. The purpose of the analysis is to estimate the annual sediment yield of the site assuming that no erosion control measures are applied during construction.

Noting the location and physical features of the project, the applicable values of independent variables in the equation are

A = 5 acres

R = 125 erosion index units per year and is obtained from Figure 7.2

K = 0.17 ton per acre per erosion index units and is obtained from Table 7.3

LS = 0.35 and is obtained from Figure 7.3

C = 1.0 and is obtained from Table 7.4

P = 1.0 and is obtained from Table 7.5

Substituting the variables into the USLE yields 37 tons of sediment per year. Assuming that all of this is deposited in a receiving storm sewer, drainage ditch, or culvert at a dry unit weight of 80 lb/ft^3, the resulting volume of sediment accumulated in an average year is 925 ft^3 or 34 cubic yards.

Note the apparent sensitivity of sediment yield from an urban construction site to various cover and other practices during construction. For example,

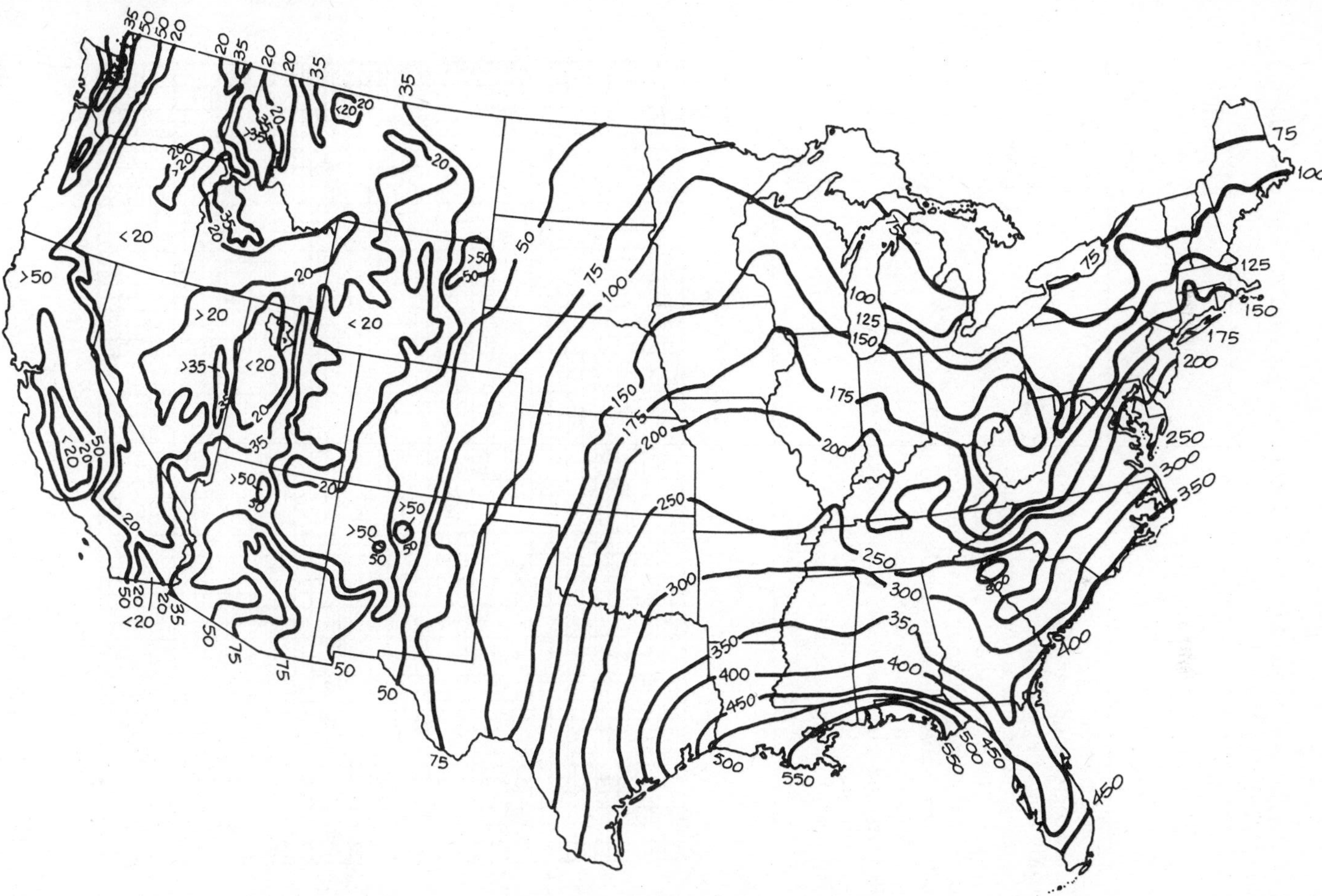

FIGURE 7.2 Rainfall factors for use with the universal soil loss equation (*Source:* Adapted from Wischmeier and Smith, 1978)

TABLE 7.3 Typical Soil Erodibility Factors for Use with the Universal Soil Loss Equation

Soil	Soil Erodibility Factor, *K* (tons/acre per erosion index unit)
Sands and gravels	0.10
Loamy coarse sand, sand and fine sand	0.15
Loamy fine sand and loamy sand	0.17
Fine sandy loam and sandy loam, mollic	0.20
Fine sandy loam and sandy loam, nonmollic	0.24
Loam, clay loam and sandy clay loam, mollic	0.28
Loam, clay loam and sandy clay loam, nonmollic	0.32
Silt loam and silty clay loam, mollic	0.32
Silt loam and silty clay loam, nonmollic	0.37
Silt loam and silty clay loam, subsoil	0.43
Clay and silty clay, less than 50 percent clay	0.32
Clay and silty clay over 50 percent clay	0.28

Source: Wylie (1979).

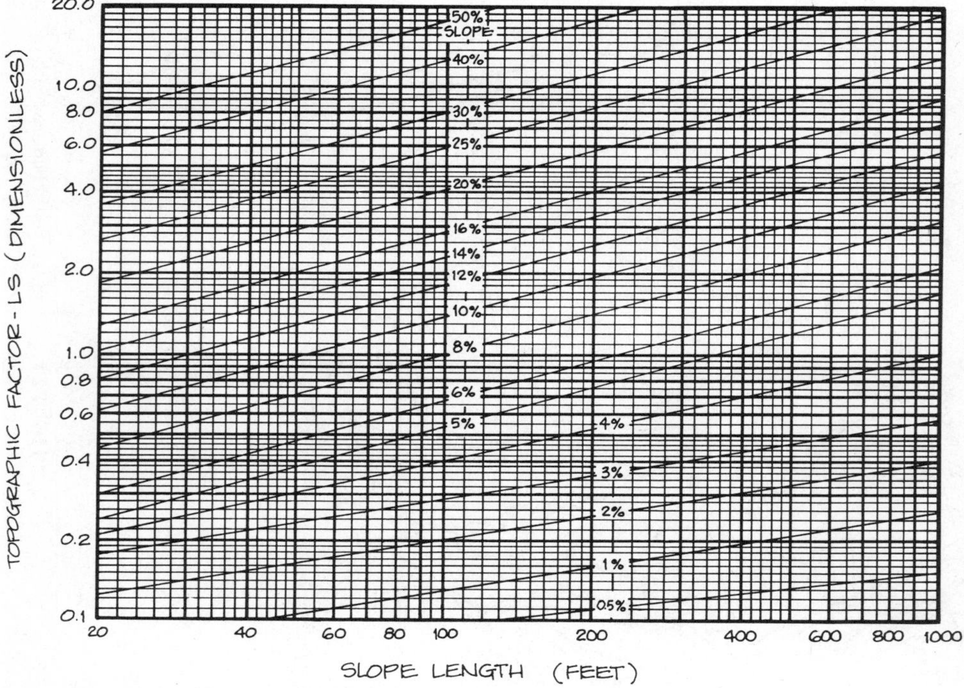

FIGURE 7.3 Topographic factor as a function of slope and slope length for use with the universal soil loss equation (*Source:* Adapted from Wischmeier and Smith, 1978)

TABLE 7.4 Cover and Management Factors for Use with the Universal Soil Loss Equation

Cover or Management Measure	Cover and Management Factor C (dimensionless)
Bare soil	1.0
Straw mulch[a]	
0.5 ton per acre	0.35
1.0 tons per acre	0.18
1.5 tons per acre	0.06
3.0 tons per acre	0.03
4.0 tons per acre	0.02
Grass	0.011

Source: Adapted from Wylie (1979).

[a]Mulch assumed to be anchored by means such as punching into soil with a disk or using mulch net.

Table 7.4 indicates that application of mulch at a rate of ½ ton/acre will reduce the cover factor by 65 percent and result in a 65 percent reduction in sediment yield. Similarly, Table 7.5 indicates that simple contouring or furrowing across the slope will reduce the practice factor by up to 50 percent and, in turn, reduce the sediment yield by up to 50 percent.

Positive Features

One positive feature of the USLE is simplicity of application. Assuming availability of basic site data, the method can be applied almost anywhere in the United States. Another advantage of the method is that it provides at least

TABLE 7.5 Erosion Control Practice Factor for Use with the Universal Soil Loss Equation

	Erosion Control Practice and Resulting Factor, P (dimensionless)		
Land Slope (percent)	No Erosion Control Practice	Contouring	Terracing
1–2	1.0	0.6	0.12
3–8	1.0	0.5	0.10
9–12	1.0	0.6	0.12
13–16	1.0	0.7	0.14
17–20	1.0	0.8	0.16

Source: Wischmeier and Smith (1978) and Wylie (1979).

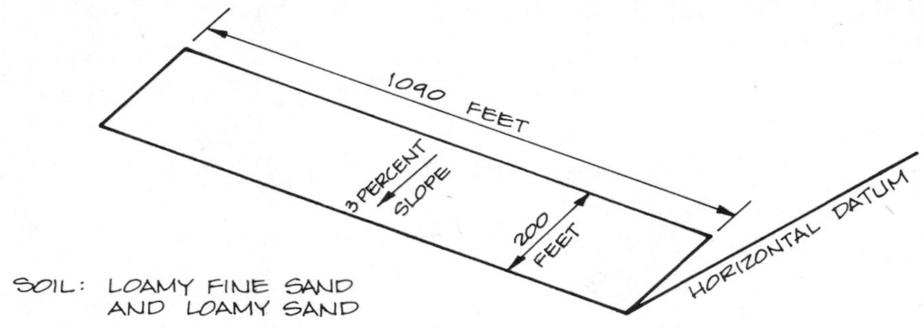

FIGURE 7.4 Parcel of land used for example application of the universal soil loss equation

an approximate indication of the impact of various procedures and practices available to control erosion in urban areas undergoing development.

Negative Features

One negative feature of the USLE is that it is confined only to one constituent, that is, suspended solids. Furthermore, the advantage of specificity to a small site is somewhat offset by lack of a means to integrate the sediment yield from many different sites in a watershed to provide an estimate of the sediment yield for the entire watershed. The average annual sediment yield of a watershed is less than the sum of the sediment yields of individual parcels as obtained with the USLE, and therefore a yield factor must be applied.

7.6 CONCENTRATION TIMES FLOW METHOD

With the concentration times flow method, dry weather and wet weather runoff and constituent concentrations are analyzed separately to obtain the average annual yield of the constituent under dry weather and wet weather conditions. The results are then combined to obtain the total average annual transport of the constituent from the watershed. Wet weather conditions are defined as those periods when direct runoff—any combination of surface runoff, interflow, and precipitation as discussed in Chapter 2—is occurring. The remaining time is defined as dry weather conditions.

Methodology

Representative dry and wet weather condition pollution concentrations are obtained by analyzing all available historic water quality data for the area under study. Concurrent meteorologic data are needed to distinguish between wet and dry weather conditions. Typically, there will be insufficient concentration data available and values for other similar watersheds or more generalized values obtained from the literature will be used. Examples of sources of data on concentration of pollutants in surface waters for wet and dry weather conditions and for various land uses are Bedient et al. (1980), Bryan (1972), Collins and Ridgway (1980), Dever et al. (1979), Howell (1979), Huber et al. (1982), Lager et al. (1977), Linsley et al. (1982), Polls and Lanyon (1980), Roesner (1982), Sheaffer et al. (1982), Terstriep et al. (1986), Torno et al. (1986), and Whipple et al. (1983).

The average annual volume of dry and wet weather runoff must also be determined. The preferable approach is to use streamflow records for the watershed under study. However, such data are typically not available for urban watersheds. Alternatives include using flows obtained from continuous simulation or other computer models or transposing the relative amounts of dry and wet weather flow from nearby similar watersheds to the study watershed.

For each NPS pollutant, the representative dry and wet weather concentrations are combined with the volumes of dry and wet weather runoff, respectively. The results are the average annual dry and wet weather pollutant yields of the watershed, which when summed give the total annual pollutant yield.

Example

The concentration times flow method was applied to the 20.24-mi^2 separately sewered area of the Kinnickinnic River watershed in the Milwaukee, Wisconsin, area (SEWRPC, 1978). The average annual yield of suspended sediment, 5-day biochemical oxygen demand (BOD_5), phosphate-phosphorus, total phosphorus, total nitrogen, and dissolved solids were estimated with the method. Dry and wet weather condition concentrations of the five water quality constituents as presented in Table 7.6 were based largely on historic monitoring in the Kinnickinnic River watershed. In analyzing the data, a water quality sample was taken as representing wet weather conditions if 0.10 in. or more of rainfall were recorded on the day of the sampling prior to the time of sampling, assuming that the time of sampling was known. If the time of sampling was not known, a sample was taken as representing wet weather conditions if 0.10 in. or more of rainfall was recorded on the day of sampling.

An estimate of the average annual hydrologic budget for the separately sewered area is presented in Figure 7.5. Average annual dry and wet weather runoff were based on the results of 13 years of continuous simulation of the watershed hydrologic–hydraulic system. On an average annual basis, 30.1 in. of rainfall is converted into 15.9 in. of evapotranspiration, 10.2 in. of direct runoff or wet weather flow, and 4.0 in. of groundwater flow. In addition, industrial discharges add 4.2 in. of runoff per year to the groundwater flow giving a total dry weather flow of 8.2 in. per year.

TABLE 7.6 Representative Dry and Wet Weather Condition Concentrations in the Kinnickinnic River Watershed

Parameters	Concentration[a] (mg/L)	
	Dry Weather Conditions	Wet Weather Conditions
Biochemical oxygen demand	5	15
Phosphate-phosphorus	0.05	0.15
Total phosphorus	0.10	0.30
Total nitrogen	1.0	1.5
Dissolved solids	500	250

Source: Adapted from Southeastern Wisconsin Regional Planning Commission (1978)
[a]Based primarily on monitoring in the Kinnickinnic River watershed under conditions representative of wet and dry weather streamflow from separately sewered areas.

Using 5-day biochemical oxygen demand as an example, the dry weather yield of BOD_5 of the watershed is

$$[5 \text{ mg/liter (L)}] \times (8.2 \text{ in. of dry weather flow}) \times (\text{conversion factor}) \\ = 60 \text{ tons/year}$$

Similarly, the wet weather yield of BOD_5 from the watershed is

$$(15 \text{ mg/L}) \times (10.2 \text{ in. of wet weather flow}) \times (\text{conversion factor}) \\ = 225 \text{ tons/year}$$

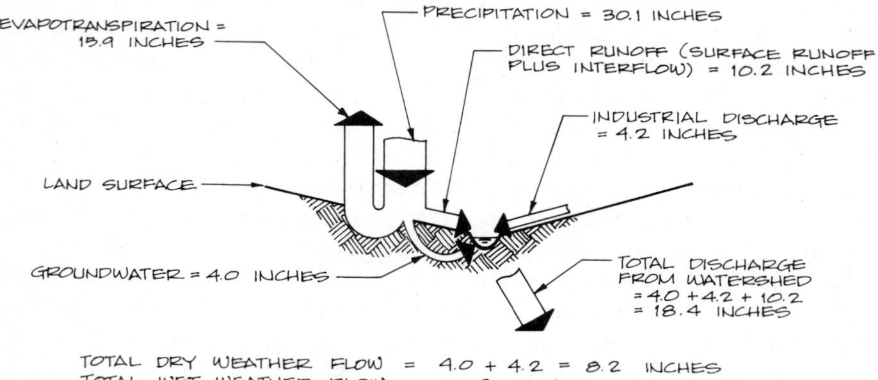

FIGURE 7.5 Average annual hydrologic budget for the separately sewered portion of the Kinnickinnic River watershed (*Source:* Adapted from Southeastern Wisconsin Regional Planning Commission, 1978)

Therefore, the total average annual BOD_5 yield of the watershed is 60 + 225 = 285 tons per year.

Using the same procedure, the average annual dry weather, wet weather, and total yields of phosphate-phosphorus, total phosphorus, total nitrogen, and dissolved solids were calculated and the results are presented in Table 7.7.

The analysis indicates that a disproportionately large amount of the BOD_5 is generated during wet weather conditions. Although only 18 percent of the days in an average year are categorized as wet weather days, about 79 percent of the BOD_5 is transported from the separately sewered areas during those days. Similar results are observed for phosphate-phosphorus, in that 80 percent is generated during wet weather days. Similarly, 65 percent of the total nitrogen yield and 40 percent of the dissolved solids yield occurs during wet weather conditions.

Positive Features

The primary strength of the concentration times flow method is that it requires the user to distinguish between dry and wet weather hydrologic conditions and dry and wet weather concentrations of potential pollutants. Accordingly, assuming that there are sufficient data to apply the method with confidence, the resulting yields of potential pollutants provide an improved understanding of the relative impact of wet and dry weather conditions as illustrated by the example.

Another advantage of the concentration times flow method is that it makes use of miscellaneous water quality data sometimes available for a watershed. Such data are often collected by different organizations or individuals under a variety of hydrologic conditions. Assuming that the conditions under which the data were collected are adequately documented, the data can be integrated and used to characterize dry and wet weather water quality conditions for use in this method.

TABLE 7.7 Estimated Nonpoint-Source Pollution Yields from the 20.24-mi^2 Separately Sewered Portion of the Kinnickinnic River Watershed Obtained from the Concentration Times Flow Method

	Yield (tons/year)		
Pollutant	Dry Weather Conditions	Wet Weather Conditions	Total
BOD_5	60	225	285
Phosphate-phosphorus	0.6	2.3	2.9
Total phosphorus	1.2	4.5	5.7
Total nitrogen	12	22	34
Dissolved solids	6000	3700	9700

Source: Adapted from Southeastern Wisconsin Regional Planning Commission (1978).

Negative Features

The main disadvantage of the concentration times flow method, relative to the preceding methods, is that it requires at least basic information on the wet and dry weather conditions average annual hydrologic budget and on representative pollutant concentrations under wet and dry weather conditions. Even when concentration data are available, they typically exhibit wide variations, especially under wet weather conditions, and therefore it is difficult to select one representative concentration for each constituent. Another disadvantage of the concentration times flow method is that it is difficult to trace pollutant yields back to source areas.

7.7 CONSTITUENT RATING CURVE–FLOW DURATION CURVE METHOD

The constituent rating curve–flow duration curve method is typically used to estimate the sediment yield of a watershed. The method can also be used to calculate the yield of other potential pollutants.

Methodology

An overview of the constituent rating curve–flow duration curve method is presented in Figure 7.6. As indicated, two relationships unique to a watershed are required to apply the method. The first is a constituent rating curve, which is the relationship between watershed discharge and transport of the constituent. This relationship is usually based on streamflow and water quality monitoring over an extended period, typically several years or more, to capture a wide range of flow and transport conditions. Although not as desirable, a constituent rating curve for a nearby similar watershed, normalized with respect to area, may be used.

The second of the two required relationships is the flow duration curve, which is the relationship between watershed discharge and the percent of time in which discharge is reached or exceeded. The flow duration curve is typically developed by monitoring streamflow over an extended period of time, usually several years or more. The curve could also be based on streamflow generated with a continuous simulation computer model or possibly on a flow duration relationship transferred from a similar watershed and adjusted for watershed size.

Information summarized in the constituent rating curve is combined with information summarized in the flow duration curve by eliminating discharge between the two relationships. The result is the yield of the particular constituent or potential pollutant. If the constituent is suspended solids, or is transported on or in suspended solids, an upward adjustment is needed to account for bedload. The bedload is that portion of the sediment which slides or rolls

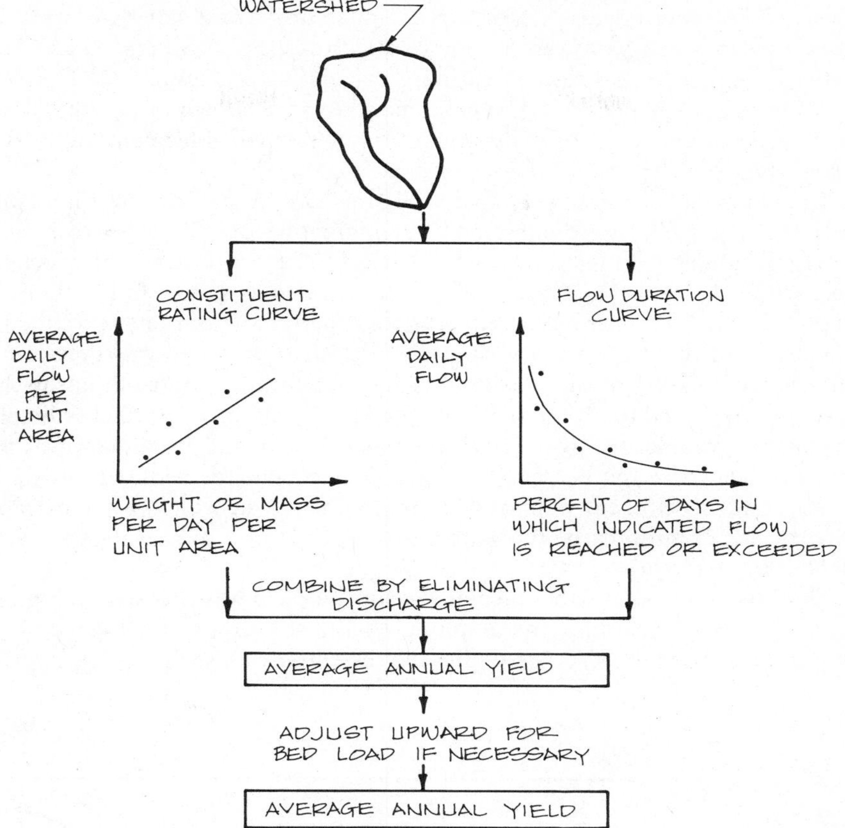

FIGURE 7.6 Constituent rating curve–flow duration curve method

along the channel bottom and is not intercepted by sediment samplers typically used to collect suspended sediment data (Hindall and Flint, 1970; Kinori and Mevorach, 1984; Linsley et al., 1982).

Example 1

The constituent rating curve–flow duration curve method was used to estimate sediment yield from the 25-mi^2 Kinnickinnic River watershed in the Milwaukee, Wisconsin area (SEWRPC, 1978). A master plan was being prepared for the watershed. The analysis and forecast phase of the master plan required an estimate of the sediment yield and the potential impact of that yield on the navigable lower reaches of the Kinnickinnic River.

Suspended sediment data based on monitoring were not available for the Kinnickinnic River watershed. However, 770 sediment transport–stream flow measurements were available for three similar urban watersheds in the Milwaukee area. These data were normalized with respect to watershed area and used to construct the sediment rating curve shown in Figure 7.7. The flow duration relationship for the Kinnickinnic watershed shown in Figure 7.8 was developed using 37 years of daily streamflow generated with a continuous simulation model.

The sediment rating curve shown in Figure 7.7 and the flow duration curve shown in Figure 7.8 were combined by eliminating discharge between them using the tabular format shown in Table 7.8. The entire range of discharge included in the flow duration curve was subdivided as shown in the first column of Table 7.8. Then, proceeding across the table from left to right, the annual sediment transport contribution of each flow range was determined and entered in the right-hand column of the table. Range contributions in that column were added to obtain the average annual yield of suspended sediment.

For the example, the average annual yield of suspended sediment was 411 tons/mi^2 per year. Ten percent was added to account for bedload, giving an average annual sediment yield of 452 tons/mi^2 per year, which, when extrapolated to the 25-mi^2 area of the watershed, resulted in a total watershed yield of 11,200 tons/year.

The reasonableness of the calculated sediment yield was checked by expressing the sediment yield in terms of annual volume of sediment delivered to the estuary portion of the river, assuming that essentially all sediment transported

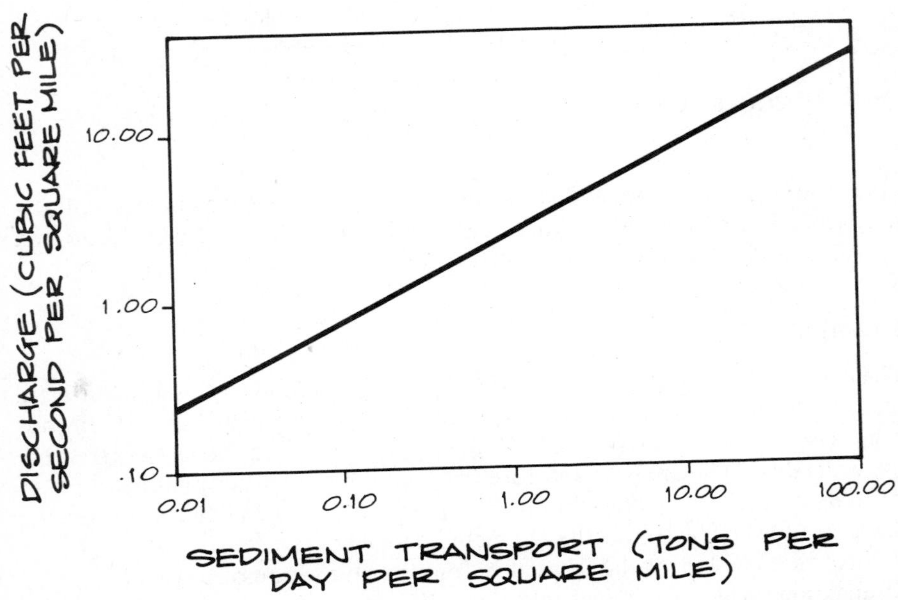

FIGURE 7.7 Sediment rating curve for urban watersheds in the Milwaukee area (*Source:* Adapted from Southeastern Wisconsin Regional Planning Commission, 1978)

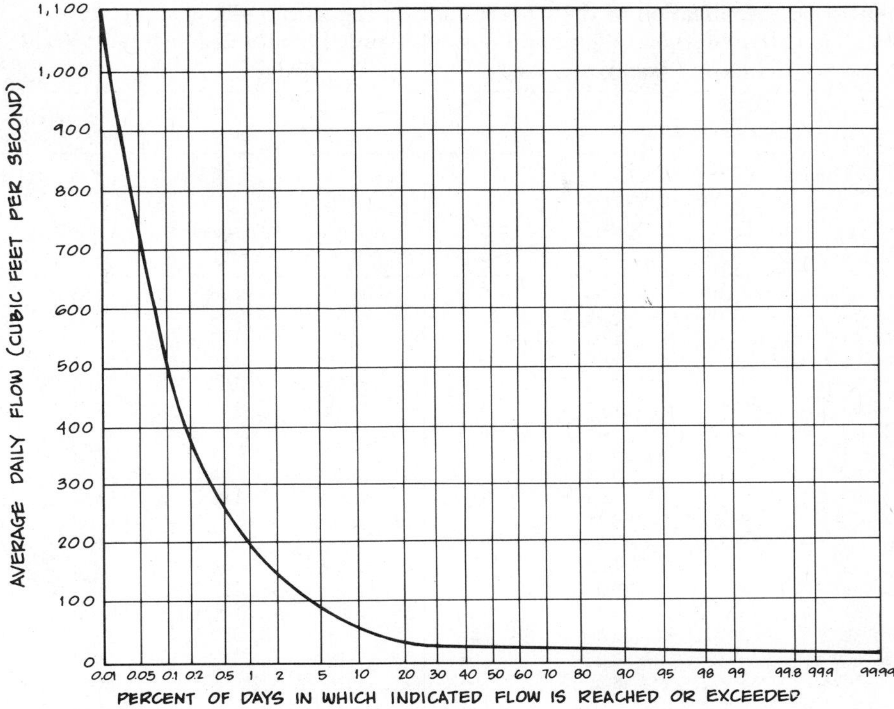

FIGURE 7.8 Flow duration curve for the Kinnickinnic River watershed (*Source:* Southeastern Wisconsin Regional Planning Commission, 1978)

from the watershed settled out in the estuary. This annual volume of sediment obtained with the constituent rating curve–flow duration curve method was compared to the annual volume of material dredged from the estuary by the Corps of Engineers to maintain navigation.

A submerged dry weight of 40 lb/ft^3 was selected. For this unit weight, the sediment yield of 11,200 tons/year converted to 20,700 cubic yards of settled sediment per year. Average annual dredging by the Corps of Engineers, based on dredging data for the period 1960–1976, indicated that an average of 26,500 cubic yards of sediment was removed per year. The approximate balance between estimated average annual sediment yield to the estuary and average annual dredging from the estuary suggests the reasonableness of the sediment yield estimate obtained with the constituent rating curve–flow duration curve method.

Example 2

Although the constituent rating curve–flow duration curve method is usually used to estimate sediment yield, it can be used to estimate the yield or transport

TABLE 7.8 Application of the Constituent Rating Curve—Flow Duration Curve Method to Estimate Average Annual Suspended Sediment Yield of the Kinnickinnic River Watershed

Average Daily Discharge[a]			Days Within Flow Range[a]		Sediment Transport	
	Representative Discharge					
Range (ft³/sec)	(ft³/sec)	(ft³/sec/ mi²)	(% of yr)	(Number of days per yr)	(Tons/day/ mi²)[b]	(Tons/mi²/ yr)
0–16	12.0	0.66	10	36.50	0.069	2.52
16–18	17.0	0.94	20	73.00	0.14	10.22
18–20	19.0	1.05	20	73.00	0.17	12.41
20–21	20.5	1.14	20	73.00	0.20	14.60
21–27	24.0	1.33	10	36.50	0.27	9.86
27–35	31.0	1.72	5	18.25	0.44	8.03
35–51	43.0	2.38	5	18.25	0.82	14.97
51–63	57.0	3.16	2	7.30	1.41	10.29
63–67	70.0	3.88	2	7.30	2.10	15.33
77–87	82.0	4.54	1	3.65	2.85	10.40
87–101	94.0	5.21	1	3.65	3.71	13.54
101–119	110.0	6.09	1	3.65	5.02	18.32
119–151	135.0	7.48	1	3.65	7.46	27.23
151–211	181.0	10.0	1	3.65	13.2	48.18
211–273	242.0	13.4	0.5	1.82	23.0	41.86
273–523	398.0	22.0	0.4	1.46	59.9	87.45
523–669	596.0	33.0	0.05	0.18	131	23.58
669–1093	881.0	48.8	0.04	0.15	280	42.00
Annual total			99.99[c]	364.96		410.79

Source: Adapted from Southeastern Wisconsin Regional Planning Commission (1978)

[a]From flow-duration curve shown in Figure 7.8.

[b]From sediment rating curve shown in Figure 7.7.

[c]For remaining 0.01 percent of year, average daily flows exceed 1093 ft³/sec.

of other constituents. For example, the constituent rating curve–flow duration curve method was used to estimate the average annual yield of several potential pollutants to Rend Lake, a 29.5-mi² artificial lake in Illinois (Kothandaraman and Evans, 1979). One purpose of the analysis was to estimate long-term nutrient loading from the 488-mi² watershed to the lake. Another purpose of the analysis was to identify portions of the watershed contributing a disproportinately large amount of nutrients. The last purpose of the nutrient yield investigation was to determine the probable effects of the nutrient loading on lake quality.

Weekly sampling was conducted for one year at six inflow locations on the periphery of the lake and used to construct rating curves (pounds per day of each nutrient versus cubic feet per second) for nitrate-nitrogen, ammonia-nitrogen, inorganic nitrogen, total phosphorus, total dissolved phosphorus, total silica, and total dissolved silica. The rating curves for each constituent at each of the six locations were combined with flow duration curves for each of the six locations to calculate average annual transport of each constituent from the watershed into the lake.

The investigation concluded that the lake received excessive nitrogen and phosphorus and was likely to experience accelerated eutrophication. The study recommended controlling the source of pollutants in those portions of the watershed contributing the disproportionately largest amounts of problematic nutrients.

Positive Features

The strength of the constituent rating curve–flow duration curve method is heavy reliance on watershed-specific data. The constituent rating curve and the flow duration curve are developed from data collected within the watershed, as is the case with Example 2, or data from within the watershed are supplemented with data from one or more similar watersheds in the immediate area, as is the case with Example 1. Contrast this watershed specificity with reliance, in other methods presented in this chapter, on parameters and coefficients generalized from many watersheds.

Another positive feature of the constituent rating curve–flow duration curve method is use of a large number of data points, regardless of whether or not they come from the watershed under study or under a similar watershed. Development of both the constituent rating curve and the flow duration curve requires many data points obtained under a wide range of hydrologic–hydraulic–water quality conditions. Over 100 data points could easily be reflected in the constituent rating curve and 1000 or more data points could be used to develop a flow duration curve. Therefore, because so many data points are used and they are obtained under widely varying conditions, the constituent rating curve and the flow duration curve each are an integrated representation of watershed phenomena.

Negative Features

The most significant weakness of the constituent rating curve–flow duration curve method is the small likelihood that sufficient watershed specific or watershed relevant data will be available to develop the constituent rating curve and the flow duration curve. Mounting a monitoring program to obtain the data is technically feasible, but time and funds may not be available.

Another negative feature of the method, which is shared with some other methods, is inability to use the method to trace the pollutant or pollutants of interest to their sources of origin in the watershed. That is, when the method is applied solely at the discharge point of the watershed, as is the case with Example 1, no direct information is provided as to the source of the potential pollutants. However, as illustrated by Example 2, with a sufficient data base the method can be applied to individual subwatersheds or subbasins within an overall watershed. These results can be used to identify those subwatersheds or subbasins contributing a disproportionate amount of the substance of interest.

7.8 OTHER NONPOINT-SOURCE POLLUTION LOAD TECHNIQUES

Discussed in detail in this chapter are five manual methods for calculating the yield of various NPS of pollutants. Variations on the methods presented in this chapter are available as well as additional techniques. For example, Lakatos and Johnson (1976) review various techniques available to calculate the yield of NPS pollutants. Included in their review are methods essentially the same as the unit load, the preliminary screening procedure, the USLE, and the concentration times flow methods presented in this chapter. In addition, the idea of estimating pollution yield as a percent of the yield of suspended solids is suggested based on the premise that some NPS substances are closely associated with suspended solids.

Whipple (1978) also reviews various methods, including methods similar to the unit load, the preliminary screening procedure, and the concentration times flow methods presented in this chapter. In addition, a mass balance technique is suggested in which the NPS contribution between two points on a stream system is calculated as the difference between the measured mass transport of the pollutant at the two points on the stream.

As an additional example of regression or empirical relationships, such as the USLE and the preliminary screening procedure, Hindall (1976) describes the use of sediment yield data from 118 gauging stations to develop sediment yield prediction equations for Wisconsin watersheds. Equations were established for various areas of Wisconsin, and independent variables were the percent of lake and marsh, average annual watershed discharge, 2-year- and 25-year-recurrence-interval discharge, an index of soil infiltration capacity, 2-year 24-hour rainfall amount, main channel slope, main channel length, 10-year runoff per unit area, mean frost depth, and the ratio of discharge that can be expected to occur at least 10 percent of the time to the discharge that can be expected to occur at least 90 percent of the time.

The U.S. Department of Agriculture (1975) suggests determining the yield of suspended solids by first determining the sediment deposited in a reservoir, retention facility, estuary, or other sediment sink and expressing it on an average annual basis. Reservoir or other trap efficiency relationships are then used to adjust the average annual sediment deposition to average annual sediment yield.

REFERENCES

Beaulac, M. N., and K. H. Reckhow, "An Examination of Land Use—Nutrient Export Relationships." *Water Resour. Bull.* **18**(6) 1013–1024, (1982).

Bedient, P. B., J. L. Lambert, and P. Machado, "Low Flow and Stormwater Quality in Urban Channels." *J. Environ. Eng. Div., Am. Soc. Civ. Eng.* **106**(EE2), 421–436 (1980).

Binder, J. J., P. B. Katz, and J. A. Ripp, "The Use of Simplified Modeling and Monitoring as a Screening Technique to Quantify Urban Runoff." *Proc. Int. Symp. Urban Storm Runoff, 1979* pp. 133–141.

Bryan, E. H., "Quality of Stormwater Drainage From Urban Land." *Water Resour. Bull.* **8**(3), 578–588 (1972).

Chen, C., "Design of Sediment Retention Basins." *Proc. Natl. Symp. Urban Hydrol., Hydraul., Sediment Control, 1975* Mini Course No. 2 (1975).

Collins, P. G., and J. W. Ridgway, "Urban Storm Runoff Quality in Southeast Michigan." *J. Environ. Eng. Div., Am. Soc. Civ. Eng.* **106**(EE1), 153–162 (1980).

Dendy, F. E., and G. C. Bolton, "Sediment Yield—Runoff Drainage Area Relationships in the United States." *J. Soil Water Conserv.* **31**(6), November-December, pp. 264–266 (1976).

Dever, R. J., Jr., L. E. Stom, and D. F. Lee, "Quality of Urban Runoff in the Twin Cities Area." *Proc. Int. Symp. Urban Storm Runoff, 1979* pp. 143–149 (1979).

Donigian, A. S., *Nonpoint Source Pollution from Land Use Activities.* Prepared by Hydrocomp, Inc., for the Northeastern Illinois Planning Commission, August 1977.

Heaney, J. P., W. C. Huber, and S. J. Nix, *Storm Water Management Model: Level I—Preliminary Screening Procedures,* EPA Rep. No. 66/2-76-275, pp. 16–21. U.S. Environ. Prot. Agency, Washington, DC, October 1976.

Heinemann, H. G., "Volume-Weight of Reservoir Sediment." *J. Hydraul. Div., Am. Soc. Civ. Eng.* **88**(HY5), 181 (1962).

Hindall, S. M., "Measurement and Prediction of Sediment Yields in Wisconsin Streams." *Water Resour. Invest. (U.S. Geol. Surv.)* **54–75,** January (1976).

Hindall, S. M., and R. F. Flint, *Sediment Yields of Wisconsin Streams,* Hydrol. Invest. Atlas HA-376. U.S. Geol. Survey, Washington, DC, 1970.

Howell, R. B., "Characterization of Constituents in Highway Runoff Waters." *Proc. Int. Symp. Urban Storm Runoff, 1979* pp. 109–114 (1979).

Huber, W. C., J. P. Heaney, D. A. Aggidis, R. E. Dickinson, K. J. Smolenyak, and R. W. Wallace, *Urban Rainfall-Runoff-Quality Data Base,* EPA-600/S2-81-238. U.S. Environ. Prot. Agency, Washington, DC, July 1982.

Kinori, B. Z., and J. Mevorach, *Manual of Surface Drainage Engineering,* Vol. II. Am. Elsevier, New York, 1984.

Kothandaraman, V., and R. L. Evans, "Nutrient Budget Analysis for Rend Lake in Illinois." *J. Environ. Eng. Div. Am. Soc. Civ. Eng.* **105**(EE3), 547–555 June (1979).

Lager, J. A., W. G. Smith, W. G. Lynard, R. M. Finn, and E. J. Finnemore, *Urban Stormwater Management and Technology: Update and Users Guide,* EPA-600/8-77-014. U.S. Environ. Prot. Agency, Washington, DC, September 1977.

Lakatos, D. F., and G. M. Johnson, "Desk-Top Analysis of Non-Point Source Pollutant Loads." *Proc. Int. Symp. Urban Storm Runoff, 1976.*

Linsley, R. K., and J. B. Franzini, *Water-Resources Engineering,* 3rd ed. McGraw-Hill, New York, 1979.

Linsley, R. K., M. A. Kohler, and J. L. H. Paulhus, *Hydrology for Engineers,* 3rd ed. McGraw-Hill, New York, 1982.

Loehr, R. C., "Agricultural Runoff—Characteristics and Control." *J. Sanit. Eng. Div., Am. Soc. Civ. Eng.* **98**(SA6), 909–925 (1972).

Marsalek, J., *Pollution Due to Urban Runoff: Unit Loads and Abatement Measures.* National Water Research Institute, Canada Centre for Inland Waters, Burlington, Ontario, October 1978.

Meyer, L. D., and M. A. Ports, "Prediction and Control of Urban Erosion and Sedimentation." *Proc. Natl. Symp. Urban Hydrol., Hydraul., Sediment Control, 1976.*

Novotny, V., and G. Chesters, *Handbook of Nonpoint Pollution,* Chapter 5, Van Nostrand Reinhold, New York, NY, 1981.

Polls, I., and R. Lanyon, "Pollutant Concentration From Homogeneous Land Uses." *J. Environ. Eng. Div., Am. Soc. Civ. Eng.* **106**(EE1), 69–80 (1980).

Roesner, L. A., "Quality of Urban Runoff." In *Urban Stormwater Hydrology* (D. F. Kibler, ed.), Water Resour. Monogr. No. 7, Chapter 6, pp. 161–187. Am. Geophys. Union, Washington, DC, 1982.

Sheaffer, J. R., K. R. Wright, W. C. Taggart, and R. M. Wright, *Urban Storm Drainage Management,* Chapter 11. Dekker, New York, 1982.

Southeastern Wisconsin Regional Planning Commission (SEWRPC), *A Comprehensive Plan for the Kinnickinnic River Watershed.* SEWRPC, Waukesha, WI, December 1978.

Terstriep, M. L., D. C. Noel, and G. M. Bender, "Sources of Urban Pollutants—Do We Know Enough?" In *Urban Runoff Quality—Impact and Quality Enhancement Technology,* Proc. Eng. Found. Conf., pp. 107–121. New England College, Henniker, NH, 1986.

Torno, H. C., J. Marsalek, and M. Desbordes (Eds.), *Urban Runoff Pollution,* NATO Adv. Sci. Inst. Ser., Ser. G, Vol. 10. Springer-Verlag, New York, 1986.

U.S. Department of Agriculture, Soil Conservation Service, *Procedure-Sediment Storage Requirements for Reservoirs,* Tech. Release No. 12. USDA, Washington, DC, January 1975 (revised).

U.S. Department of Agriculture, Soil Conservation Service, *Procedure for Computing Sheet and Rill Erosion on Project Areas,* Tech. Release No. 51 (Rev. 2). USDA, Washington, DC, September 1977.

Uttormark, P. D., J. D. Chapin, and K. M. Green, *Estimating Nutrient Loadings of Lakes From Non-Point Sources,* EPA-660/3-74-020. U.S. Environ. Prot. Agency, Washington, DC, August 1974.

Whipple, W., *Estimating Runoff Pollution From Large Urban Areas—The Delaware Estuary.* Water Resour. Res. Inst., Rutgers, The State University of New Jersey, New Brunswick, July 1978.

Whipple, W., N. S. Grigg, T. Grizzard, C. W. Randall, R. P. Shubinski, and L. S. Tucker, *Stormwater Management in Urbanizing Areas,* Chapter 4. Prentice-Hall, Englewood Cliffs, NJ, 1983.

Wischmeier, W. H., "Use and Misuse of the Universal Soil Loss Equation." *J. Soil Water Conserv.* January-February, pp. 5–9 (1976).

Wischmeier, W. H., and D. D. Smith, "Predicting Rainfall Erosion Losses—A Guide to Conservation Planning." *U.S., Dep. Agric., Agric. Handb.* **537,** December (1978).

Wischmeier, W. H., C. B. Johnson, and B. V. Cross, "A Soil Erodibility Nomograph for Farmland and Construction Sites." *J. Soil Water Conserv.* September-October, pp. 189–193 (1971).

Wylie, L. J., *Storm Water Sediment and Erosion Control.* Presented at Storm Water Detention Systems, University of Wisconsin-Extension Course, Madison, April 17-18, 1979.

8

PLANNING AND DESIGNING DETENTION/RETENTION FACILITIES

As explained in Chapter 1, there are two fundamentally different but not mutually exclusive approaches to controlling the quantity and quality of surface water runoff. One is the conveyance-oriented approach and the other is the storage-oriented approach.

This chapter addresses the latter by presenting a comprehensive, stepwise approach to the planning and design of multipurpose detention/retention (D/R) facilities. The stepwise approach recognizes that a detailed cookbook methodology is neither desirable nor possible. However, it is feasible and useful to chart a course that, based on experience, identifies the essential technical and nontechnical aspects of D/R planning and design and shows ways in which they can be addressed. The subject of sedimentation basin (SB) planning and design is introduced but detailed discussion is deferred to Chapter 9.

This chapter begins with definitions of commonly used D/R terms. An overview of the D/R planning and design process is presented, followed by a description and explanation of each step in the process, which constitutes most of the chapter. Discussions of multiple D/R facilities and use of D/R for enhancement of water quality are then presented. The chapter concludes with a treatment of the negative aspects and misuse of D/R facilities.

8.1 DEFINITION OF TERMS

Terms such as "detention facility" and "retention facility" have varied and inconsistent meanings and interpretations in the literature and in common usage. A workable and consistent set of definitions of three basic terms is established for use in this book.

Detention Facility. A surface water runoff storage facility that is normally dry but is designed to hold (detain) surface water temporarily during and immediately after a runoff event. Examples of detention facilities are: natural swales provided with crosswise earthen berms to serve as control structures, constructed or natural surface depressions, subsurface tanks or reservoirs, rooftop storage, and infiltration or filtration basins.

Retention Facility. A surface water runoff storage facility that always contains (retains) a substantial volume of water to serve recreational, aesthetic, water supply, or other functions. Surface water is temporarily stored above the normal stage during and immediately after runoff events. Examples of retention storage facilities include ponds and small lakes in residential and commercial developments and in public areas.

Sedimentation Basin. A surface water runoff storage facility intended to trap suspended solids, suspended and buoyant debris, and adsorbed or absorbed potential pollutants which are carried by surface water runoff. The SB may be part of an overall multipurpose D/R facility.

8.2 PLANNING AND DESIGN PROCEDURE: AN OVERVIEW

Planning and designing D/R facilities is much more than an exercise in hydrology and hydraulics. Technical factors including, but not limited to, hydrology and hydraulics must be considered. Examples are: (1) determining the steepest slope at which in situ soils at a potential D/R site can be excavated to provide for water storage or can be placed to construct a berm across a swale to provide for water storage; (2) estimating the annual suspended solids yield of the tributary watershed and determining if a sedimentation basin or other means of sediment control will be required; and (3) selecting grass varieties consistent with occasional flooding of several hours to several days duration.

In addition, nontechnical constraints and needs must be addressed. Examples are: (1) determining or forecasting community recreation and other needs in anticipation of developing a multipurpose D/R facility; (2) assessing potential safety hazards and making provisions to mitigate them; and (3) seeking creative and equitable ways to finance land acquisition, construction, and maintenance.

One way of encouraging a thorough consideration of all pertinent technical and nontechnical factors involved in the planning and design of D/R facilities is to use the following eight-step process.

1. Collect and analyze watershed data.
2. Identify potential sites.
3. Select the design discharge–probability condition.
4. Perform preliminary hydrologic design.
5. Accommodate lateral and vertical constraints.

6. Perform refined hydrologic–hydraulic design.
7. Finalize the design.
8. Begin implementation.

The eight steps will not apply to all situations—special circumstances will arise. Each of the eight steps is presented and illustrated in the subsequent sections of this chapter.

8.3 STEP 1: COLLECT AND ANALYZE WATERSHED DATA

If one or more D/R facilities are being considered as an alternative surface water management measure for an urbanizing watershed, certain watershed data collection and analysis will typically be required. Much of the data collection and analysis effort is identical to that required for planning and designing a conveyance-oriented system.

Delineate the Watershed, Subwatersheds, and Subbasins

Guidelines for the delineation of watersheds, subwatersheds, and subbasins are presented in Chapter 3. To the extent that potential D/R sites are known, subbasins should be delineated to discharge at those points.

Determine Existing and Future Land Use

Chapter 2 discusses the potential significant impact of urbanization and other changes in land use on the quantity and quality of surface water runoff. D/R facilities should be sized to accommodate runoff for the most critical land use condition, which for quantity purposes will usually be a future condition. The most critical condition for some water quality purposes, such as suspended solids, may be neither the existing nor the future condition but rather the condition when the land is being developed, that is, is under construction.

Although the D/R facility will usually be planned and designed for a critical future land use and land cover condition, the existing land use cover should also be determined. Establishment of existing and future conditions, and subsequent calculation of existing and future hydrologic, suspended solids, and other loads, will help decision makers understand the impact of urbanization and reinforce the need for recommended D/R facilities and other surface water management measures.

Land use should be determined on a subbasin-by-subbasin basis. Potential data sources for existing land use include land use maps, which are rarely available; aerial photographs; and field reconnaissance. Likely future land use may be developed from sources such as land use plans, plans for proposed development, zoning maps, and discussion with local developers and public officials. Assuming that some condition other than the existing is the most critical, a crude forecast of future land use is much more useful in planning and designing D/R facilities than an accurate accounting of existing land use.

Zoning maps should be used with caution. Although they can be a useful indicator of probable future land use, zoning can change dramatically. For example, certain zoning categories such as agricultural or very large lot residential, are sometimes used as interim zoning categories to prevent scattered pockets of urban development or to discourage it until such time as the necessary utilities and facilities can be provided.

Quantify Hydrologic–Hydraulic Parameters

Having delineated subbasins and established existing and future land use patterns for each subbasin, subbasin hydrologic–hydraulic parameters should be determined. Examples of such parameters, which should be determined for each subbasin under existing and future land use conditions, are:

1. Dominant soil type, such as sand, silt, clay, or combinations of soil types and possible erosion susceptibility and other soil characteristics
2. Hydrologic soil group A, B, C, and D, as defined by the Soil Conservation Service and as determined from county soil survey reports
3. Nominal slope
4. Percent imperviousness based on aerial photographs or other sources
5. Runoff coefficient, such as is used in the rational method and modified rational method
6. The runoff curve number as used in SCS hydrologic methods and derived on the basis of land use and hydrologic soil group
7. Time of concentration and flow through time as needed for techniques such as the rational method and for some computer models

The specific hydrologic–hydraulic parameters to be determined depend on the hydrologic–hydraulic techniques to be used. Given that at least two techniques should be used, as recommended in Chapter 3, a relatively large number of parameters may be needed for a given D/R planning and design project. Guidance on and aids for the determination of hydrologic parameters are presented in Chapter 3.

Table 8.1 is an example of subbasin hydrologic–hydraulic data and a format for presenting the data. The project for which this table was prepared (Donohue & Associates, 1980) used several hydrologic methods, including the rational and SCS TR55 methods for estimating peak flows and the modified rational method and the SCS TR55 method for estimating D/R facility volumes.

Note how the existing and future condition subbasin hydrologic parameters in Table 8.1 provide an indication of or an index to the probable impact of urbanization on surface water runoff and volumes. For example, the future land use condition area-weighted imperviousness is over three times the existing condition area-weighted imperviousness. The future condition area-weighted runoff coefficient is twice the existing condition value, and the future condition time of concentration for the entire watershed is one-third of the existing condition value. Note also how watershed size increases under future conditions with the enlargement of subbasin I. Storm sewer plans for the future residential development in this area would have the effect of draining additional areas into subbasin I, which, prior to the storm sewers, were outside subbasin I and the watershed.

Assemble Precipitation Data

Two categories of precipitation data are typically needed in the planning and design of D/R facilities. The first is intensity–duration–frequency curves or tables which are often used for the sizing of D/R facilities using methods described in Chapter 3.

The second category of useful precipitation data are hyetographs or other data on historic storms. More specifically, data should be gathered on recent major storms widely recognized by area residents to have caused flooding problems. After one or more D/R facilities have been sized, the historic precipitation data should be used to demonstrate how the recommended D/R facilities would prevent or substantially reduce flood damage if the historic storms, or storms similar to them, were to occur again. Citizens and elected officials are most likely to understand the intended function of surface water management facilities if that function is demonstrated using rainfall data from easily recalled historic storms.

Search for Groundwater-Level Data

The position of the water table can have a significant effect on the construction and operation of a D/R facility. If, for example, a retention facility is desired for recreational or aesthetic purposes, a water table located above the bottom of the retention facility may be necessary to provide for a permanent pool independent of sources of surface inflow. In contrast, the construction of a detention facility may be complicated and the operation of a detention facility may be compromised if the local water table is positioned well above the intended dry bottom of the detention facility.

Although a special geotechnical investigation, including water table logging, is often required for the design of a particular D/R facility, existing watershed-wide groundwater data should be examined during the planning stage as an aid in screening potential D/R sites. Possible sources of groundwater data include existing ponds or wetlands in the watershed; nonpumping

TABLE 8.1 Subbasin Hydrologic-Hydraulic Parameters for Hahns Creek Watershed

Sub-basin Characteristic	Units	Subbasin A	B	C	D	E	F	G	H	I	Total or Area-Weight Value
Nominal slope[a]	percent	1.8	1.4	1.0	1.0	1.1	0.7	1.3	1.1	1.1	1.1
Dominant soil type	—	Loam	Silt loam	Loam	Silt loam	Silt loam	Silt loam	Silt loam	Silt loam	Loam	—
Hydrologic soil type	—	C	C	C	B	B	C	B	B	C	—
Existing conditions											
Dominant land use	—	Open space with vegetation	Open space with vegetation	Cropland	Open space with vegetation	Open space with vegetation	Park	Cropland	Cropland	Cropland	—
Area	acres	26	53	59	42	38	7	55	58	57	395
Imperviousness	percent	15	8	10	28	32	8	11	9	9	14
Runoff coefficient	—	0.24	0.26	0.20	0.30	0.32	0.19	0.18	0.16	0.20	0.22
Runoff curve number	—	82	81	77	82	81	76	79	79	83	80
Time of concentration	minutes	18	27	60	42	28	20	46	58	69	—
Flow through time	minutes	—	21	47	—	38	—	45	58	—	—

Cumulatative time of concentration	minutes	18	39	86	—	124	—	169	227	—	227
		Ultimate development									
Dominant land use	—	Quasi-public (open space with vegetation)	Quasi-public (open space with vegetation)	Quasi-public (open space with vegetation)	Quasi-public (open space with vegetation)	Quasi-public (open space with vegetation)	Conservancy (park)	Commercial	Park/storm water retention	Multi-family residential	
Area	acres	26	53	59	42	38	7	55	58	115	453
Imperviousness	percent	15	34	52	45	42	8	74	32	62	48
Runoff coefficient	—	0.24	0.36	0.46	0.40	0.38	0.19	0.60	0.32	0.54	0.44
Runoff curve number	—	82	86	98	86	85	76	93	36	87	87
Time of concentration	minutes	18	13	14	15	10	20	11	14	22	—
Flow through time	minutes	—	8	11	—	13	—	10	14	—	—
Cumulative time of concentration	minutes	18	26	37	—	50	—	60	74	—	74

Source: Adapted from Donohue & Associates (1980).

[a]Along the principal hydraulic flow path through the subbasin.

wells or active wells that could be temporarily shut off and used as piezometers; recent experience with construction of sanitary sewers, major buildings, and other utilities and facilities; and special reports prepared by federal agencies, such as U.S. Geological Survey and their state or local counterparts.

Establish Land Ownership

In anticipation of screening parcels of land for potential D/R sites, land ownership should be determined for large or otherwise significant parcels of land. Large tracts of undeveloped publicly owned land are most desirable, followed by undeveloped, privately held land. Fully or partially developed public or private parcels in need of redevelopment may also offer opportunities for siting of a D/R facility.

The 27-acre former Porter County, Indiana, fairgrounds site shown in Figure 8.1 was a prime location for an excavated detention facility. It was also the last feasible detention location in a heavily developed area within the city of Valparaiso, Indiana, the county seat of Porter County. At the time a flood control plan was being developed by the city, the fairgrounds site was owned by the county. However, it was no longer used for the county fair because the county had just opened a new fairgrounds outside the city in a rural area. The

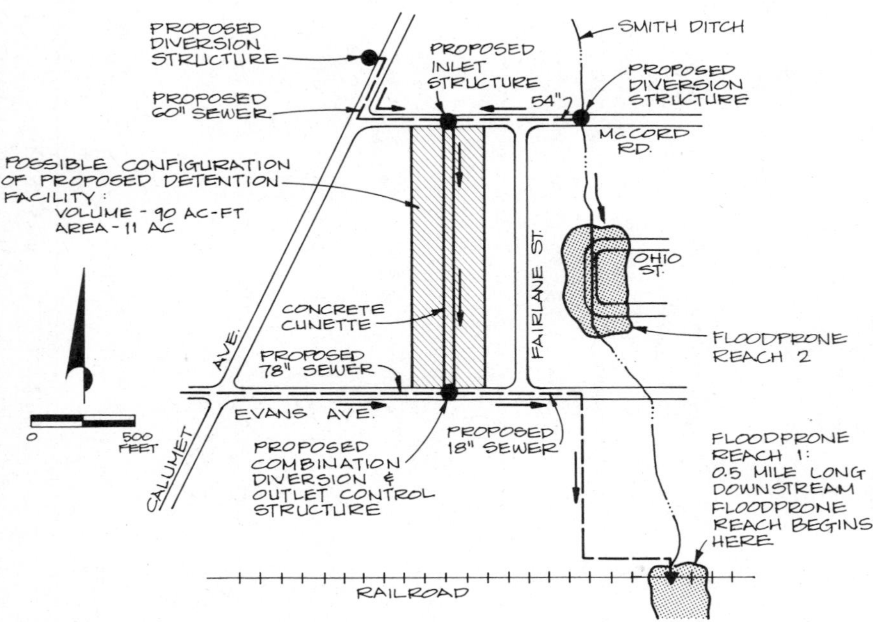

FIGURE 8.1 Conceptual detention plan for fairgrounds site (*Source:* Adapted from Walesh et al., 1987, 1988)

city eventually negotiated with the county to buy the old fairgrounds and to redevelop the site by converting it into a multipurpose flood control–recreation complex (Walesh et al., 1987, 1988).

In summary, the ownership characteristics of larger parcels of land in the watershed, particularly potential D/R sites, should be determined as early as possible in the planning process. Careful identification of current ownership and intended use, in combination with an assessment of recreational and other needs of a community, can lay the groundwork for successful negotiation for purchase and development or redevelopment of property for flood control possibly in combination with other uses.

Determine Applicable Surface Water Management Regulations

Local regulations affecting surface water management should be reviewed because they establish the framework within which D/R facilities will be planned and designed by the engineer and received and reviewed by local officials and area citizenry. Some municipalities and other government entities have specific D/R requirements, whereas other communities have very general provisions which simply require that recommended surface water management facilities receive the approval of one or more local officials.

In the latter, loosely structured situation, the engineer responsible for the planning and design of D/R and other surface water management facilities will have more latitude to be creative in seeking a solution to flooding problems but may find less than enthusiastic support by local officials and citizenry. In the former, highly regulated situation, although the engineer may be subject to more planning and design constraints, local officials and citizens are more likely to be receptive to recommendations for D/R facilities.

Special attention should be given to recurrence interval or probability criteria included in local regulations in anticipation of possibly recommending an expansion of or modification to the criteria. For example, many communities have D/R regulations which specify that for a flood of specific recurrence interval the peak discharge after development shall not exceed the peak discharge before development. The intent of such provisions is conceptually sound. However, their implementation can, as discussed later in this chapter, lead to less than optimum facilities which exert relatively little control over flood events of magnitude less than or greater than that set forth in the regulations. In these situations, consideration should be given to modifying the D/R regulations to incorporate the continuous probability criterion presented later in this chapter.

Identify Related Problems and Opportunities

When seeking the solution to one or more actual or potential flooding problems in a watershed, it is important to view those problems in the context of other, possibly related problems and needs within the watershed or within the larger community or area containing the watershed. Most communities have several needs. Therefore, communities with flooding problems may be expected to have needs in one or more other areas of community services, such as preserving open space, providing recreation facilities, abating water pollution, improving water supply, upgrading transportation, and redeveloping blighted areas. These needs may be relevant to the surface water management effort and, therefore, should be noted and understood in the early stages of planning and designing D/R and other surface water management facilities.

A comprehensive, creative approach to urban flood mitigation and prevention aided by a fortuitous set of circumstances can sometimes lead to the cost-effective solution of two or more problems. That is, the existence of two or more community problems, one of which is a flood problem, can present an opportunity for the economic solution of all the problems.

For example, the 1984 identification by one consultant of the former fairgrounds site in Valparaiso, Indiana, as shown in Figure 8.1, as a prime location for a detention facility coincided with the selection of that site by another consultant for a major outdoor recreation complex. These two events coincided with the growing need to obtain fill material for a bypass highway ready for construction 1.6 mi from the site (Walesh et al., 1987, 1988).

In this case, the early identification of three coincident community problems or needs eventually led to recommendations and implementation of a plan which simultaneously met all needs in a cost-effective fashion. Some aspects of implementation of the fairgrounds project are presented in Step 8 of this D/R planning and design process.

8.4 STEP 2: IDENTIFY POTENTIAL SITES

Potential sites are screened in this step before doing a detailed analysis for any given location. Factors to consider in the screening process are presented in the approximate order in which they should be considered.

Assess Proximity to Flood-prone Areas

A primary consideration for a D/R facility is that it be located upstream of and as close as possible to an area requiring flood protection. The nearer the storage site is to the flood-prone area, the greater the portion of the tributary area that will be controlled by the site.

Consider, for example, the detention site shown in Figure 8.1. The site was immediately upstream of an open channel reach of Smith Ditch, which had experienced serious overbank flooding (flood-prone reach 1 in Figure 8.1). Temporary storage at the site with slow release offered the opportunity to mitigate downstream flooding. Furthermore, the site was adjacent to a sewered segment of Smith Ditch, which, because of inadequate capacity, had surcharged and caused localized surface flooding (floodprone reach 2 in Figure

8.1). Diversion of excess flow from the upstream end of the sewered reach into the detention facility offered the opportunity to relieve the surcharging problem.

Determine If Site Size Is Adequate

A potential site should have, in an approximate sense, adequate size as determined by the areal extent of the site and the volume of water that could be stored temporarily on the site. Given the size and characteristics of the watershed area tributary to the site, a rough estimate can be made of the volume of surface water runoff that will be delivered to the site. The rough estimate could be made by beginning with the 100-year 24-hour rainfall depth; estimating an area-weighted runoff curve number or runoff coefficient for the watershed; and converting rainfall volume to surface water runoff at the potential D/R site.

As discussed under Step 1, the collection and analysis of watershed data should determine ownership of and plans for large or otherwise significant parcels of land. Furthermore, and as also treated in Step 1, potential related community problems and opportunities should be identified. Given the preceding information, a preliminary decision can be made as to whether or not a particular parcel is available or could be made available for a D/R facility.

Evaluate Likelihood of Gravity Inflow and Outflow

As discussed in Chapter 5, gravity-driven inflow and outflow are most desirable. Potential D/R sites should be examined to determine if gravity-driven inflow and outflow, or at least gravity inflow, can be accomplished through judicious site layout and vertical positioning of inflow and outflow works.

In the case of the former fairgrounds site shown in Figure 8.1, the screening suggested that an off-channel, excavated detention facility could have gravity inflow and gravity outflow. This would be possible because of the relatively steep slope of the adjacent southerly-flowing Smith Ditch.

Evaluating the likelihood of gravity inflow and outflow at a given potential D/R site includes considering the means by which flow would be conveyed into and out of the D/R facility. For example, screening of the off-line fairgrounds detention facility shown in Figure 8.1 indicated that special and costly inflow and outflow sewers would be needed. Less elaborate and therefore less costly inlet and outlet works are usually needed for on-line D/R facilities, such as that shown in Figure 8.2.

In a related matter, it is preferable to have a site that is already topographically low, thus minimizing the excavation and earthwork needed to achieve the desired storage volume. Sites with favorable topographic features are most likely to be found in undeveloped or newly developing areas. Such sites are less likely to exist in developed areas because existing development preempts the topographically low sites or results in their being filled in. Therefore, ex-

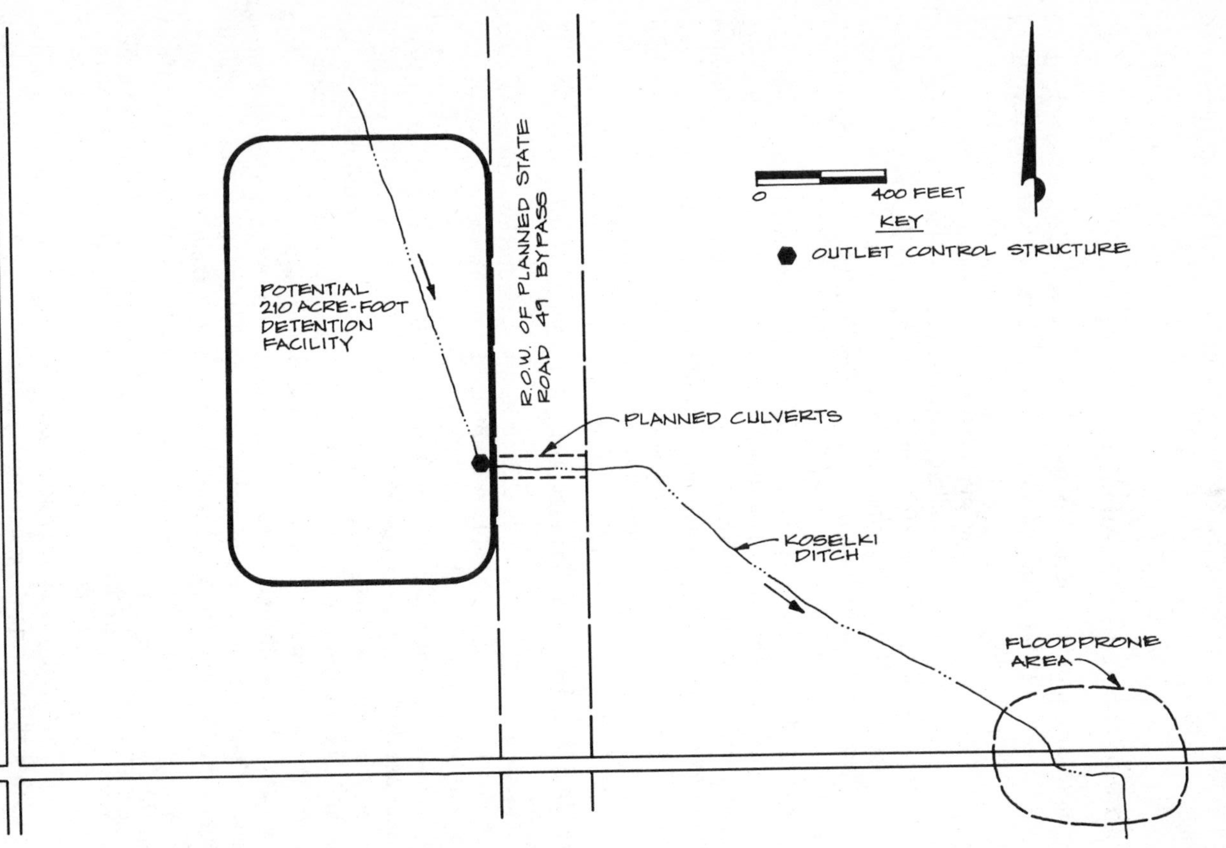

FIGURE 8.2 Detention plan for proposed highway site (*Source:* Adapted from Donohue & Associates, 1984)

tensive excavation is common when retrofitting D/R facilities into urbanized areas.

Prior to construction, ground grades at the fairgrounds site shown in Figure 8.1 were coincident with that of adjacent areas. At this site, which is in an urbanized area, all storage was eventually developed by excavation. In the case of the detention facility illustrated in Figure 8.2, which is outside the urban area, the site was topographically low, but supplemental excavation was required to provide the necessary volume of storage.

Consider Other Factors

Other relevant factors should be addressed, at least in a preliminary fashion, in the site screening process, depending on the circumstances. For example, some potential D/R sites may already serve at least a partial surface water runoff storage function. A hypothetical example is a topographically low area on the upstream side of a roadway embankment penetrated by a small culvert. If the site is selected for further investigation, subsequent hydrologic–hydraulic analysis should determine the downstream benefit of the incremental storage, not the total storage, which could be developed at the site.

Some potential D/R sites may be isolated from public roadways or public right-of-way, thus presenting potential access problems for inspection, maintenance, and repair. Regardless of the ultimate ownership of a D/R facility, access must be provided for government officials and work crews either from contiguous publicly owned land or by means of an access easement.

8.5 STEP 3: SELECT THE DESIGN DISCHARGE–PROBABILITY CONDITION

If not dictated by local regulations, the design discharge–probability condition must be established for D/R and other surface water management facilities. Selection of this design condition should include consideration of two factors: (1) determination of the most severe flood condition, in terms of probability or recurrence interval, to be accommodated; and (2) determination of what control, if any, should be achieved over smaller flood events.

Selection of the Most Severe Event

In determining the most severe flood condition to be accommodated, consistency with other parts of the community should be the goal. Equity suggests that all residents and all similar properties should be provided with the same level of flood protection.

Such areal or community-wide consistency in the level of flood protection is often lacking. For example, as illustrated schematically in Figure 8.3, properties in or near floodplains are usually protected up to the 1 percent flood

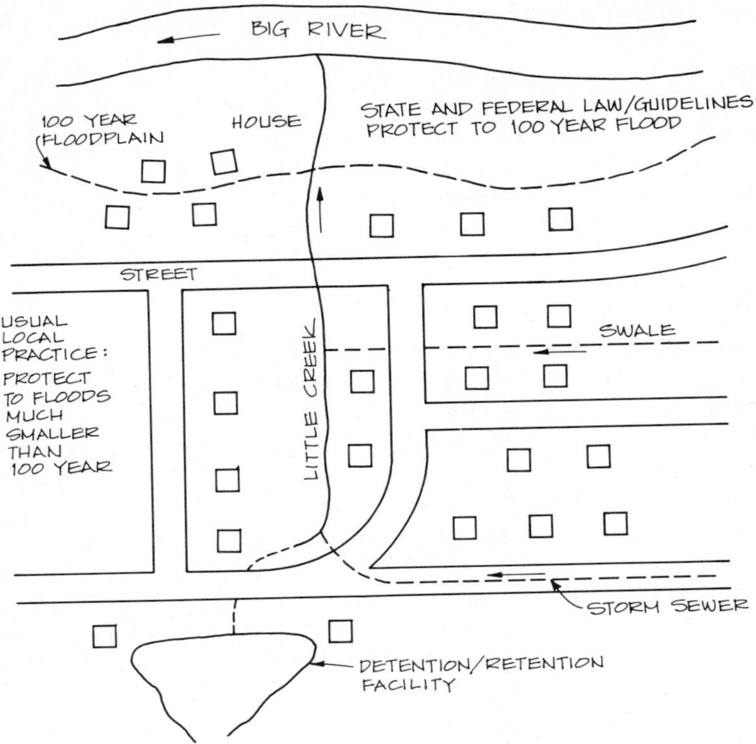

FIGURE 8.3 Inconsistent levels of flood protection

level, whereas properties away from the river or floodplain are often explicitly protected only up to a much lesser flood, such as the 10 percent event. Explicit protection to a lesser flood level usually means that storm sewers, and in some cases D/R facilities, are designed to convey or store safely flows or volumes generated by the lesser flood.

The apparent lack of consistency and equity in flood protection may be somewhat ameliorated by the availability during more severe events of conveyance and storage capacity on streets and in low-lying areas, even though such capacity was not explicitly designed into the system. That is, the emergency system as defined in Chapter 1 will function to some extent even though it was not designed.

The basis for the higher, usually 1 percent flood, level of protection provided to floodplain properties can usually be traced to federal or state programs or regulations. These agencies have typically used the 1 percent flood event for planning and design purposes. Local jurisdictions, when free of federal and state constraints, have usually selected less stringent levels of protection against flood damage.

Regardless of the reasons for the typical inconsistencies and levels of protection, equity suggests that a consistent upper limit of protection be applied to the entire community. Given that the federal or state 1 percent criterion will prevail in many communities, it follows that the 1 percent event will often be the most severe condition to be accommodated by all D/R and other surface water management facilities.

Accommodating Lesser Events

Having selected the most severe flood event to be used for planning and design purposes, attention should be directed to the matter of what control, if any, which should be exercised over lesser events. It is not unusual for regulations and design criteria to be silent on this matter. For example, the Metropolitan Sanitary District of Greater Chicago (1972) requires that the 100-year discharge from a D/R facility after development be less than or equal to the 3-year discharge for natural conditions. However, there are no explicit criteria for other recurrence intervals.

Accordingly, some D/R and other surface water management facilities are designed on the basis of a single probability or recurrence interval criterion. That is, the design does not explicitly address performance under other more or less severe flood events.

The potential fallacy of the single probability or single recurrence interval criterion is illustrated in Figure 8.4. With this criterion, and in an attempt to prevent an increase in peak flows that could result from urbanization (case 2 in Figure 8.4), a D/R facility could be designed such that peak flows for the specified probability after land development (cases 4 and 5) are no greater than before development (case 1). However, for other than the specified single design probability, future development peak flows (cases 4 and 5) could be less than or greater than the existing condition peak flows (case 1). Refer to Section 10.7 for results of a simulation study of the performance of a detention facility over a wide range of recurrence intervals.

Whether peak flows increase or decrease for other than the specified probability depends on the characteristics of the outlet structure used to achieve the desired "no increase" performance at the specified single probability. Typically, the performance of the outlet structure is not checked under hydrologic conditions other than the one for which it is designed.

The haphazard performance over a range of storm events typical of the single probability approach seems wasteful of resources. That is, a major facility is designed, constructed, and operated to control explicitly only one rare, although typically severe, hydrologic condition. Any favorable control exercised over other events is achieved by chance, not design.

With a fundamentally different design criterion, additional design effort, and possibly, but not necessarily, additional construction and operation costs, flood events over a wide range of severity may be controlled. The fundamentally different design criterion is the continuous probability or continuous re-

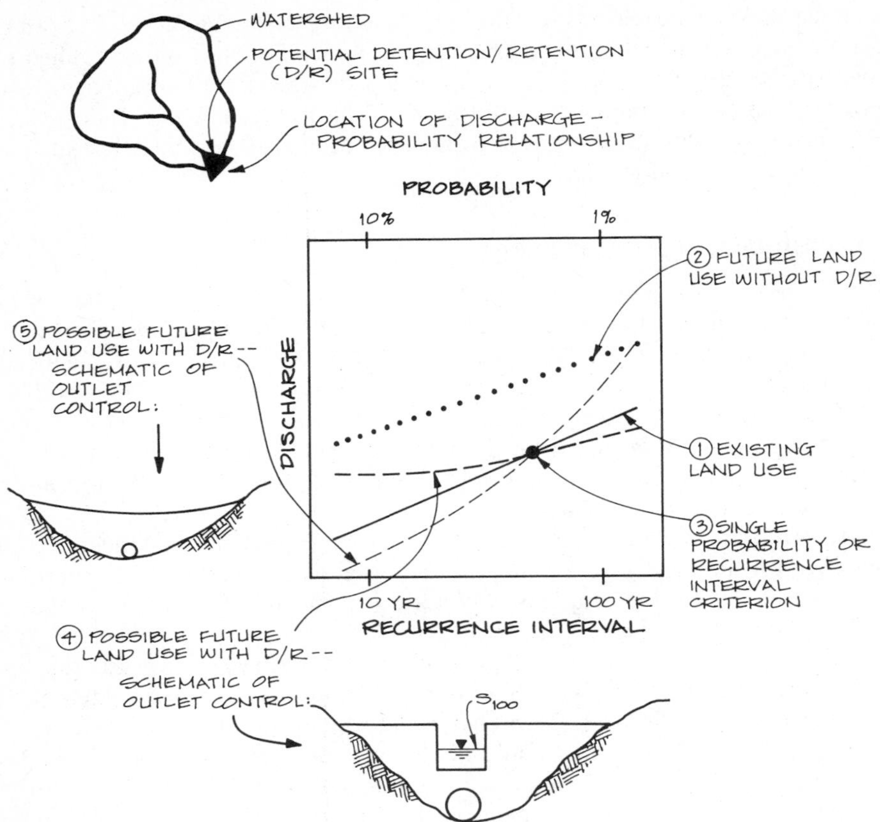

FIGURE 8.4 Single probability detention/retention criterion and possible outcomes

currence interval criterion illustrated in Figure 8.5. The idea is to require that for all storm events up to and including the most severe event peak downstream flows for the developed condition with the D/R facility in place will not exceed peak downstream flows for existing development conditions without the D/R facility in place. For example, the surface water management guidelines for Bettendorf, Iowa (Donohue & Associates, 1985), specify that there shall be no adverse shift in the discharge-probability relationship. A design procedure consistent with the continuous probability or continuous recurrence interval criterion is presented later in this chapter under Step 6 of the planning and design procedure.

Maximum Allowable Release Rate or Rates

The maximum allowable release rate or rates from a D/R facility will be established as a consequence of selecting or specifying the design discharge-proba-

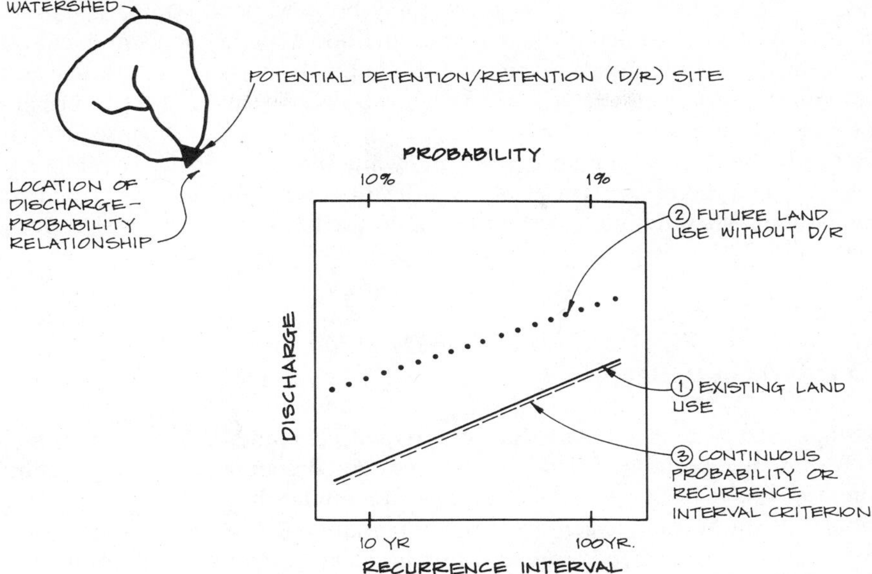

FIGURE 8.5 Continuous probability or recurrence interval detention/retention criterion

bility condition. Assume, for example, that the single probability or recurrence interval criterion applies. Then the maximum allowable release rate is the predevelopment or existing condition peak flow corresponding to the specified probability. This flow is illustrated by the single point on Figure 8.4.

If, on the other hand, the continuous probability or recurrence interval criterion applies, there will be a series of maximum allowable release rates, with each rate corresponding to one probability in a range of probabilities. These flows are illustrated by the case 3 discharge–probability relationship shown in Figure 8.5, which is coincident with the case 1 relationship.

Regardless of the value of the maximum allowable release rate or rates based on the design criterion, the flow or flows should also be evaluated in the context of the capacity of the downstream swale, channel, or sewer. In general, the more stringent condition should govern and be used to determine the maximum allowable release rate or rates.

There may be exceptions, however, such as when the downstream channel has a very large capacity—much larger than the greatest flow that would be allowed by the design discharge–probability condition. In such a situation, consideration should be given to seeking a waiver from the applicable regulations. Such waivers can sometimes be granted at the discretion of the municipal engineer or other local officials on presentation of supporting technical evidence.

Regardless of how it is determined, the maximum allowable release or bypass rate for a particular D/R facility may be very small compared to discharges such as the uncontrolled 1 percent flood flow at the same location. Accordingly, the engineer may be tempted simply to set the release rate to zero to simplify subsequent analysis and design. This is generally ill advised because the release or bypass rate, although it may appear deceptively inconsequential, may function over a long period of time while the D/R facility is filling and emptying and therefore represents a significant volume. Failure to account for this volume when designing an on-line D/R facility may result in significant oversizing.

8.6 STEP 4: PERFORM PRELIMINARY HYDROLOGIC DESIGN

Assume that watershed data have been collected and analyzed, potential D/R sites have been identified and screened, and design discharge–probability conditions have been established. Step 4, preliminary hydrologic design, contributes to the further screening of the potential D/R sites. More specifically, the objective of this step is to make a preliminary estimate of the storage volume required at a potential D/R site to determine if adequate volume is available or can be developed. The step is preliminary in that it does not require detailed, time-consuming hydrologic–hydraulic analysis and yet provides information sufficiently accurate to assist with the screening process.

One approach to preliminary hydrologic design is illustrated schematically in Figure 8.6 for both on-line and off-line D/R facilities. The variables shown on Figure 8.6 are defined as follows:

R = design rainfall volume in inches over the tributary area.
d_R = duration of the design rainfall in hours
t_c = time of concentration of the tributary area in hours
Q_{in} = the inflow hydrograph for the on-line and off-line D/R facility
Q_{out} = the outflow hydrograph for the on-line D/R facility

The idea is first to convert a design rainfall volume to a runoff volume. Then the required storage volume is calculated as the runoff volume minus the concurrent release volume, in the case of an on-line D/R facility, or the concurrent bypass volume, in the case of an off-line D/R facility.

Simplifying assumptions include setting the base of the inflow hydrograph equal to the sum of d_R and t_c, linear rising and falling limbs on the outflow hydrograph for the on-line D/R facility, and a uniform diversion rate for the off-line D/R facility. Hyetograph and hydrograph shapes are not needed.

As an example of the application of the preliminary hydrologic design method, consider the off-line D/R situation illustrated in Figure 8.7. Selected

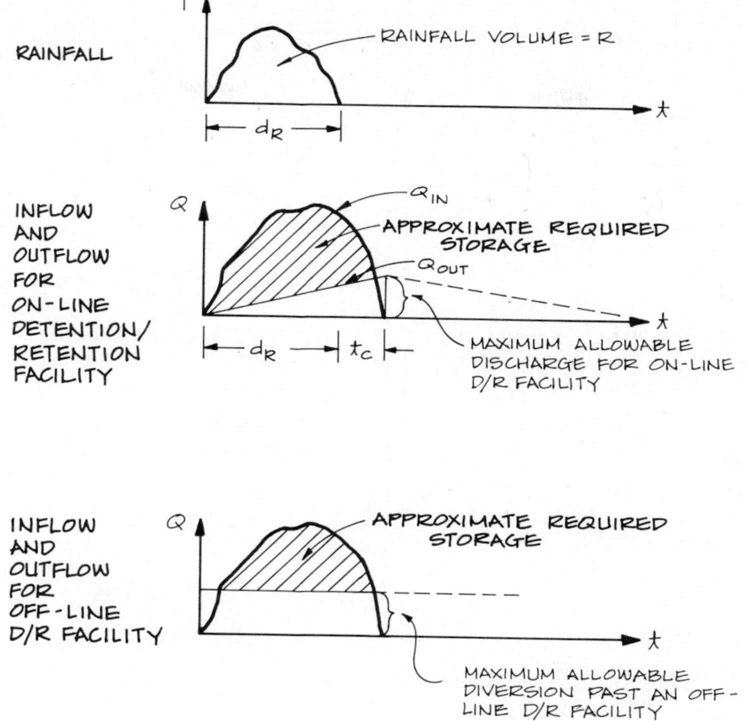

FIGURE 8.6 Simple inflow–outflow models for on-line and off-line detention/retention facilities

design rainfall data for this Valparaiso, Indiana, site are (Donohue & Associates, 1984):

1. 100-year 3-hour = 4.11 in.
2. 100-year 6-hour = 4.62 in.
3. 100-year 12-hour = 5.04 in
4. Largest 24-hour rainfall in period of record (1915–1983) = 7.34 in. (occurred in 1915 and in 1983)

The design–discharge–probability condition applicable to this situation was to control all flood events up to the 1 percent event without exceeding downstream sewer and channel capacities. The first three design rainfall conditions were selected as being representative, in terms of duration, of the range of possible 1 percent events. The fourth event was selected because it represented a recent flood which inflicted considerable damage and disruption and there-

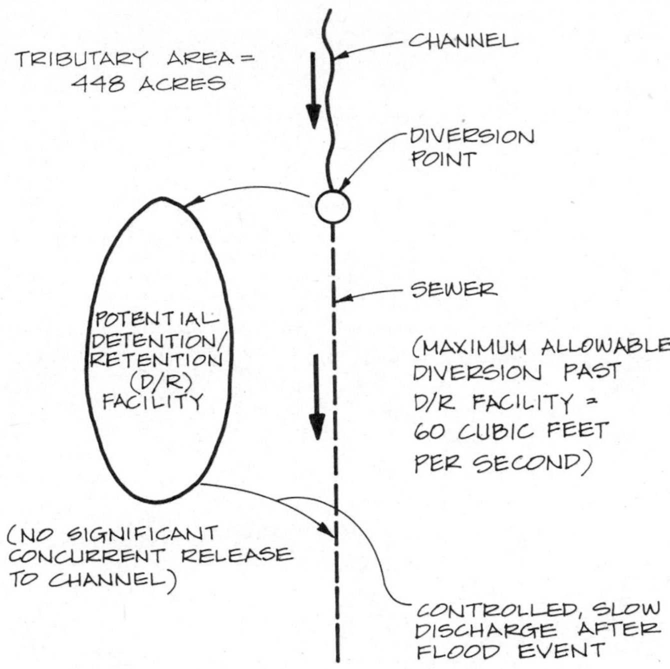

FIGURE 8.7 Schematic of off-line detention/retention facility

fore was readily remembered by and was of great interest to local officials and citizens.

The calculation of required storage for each of the four design rainfalls proceeds as follows:

1. For 100-year 3-hour rainfall of 4.11 in.:
   (a) Runoff (from SCS rainfall–runoff table—Table 3.6 in this book—using the given RCN of 86) = 2.65 in.
   (b) Runoff volume = (2.65 in.)(ft/12 in.)(448 acres) = 98.9 acre-ft
   (c) Concurrent bypass volume = (60 ft^3/sec)(4 hours)(3600 sec/hr)(acre-ft)/(43,560 ft^3) = 19.8 acre-ft
   (d) Required storage = 98.9 − 19.8 = 79.1 acre-ft
2. For 100-year 6-hour rainfall of 4.62 in:
   (a) Runoff = 3.11 in.
   (b) Runoff volume = (3.11)(1/12)(448) = 116 acre-ft
   (c) Concurrent bypass volume = (60)(7)(3600)(1/43,560) = 34.8 acre-ft
   (d) Required storage = 116 − 35 = 81 acre-ft

3. For 100-year 12-hour rainfall of 5.04 in.:
   (a) Runoff = 3.50 in.
   (b) Runoff volume = (3.50)(1/12)(448) = 131 acre-ft
   (c) Concurrent bypass volume = (60)(13)(3600)(1/43,560) = 64.5 acre-ft
   (d) Required storage = 131 − 65 = 66 acre-ft
4. For 24-hour rainfall of record of 7.34 in.:
   (a) Runoff = 5.69 in.
   (b) Runoff volume = (5.69)(1/12)(448) = 212 acre-ft
   (c) Concurrent release volume = (60)(25)(3600)(1/43,560) = 124 acre-ft
   (d) Required storage = 212 − 124 = 88 acre-ft

Using the three 1 percent design rainfall events, the preceding analysis indicates that the required D/R volume is in the approximate range 66 to 81 acre-ft.

It is of value to note that preliminary design techniques, such as the preceding, show in an approximate way the manner in which required D/R storage may vary as a function of design storm duration. More specifically, the example indicates that for a particular probability or recurrence interval, the required D/R storage does not necessarily increase as design storm duration, and therefore design storm volume, increases. As expected, design storm runoff volume increases with increasing duration. However, concurrent D/R facility bypass or release volume increases and then decreases with increasing duration. Therefore, as is the case in the example presented, the incremental runoff volume associated with increased design storm duration may be more than offset by the incremental bypass or release volume.

In summary, an increase in the duration of the design storm does not necessarily mean an increase in the required volume of storage. This conclusion can be safely drawn from experience with preliminary hydrologic design procedures such as that described here. The conclusion is also borne out by more detailed hydrologic–hydraulic analyses of D/R facilities.

8.7 STEP 5: ACCOMMODATE LATERAL AND VERTICAL CONSTRAINTS

Assume that volume and other previous considerations indicate that it is still feasible to place a D/R facility at a given site. Then the next step is to carefully identify the site-specific lateral and vertical constraints and to seek ways to accommodate them. Stated differently, the purpose of step 5 is to determine how the required storage volume can be tailored to fit the site.

Lateral Constraints

Examples of lateral or planimetric constraints are existing and proposed streets, utilities, buildings, and other facilities contiguous with or within the potential D/R site. Recreation facilities such as softball and soccer fields and tennis and basketball courts to be located within D/R facilities are planimetric constraints in that they define minimum areas and dimensions of portions of the D/R facility which must be relatively flat.

Sometimes planimetric constraints can be accommodated in such a way as to enhance the desirability of a potential D/R facility. For example, the Hahns Creek detention facility site in Manitowoc, Wisconsin (Donohue & Associates, 1980), was bounded on the north by city-owned land designated for a school and on the west by privately owned land planned for multifamily residential development. These two planned uses of large portions of the undeveloped site significantly limited the portion of the site available for a D/R facility. Accordingly, the two planned land uses were accommodated by designing a multipurpose recreation–flood control facility which used land efficiently by being oriented, in part, to providing recreational areas and open space amenities for the future contiguous school and residential development.

Vertical Constraints

One category of vertical constraints are existing and proposed street and ground grades in contiguous areas, coupled with the freeboard to be provided above the design high water level. Other examples are the grades of existing or future incoming storm sewers or channels and the ground grades of areas served by them, the grade of the receiving storm sewer or channel, and the water table elevation. The preferred result is to accommodate vertical constraints without resorting to pumped inflow or outflow.

Lateral and Vertical Constraints: Example

A progressive, sometimes iterative process is often needed to fit the required storage volume to a site while simultaneously accommodating lateral and vertical constraints. The initial design for flood control–recreation development of the fairgrounds site shown in Figure 8.1 is illustrated in Figure 8.8. Because a park planning project was conducted parallel to and with knowledge of the flood control planning project, the park master plan recommended that a major recreation facility be developed at the fairground site and that it be configured to permit joint use for flood control. Recreation facilities included four softball diamonds, three or four soccer fields, a general-purpose athletic field utilizing the existing grandstand for spectator viewing, picnic and playground areas, an ice skating rink, a circumferential jogging–walking–biking trail, parking areas, and public use and service buildings. The initial concept envisioned incorporating flood control by excavating the approximate eastern

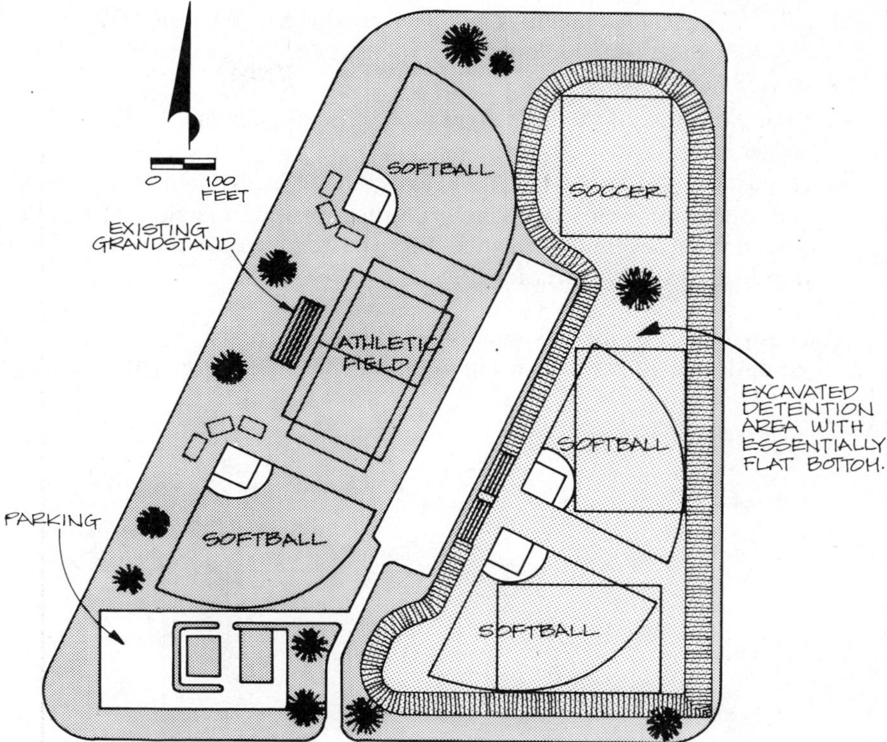

FIGURE 8.8 Initial flood control–recreation plan for fairgrounds site (*Source:* Adapted from Walesh et al., 1987, 1988)

third of the site to one depressed level thus providing the necessary 90 acre-ft of storage.

As the flood control–recreation project moved from planning to designing, a refined plan and final design were developed. The project team, consisting of city engineering and city park personnel, flood control engineers, and park and recreation designers, considered and resolved many factors during design, such as:

1. Flattening of side slopes on all excavated areas for ease of maintenance and for aesthetic quality
2. Grading of all recreation field surfaces to provide positive and rapid surface and subsurface drainage
3. Raising the minimum outlet elevation of the detention facility based on a field survey of local storm sewers specially conducted for the design
4. Elimination of the old, unsightly fairground bleachers and rearrange-

ment of recreation facilities, particularly the four softball fields, to provide improved function and less costly concession, rest room, and other service facilities

5. Saving of most of the large trees along the east side and in the northeast corner of the site
6. Using terraces—relatively flat areas at slightly different elevations—to add topographic variety and interest, in plan and in section, to an otherwise flat or severely excavated site and to decrease the frequency of flooding of some sports facilities

The design process, including consideration of the preceding factors, led to the final design as illustrated in Figure 8.9. A set of terraces covers the site.

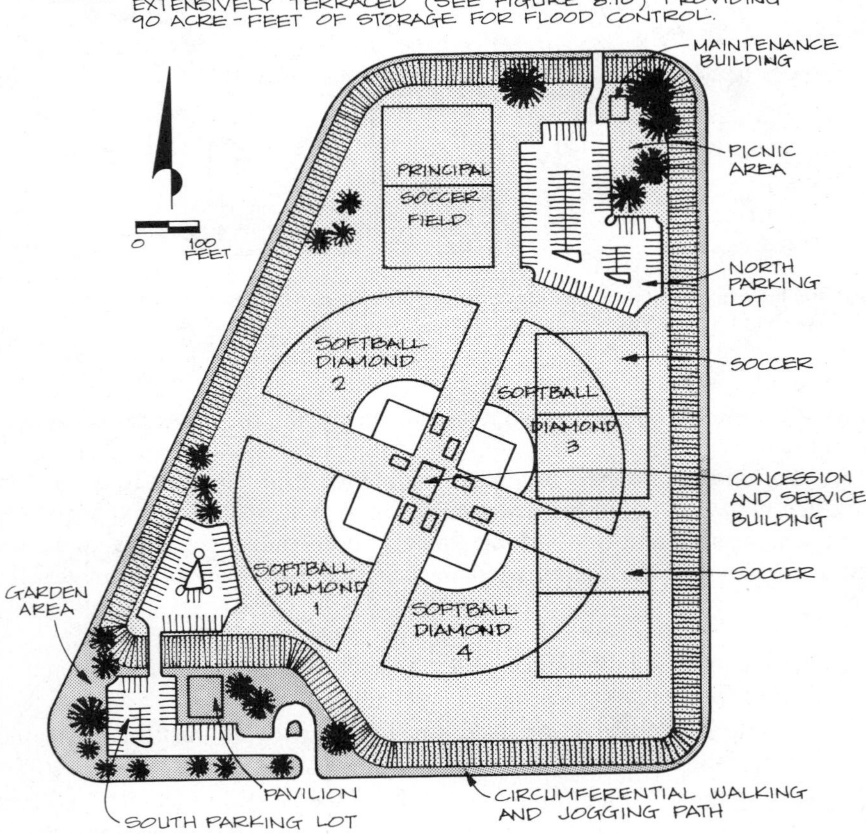

FIGURE 8.9 Final flood control–recreation design for fairgrounds site (*Source:* Adapted from Walesh et al., 1987, 1988)

As a result of excavation and terracing, the originally flat site now has a total relief of 13.0 ft, with the lowest area occurring at the outlet in the southeast corner.

Least used portions of the facility, such as one of the overflow soccer areas, are on the lowest levels. In contrast, frequently used areas, such as lighted softball fields, are on the highest terraces. Frequency and duration of stormwater storage for all areas of the site are presented in Figure 8.10. Indicated frequency and duration assume ultimate watershed development corresponding to the greatest stormwater loads. As indicated in Figure 8.10, the southeast or lowest corner of the facility will be inundated at least once per year for a duration of about 2 days. In contrast, the softball diamonds will be inundated for a fraction of a day once every 10 or more years, depending on their specific vertical position.

8.8 STEP 6: PERFORM REFINED HYDROLOGIC–HYDRAULIC DESIGN

The purpose of this step is to determine the D/R facility storage versus release rate or bypass rate relationship required to satisfy the design discharge–probability condition and then to do the final geometric design of the D/R facility and the final hydraulic design of the outlet works. The description of this step assumes that the continuous probability or recurrence interval criterion rather than the single probability or recurrence interval criterion is being used. In the event that the latter is being used, the generic procedure presented here can be simplified accordingly.

Overview

The five-step process used to determine the storage versus release rate or bypass rate relationship is first presented in a generic fashion. Three examples follow to illustrate application of the process through final site geometry and outlet works hydraulics.

Step A: Generate Hydrographs. For a given duration, generate inflow hydrographs for a full range of probabilities, such as 50, 10, 4, 2, and 1 percent. Inflow hydrographs are illustrated in Figures 8.11 and 8.12. Inflow hydrographs can be developed using many methods, including, but not limited to, manual techniques such as the MRM (modified rational method) and the Soil Conservation Service TR55 method, both of which are described in Chapter 3, and computer programs such as HEC-1 and ILUDRAIN, which are described in Chapter 10.

Step B: Establish Release or Bypass Rates. For each probability, note the corresponding maximum allowable release rate (for an on-line D/R facility) or

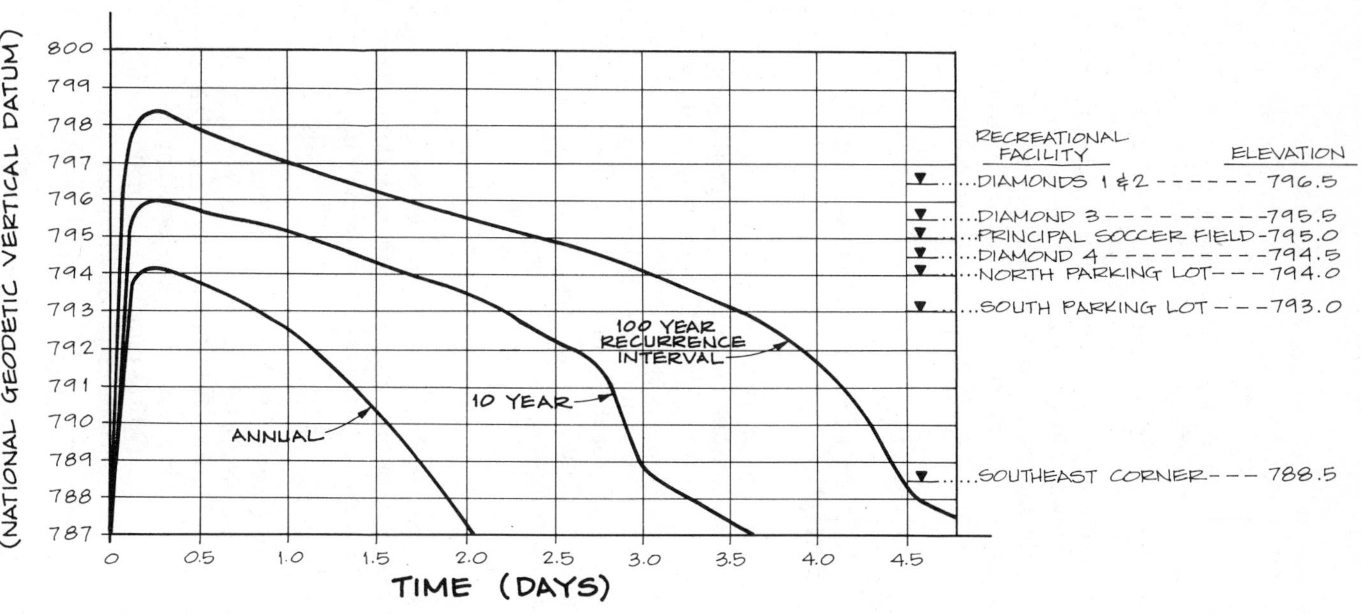

FIGURE 8.10 Frequency and duration of stormwater storage for fairgrounds flood control recreation project (*Source:* Adapted from Walesh et al., 1987, 1988)

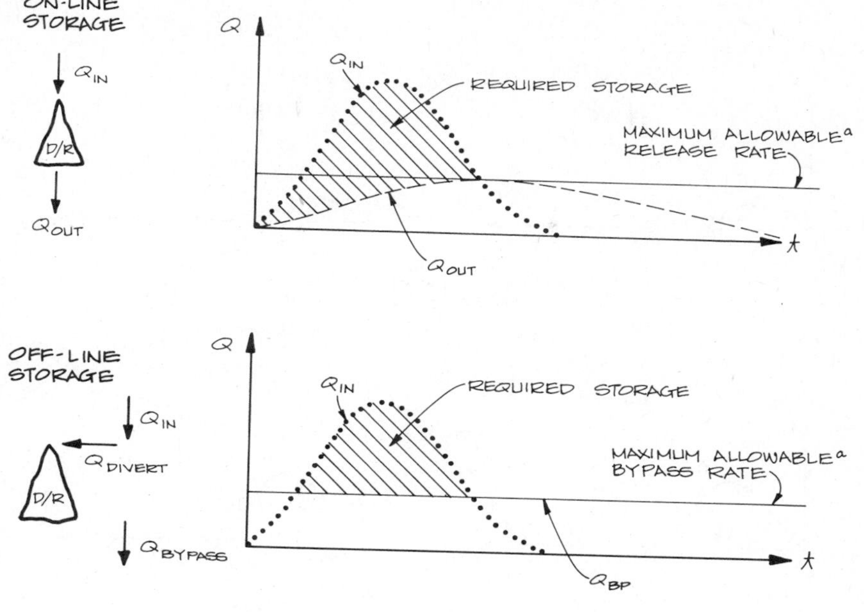

FIGURE 8.11 Inflow–outflow hydrographs for on-line and off-line detention/retention facilities

maximum allowable bypass rate (for an off-line facility). Maximum allowable release and bypass rates are illustrated in Figures 8.11 and 8.12.

Step C: Determine Storage Volumes. Route each hydrograph through the D/R facility while satisfying the appropriate release or bypass rate, and for each routing, determine the required storage volume. The manner in which the required storage volume will be obtained depends on the method used to generate and route the hydrograph. For example, if the MRM is used, the volume will be obtained from a tabulation of rainfall averaging period and corresponding storage volumes. If TR55 is utilized, storage volume will be taken from a graph in the SCS report. When computer models such as HEC-1 and ILUDRAIN are used, the required storage volume is read directly from the output. At this point, the process shown on Figure 8.12 will be completed in that pairs of release, or bypass rates, and storages will have been generated.

Step D: Construct Desired Storage versus Rate Graph. Use the release rates or bypass rates and the corresponding calculated storage to construct a storage versus release rate or a storage versus bypass rate graph as shown schematically in Figures 8.13 and 8.14. That is, graphically summarize the results of the hydrograph generation and routing process by plotting a storage–discharge

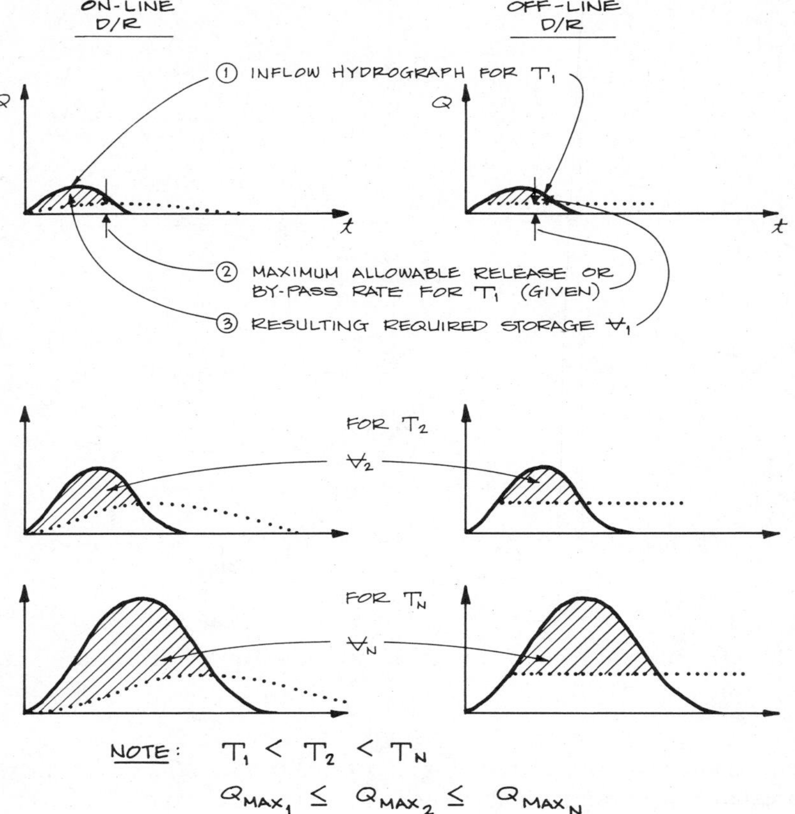

FIGURE 8.12 Process used to develop storage versus release rate or storage versus bypass rate relationship

point for each probability or recurrence interval. Figure 8.13 is a conceptual storage versus release or bypass rate relationship. Figure 8.14 is more specific in that it shows the general form of storage versus release or bypass rate relationships for two different release rate conditions.

Step E: Determine D/R Geometry and Outlet Hydraulics. Use the storage versus rate relationship to design the D/R facility final geometry and the outlet works hydraulics. If the geometry of the site is to remain fixed, such as when a D/R facility is formed by damming a natural valley, the required storage versus rate relationship must be achieved entirely through the hydraulic design of the outlet works. In contrast, if the geometry of the D/R site can be altered, such as when a relatively flat site is excavated to provide storage, the required storage versus rate relationship can be accomplished through the integrated design of site geometry and outlet works hydraulics.

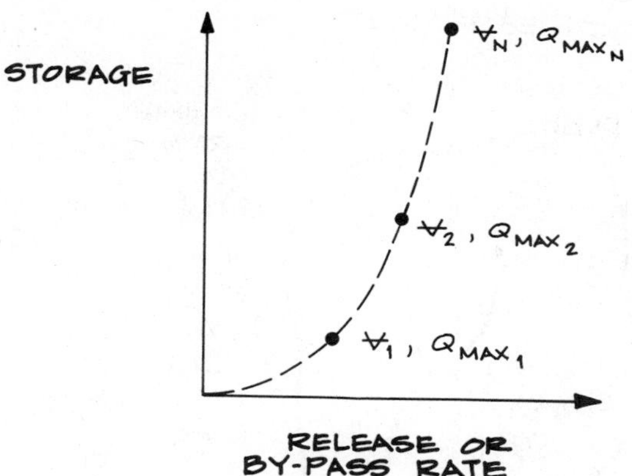

FIGURE 8.13 Conceptual storage versus release or bypass rate relationship

Three examples of the foregoing five-step process follow to reinforce the concepts and illustrate, in detail, the process and the calculations. TR55 is the hydrologic method used for the examples. However, any hydrologic method that generates runoff hydrographs and determines storage volumes could be used.

Example 1

The watershed and its basic hydrologic parameters are shown in Figure 8.15. Both predevelopment and future condition watershed data are pertinent to this example. A continuous probability condition is imposed. The on-line D/R facility must allow no increase in downstream flows under future conditions

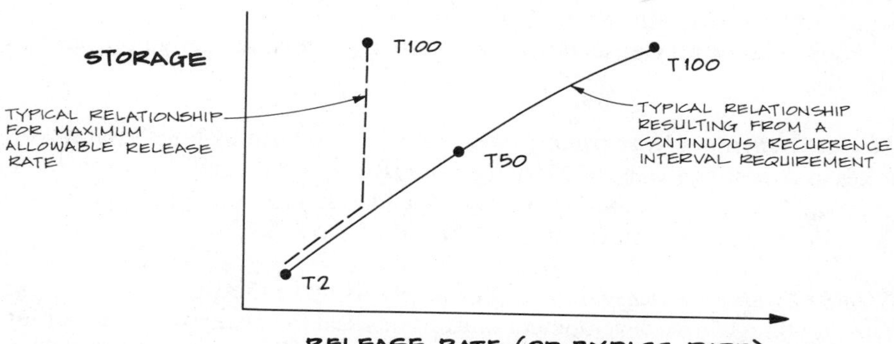

FIGURE 8.14 Typical storage versus release or bypass rate relationships

LOCATION: NORTHWESTERN INDIANA

AREA: 0.80 SQUARE MILES

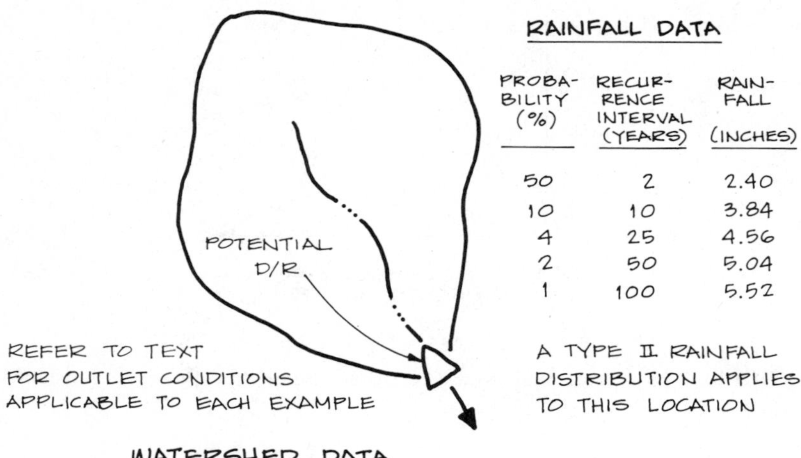

WATERSHED DATA

PARAMETER	EXISTING CONDITION	DEVELOPED CONDITION
RUNOFF CURVE NUMBER	77	84
TIME OF CONCENTRATION (HOURS)	1.7	0.9

FIGURE 8.15 Watershed used for Examples 1, 2, and 3

relative to predevelopment conditions for flood events up to and including the 1 percent event. A 24-hour design storm duration is used and the hydrology will be performed with the TR55 method.

Following the previously described five-step process, the design proceeds as follows:

Step A: Generate Hydrographs. Use TR55 to establish 24-hour rainfall volumes and existing (RCN = 77) and future (RCN = 84) condition runoff volumes and peak discharges. The process and results for existing and future conditions are shown in the step A portion of Table 8.2.

Step B: Establish Release Rates. For all events up to the 1 percent event, the peak discharge after development must not exceed the peak existing condition discharge. The target (i.e., existing condition) peak discharges for representative probabilities are shown in the step B portion of Table 8.2.

TABLE 8.2 Summary of Calculations for Examples 1 and 2 Using TR55

(1)	(2)	(3)	(4)	(5)	(6)	(7)	(8)	(9) Step B: Establish Release Rates	(10)	(11)	(12)	(13)
Design Rainfall Event		Step A: Generate Hydrographs							Step C: Determine Storage Volumes			
Probability[a] (percent)	Rainfall[a] (in.)	Runoff[b] (in.)	Time of Concentration[a] (hr)	Initial Abstraction[c] (in.)	(Initial Abstraction)/(Rainfall)[d]	Unit Peak Discharge[e,f]	Peak Discharge from Watershed without D/R[g] (ft³/sec)	Peak Discharge from Watershed with D/R in Place[h] (ft³/sec)	(Peak Outflow Discharge)/(Peak Inflow Discharge)[i]	(Storage Volume)/(Runoff Volume)[j]	Storage Volume[k] (in.)	(Acre-ft)
				Existing Conditions:								
50	2.40	0.68	1.7	0.597	0.25	220	120					
10	3.84	1.68	1.7	0.597	0.16	235	315					
4	4.56	2.26	1.7	0.597	0.13	240	435					
2	5.04	2.65	1.7	0.597	0.12	240	510					
1	5.52	3.06	1.7	0.597	0.11	240	590					
				Developed Conditions:								
50	2.40	1.04	0.9	0.381	0.16	370	310	120	0.39	0.325	0.338	14.4
10	3.84	2.24	0.9	0.381	0.10	380	680	315	0.46	0.290	0.650	27.7
4	4.56	2.87	0.9	0.381	0.08	380	870	435	0.50	0.275	0.789	33.7
2	5.04	3.31	0.9	0.381	0.08	380	1010	510	0.50	0.275	0.910	38.8
1	5.52	3.75	0.9	0.381	0.07	380	1140	590	0.52	0.270	1.013	43.2

[a]Given.
[b]From Table 3.6 using runoff curve number of 77 for existing conditions and runoff curve number of 84 for developed conditions.
[c]From Table 3.8 using runoff curve number of 77 for existing conditions and runoff curve number of 84 for developed conditions.
[d](Column 5)/(column 2).
[e]From Figure 3.17.
[f]Cubic feet per second per square mile per inch of runoff.
[g](Column 3) × (column 7) × (watershed area in square miles).
[h]Use column 8, existing conditions, because peak discharges after development must not exceed peak discharges before development.
[i](Column 9)/(column 8).
[j]From Figure 3.18 for type 2 rainfall.
[k](Column 11) × (column 3).

Step C: Determine Storage Volumes. TR55 is used to determine the volume of storage required for each of the five probabilities as a function of the release rate for each probability. The process and the results are shown in the step C portion of Table 8.2.

Step D: Construct Desired Storage versus Rate Graph. The prescribed release rates for each of the five probabilities and the corresponding calculated storages are used to prepare the storage versus release rate graph shown in Figure 8.16.

Step E: Determine D/R Geometry and Outlet Hydraulics. Assume that a detention facility will be placed at the watershed outlet in the natural valley shown in Figure 8.17. An earth berm will be constructed across the valley, resulting in the stage–volume relationship shown in Figure 8.18. Outlet control will be provided by a series of orifices in a concrete headwall, with the top of the headwall serving as a weir for control of the largest discharge.

Perform trial-and-error calculations to determine orifice positions and diameters and weir length and crest elevation.

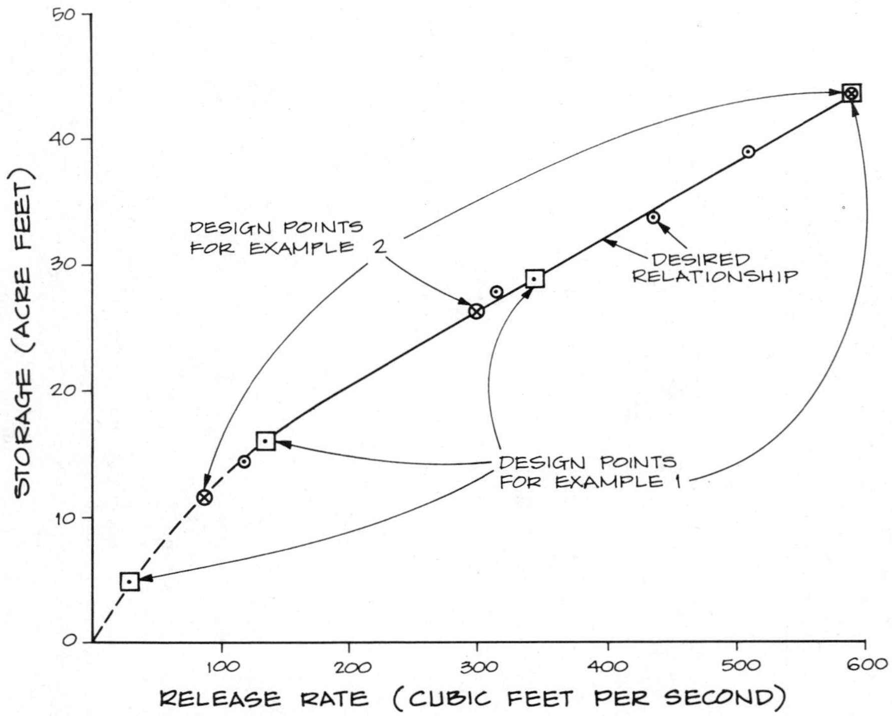

FIGURE 8.16 Storage versus release rate for Examples 1 and 2

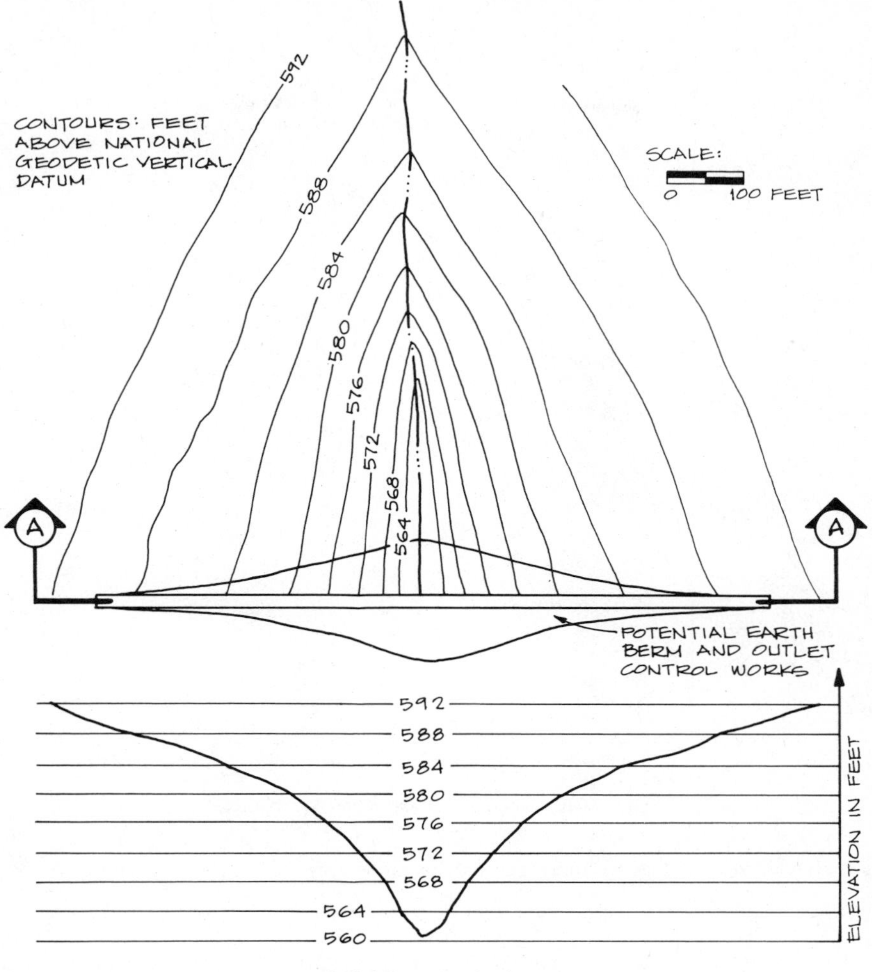

FIGURE 8.17 Detention site for Example 1

(a) The orifice equation as presented in Chapter 5 is

$$Q = C_d A(2gh)^{1/2} = \frac{C_d \pi D^2 (2gh)^{1/2}}{4}$$

Substitute $C_d = 0.8$ for a sharp-edged orifice in a concrete headwall and then $Q = (5.04)(D^2)(h^{1/2})$.

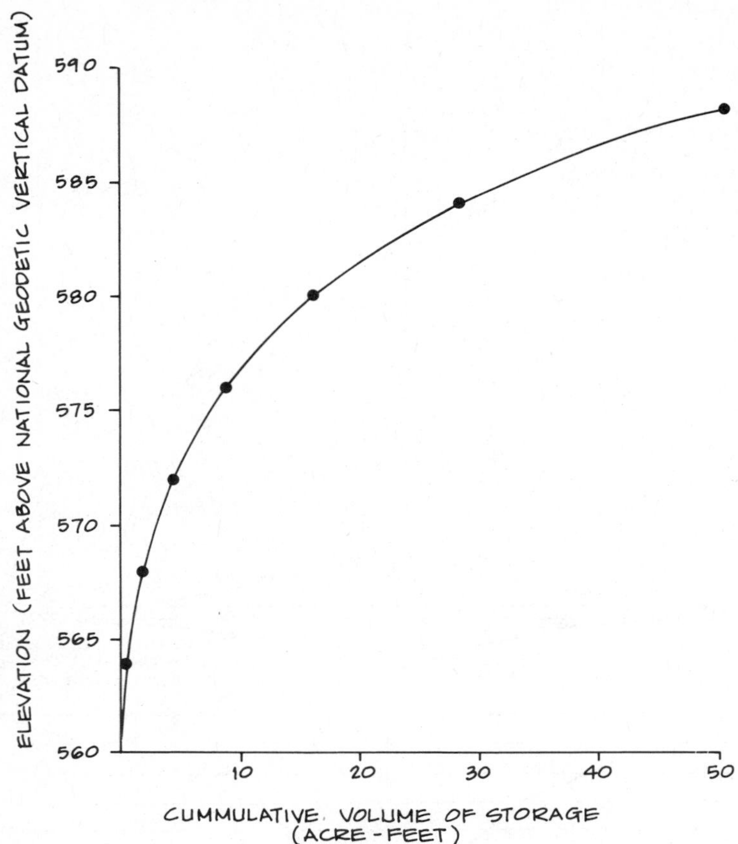

FIGURE 8.18 Stage–volume relationship for detention site used in Example 1

(b) Set the stage at an elevation of 572.0 ft National Geodetic Vertical Datum (NGVD). Enter Figure 8.18 with this stage and obtain $V = 5.0$ acre-ft. For this volume, Figure 8.16 indicates that the desired release rate is 30 ft^3/sec.

Try one orifice at the bottom of the facility. For $Q = 30$ ft^3/sec and $h =$ 12 ft, calculate $D = 1.72$ ft. This is acceptable. *Therefore, place one 1.72-ft-diameter orifice at the bottom of the outlet control with its center at elevation 560 ft NGVD, as shown in Figure 8.19. A similar-sized orifice* equipped with a sluice gate and normally kept closed might be placed at the bottom of the outlet control to provide for drainage of the facility if the open orifice becomes blocked with debris.

(c) Set the stage at elevation 580.0 ft NGVD. Enter Figure 8.18 with this stage and obtain $V = 16$ acre-ft. For this volume, Figure 8.16 indicates that the desired release rate is 135 ft^3/sec.

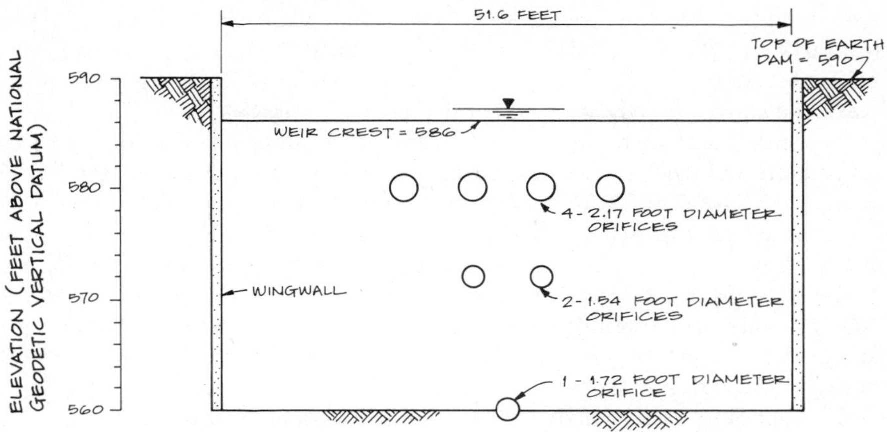

FIGURE 8.19 Hydraulic features of outlet control designed for Example 1

At a stage of 580.0 NGVD, the single orifice at the bottom of the detention facility will pass $Q = (5.04)(1.72^2)(20^{1/2}) = 66.7$ ft^3/sec. This is 135 − 66.7 = 68.3 ft^3/sec less than required when the stage is at 580.0 ft NGVD.

Therefore, try one orifice at elevation 572.0 ft NGVD. For $Q = 66.7$ ft^3/sec and $h = 8.0$ ft, calculate $D = 2.18$ ft. This may be too large. Try two orifices at elevation 572.0 ft NGVD. For $Q = 66.7/2 = 33.3$ ft^3/sec per orifice, calculate $D = 1.54$ ft. This is acceptable. *Therefore, place two 1.54-ft-diameter orifices at elevation 572.0 ft NGVD, as shown in Figure 8.19.*

(d) Set the stage at elevation 584.0 ft NGVD. Enter Figure 8.18 with this stage and obtain $V = 28.5$ acre-ft. For this volume, Figure 8.16 indicates that the desired release rate is 345 ft^3/sec.

At a stage of 584.0 ft NGVD, the single orifice at the bottom of the detention facility will pass $Q = (5.04)(1.72^2)(24^{1/2}) = 73.0$ ft^3/sec. The two orifices at elevation 572.0 ft NGVD will discharge $Q = (2)(5.04)(1.54^2)(12^{1/2}) = 82.8$ ft^3/sec. The total discharge through the three orifices is 345 − 73.0 − 82.8 = 189.2 ft^3/sec less than required when the stage is at elevation 584 ft NGVD.

Therefore, try two orifices at elevation 580.0 ft NGVD. For $Q = 189.2/2 = 94.6$ ft^3/sec per orifice, calculate $D = 3.06$ ft. This is too large. Try four orifices at elevation 580 ft NGVD. For $Q = 189.2/4 = 47.3$ ft^3/sec per orifice, calculate $D = 2.17$ ft. This is acceptable. *Therefore, place four 2.17-ft-diameter orifices at elevation 580.0 ft NGVD, as shown in Figure 8.19.*

(e) Set the stage at 587.0 ft NGVD. This stage corresponds to the 100-year-recurrence-interval volume of 43.2 acre-ft and discharge of 590 ft^3/sec, as indicated by Figure 8.16.

At the stage of 587.0 ft NGVD, the single orifice at the bottom of the detention facility will pass $Q = (5.04)(1.72^2)(27^{1/2}) = 77.4$ ft^3/sec. The two orifices at elevation 572.0 ft NGVD will discharge $(2)(5.04)(1.54^2)(15^{1/2}) = 92.6$ ft^3/

sec. Finally, the four orifices at elevation 580.0 ft NGVD will pass $(4)(5.04)(2.17^2)(7^{1/2}) = 251.2$ ft^3/sec. The total discharge through the seven orifices is $590.0 - 77.4 - 92.6 - 251.2 = 168.8$ ft^3/sec less than required when the stage is at elevation 587.0 ft NGVD.

Therefore, size a weir at the crest of the headwall to provide for the necessary additional discharge. The weir equation as presented in Chapter 5 is $Q = C_w L h^{1.5}$. Use $C_w = 3.27$ for a high, relative to head, sharp-crested weir.

Assume that the weir crest at elevation 586.5 ft NGVD. Substitute $Q = 168.8$ ft^3/sec, $C_w = 3.27$, and $h = 0.5$ ft in the weir equation and calculate $L = 146$ ft. This is too long. Assume that the weir crest is at elevation 586.0 ft NGVD, which corresponds to $h = 1.0$ ft and compute $L = 51.6$ ft. This is acceptable. *Therefore, place a 51.6-ft-long weir at elevation 586.0 ft NGVD, as shown in Figure 8.19.*

Design points used to arrive at the resulting outlet control structure as illustrated in Figure 8.19 are shown on the required storage versus release rate relationship in Figure 8.16.

Figure 8.19 presents the necessary hydraulic features—orifice and weir locations and sizes—of the outlet works. The construction materials and physical configuration used to achieve the desired hydraulic features would depend on site conditions. A reinforced-concrete headwall might be used, as suggested by Figure 8.19, with adequate provision for energy dissipation and erosion protection immediately downstream of the structure. Orifices might be rectangular rather than circular, for constructability purposes. As an alternative, the multiple-orifice effect might be achieved by using a series of hydraulically short conduits, probably corrugated metal pipes, passing through an earthen embankment as shown schematically in Figure 8.20. The pipes would be provided with seepage collars and discharged at the toe of the earthen embankment.

Example 2

This example is the same as Example 1 except for the nature of the detention facility. In Example 1, the detention facility was formed by building an earth

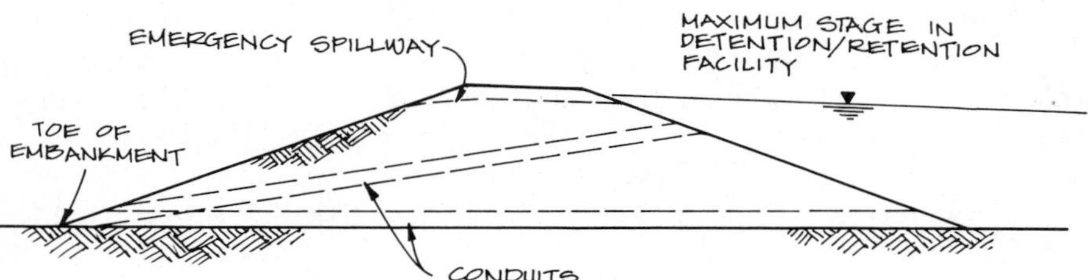

FIGURE 8.20 Concept of using multiple conduits in an earthen dam to achieve desired hydraulic results

dam across a small natural valley. Example 2 uses an excavated detention facility. Accordingly, steps 1 through 4 from Example 1 are completely applicable to Example 2. Similarly, Figures 8.15 and 8.16 and Table 8.2 are entirely applicable to Example 2. The two examples differ beginning with step E.

Step E: Determine D/R Geometry and Outlet Hydraulics. Assume that the D/R facility will be a rectangular excavation and that a series of orifices will be used in the outlet with an emergency spillway at the stage of the 1 percent storage volume.

Assume further that the rectangular basin has sloped side walls (4 horizontal to 1 vertical) and a maximum depth of 12 ft when storing the 100-year volume of 43.2 acre-ft (1,881,792 ft^3). Therefore, the plan area of the D/R facility is approximately $43.2/12 = 3.6$ acres.

Use a plan proportion of 3 to 1 with the long axis being along the existing drainageway. Note that this is one of an infinite number of D/R configurations that could be used. Refer to Figure 8.21 for D/R facility geometry and equations.

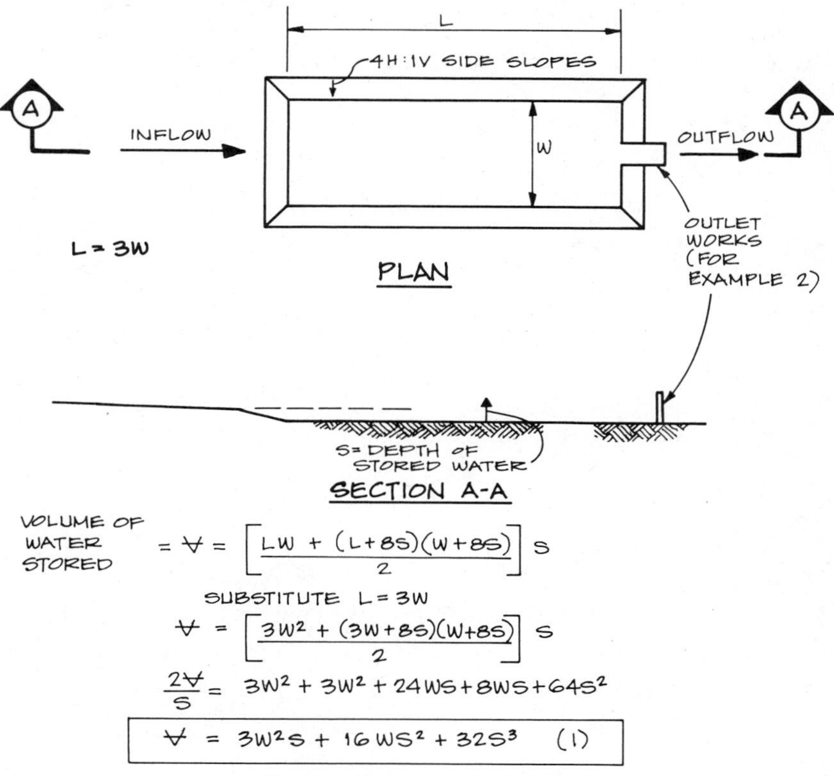

FIGURE 8.21 Geometry and equation for Examples 1 and 2

Substitute $V = 1{,}881{,}792\ \text{ft}^3$ and $S = 12$ ft into equation (1) of Figure 8.21 and obtain $W = 195.5$ ft. Therefore, $L = 3W = 586.5$ ft and bottom area = 2.63 acres.

Substitute $W = 195.5$ ft into equation (1) and obtain $V = 114{,}661S + 3128S^2 + 32S^3$. This volume–stage relationship is equation (2) for this example and is analogous to Figure 8.18 in Example 1.

Perform a trial-and-error calculation of orifice positions and orifice diameters. Orifice position is defined relative to the bottom of the D/R facility.

(a) Select a stage of 4.0 ft. Substitute $S = 4.0$ ft in equation (2) and obtain $V = 11.7$ acre-ft. For this volume, Figure 8.16 indicates that the desired release rate is 90 ft^3/sec.

(b) Size one orifice at the bottom of the D/R facility. For $Q = 90$ ft^3/sec and $h = 4.0$ ft, calculate $D = 2.98$ ft. This is too large compared to the head.

Size three orifices at the bottom of the D/R facility. For $Q = 90/3 = 30.0$ ft^3/sec per orifice and $h = 4.0$ ft, calculate $D = 1.72$ ft. This is acceptable. *Therefore, place three 1.72-ft-diameter orifices at $S = 0.0$ ft, as shown in Figure 8.22.*

(c) Select a stage of 8.0 ft. Substitute $S = 8$ ft in equation (2) and obtain $V = 26.0$ acre-ft. For this volume, Figure 8.16 indicates that the desired release rate is 300 ft^3/sec.

At $S = 8.0$ ft, the three orifices at the bottom of the D/R facility will pass

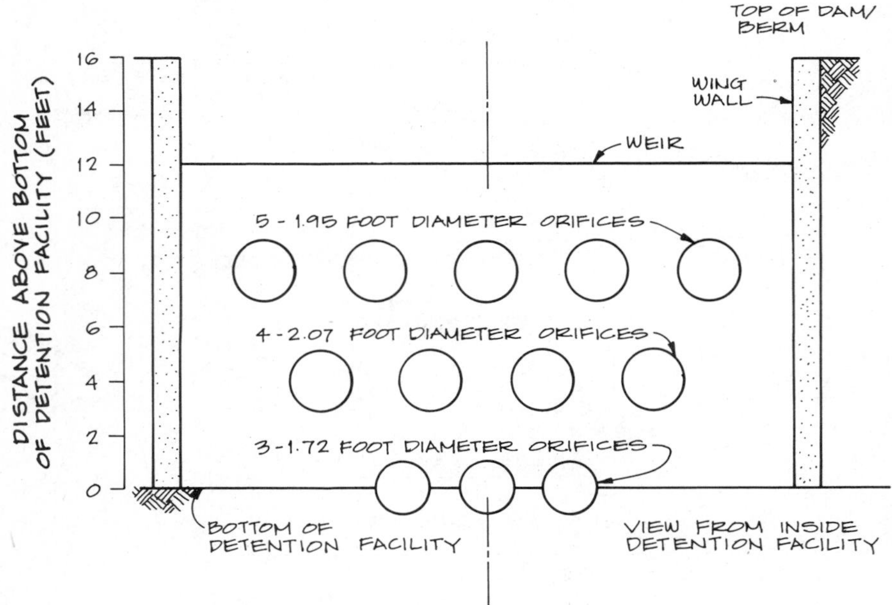

FIGURE 8.22 Outlet control designed for Example 2

$Q = (3)(5.04)(1.72^2)(8^{1/2}) = 127$ ft³/sec. This is $300 - 127 = 173$ ft³/sec less than required.

Therefore, place one or more orifices at $S = 4.0$ ft, which will convey 173 ft³/sec when the stage is at $S = 8.0$ ft. Size four orifices at $S = 4.0$ ft. For $Q = 173/4 = 43.3$ ft³/sec and $h = 4.0$ ft, calculate $D = 2.07$ ft. This is acceptable. *Therefore, place four 2.07-ft-diameter orifices at S = 4.0 ft, as shown in Figure 8.22.*

(d) Select a stage of 12.0 ft. At $S = 12.0$ ft the detention facility will be full and storing 43.2 acre-ft, and the desired release rate is 590 ft³/sec.

At $S = 12.0$ ft, the three orifices at the bottom of the D/R facility will pass $Q = (3)(5.04)(1.72^2)(12^{1/2}) = 155$ ft³/sec. Also at $S = 12.0$ ft, the four orifices at $S = 4.0$ ft will pass $Q = (4)(5.04)(2.07^2)(8^{1/2}) = 244$ ft³/sec. Therefore, at $S = 12.0$ ft the total discharge is $155 + 244 = 399$ ft³/sec. This is $590 - 399 = 191$ ft³ less than desired.

Therefore, place one or more orifices at $S = 8.0$ ft, which will convey 191 ft³/sec when the stage is at $S = 12.0$ ft. Size five orifices at $S = 8.0$ ft. For $Q = 191/5 = 38$ ft³/sec and $h = 4.0$ ft, calculate $D = 1.95$ ft. This is acceptable. *Therefore, place five 1.95-ft-diameter orifices at S = 8.0 ft, as shown in Figure 8.22.*

Design points used to arrive at the resulting outlet control structure as illustrated in Figure 8.22 are shown on the required storage versus release rate relationship in Figure 8.16.

Example 3

The watershed and its basic hydrologic parameters are the same as for Examples 1 and 2 as presented in Figure 8.15. Only the future watershed conditions are pertinent to this example. The potential on-line D/R facility must control all floods up to the 1 percent event such that the maximum discharge is 150 ft³/sec per second. The desired storage versus release rate relationship is shown in Figure 8.23. A 24-hour design storm duration is used and the hydrology will be performed with the TR55 method.

Following the previously described five steps the design proceeds as follows:

Step A: Generate Hydrographs. Use TR55 to establish future condition (RCN = 84) 24-hour rainfall volumes and runoff volumes for a range of probabilities. The results are shown in columns 3 through 8 of Table 8.3.

Step B: Establish Release Rate. The potential D/R facility must control all floods up to and including the 1 percent event such that the maximum discharge is 150 ft³/sec.

Step C: Determine Storage Volumes. Use TR55, with a release rate of 150 ft³/sec, to determine storage volumes for the selected range of probabilities. The process and results are shown in columns 10 through 13 of Table 8.3.

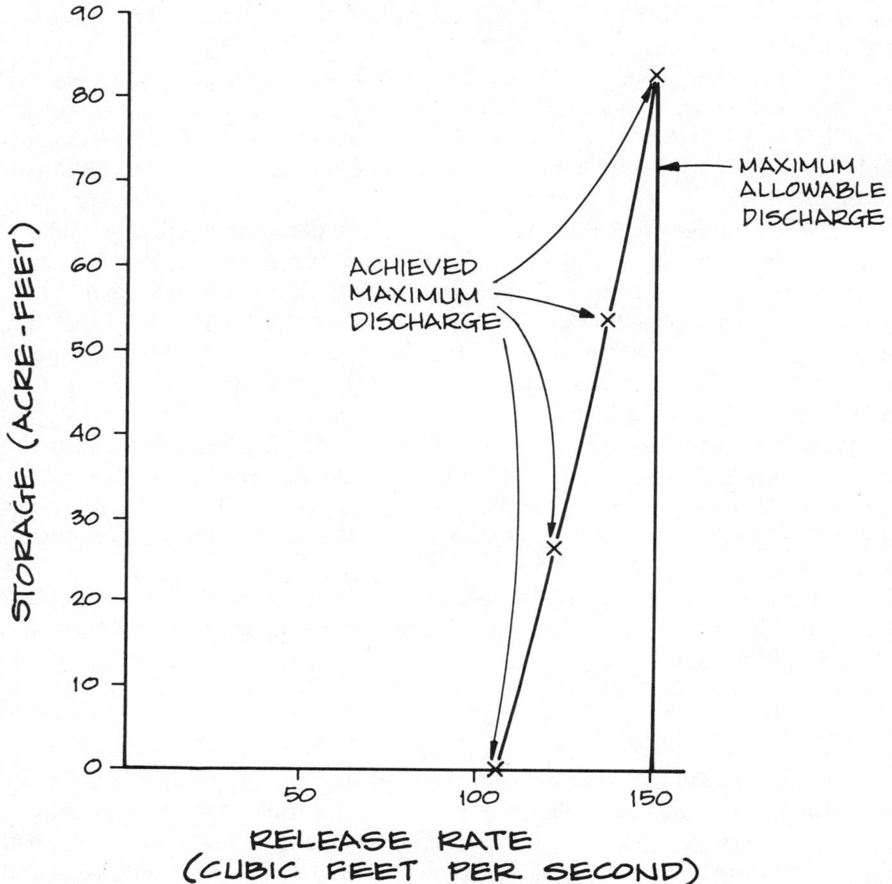

FIGURE 8.23 Storage versus release rate for Example 3

Step D: Construct Desired Storage versus Rate Graph. Because of the prescribed 150 ft^3/sec maximum release rate for events up to and including the 1 percent event, the graph is the simple linear relationship shown on Figure 8.23.

Step E: Determine D/R Geometry and Outlet Hydraulics. This is a geometric and hydraulic challenge because, in a strict sense, the outlet control structure must maintain a uniform flow over a range of storage and, therefore, stage. Assume that some form of orifice will be used with an emergency spillway at the stage of the 100-year storage volume.

To minimize the stage range, a large (in plan), shallow D/R facility should be used in contrast with a small, deep facility. Assume a rectangular basin with sloped sidewalls (4 horizontal to 1 vertical) and a maximum depth of 6.0 ft

TABLE 8.3 Summary of Calculations for Example 3 Using TR55

(1)	(2)	(3)	(4)	(5)	(6)	(7)	(8)	(9)	(10)	(11)	(12)	(13)
Design Rainfall Event		Step A: Generate Hydrographs						Step B: Establish Release Rates	Step C: Determine Storage Volumes			
Probability[a] (percent)	Rainfall[a] (in.)	Runoff[b] (in.)	Time of Concentration[a] (hr)	Initial Abstraction[c] (in.)	(Initial Abstraction)/(Rainfall)[d]	Unit Peak Discharge[e,f]	Peak Discharge from Watershed without D/R[g] (ft³/sec)	Peak Discharge from Watershed with D/R in Place[h] (ft³/sec)	(Peak Outflow Discharge)/(Peak Inflow Discharge)[h]	(Storage Volume)/(Runoff Volume)[i]	Storage Volume[j] (in.)	(Acre-ft)
				Developed Conditions:								
50	2.40	1.04	0.9	0.381	0.16	370	310	150	0.48	0.285	0.296	12.6
10	3.84	2.24	0.9	0.381	0.10	380	680	150	0.22	0.435	0.974	41.6
4	4.56	2.87	0.9	0.381	0.08	380	870	150	0.17	0.480	1.378	58.8
2	5.04	3.31	0.9	0.381	0.08	380	1010	150	0.15	0.500	1.655	70.6
1	5.52	3.75	0.9	0.381	0.07	380	1140	150	0.13	0.520	1.950	83.2

[a]Given.
[b]From Table 3.6 using runoff curve number of 84 for developed conditions.
[c]From Table 3.8 using runoff curve number of 84 for developed conditions.
[d](Column 5)/(column 2).
[e]From Figure 3.17.
[f]Cubic feet per second per square mile per inch of runoff.
[g](Column 3) × (column 7) × (watershed area in square miles).
[h](Column 9)/(column 8).
[i]From Figure 3.18 for type 2 rainfall.
[j](Column 11) × (column 3).

when storing the 100-year volume of 83.2 acre-ft. Therefore, the plan area of the D/R facility is approximately 83.2/6 = 13.8 acres. Use plan proportions of 3 to 1 with the long axis being along the existing drainageway. Refer to Figure 8.21 for D/R facility geometry and equations. Note that this is one of an infinite number of D/R configurations that could be used.

To "move up" on the typical stage–discharge relationship for an orifice, and thus decrease flow sensitivity to stage, depress the outlet orifice well below the bottom of the D/R facility. The concept is shown in Figure 8.24. Another option would be to use one or more commercial or specially fabricated flow regulators. Flow regulators might produce the desired storage–volume relationship in lieu of the depressed outlet orifice.

Substitute V = 3,624,192 ft^3 (equivalent to 83.2 acre-ft) and S = 6 ft into equation (1) of Figure 8.21 and obtain W = 433 ft. Therefore, $3W$ = 1299 ft and bottom area = 12.9 acres. Substitute W = 433 ft into equation (1) and obtain $V = 562{,}467S + 6928S^2 + 32S^3$, which is equation (2) for this example.

Perform a trial-and-error calculation of orifice position, in a vertical sense relative to the bottom of the D/R facility, and diameter.

(a) Assume that k = 4.0 ft. Then h = 10 ft for T = 100 and Q_{100} = 150 ft^3/sec. Calculate D = 3.07 using the orifice equation.

(b) Determine Q when the D/R facility storage drops to zero. Then h = 4.0 ft and $Q = (5.04)(3.07^2)(4.0^{1/2}) = 95.0$ ft^3/sec. This is too small compared to 150 ft^3/sec; therefore, increase k.

(c) Assume that k = 6.0 ft. Then h = 12 ft for T = 100 and Q_{100} = 150 ft^3/sec. Calculate D = 2.93 ft.

(d) Determine Q when storage drops to zero. Then h = 6.0 ft and $Q = (5.04)(2.93^2)(6.0^{1/2}) = 106$ ft^3/sec. This is acceptable compared to 150 ft^3/sec.

(3) Calculate Q at two intermediate stages. At S = 2.0 ft, h = 8.0 ft and $Q = (5.04)(2.93^2)(8^{1/2}) = 122$ ft^3/sec. Also, at S = 2.0 ft, V = 26.5 acre-ft. At S = 4.0 ft, h = 10.0 ft and $Q = (5.04)(2.93^2)(10^{1/2}) = 137$ ft^3/sec. Also, at S = 4.0 ft, V = 54.2 acre-ft.

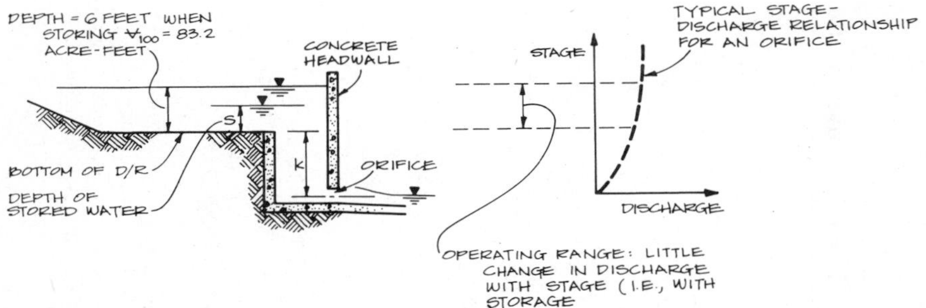

FIGURE 8.24 Depressed outlet orifice for Example 3

Refer to Figure 8.23 for a comparison of the achieved storage–release rate relationship to the desired relationship.

Besides illustrating the continuous probability approach to the design of D/R facilities, the three examples begin to suggest the variety of outlet works configurations potentially available to satisfy particular situations. Additional examples of specially designed outlet works configurations, including roof drain controls, are provided by American Society of Civil Engineers (1985) and American Public Works Association (1974, 1981).

8.9 STEP 7: FINALIZE THE DESIGN

Chapter 5 identifies and discusses analysis and design guidelines many of which are applicable to the final design of D/R facilities. Included are important areas such as gravity operation, backwater effects, safety, aesthetics, low-flow control, antiplugging measures, vortex protection, antivandalism provisions, drainage of recreation areas, maximum slopes, erosion and sedimentation control, and inspection and maintenance. These guidelines should be considered explicitly at this point in the design process because they help to identify design details.

To illustrate the importance of attending to final design details, refer to Figure 8.25, which shows the on-site drainage for the previously cited fairgrounds flood control–recreation project. Rapid drainage and drying of recreation areas was an important aspect of the multiple-purpose project. Accordingly, very close attention was given to the hydraulic function of the facility. The diversion structure on Smith Ditch will divert all flows above 60 ft^3/sec into the detention facility. On-site drainage is shown in Figure 8.25. Flows of less than 6 ft^3/sec which enter the facility from the north will be carried to the outlet control structure in an 18-in.-diameter southerly flowing sewer unobtrusively located along the eastern edge of the facility. Flows below 6 ft^3/sec from the west will bypass the facility and go directly to Smith Ditch. Excess flows from the west will be diverted into storage temporarily.

An extensive network of 4- to 18-in.-diameter corrugated perforated polyethylene pipe underdrains and sewers provides subsurface drainage of all recreation areas. All recreation surfaces have a surface slope of at least 1 percent to encourage rapid drainage after rainfall or flooding. Because of the careful attention given to grading and subsurface drainage, even the lower recreation facilities on this site will probably be available for use more often than other single-purpose recreation facilities in the community.

8.10 STEP 8: BEGIN IMPLEMENTATION

As a result of being intimately involved in the planning and design process, members of the planning and design team are often in a very good position to

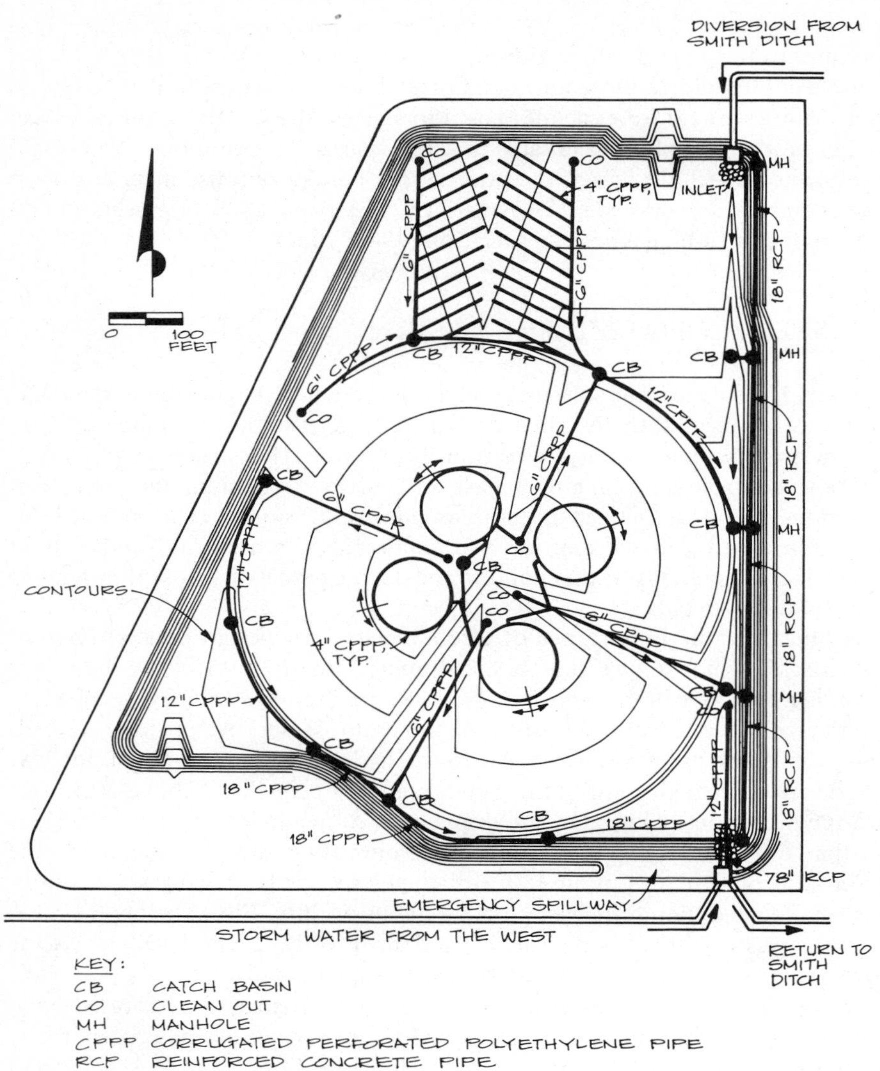

FIGURE 8.25 On-site drainage for the fairgrounds flood control–recreation project (*Source:* Adapted from Walesh et al., 1987, 1988)

make suggestions and take actions that will lead to the early construction of the designed facilities. Besides being very knowledgeable about all aspects of the D/R facility, design team members should have become familiar with related problems and needs within the watershed and within the larger community. There may be an opportunity to solve some of these problems and satisfy some of these needs while advancing the implementation of the designed D/R facilities.

Specific creative means of implementing D/R facility projects are highly dependent on local conditions and, therefore, cannot be enumerated. However, the range and types of possibilities may be suggested by noting the efforts that enabled early implementation of the fairgrounds multiple-purpose flood control–recreation facility in Valparaiso, Indiana (Walesh et al., 1987, 1988). These efforts, most of which are illustrated in Figure 8.26 and took place in 1985, are:

1. During the planning process, the solution to a major flooding problem was tied to satisfying a major recreation need by planning and designing a multipurpose recreation–flood control project. The multipurpose nature of the project subsequently engendered widespread, sustained community support because many more citizens would benefit from a multipurpose project than from a single-purpose project.

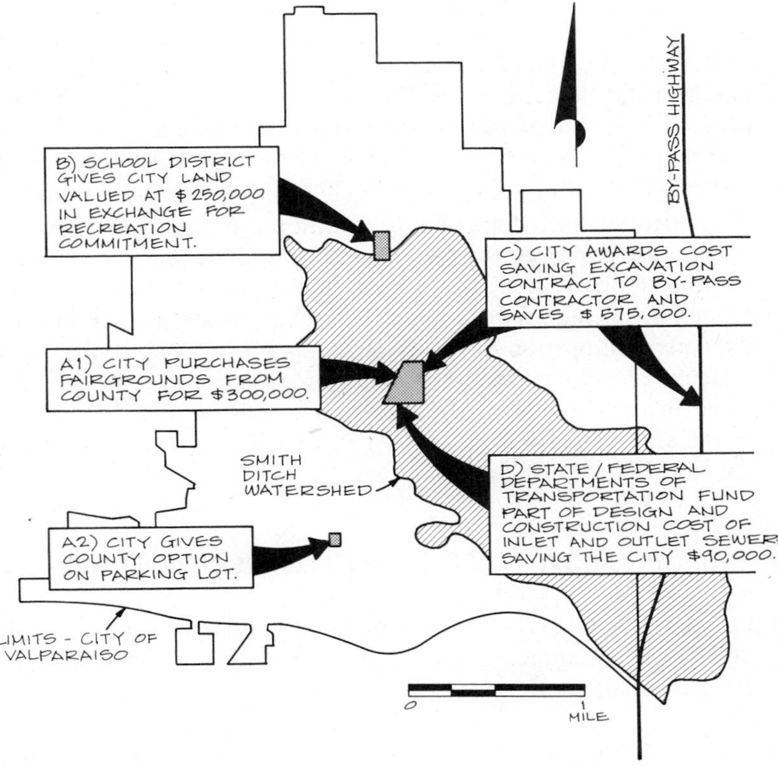

FIGURE 8.26 Financing the fairgrounds flood control–recreation project (*Source:* Adapted from Walesh et al., 1987, 1988)

2. The city obtained the 27-acre old fairgrounds site from the county, the considerations being cash (A1 in Figure 8.26) and an option on a parcel of city-owned land that the county needed for a parking lot within the city (A2 in Figure 8.26). As soon as the city obtained ownership of the fairgrounds site, the city entered into an agreement with the area-wide school district, in which the district would get use of recreation facilities to be developed on the fairgrounds site and elsewhere (B in Figure 8.26). In return, the school district gave the city 11 acres of prime land it was holding elsewhere in the city for future school-related recreation development (B in Figure 8.26). At the time of the transaction, the land was valued at $250,000 and could be sold by the city. The effect of this exchange was to offset most of the cost incurred by the city in purchasing the fairgrounds inasmuch as the city could sell the land acquired from the school district as a means of recovering the fairgrounds purchase price or other costs as may be required to provide the additional recreation services.
3. Construction was beginning on a major highway project 1.6 mi from the recreation–flood control project, as shown in Figure 8.27, thus creating a demand for embankment fill material. As a result, the city took the initiative and negotiated an agreement with a highway project contractor in which the city paid only $1.02 per cubic yard to have 180,500 cubic yards of material excavated and hauled from the site (C in Figure 8.26). The savings to the city was $574,000 based on the prevailing excavation and haul rate of approximately $4.20 per cubic yard.
4. Coincident with the fairgrounds detention project, the city was working with the Federal Highway Administration and the Indiana Department of Highways on improvements to roadways contiguous with the fairgrounds, which are partly under federal and state jurisdiction. Because the detention portion of the fairgrounds project would help to control surface water runoff from the roadways, the federal government participated in the funding of the design and construction of outlet control works and related sewers, resulting in additional savings of $90,000 to the city (D in Figure 8.26).

8.11 MULTIPLE DETENTION/RETENTION FACILITIES

Multiple D/R facilities refers to two or more D/R in a watershed. Multiple D/R facilities are becoming common in some parts of the United States for one of two reasons:

1. Detention or retention is mandated and therefore numerous D/R facilities are being constructed, often on a development-by-development ba-

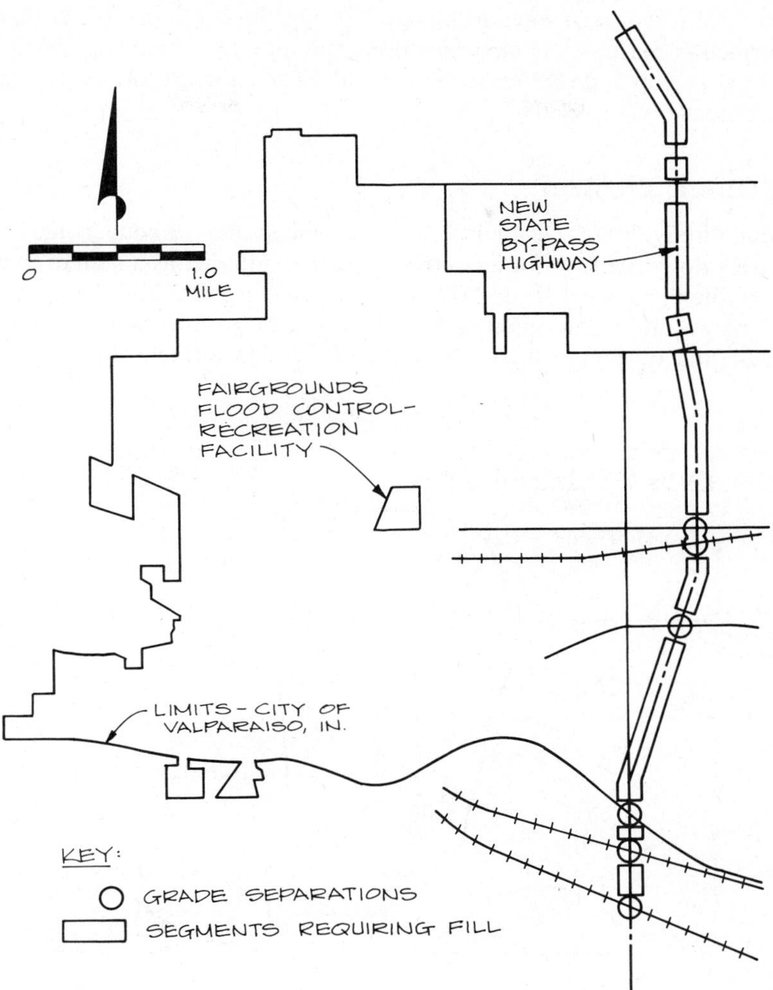

FIGURE 8.27 Highway construction project coincident with the fairgrounds flood control–recreation project (*Source:* Adapted from Walesh et al., 1987, 1988)

sis. That is, the planning and design of the D/R facilities often, if not usually, does not include consideration of the probable interactions between two or more D/R facilities in a watershed.

2. Surface water management systems are being planned and constructed on their merits for individual developments in an urbanizing watershed, and sometimes the result is multiple, uncoordinated D/R facilities.

The proliferation of uncoordinated D/R facilities is cause for concern for two principal reasons: (1) unexpected and potentially damaging timing problems can develop and (2) local officials and area residents may develop a false sense of security.

The Timing Problem

Although individual D/R facilities in a watershed may provide local (i.e., immediately downstream) flood control, they may interact as a system to aggravate flooding problems around the watershed. Figure 8.28 illustrates schematically one way in which one new D/R facility can produce a timing problem. As suggested by Figure 8.28, construction of a D/R facility on a tributary

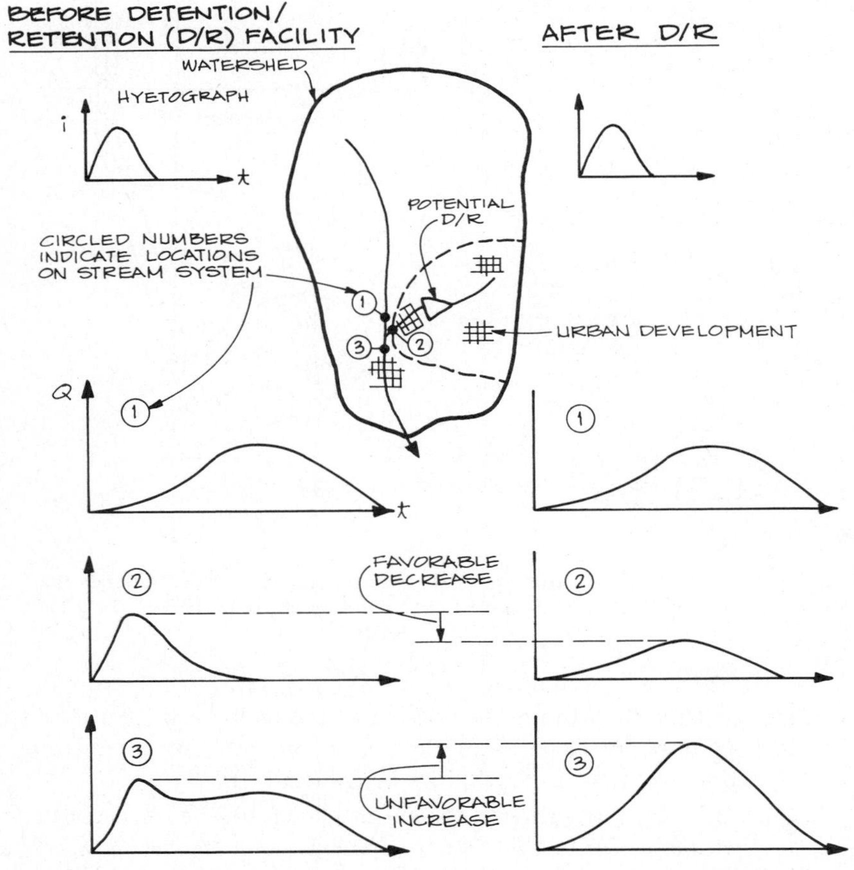

FIGURE 8.28 Adverse interaction of hydrographs as a result of detention/retention facility

could have the favorable effect of mitigating flooding on that tributary while having the unfavorable effect of causing or aggravating flooding downstream on the principal stream.

As another example of the potential timing problem, consider McCuen's (1979) computer modeling study of the downstream effect of a retention facility constructed at the outlet of a 0.07-mi^2 subbasin located about halfway up a 2.12-mi^2 watershed. The retention facility was designed to limit the 10-year-recurrence-interval discharge after development of the subbasin to the pre-development level. However, as a result of changes in timing of flow from the subbasin and the single-recurrence-interval design, downstream flood flows will increase rather then remain unchanged. For example, a 1-mi-long reach of the main channel in the watershed will have 100-year-recurrence-interval flood flows after development of the subbasin with retention that are greater then before development without the retention.

Mein (1980) outlines a D/R facility design procedure that keys in on maximum allowable discharge as the principal design objective and recommends use of a range of storm durations. He concludes that one D/R facility is more effective in controlling discharge than two in series, which, in turn, are more effective than three in series, and so on.

The False Sense of Security Problem

A system of D/R facilities in which each facility is sized on the basis of the single-probability criterion may have no significant effect on reducing flows from much larger or much smaller runoff events. This is one aspect of the "false sense of security" problem. That is, a system of one or more D/R facilities in a watershed may, by design, effectively control only a very small portion of the probability spectrum, leaving watershed areas vulnerable under other flood conditions.

Mein (1980) indicates that although a set of D/R facilities in series and parallel are not likely to increase significantly the peak discharge from a watershed, they may also interact so as not to reduce significantly the peak discharge from the watershed. This is another aspect of the "false sense of security" problem. That is, in a system of D/R facilities, the facilities may unexpectedly interact to nullify each other for certain downstream areas of the watershed.

Avoiding the Problems

There is no simple, widely applicable method for avoiding the two problems discussed above, which can occur when multiple D/R facilities are used. If there is concern with the potential adverse effects of one or more D/R facilities, a watershed-wide system analysis of all existing or planned D/R facilities is clearly required.

8.12 NEGATIVE ASPECTS AND MISUSE OF DETENTION/ RETENTION FACILITIES

D/R facilities are one of many structural and nonstructural measures available for managing the quantity and quality of urban storm water runoff. They are not a panacea, and mandatory use of D/R facilities instead of the application of engineering design and judgment is likely to lead to numerous and varied problems.

Debo and Ruby (1982) discuss the negative aspects of D/R facilities based on about 10 years of experience in the Atlanta metropolitan area. Problems cited include:

1. Improper maintenance or lack of maintenance, particularly when facilities are the responsibility of adjacent or nearby landowners. Possible solution: assign maintenance responsibility to municipalities, except possibly in those cases where commercial or industrial developments could handle the responsibility adequately.
2. Difficult access for maintenance purposes. Possible solution: facilities should be contiguous with public land or right-of-way, or an easement should be obtained.
3. Steep banks, causing difficulty in establishing and maintaining vegetation.
4. Outlet structures that are so small that they are prone to clogging, resulting in improper operation and stagnant water pools subsequent to operation during storm events.
5. Weed control problems, particularly in facilities that have permanent pools by design or as a result of poor maintenance.
6. Mosquito breeding, rodents, and other pests.
7. Safety hazards, particularly to children, created by great depth and high velocity.
8. Absence of downstream flood control or actually increasing downstream flood problems as a result of the unexpected adverse interaction between two or more basins in a watershed. Solution: watershed-based planning and design.
9. Erosion problems immediately below the outlet structures. Possible solution: adequate armoring.
10. Increased stream channel erosion, possibly as a result of longer flow time within the channel as compared to pre-D/R facility conditions.
11. Little or no effect on floods other than those for which the facility was designed.
12. "Wasteful" placing of the facilities near the outlet of areas that drain directly into lakes, rivers, or other surface water bodies not affected by

increased runoff in small tributary areas. Facilities were installed only to comply with local regulations.

13. Cost in terms of significant amount of land not available for development, coupled with high and continuous maintenance costs.
14. Missed opportunities for downstream flood control. That is, many facilities are, according to regulations, designed not to increase downstream flood flows but do not facilitate decreasing downstream flood flows, particularly in flood-prone areas. The solution: watershed-wide analysis of alternatives.

REFERENCES

American Public Works Association, *Practices in Detention of Urban Stormwater Runoff,* Spec. Rep. No. 43. APWA, Chicago, IL, 1974.

American Public Works Association, *Urban Stormwater Management,* Spec. Rep. No. 49. APWA, Chicago, IL, 1981.

American Society of Civil Engineers, *Stormwater Detention Outlet Control Structures.* Report of the Task Committee on the Design of Outlet Control Structures of the Committee on Hydraulic Structures of the Hydraulics Division, 1985.

Debo, T. N., and H. Ruby, "Detention Basins—An Urban Experience." *Public Works,* January, pp. 42–43 (1982).

Donohue & Associates, Inc., *Upper Hahns Creek Watershed—Detention/Retention Facility,* City of Manitowoc, WI, November 1980.

Donohue & Associates, Inc., *Smith Ditch Lagoon No. 1 and Hotter Lagoon Investigation,* City of Valparaiso, IN, June, 1984.

Donohue & Associates, Inc., *Stormwater Management Guidelines for the City of Bettendorf, IA,* Final Draft, June 1985.

McCuen, R. H., "Downstream Effects of Stormwater Management Basins." *J. Hydraul. Div., Am. Soc. Civ. Eng.* **105**(HY11), 1343–1356 (1979).

Mein, R. G., "Analysis of Detention Basin Systems." *Water Resour. Bull.* **16**(5), 824–829 (1980).

Metropolitan Sanitary District of Greater Chicago, *Sewer Permit Ordinance.* MSDGC, IL, October 1972.

Walesh, S. G., J. A. Hardwick, and D. H. Lau, *Cost Savings Through Multi-Purpose Flood Control-Recreation Facility: Case Study.* Presented at the 14th Annual Water Resources Planning and Management Conference, American Society of Civil Engineers, Kansas City, MO, March 1987.

Walesh, S. G., J. A. Hardwick, and D. H. Lau, *Multi-Purpose Flood Control-Recreation Facility: Case Study.* Proceedings of the 9th Annual Water Resources Symposium, Indiana Water Resources Association, Greencastle, IN, June 8–10, 1988, edited by K. E. Bobay, pp. 126–137.

9

SEDIMENTATION BASIN DESIGN

Temporary or permanent sedimentation basins (SB) may be important components of an urban surface water system. Such facilities can control much of the suspended solids and other debris typically conveyed by stormwater runoff. By removing solids and debris, SBs also remove potential pollutants adsorbed onto or absorbed into the solids and debris.

Concepts and definitions pertinent to SBs are presented and discussed at the beginning of this chapter. An SB design procedure is then presented and comprises most of the chapter.

9.1 OVERVIEW

Definitions

A SB is a facility intended to trap suspended and buoyant solids and debris carried by stormwater runoff. Another purpose of SBs is removal from stormwater of adsorbed and absorbed potential pollutants. For comparison purposes, schematics of a SB, a retention facility, and a detention facility are presented in Figure 9.1. The terms "detention facility" and "retention facility" are defined in Chapter 8.

Settling efficiency and similar terms, any one of which may be used in association with SBs, have varied and inconsistent meanings and interpretations in erosion–sedimentation literature. Malcom and New (1975) draw a useful distinction between these terms. They define settling efficiency as the fraction of particles of a prescribed size trapped in a SB under design conditions at peak outflow.

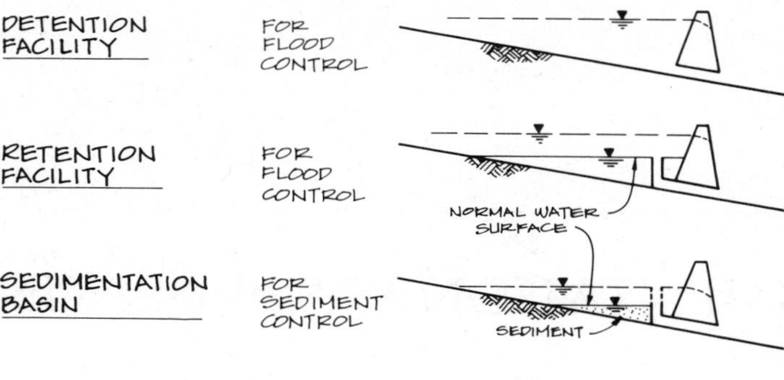

FIGURE 9.1 Schematics of detention facility, retention facility, and sedimentation basin

Trap efficiency is defined by Malcom and New as the fraction of all material removed by the SB under a variety of flow conditions and over an extended period. Trap efficiency is expected to exceed settling efficiency significantly because most runoff events will be less severe than the design event. The concepts supporting the terms ''settling efficiency'' and ''trap efficiency'' could also be used with respect to pollutants other than suspended solids.

Historic Note

Traditionally, SBs have been used as part of erosion and sedimentation control systems in agricultural areas. Recently, as water quality impacts caused by water erosion and sedimentation have been observed in urban and urbanizing areas, SBs have found increased use in those areas. In agricultural settings, SBs tend to be separate, single-purpose facilities. However, and as discussed and illustrated in Chapter 11, a SB may be one part of a multipurpose stormwater D/R facility. Combining SBs with other D/R-type facilities is one feature of the more recent use of SBs in urban and urbanizing areas.

In these multipurpose applications, the SB is sometimes placed in series with and upstream of the D/R facility and is used to remove suspended solids and associated pollutants from frequent, small stormwater runoff events. The D/R facility is intended primarily to reduce peak downstream flows during infrequent but major stormwater runoff events.

Sedimentation will occur in most D/R facilities even though they may not be explicitly designed to serve a sedimentation function. Unexpected sedimentation may have adverse impacts on D/R facilities, such as loss of storage caused by sediment accumulation, unsightly accumulation of suspended solids and other debris near the inlet, and partial obstruction of outlet works, primarily by buoyant debris. Accordingly, designers of D/R facilities should make at least a preliminary assessment of likely sediment loads to the facility using one or more of the techniques presented in Chapter 7, roughly determine probable quantities of sediment that will be retained by the facility, and identify possible adverse effects of the sedimentation. If significant adverse effects are expected, preventive measures, such as a SB, should be considered.

Temporary and Permanent Installations

A SB may be a temporary measure at a construction or development site intended to retain soil and other materials on the site during construction. Later, this type of SB is typically converted to some other use, such as open space or filled to provide land for additional development.

In contrast, a SB may also be a permanent measure in an urban and urbanizing area intended to prevent suspended solids, buoyant debris, and adsorbed/absorbed pollutants from entering environmentally sensitive areas such as lakes, rivers, and wetlands. A permanent SB could be constructed, at least in rough form, prior to the initiation of land development and initially serve the same purpose as a temporary SB.

Surface and Subsurface Installations

Most SB are surface facilities; that is, they are constructed on the land surface and are readily visible. Some SBs are constructed below ground, a common example being the sumps in standard stormwater system catch basins or in special catch basin installations such as that shown in Figure 11.6. Although not common, large subsurface SBs have been designed and constructed. For example, and as shown schematically in Figure 11.7, Skokie, Illinois, constructed eight large subsurface stormwater detention facilities, each of which includes a separate sedimentation section (Walesh, 1986; Walesh and Schoeffmann, 1984, 1985).

9.2 COMPONENTS OF A SEDIMENTATION BASIN

The possibility of permanent and temporary installations, single- and multipurpose arrangements, rural and urban sites, and surface and subsurface construction suggest many possible SB configurations. Although many configurations are possible, most SBs will have four basic components: an inlet control zone, a settling zone, a sediment storage zone, and an outlet control zone. The four components common to most SBs are illustrated in Figure 9.2 and each is described in the following.

1. Inlet Control Zone. The purpose of this component is to intercept incoming flow and distribute it laterally over the SB. The objective is to utilize fully the subsequent settling zone by minimizing short circuiting.

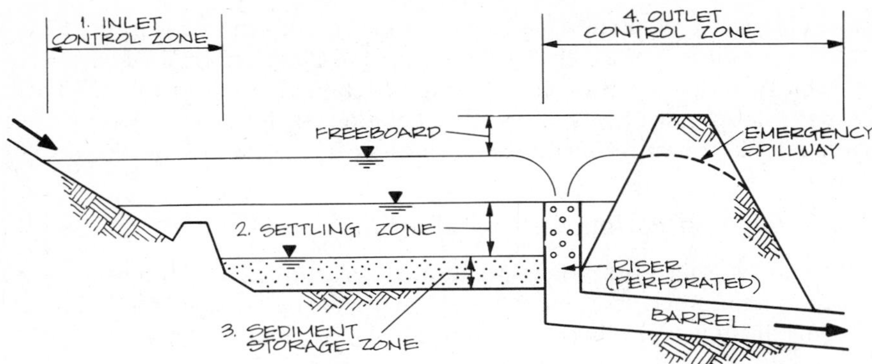

FIGURE 9.2 Sedimentation basin components (*Source:* Adapted from Malcom and New, 1975)

Depending on the particular facility, lateral flow distribution may be accomplished with a submerged berm or weir, baffles, or a silt fence.

2. Settling Zone. Particle settling occurs in the settling zone, a shallow, relatively tranquil portion of the SB. To minimize short circuiting further, the settling zone is generally long and narrow, with flow entering and leaving at the opposite, narrow ends. Design flow rate, design particle size, and design settling efficiency determine the plan area of the settling zone.
3. Storage Zone. Positioned beneath the settling zone, the sediment storage zone stores settled solids. The size of this zone is a function of the annual sediment load to the SB, the trap efficiency of the SB, and the expected frequency of sediment removal.
4. Outlet Control Zone. The purpose of this zone is to control the water level in and rate of discharge from the SB. In addition, the outlet must provide a smooth transition of flow from the settling zone, including holding velocities to values that will not disturb solids in the storage zone. Velocity control might be accomplished with a submerged weir or berm, baffles, or with a silt fence. A means of trapping buoyant debris and preventing it from being carried downstream should also be included in the outlet control zone.

   Finally, the outlet control zone should be hydrologically and structurally designed to accommodate, without failure, flood flows well in excess of those used to design the settling function. For this reason, an emergency spillway is normally included as part of the outlet control zone. Various combinations of weirs, pipes, drop inlets, service spillways, emergency spillways, and infiltration areas are typically used to form the outlet control zone. The types and sizes of these units are a function of the SB design criteria and flood flow design criteria.

9.3 DESIGN PROCEDURE: OVERVIEW

The design of an SB, excluding structural and geotechnical considerations, may be viewed as consisting of the following nine-step process:

1. Collect and analyze watershed data.
2. Identify potential sites.
3. Select design recurrence intervals.
4. Select design particle size and settling efficiency.
5. Determine design flows and the volume of the settling zone.
6. Design the inlet zone.
7. Determine settling zone geometry.
8. Determine volume of the sediment storage zone.
9. Design the outlet zone.

The preceding process, in an appropriately abbreviated and more general form, is also applicable to planning of one or more SBs. The nine-step SB design process is presented in the following section of this chapter, where a simplification of an actual designed and constructed permanent SB is used to illustrate some aspects of the design process.

The example SB, the west SB shown in Figure 11.13, was designed for and subsequently constructed in the town of Madison, Wisconsin (Raasch, 1982; Walesh, 1986). The primary purpose of the SB was to intercept suspended solids and buoyant debris from a 225-acre urban and urbanizing area and thereby prevent the materials from entering and being deposited in an arboretum, a publicly owned, environmentally sensitive natural area. Construction of the SB was completed in 1982. As also shown in Figure 11.13, the SB is part of a multipurpose sedimentation basin–flood control system consisting of two SBs and a detention facility. Additional information about the system is presented in Chapter 11.

9.4 STEP 1: COLLECT AND ANALYZE WATERSHED DATA

The intent of this step is to gather the data that will be needed to calculate hydrologic loads and sediment loads from the watershed to the SB. This step is essentially identical to step 1 in the design of D/R facilities as presented in Chapter 8; therefore, the information is not repeated here.

For the example SB, step 1 indicated that the tributary area was 225 acres, watershed soils were loam and silt loam, hydrologic soil group B was dominant, and land slopes were generally in the 3 to 4 percent range. The design future development condition was primarily residential, with some commercial land use, all of which would be served by curb, gutter, and storm sewers.

The corresponding watershed imperviousness was 49 percent and the time of concentration was 39 minutes.

9.5 STEP 2: IDENTIFY POTENTIAL SITES

In this step, potential sites are screened prior to doing a more detailed analysis on any given location. The screening process is similar to step 2 in the design of D/R facilities, as presented in Chapter 8.

Assess Proximity to Sensitive Area

The SB must be upstream of and should be as close as possible to the sensitive area that is to be protected. The sensitive area may be a lake, reservoir, stream, wetland, or similar feature. The closer a site is to a sensitive area, the greater the portion of the tributary area that will be controlled by the site. For the example, the site selected was on the upstream edge of the environmentally sensitive arboretum.

Determine If Site Size Is Adequate

One consideration is adequate size, that is, the areal extent of a potential site, with the determination being based primarily on judgment. For the example SB, sufficient space was available at several locations along the upstream edge of the arboretum.

Determine Likely Availability of Site

Undeveloped land not committed to conflicting development is desired. In some situations, a parcel of land with other economically more productive uses will have to be dedicated for a SB site, as might be the case if a new housing development is to be constructed close to and upstream of a lake or reservoir. In other instances, a SB may be placed on a location not readily suited for other development, such as a low-lying wetland area. In the example, the criterion of undeveloped land not committed to conflicting development was easily satisfied because the arboretum was zoned open space and was publicly owned.

Assess Topographic Position

A SB site should be topographically low to minimize excavation and other earth work. This consideration will be automatically accounted for to the extent that most potential sites are on swales, channels, and other drainageways. For the example, the site selected was naturally low because it was on a channel, and therefore, excavation and earth work were minimized.

Evaluate Likelihood of Gravity Inflow and Outflow

Gravity-driven inflow and outflow are obviously very desirable, as they are with D/R facilities and with stormwater facilities in general. Pumped discharge may be feasible for some temporary SBs, such as an SB at the downstream end of a construction site, where effluent is pumped into a nearby storm sewer.

For the example, the SB was positioned on a channel having sufficient grade to provide gravity inflow and outflow. However, the sediment storage zone was excavated below the original channel grade, and therefore contained standing water after runoff events.

Evaluate Access

An SB must be easily accessible for inspection and maintenance. An important aspect of the latter is occasional removal of accumulated sediment and other materials; therefore, the site must be accessible with a truck, front-end loader, or similar equipment. A site contiguous with a public roadway is preferred, but an access easement across private land is feasible. In the example, the site selected was contiguous with a then-proposed public roadway.

Access Consequences of Failure

Failure of the SB, such as overtopping and washout of an earthen berm as a result of a major flood, could cause downstream damage. An SB facility, in contrast with a D/R facility, is usually intended to protect a natural or seminatural area rather than a residential, commercial, or industrial area, or is located immediately upstream of a D/R facility. Furthermore, at the time of imminent failure, a SB will generally contain much less water than a D/R facility serving the same-sized tributary area. Therefore, the consequences of a failure of a SB facility are likely to be markedly less than for a D/R facility. For the example situation, a large natural area and stormwater detention facility were located immediately downstream of the SB, providing a buffer between the SB and the nearest downstream urban development.

Consider Multiple-Use Possibility

An SB, in contrast with a D/R facility, is not usually suitable for multiple uses, primarily because most of the bottom of the SB will be covered with objectionable sediment and other debris, and because SBs typically cover a small area relative to D/R facilities. However, a SB may facilitate or enhance the multiple use of an associated D/R facility.

For example, placing a separate SB immediately upstream of a detention facility may enhance the cleanliness and desirability of a park or recreation area developed in or contiguous with the detention facility. Similarly, construction of a SB at the upstream end of a retention facility may protect water quality and prevent excessive sediment accumulation in the retention facility.

In the example, the SBs were used to minimize the entry of suspended solids and debris into the sensitive arboretum. This function was critical because the detention facility, with its inherent settling capability, was to be constructed in the arboretum. That is, construction of a detention facility in the arboretum would have, in the absence of the SB, solved the downstream flooding problem but introduced an aesthetic and ecologic problem in the arboretum.

9.6 STEP 3: SELECT DESIGN RECURRENCE INTERVALS

SB design requires the selection of two recurrence intervals. The first is the recurrence interval used for designing the sedimentation features of the facility. This sedimentation recurrence interval enters into the calculation of the area and depth of the settling zone and the sizing of the outlet works. The sedimentation recurrence interval is typically for a frequent or small hydrologic event, such as 1 or 2 years, because most of the total load of suspended solids and other debris is transported by stormwater runoff from smaller, more frequent storms. For temporary SBs, such as may be used at a construction site, the sedimentation recurrence interval may be even smaller because of the reduced probability that a 1- or 2-year hydrologic event will occur during the construction period.

The second recurrence interval is used for designing the emergency spillway in the outlet zone. This emergency spillway recurrence interval is usually that of an infrequent or large hydrologic event, such as in the 10- to 100-year range, with the value being selected based on the downstream consequences of failure of the outlet zone. For example, if the only significant adverse impact of washout of the berm that contains the SB is the need to rebuild the damaged portion of the berm, the emergency spillway recurrence interval might be set at 10 years. In contrast, if berm failure could endanger downstream property or residents, a 100-year recurrence interval should be selected, to be consistent with flood control practice. Local, state, or other dam design criteria may also affect the selection of recurrence interval.

In the example, the sedimentation interval was set at 1 year. The emergency spillway interval was set at 10 years, meaning that the emergency spillway begins to function at the discharge approximating a 10-year-recurrence-interval event. Emergency spillway recurrence interval selection reflected the limited consequences of failure—the need to rebuild the berm.

9.7 STEP 4: SELECT DESIGN PARTICLE SIZE AND SETTLING EFFICIENCY

Settling zone area and configuration are dependent on the size, that is, nominal diameter of the smallest particle to be removed by sedimentation, and by the expected effectiveness, that is, settling efficiency. Complete removal of particles of all sizes is physically and economically impractical because as design particle size is decreased and desired particle settling efficiency is increased, settling zone volume and SB cost increase.

The definitions of and distinction between settling efficiency and trap efficiency were presented earlier in this chapter. Settling efficiency, rather than trap efficiency, is used as a design parameter because of the existence of the analytic relationship between settling zone plan area and other design parameters, including settling efficiency. This relationship is based on Hazen's theory of sedimentation (Malcom and New, 1975) and is borrowed from the field of sanitary engineering, where it is used to design settling tanks in water and wastewater treatment plants.

Hazen's equation is

$$A = \frac{NQ}{V_0}[(1 - E)^{-1/N} - 1]$$

where, as illustrated in Figure 9.3,

A = plan area of the settling zone in square feet
N = effective number of cells
Q = design flow in cubic feet per second
V_0 = design particle settling velocity in feet per second
E = settling efficiency of design particle expressed as a decimal

The resulting trap efficiency is likely to be significantly greater than the prescribed settling efficiency, because smaller, more frequent rainfall events

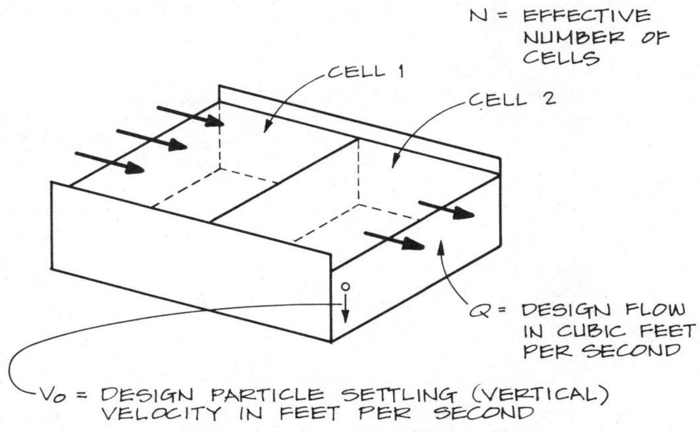

FIGURE 9.3 Schematic representation of a sedimentation basin

transport most of these suspended solids and most of these events will be smaller—lower recurrence interval—than the recurrence interval selected for sedimentation. Therefore, the long-term trap efficiency of the SB, which is the integrated effect of the settling efficiencies of all rainfall–runoff events, will be significantly greater than the design settling efficiency.

Design particle size should be selected based on the probable size distribution of suspended solids entering the SB. Particle size ranges may be approximated by knowing the soil types in the watershed and using soil separates size data such as those presented in Table 9.1.

Experience suggests that whereas it is usually physically practical and economically feasible to design for removal of sands, removal of silts and clays is likely to be physically and economically prohibitive. As illustrated by Table 9.1, silts are one order of magnitude smaller than sand. Because settling velocity is inversely proportional to particle size to the second power, as expressed by Stokes' law, with typical results shown in Table 9.2, silts have settling velocities at least two orders of magnitude smaller than sands. Finally, because settling zone area is inversely proportional to settling velocity to the first power, as indicated by Hazen's equation, it follows that for the same settling efficiency, an SB designed for silt removal would require an area at least two orders of magnitude greater than the area of an SB designed for sand removal. An analogous argument for clay leads to the conclusion that an SB designed for clay removal would require an area at least five orders of magnitude greater than that of an SB designed for sand removal.

Unfortunately, receiving water turbidity—a readily visible indicator of erosion—is caused primarily by clay-sized particles. Therefore, although SBs can be effective in controlling unwanted sedimentation, they are not likely to be highly effective in preventing turbidity, a highly visible phenomenon.

For the example, the design particle size was set at 40 microns (μm) based on watershed soil characteristics. The settling efficiency was set at 70 percent.

TABLE 9.1 Soil Separates and Corresponding Particle Size

	Size (Equivalent Diameter)	
Soil Separates	Millimeters	Microns[a]
Very coarse sand	2.0–1.0	2000–1000
Coarse sand	1.0–0.5	1000–500
Medium sand	0.5–0.25	500–250
Fine sand	0.25–0.10	250–100
Very fine sand	0.10–0.05	100–50
Silt	0.05–0.002	50–2
Clay	<0.002	<2

Source: Adapted from U.S. Department of Agriculture (1981).

[a]1 micron = 1 micrometer = 0.001 mm or about 1/25,000 of an inch.

TABLE 9.2 Settling Velocities[a]

Particle Diameter (μm)	Settling Velocity (ft/sec)
8	0.00019
10	0.00029
12	0.00042
15	0.00066
18	0.00095
20	0.0012
25	0.0018
30	0.0027
40	0.0047
50	0.0074
60	0.011
80	0.019
100	0.029

Source: Malcom and New (1975).

[a]Calculated with Stokes' law for spherical particles of specific gravity 2.65 settling in quiescent water at 68°F.

A sensitivity analysis presented later in this chapter illustrates the impact of variations in design, particle size, and settling efficiency on settling zone area.

9.8 STEP 5: DETERMINE DESIGN FLOWS AND VOLUME OF THE SETTLING ZONE

Two design flows are required. The first is the peak outflow rate corresponding to the sedimentation recurrence interval selected in step 3. Various methods available for determining watershed runoff corresponding to the specified recurrence interval are presented in Chapter 3. In the design used as the basis for this example, the computer program ILLUDAS was used. The peak inflow to the SB corresponding to the selected sedimentation recurrence interval of 1 year was 100 ft^3/sec.

Routing procedures can be used to determine the peak outflow rate corresponding to the peak inflow rate. For the example, a trial-and-error procedure was used in which various settling zone sizes and outlet works configurations were selected and the inflow hydrograph was then routed through the SB. The resulting 1-year recurrence interval peak outflow rate was determined to be 30 ft^3/sec. A more conservative approach would be to assume that the peak outflow rate approximates the peak inflow rate because of the small volume of storage often available in an SB.

The volume of the settling zone is the maximum storage that will occur under the condition of the sedimentation recurrence interval. Given the inflow hydrograph for the sedimentation recurrence interval, the maximum storage can be determined by methods such as routing the hydrograph through the SB, the TR55 graphical procedure, or a linear approximation of an outflow hydrograph superimposed on the inflow hydrograph. For the example, the maximum volume of water in storage under the 1-year sedimentation recurrence interval was determined to be 124,000 ft^3.

The second design flow for a SB is the peak outflow, corresponding to the emergency spillway recurrence interval selected in step 3. Again, many methods are available for determining peak flow from a watershed for a prescribed recurrence interval. In the case of the example, ILLUDAS was used. Peak inflow to the SB, corresponding to the selected emergency spillway recurrence interval of 10 years, was 150 ft^3/sec. Because available storage is small, the peak outflow was also assumed to be 150 ft^3/sec.

9.9 STEP 6: DESIGN THE INLET ZONE

Inflow to the SB should occur without erosion at or near the inlet. Riprap, gabions, or some other form of armoring may be required. Furthermore, incoming flow should be spread across the full width of the SB to introduce minimal turbulence into the settling zone. Distribution of flow may be achieved with a submerged berm or weir, a suspended baffle, or a silt fence, any of which would be positioned perpendicular to the incoming flow and extend across the width of the SB. In the example, riprap was placed on the bottom and sides of the inflow channel to control erosion.

9.10 STEP 7: DETERMINE SETTLING ZONE GEOMETRY

Given the design particle size, the quiescent condition settling velocity is determined from Table 9.2, which assumes a specific gravity of 2.65, or the settling velocity can be calculated directly from Stokes' law. For the example, the design particle size of 40 μm corresponds to a settling velocity of 0.0047 ft/sec.

Hazen's equation is used to determine the plan area of the settling zone. Substituting $V_0 = 0.0047$ ft/sec, $E = 0.7$, $Q = 30$ ft^3/sec, and $N = 1$ into Hazen's equation yields a settling zone plan area of 14,900 ft^2, as shown in Table 9.3.

Sensitivity of plan area to design particle size, number of cells, and settling efficiency, is shown in Table 9.3. For example, reducing design particle size by one-half quadruples the plan area, whereas doubling design particle size reduces the plan area by three-fourths.

Plan area is also sensitive to settling efficiency. Increasing efficiency by about 30%, from 0.7 to 0.9, results in quadrupling of the plan area. Reducing efficiency by about 30 percent, from 0.7 to 0.5, reduces the plan area by more than half.

TABLE 9.3 Sensitivity of Plan Area to Design Particle Size, Effective Number of Cells, and Settling Efficiency for the Example Sedimentation Basin

Sensitivity to Particle Size[a]

Size (Equivalent Diameter) (μm)	Settling Velocity V_0, (ft/sec)	Plan Area, A (ft^2)
20	0.0012	58,400
40[b]	0.0047[b]	14,900[b]
80	0.019	3,700

Sensitivity to Effective Number of Cells (N)[c]

Effective Number of Cells, N	Plan Area, A (ft^2)
1[b]	14,900[b]
2	10,500
3	9,450

Sensitivity to Settling Efficiency (E) of Design Particle[d]

Settling Efficiency, E	Plan Area, A (ft^2)
0.5	6,400
0.7[b]	14,900[b]
0.9	57,400

[a]For all three particle sizes, $E = 0.7$, $Q = 30$ ft^3/sec, and $N = 1$ in Hazen's equation.
[b]Design conditions used in example.
[c]For all three effective number of cells, $V_0 = 0.0047$ ft/sec, $E = 0.7$, and $Q = 30$ ft^3/sec.
[d]For all three settling efficiencies, $V_0 = 0.0047$ ft/sec, $Q = 30$ ft^3/sec, and $N = 1$.

Increasing the number of cells to two or three decreases the plan area by 30 and 37 percent, respectively. The number of cells can be increased by means such as construction of submerged weirs or berms and installation of filter fabric fences.

The corresponding depth is then calculated to check for reasonableness. Settling zone depth is the quotient of the volume of the settling zone as determined from step 5 and the area of the settling zone. For the example, the settling zone depth is 124,000 ft^3 divided by 14,900 ft^2, or 8.3 ft. A depth of 8.0 ft was used in the design.

9.11 STEP 8: DETERMINE VOLUME OF THE SEDIMENT STORAGE ZONE

As noted earlier in this chapter, the volume of the storage zone is determined by the annual sediment load to the SB, the trap efficiency of the SB, and the expected frequency of sediment removal. Average annual sediment load, which is the same as the annual sediment yield of the tributary watershed, may be determined using one or more of the sediment yield methods presented in Chapter 7.

For the example, the universal soil loss equation was used with a sediment delivery ratio of 1. The estimated average annual sediment yield was determined to be 53 tons. Assuming a submerged unit weight of solids of 70 lb/ft^3, the average annual volume of settleable solids was 1500 ft^3. Trap efficiency may be conservatively assumed to be 100 percent. This assumption is suggested because trap efficiency typically exceeds design settling efficiency and because of the difficulty of estimating trap efficiency accurately. Trap efficiency was assumed to be 100 percent for the example.

Frequency of sediment removal is established based on several factors. One is the need, if any, to minimize the size of the storage zone. Increasing the frequency of sediment removal reduces the volume of the storage zone. A second factor is the ease of access and availability of special equipment that may be needed. For example, a small unit of government may need to contract for SB cleaning, and therefore, may wish to increase the time between cleaning. Finally, the designer must recognize that as the time period between sediment removal is increased, there is an increased probability of occurrence of a major storm and flushing settled solids downstream, thus at least partly negating the intended function of the SB. For the example, the cleaning frequency was conservatively set at 10 years. Therefore, the maximum volume of the storage zone is 15,000 ft^3.

The configuration of the storage zone, that is, the length, width, and depth, is determined next. Recognizing that the storage zone lies beneath and is conterminous with the settling zone, the plan area and the length and width of the storage zone are the same as the settling zone. Therefore, for the example, the areas for both the settling zone and the storage zone are 14,900 ft^2.

Storage zone depth is the quotient of the volume of the storage zone and the area of the storage zone. For the example, the storage zone volume is 15,000 ft^3, which, when divided by the sediment storage zone plan area of 14,900 ft^2, is 1 ft. Storage zone depth was conservatively set at 2 ft.

9.12 STEP 9: DESIGN THE OUTLET ZONE

The two principal hydraulic components of the outlet zone are the outlet control for the sediment function and the emergency spillway. The idea is to have the former component function for hydrologic events having recurrence intervals up to and including the sedimentation recurrence interval. The emergency spillway functions for extreme events, that is, those exceeding the emergency spillway recurrence interval.

The outlet control for the sedimentation function often takes the form of a riser—unperforated or perforated in the upper portion—connected to an approximately horizontal barrel, as illustrated in Figure 9.2. An unperforated riser is used if a permanent pool is desired, and a riser perforated above the level of the sediment storage zone is used if the SB is to be dry between runoff events. For the example, a perforated riser was used.

Riser design begins with designing its diameter. The critical condition is that at which the flow through the riser approaches the emergency spillway condition. This is the maximum flow that must be carried by the riser because higher flows are conveyed over the emergency spillway. The top edge of the riser is analyzed as a weir. A trial-and-error procedure is used to determine technically feasible riser diameter–head combinations. Figure 9.4 illustrates the process used to determine riser diameter. Note the allowance for concurrent flow of 30 ft³/sec or more through the riser perforations.

SKETCH:

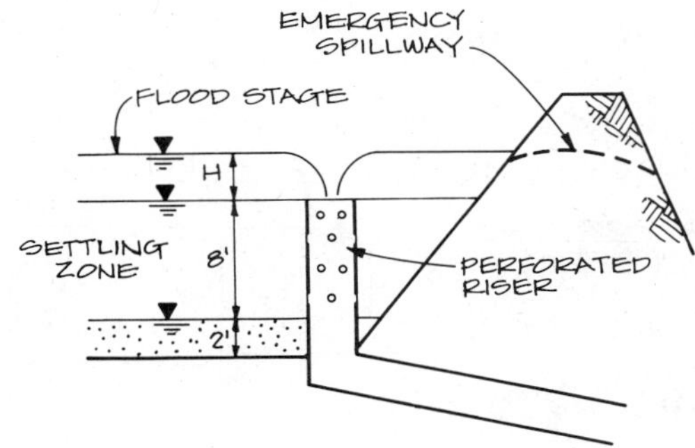

DESIGN Q:

$Q_{10_{IN}}$ = 150 CUBIC FEET PER SECOND

ASSUME NO ATTENUATION BUT ACCOUNT FOR CONCURRENT OUTFLOW OF 30 CFS OR MORE THROUGH PERFORATIONS IN RISER

$\therefore$ $Q_{10_{OUT}}$ = 120 CUBIC FEET PER SECOND OR LESS

CALCULATE DIAMETER OF RISER:

$L = \dfrac{Q}{C_W H^{3/2}}$ $\begin{cases} Q = 120 \text{ CUBIC FEET PER SECOND} \\ C_W = 3.0 \end{cases}$

H (FEET)	L (FEET)	D (FEET, INCHES)	
1	40	12.7	152
2	14.1	4.50	54
3	7.7	2.45	29

USE H = 3 FEET AND 30 INCH DIAMETER RISER

FIGURE 9.4 Determination of riser diameter

Given the riser diameter, riser perforations are sized and positioned. This trial-and-error procedure involves use of the orifice equation. The perforation design is illustrated in Figure 9.5.

Provision should be made to trap floating debris. This can be accomplished by mounting a concentric baffle ring on the riser and by placing a trash rack over the top of the riser. These trash racks can also serve as safety devices to discourage children from entering the riser-barrel assembly.

The vertical riser, because of its position away from the edge of the SB, may induce vortexing. This phenomenon is a swirling or circling motion of water in a horizontal plane as it moves toward and into the riser. Vortexing and the associated air entrainment can significantly reduce the capacity of the riser and may also induce unwanted or even damaging erosion to the bottom and sides of the SB. If vortexing is likely, preventive measures such as vertical vanes mounted on the riser, should be used. In the case of the example, vertical vanes were used.

SIZE HOLES IN RISER PIPE

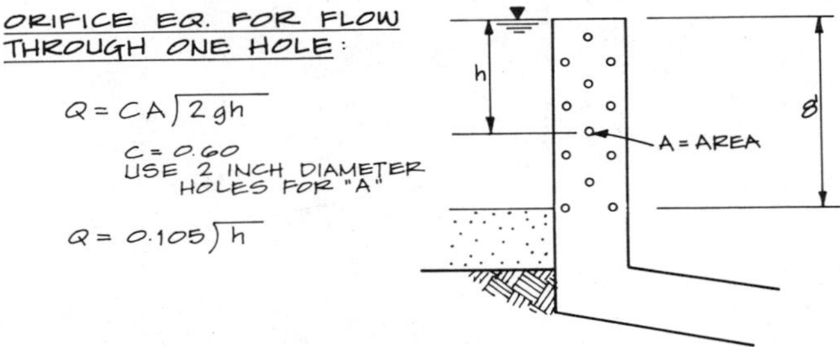

OBJECTIVE: SIZE AND SPACE HOLES SO INDICATED WATER LEVEL CAUSES FLOW OF 30 CFS INTO RISER

SOLUTION:

h (FEET)	Q FOR ONE HOLE (CFS)
0.50	0.074
1.25	0.117
2.00	0.148
2.75	0.174
3.50	0.196
4.25	0.216
5.00	0.234
5.75	0.252
6.50	0.268
7.25	0.283
8.00	0.296
TOTAL	2.26 CFS

$\frac{30}{2.26} = 13.2$ ∴ PUT 13 HOLES AT EACH ELEVATION

SPACING ON CIRCUMFERENCE WILL BE:

$\frac{\pi\, 30''}{13} = 7.2$ INCHES

FIGURE 9.5 Design of riser perforations

The emergency spillway should be designed to accommodate flows at and above those corresponding to the spillway recurrence interval. A conservative approach is to assume that the riser and barrel are not functioning once the emergency spillway begins to operate. This may occur, for example, as a result of blockage of the riser and barrel by flood-borne debris.

In the case of the example, if the riser and barrel are blocked, the emergency spillway would have to carry 150 ft^3/sec under a 10-year emergency spillway recurrence interval condition. A 120-ft-long turf-covered emergency spillway (weir coefficient of 3.6) would convey 150 ft^3/sec at a head (depth on the spillway crest) 0.5 ft. The velocity would be 2.5 ft/sec, a noneroding, and therefore acceptable, value.

REFERENCES

Malcom, H., and V. New, *Design Approaches for Stormwater Management in Urban Areas,* Part 4. North Carolina State University, Raleigh, May 1975.

Raasch, G. E., "Urban Stormwater Control Project in an Ecologically Sensitive Area." *Proc. Int. Symp. Urban Hydrol., Hydraul., Sediment Control, 1982,* pp. 187–192.

U.S. Department of Agriculture, Soil Conservation Service, *Soil Survey of Porter County, Indiana.* USDA, Washington, DC, February 1981.

Walesh, S. G., "Case Studies of Need-Based Quality-Quantity Control Projects." In *Urban Runoff Quality—Impact and Quality Enhancement Technology,* Proc. Eng. Found. Conf. New England College, Henniker, NH, edited by B. Urbonas and L. A. Roesner, 1986, pp. 423–437.

Walesh, S. G., and M. L. Schoeffmann, *Surface and Subsurface Detention in Developed Urban Areas: A Case Study.* Presented at the American Society of Civil Engineers Water Resources Planning and Management Division Conference, Baltimore, MD, May 1984.

Walesh, S. G., and M. L. Schoeffmann, "One Alternative to Flooded Basements." *Am. Public Works Assoc., Rep.* **52**(3), pp. 6–7, March (1985).

10

COMPUTER MODELING

A fundamental requirement of almost any urban surface water management (SWM) engineering or planning effort is a quantitative analysis of watershed hydrology and hydraulics, and sometimes water quality and economics, under existing and potential future conditions. Computer modeling techniques developed within the water resources engineering field during the last two decades facilitate the desired quantitative analysis. Detailed and comprehensive watershed analysis and design are now technically and economically feasible and offer great potential. As stated anonymously:

> The computer is incredibly fast, accurate, and stupid. Man is unbelievably slow, inaccurate, and brilliant. The marriage of the two is a challenge and opportunity beyond imagination.

This chapter provides a comprehensive and in-depth treatment of computer modeling in urban SWM. The chapter begins with a discussion of system and modeling concepts, emphasizing that a watershed is a system and that the system is best simulated with digital computer models. Key aspects of digital computer models are discussed, including the basic components comprising a model; algorithms; the way in which computer models facilitate the application of long-established hydrologic, hydraulic, and water quality principles; and the distinction between discrete and continuous models. Factors to consider in model selection are presented followed by an introduction to some widely used models and examples of model packages used in actual projects. Model calibration and verification are discussed in detail. Numerous examples of the application of watershed models are presented. The chapter concludes with a brief discussion on the role of the computer model in the work of the practicing engineer.

Mention of specific software products in this chapter is for illustrative purposes. Reference to software is not necessarily an endorsement of either the software or the organization that provides or supports the software.

10.1 SYSTEM AND MODELING CONCEPTS

The Watershed as a System

A system may be defined as a set of interdependent physical units and processes that function in a regular manner. A watershed is a system in that it includes, as shown in Figure 10.1, various units which, as indicated in Table 10.1, each has one or more processes associated with it. Furthermore, the units and processes function in a regular manner and are interrelated. For example, the hydrograph from an upstream basin affects hydrograph shape in a downstream reach. Nutrients discharged from a wastewater treatment plant affect the nutrient levels in downstream surface waters. Also, the hydraulic restriction of a bridge affects the flood stage profile in the reach upstream of a bridge. Conceptualizing and analyzing a complex watershed as a system composed of well-defined units and processes facilitates the development of a model of a watershed.

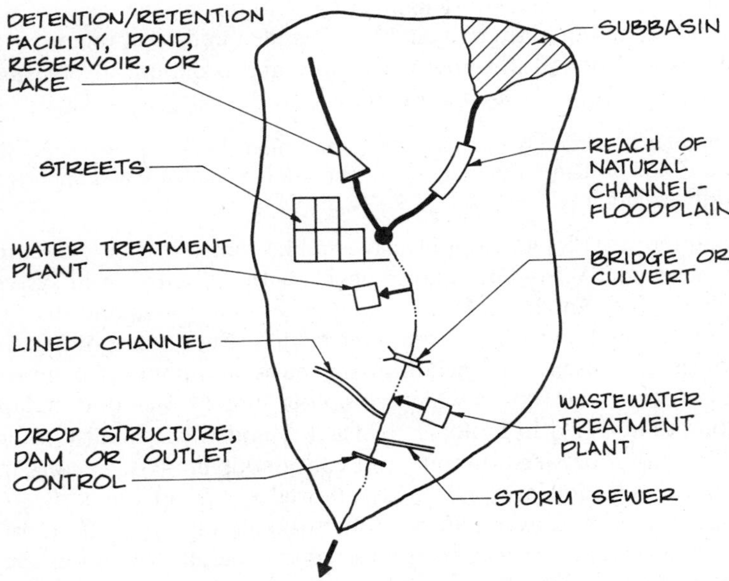

FIGURE 10.1 Representative units in a watershed system

TABLE 10.1 Representative Units and Processes in a Watershed

Units	Example Processes
Subbasin	Conversion of precipitation to runoff Accumulation and washoff of pollutants
Reach of natural channel–floodplain	Attenuation of flow Erosion and sedimentation Dilution and dispersion of pollutants
Lined channel	Attenuation of flow Hydraulic jumps at transitions
Storm sewer	Attenuation of flow Surcharging
Streets	Ponding Attenuation of flow Gutter and inlet flow
Detention/retention facility; reservoir, pond, or lake	Attenuation of flow Sedimentation Shoreline wave action and erosion
Outlet control structure, dam, or drop structure	Drawdown effect Erosion immediately downstream
Bridge or culvert	Backwater effect Scour and erosion Attenuation of flow
Confluence	Addition of flows
Wastewater treatment plant	Contribution of flow to the surface water system Input of pollutants

Two Fundamental Aspects of the Watershed System

Two fundamental questions must typically be addressed in all but the most trivial urban projects in SWM. First, how does the existing system function, that is, what is the cause of the water-related problems such as flooding, pollution, and erosion in the watershed, and how severe are they? This, the problem definition phase, must look beyond symptoms, such as frequent roadway overertopping, recurring basement flooding, frequent inundation of low-lying areas, and periodic algae blooms on urban lakes. Instead, problem definition focuses on addressing the problem and its cause or causes. A clear understanding of the cause of a problem tends to lead to its solution.

The second fundamental question is: How can the watershed system be modified or altered to mitigate existing problems significantly and to prevent similar or new problems from occurring in the future? Or, more pragmatically, will proposed solutions be successful? From a technical viewpoint, this is the problem-solving phase.

Monitoring: The Ideal Solution?

The ideal way to determine the behavior of a watershed system is to make direct observations or measurements for each of the units of interest and related processes in the watershed. This direct approach is generally not feasible, for three reasons. First, the cost of installing, operating, and maintaining an adequate network of precipitation gauges, streamflow gauges, water quality monitoring stations, and other monitoring equipment is usually prohibitive. Second, even if an ideal data-monitoring system is established in a watershed, it is highly improbable that the available sampling period would include critical events such as the extreme high-flow periods required for flood control planning and design purposes or the extreme low-flow periods required for some water quality planning and design purposes. Third and last, even if an elaborate watershed monitoring system is established, it would be of limited value because measurements and observations would only reflect existing conditions and could not be used to determine probable watershed behavior under hypothetical future conditions.

Computer Modeling: The Best Approach

Therefore, achieving the necessary understanding of the spatial and temporal fluctuations in surface water quantity and quality under existing and hypothetical future conditions requires application of some technique that can supplement and build upon a necessarily limited data base and be accomplished at an acceptable cost. The technique must quantify watershed behavior under existing and alternative future conditions with sufficient accuracy to permit decisions to be made concerning the location, type, and size of usually costly water control structures and facilities.

Hydrologic–hydraulic–water quality–economics simulation accomplished with the set of interconnected digital computer programs has proven to be a very effective planning and design approach. While it is true that watershed systems could be simulated by physical models and electric analogs, digital computer simulation has been by far the most widely used modeling approach.

Beginning in the 1960s, private consulting firms and government agencies began to gain access to the computer programs and the digital computers required to use digital computer simulation effectively. With the advent of the personal computer and suitable applications software in the 1980s, computer modeling in urban SWM was given added impetus. As a result of the dramatic decline in the capital cost of hardware and well-supported software, computer simulation is now within the reach of the entire spectrum of private and public organizations working in urban SWM. In addition, new civil engineering graduates, some of whom will serve as entry-level engineers working in urban SWM, have basic computer literacy.

10.2 NATURE OF WATERSHED DIGITAL COMPUTER MODELS

The Basics of a Computer Model

In its simplest form, a watershed computer model, like almost any of those used in urban SWM, can be viewed (Figure 10.2) as consisting of three components: input, the computer program itself, and output. Besides algorithms and input/output control, the computer programs may also include other data management routines, graphical routines, and statistical routines.

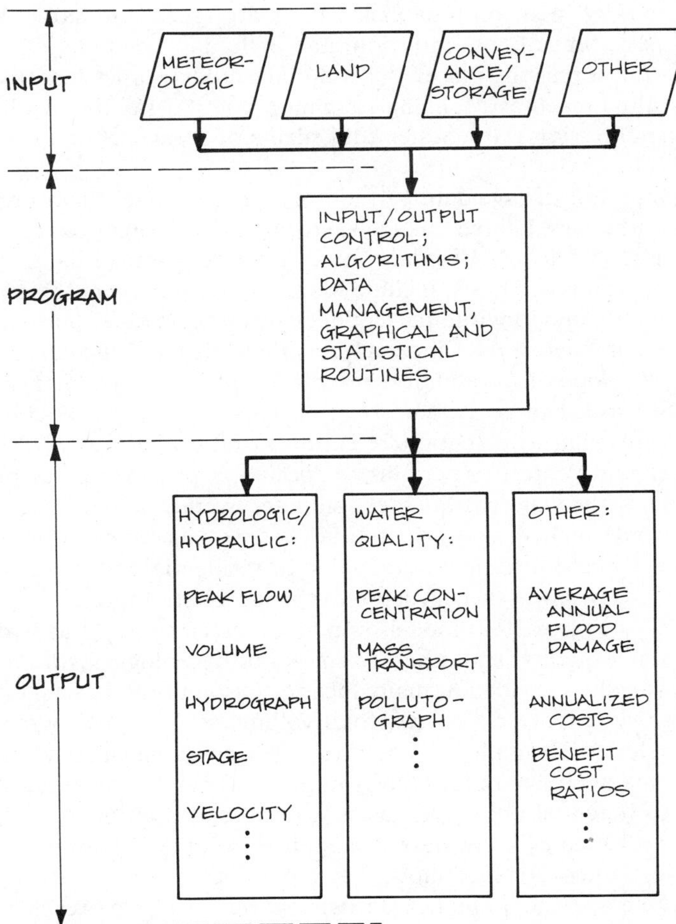

FIGURE 10.2 Basic components of a watershed computer model

Input usually includes meteorologic data, land data, and conveyance/storage data and occasionally includes other types of data, depending on the nature and use of the model. Examples of meteorologic data which are usually input for the entire watershed but sometimes for individual subbasins or groups of subbasins in the watershed are rainfall hyetographs, rainfall volume and duration, and antecedent moisture conditions. Depending on the sophistication of the model, other types of meteorologic data, such as radiation, potential evaporation, temperature, wind movement, and cloud cover, may be required.

Land data are broadly interpreted to encompass those characteristics of the land surface and subsurface that are likely to influence runoff volumes and rates in response to meteorologic events and conditions. Examples of land data which are usually input on a subbasin basis are area, land slope, soil type, imperviousness, time of concentration, and pollutant accumulation rates. Depending on the sophistication of the computer model, other land parameters may be required, such as Manning roughness coefficients for overland flow, groundwater recession rates, water equivalents of snowpack, and infiltration rates.

Conveyance and storage data pertain to swales, channels, and floodplains, armored or otherwise altered channels, storm sewers, detention/retention facilities, ponds, and lakes. Examples of conveyance/storage data, which are usually input on a reach basis, include Manning roughness coefficients, cross-section geometry, and longitudinal slope. Depending on the sophistication and purpose of the model, conveyance/storage input may also include parameters such as lateral slope of floodplains, reaeration rates, and geometry of special conduits. Stage–discharge–volume relationships are a type of conveyance/storage data peculiar to detention/retention facilities.

The other input category pertains to special-purpose computer programs. For example, if the watershed model computes average annual flood damage in addition to doing hydrologic–hydraulic–water quality simulation, examples of other input might include market value of residential and other structures and the vertical position of a structure or group of structures.

As shown in Figure 10.2, model output can be categorized as hydrologic–hydraulic, water quality, and other. Examples of hydrologic–hydraulic output which are typically provided at many locations around the watershed conveyance/storage system are peak flow, total volume of flow, hydrograph shape, stage, and velocity. Examples of water quality output, which are also typically provided at many points on the conveyance/storage system, include peak concentration of potential pollutants, mass transport of pollutants, and pollutographs. The other category of output is applicable for special-purpose models. For example, if the watershed model has economics capability, the output examples may be average annual flood damage on a reach-by-reach basis, annualized capital and operation and maintenance costs for various flood mitigation alternatives, and corresponding benefit cost ratios.

Algorithms

Each watershed unit and its associated processes are represented by algorithms. An algorithm is a set of mathematical equations or expressions within the computer program that simulates the unit and each of its processes. For example, watershed models typically include a reservoir routing algorithm and an algorithm that converts rainfall on a subbasin to runoff from the subbasin. Equations used in the algorithms are usually solved by analytic or numerical methods.

Algorithms are linked or interconnected within the computer program. The output from one algorithm becomes the input to another to represent the integrated behavior of the watershed system. For example, watershed models typically include an algorithm to determine the storage effect of a stream reach on the shape of a hydrograph that passes through the reach. Simulation of this unit and its processes is accomplished by mathematically expressing the change in hydrograph shape as a function of reach geometry and hydraulic parameters. Furthermore, the hydrograph that enters the reach is a function of all watershed hydrologic and hydraulic characteristics upstream of the reach.

Event and Continuous Models

One way to classify computer models available for use in urban SWM is event versus continuous. This event versus continuous categorization may be very important because it significantly affects the data collection costs that are likely to be incurred and the usefulness of the results. In general, continuous models cost more to use, but they produce technically superior results.

Figure 10.3 shows in schematic form the essential features of event and continuous hydrologic–hydraulic models. As is the case with all rainfall-runoff models, both event and continuous models have three components: input, the computer program, and output.

The principal distinction between event and continuous models is the type and amount of meteorologic data used as input and the manner in which the data are processed within the computer program (Crawford and Linsley, 1966; Linsley and Crawford, 1974). Event models typically require as input one or a few design storms in the form of a synthetic hyetograph or one or a few historic storm hyetographs. These few hyetographs are entered into the model, and output consists of a hydrograph and perhaps a pollutograph for each hyetograph. If a recurrence interval was attached to the input hyetograph, the usual assumption is that the resulting output hydrograph and pollutograph have the same recurrence interval.

A fundamentally different approach is used with continuous models, as exemplified by the Stanford Watershed model (Crawford and Linsley, 1966), which was the first continuous hydrologic–hydraulic model in the United States. The entire meteorologic record, usually consisting of decades of avail-

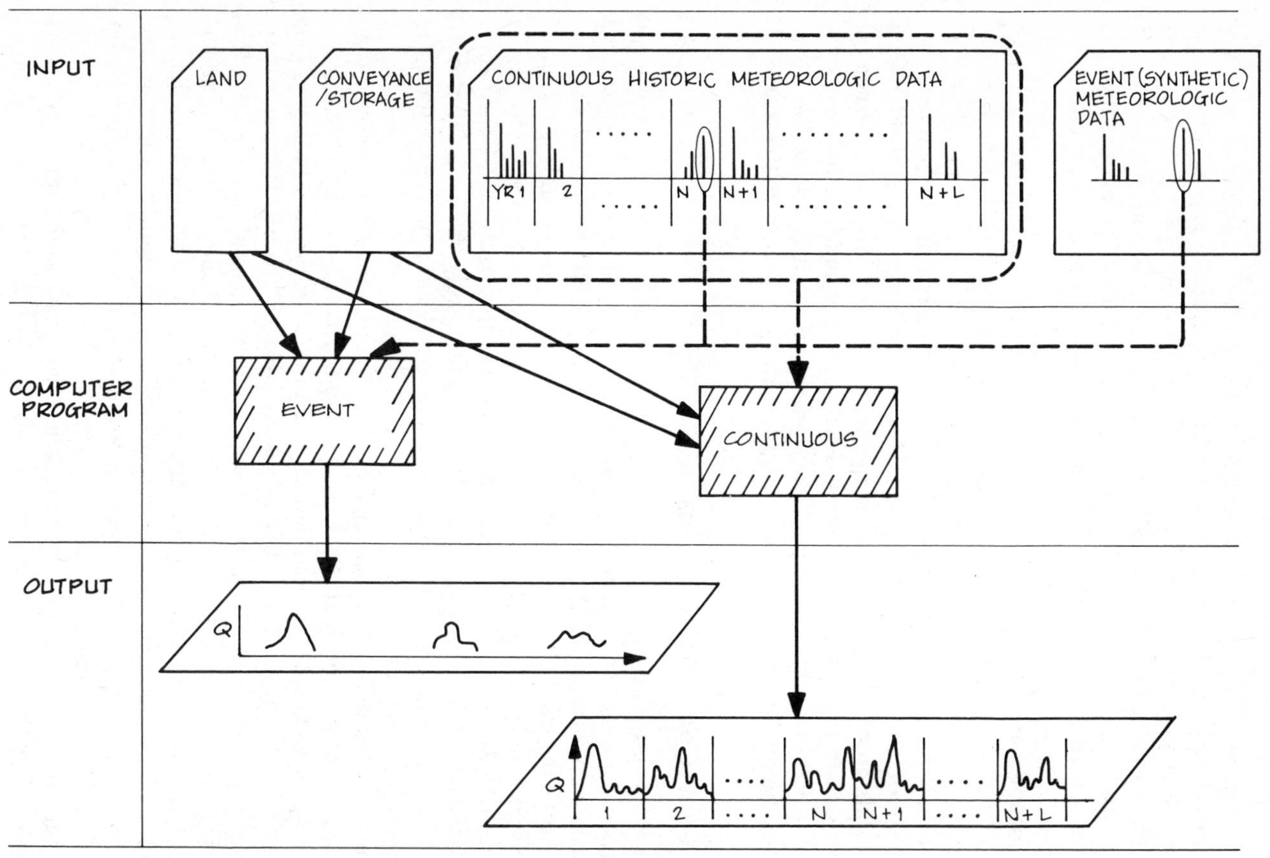

FIGURE 10.3 Comparison of event and continuous hydrologic–hydraulic models (*Source:* Adapted from Walesh, 1981)

able precipitation, temperature, and other similar data, is used as input. The computer processes all the continuous meteorologic input and produces, as output, a continuous hydrograph or pollutograph. The usual procedure is to analyze the continuous output in the same fashion as one would analyze a long streamflow or water quality record. Typical basic statistical analyses performed on the output from a continuous model are discharge–frequency relationships, volume–frequency relationships, flow–duration relationships, quality–duration relationships, and seasonal and annual mass transport quantities.

Various aspects of event and continuous models are summarized in Table 10.2. The principal advantage of the continuous model is that it eliminates the design storm dilemma and instead produces a result similar to a long-term streamflow and water quality record on which a variety of useful statistical

TABLE 10.2 Selected Characteristics of Event and Continuous Models

Event Models	Continuous Models
Simulate hydrograph or pollutograph in response to discrete rainfall event or rainfall–snowmelt event.	Simulate continuous hydrologic, hydraulic, and water quality characteristics in response to continuous meteorologic data.
Antecedent moisture and other conditions must be set by user.	Antecedent moisture and other conditions are determined by model (after initial startup).
Output not well suited for basic statistical analyses. Must use recurrence interval of output (e.g., peak discharge) equals recurrence interval of input (e.g., rainfall event) concept.	Output for long term well suited for basic statistical analyses, such as discharge-frequency relations, volume-frequency relations, flow-duration relations, and quality-duration relations.
Output not well suited for long-term (e.g., seasonal or annual) pollutant transport calculations.	Output options include calculation of pollutant transport for various periods.
Minimal use of historic streamflow, stage, and water quality data during calibration.	Extensive use of historic streamflow, stage, and water quality data during calibration.
Modest data requirements.	Volumnious data requirements—primarily because of long term (several decades) of meteorologic data.
Large computer capacity but short run time.	Similar computer capacity but long time.
Examples: TR20 HEC-1 ILUDRAIN SWWM	Examples: Stanford Watershed Model STORM HSPF

analyses may be performed. The principal disadvantage of the continuous model relative to the event model is that the former requires a much larger effort to assemble the historic meteorologic data. One way to reduce the cost of continuous simulation is to share meteorological data sets once they have been developed for a given geographic area.

Prior to the availability of personal computers with their negligible run costs, continuous models also had the disadvantage of much higher costs for computer runs. As a result, research conducted in the 1970s led to the development of methods to reduce the cost of continuous simulation (e.g., Geiger et al., 1976; Walesh and Snyder, 1979). Although the need for these cost-saving techniques has been diminished with the advent of personal computers and compatible versions of continuous models, users of continuous models may still find that these techniques are useful.

Another approach is a compromise between the typical use of continuous and event models. It capitalizes on the wider availability of event models while approximating the favorable historic storm approach of continuous models. With this approach, a series of major historic storms carefully selected from a long period—several decades—is used as input to an event model followed by a discharge–probability or volume–probability analysis of the model output. Initial indications are that this pseudo-continuous simulation produces results significantly different than, and probably more accurate than, those obtained by design storm methods, whether they are applied manually or with event models (Marsalek, 1978; Walesh et al., 1979; Wenzel and Voorhees, 1978, 1979). An example of this method is presented later in the applications section of this chapter.

Potential users of computer programs in urban SWM should be aware of and give due consideration to the distinction between event and continuous models. After weighing the objectives of the modeling and the resources available to perform it, a choice should be made between the use of event models with the traditional design storm approach, application of event models using a series of major historic events, or the use of continuous models.

What Is New or Unique about Computer Models?

Most of the algorithms in watershed models are not new. For example, the Manning open channel flow equation which is used in most models goes back to at least 1770 (Williams, 1970). Rational method concepts, such as time of concentration, used in computer models can be traced back at least a century (McPherson, 1969). Unit hydrograph theory, which is used in many watershed computer models, was well established as early as 1932 (Sherman, 1932). The method used in some computer models to compute average annual flood damage by combining discharge, stage, probability, and damage data for stream reaches was used prior to the 1960s (Franzini, 1961). Therefore, in general, the algorithms within the computer models, and more specifically, the theories and techniques on which they are based, are generally not new. Their origins can typically be traced back one or more decades or even centuries.

Furthermore, the mathematical computations and logic decisions completed during the operation of a watershed computer model are no more and no less sophisticated or valid than they would be if accomplished manually by sufficient technical personnel. However, the labor costs of such manual computations would be prohibitive.

The most important feature of watershed computer models is that they provide the ability to link the algorithms representing many hydrologic–hydraulic–water quality–economic processes and to make the necessary voluminous computations at a reasonable cost. The labor and other costs or manual calculations are easily one to two orders of magnitude greater than the labor, software, hardware, and other costs of computer calculations. The practical application of hydrologic and hydraulic knowledge was not fully realized until the development of the digital computer. The digital computer facilitates voluminous repetitive calculations and logic operations at costs that are very small compared to the usefulness of the resulting information.

Since the advent of the digital computer and suitable watershed modeling software, the same investment in engineering effort can be used to obtain a more thorough diagnosis of watershed problems and a more thorough evaluation of alternatives. The combination of more thorough problem diagnosis and alternative evaluation should improve the overall quality of the resulting recommendations and actions.

10.3 MODEL SELECTION GUIDELINES

Many computer models are available in the urban SWM field. The labor and other costs of designing, writing, testing, and documenting high-quality software are high. Therefore, it is becoming increasingly unlikely that any organization will write the basic software for an urban SWM project. Approximately 95 percent of the water resources engineering organizations responding to a recent survey (Walesh, 1987, 1988b, 1988c) reported that less that 20 percent of their water resources engineering staff hours are devoted to writing software. About 60 percent of the respondents reported less than 5 percent of the hours devoted to writing software.

Assume that a decision has been made to use computer modeling on an urban SWM project and that one or more existing models will be used. Selection of specific computer programs should be made after carefully enumerating technical needs, identifying available resources, and reviewing available models.

Technical Needs

Enumeration of the technical capabilities of the computer model or models to be used on a particular project follows from familiarity with the hydrologic, hydraulic, water quality, economic, and other problems of the watershed and the overall objective of the project. Examples of technical capabilities which may be needed in a model for a given urban SWM project are:

1. Simulate urban and rural hydrologic, hydraulic, and water quality processes.
2. Compute 100-year-recurrence-interval flood discharges and stages with sufficient accuracy for regulatory purposes.
3. Calculate a range of flood discharges and stages for use in flood insurance studies and for input to flood damage computations.
4. Determine the impact of detention/retention facilities on downstream peak discharges and peak stages.
5. Simulate the effects of bridges, culverts, encroachments, drop structures, dams, channel modifications, dikes, floodwalls, levees, and other hydraulic structures on flood stage profiles.
6. Determine the hydrologic, hydraulic, and water quality effects of urbanization.
7. Compute average annual flood damage.
8. Calculate the costs and benefits of alternative structural and nonstructural flood control management measures.
9. Determine the impact on surface water quality of increases in and reductions to point and nonpoint sources of pollution.
10. Assess the effect on surface water quality of municipal and industrial point sources of pollution.
11. Compute the depth and duration of street flooding when the storm sewer capacity is exceeded.
12. Simulate storm sewer flow under open channel and surcharged conditions.

Resources Available

Competent professional and other staff are the most important resource for any computer modeling effort. These personnel must have many attributes, including an understanding of hydrologic, hydraulic, and other processes inherent in an urban watershed and how these processes are represented in computer models.

A suitable data base consisting of meteorologic, land, conveyance, or storage, and other data, or the financial and other resources to obtain the data base are necessary. The model or models selected must be consistent with the existing data base or the data base that can be realistically developed.

The type and extent of data available for a given community, particularly land, conveyance, and storage data, may be deficient to support even the simplest computer models. A side benefit of computer modeling is that the input requirements force the development of an accurate, complete representation of the surface water system. This data base may subsequently be very useful in the day-to-day operations of the city, village, county, district, or other entity. For example, a by-product might be a computerized list of all storm water facilities in the system which could be used for scheduling inspection and maintenance work, responding to calls for service, and preparing an annual report.

The third key resource is availability of or access to computer hardware and peripherals that support the computer models being considered. As recent as the early 1980s, hardware availability was often an obstacle to many potential model users. However, with the advent of low-cost, powerful personal computers, and the dominance of the MS DOS operating system in the water resource field, hardware has become much less of a consideration.

Although hardware costs have plummeted in recent years, the annual costs that will be incurred by a private or public organization in developing and maintaining computer modeling capability will probably exceed the initial purchase price. Writing about microcomputer costs, Orenstein (1986) argues that the annual cost of maintenance, supplies, training, software purchase, software development, and occasional additional hardware acquisition is likely to be several times the initial purchase price of the basic personal computer.

A final consideration in the selection of a model or models for a particular project is the extent to which the model or portions of it will be used in successive or parallel projects by the government agency or private organization considering modeling. The initial cost of acquiring the necessary personnel and knowledge, the data base, and the software and hardware can be very high. As noted, annual support costs can be very large. These initial investments and ongoing costs may be acceptable only if they can be allocated to or absorbed by two or more parallel or successive computer modeling projects.

For example, in 1975, the Southeastern Wisconsin Regional Planning Commission, which had responsibility for water resources planning in a 2700-mi^2 urban and urbanizing area, made personnel and other commitments to acquire and develop a sophisticated hydrologic–hydraulic–water quality–economics model (SEWRPC, 1976b). The commitment was based on current and anticipated water resources planning efforts, including several comprehensive watershed plans, a major monitoring and modeling research effort on one watershed, and an area-wide water quality planning and management program. The set of computer programs developed in the mid-1970s are still being used, with appropriate updates, in the 1980s.

Models Available

As early as 1977, Brandstetter identified and published a comparison of computer models potentially applicable to urban SWM. More recently, a task committee of the American Society of Civil Engineers identified and compared the features of 28 models applicable to urban areas (Task Committee, 1985). The number of computer models or variations on models has increased significantly in the past decade, driven in part by the advent of the personal computer. For example, about 75 computer programs are available from and supported by the Hydrologic Engineering Center (HEC) of the U.S. Army Corps of Engineers. Almost one-third of these are operable on microcomputers and the conversion to microcomputer capability continues. HEC program categories includes river hydraulics, surface water hydrology, statistical hydrology, groundwater hydrology, reservoirs, water quality, economic analysis, and spatial data management (U.S. Army Corps of Engineers, 1986b).

Although a large number of models are now available, a recent survey (Walesh, 1987, 1988b, 1988c) of leading private and public water resources organizations suggests that only a relatively small number of computer programs are widely and frequently used. Of the 21 leading organizations surveyed, 19 responded and reported using at least 40 different microcomputer and mainframe programs ranging from readily available and supported programs, such as the HEC series, to proprietary in-house programs. Heavy use was made of computer models that are readily available, usually from government agencies at low cost, and are well supported. Over half of the respondents reported using all of the following models on contract projects in the past few years: HEC-1, HEC-2, TR20, SWMM, ILLUDAS, and ILUDRAIN. All of these heavily used programs, with the exception of ILUDRAIN, are available from and supported by public agencies. All of the models, variations on them, or supplements to them are also available from one or more private service organizations. Some of these programs, plus a few other programs known to be in moderate use in urban SWM, are discussed briefly in the next sections of this chapter to provide a further introduction to available computer programs.

In evaluating available models, urban SWM planning and design projects, the operational test suggested by Bedient and Huber (1988) has merit. A model is defined as being operational, that is, suitable for production as opposed to research use, if it satisfies three criteria. First, documentation must be provided in the form of a user's manual and a description of the theory and algorithms in the model. Second, support must be available as needed via telephone or correspondence. Finally, a model should have a history of many and varied successful uses by other practitioners.

Confer with Other Users

Having enumerated technical needs, identified resources, and reviewed available models, the engineer should be in a good position to select one or more computer models for a given project. The final decision might include conversations with colleagues in government agencies and consulting firms who are using the models under final consideration. Matters to discuss include model technical capabilities and limitations, adequacy of documentation, and extent and quality of support provided by the supplier of the model.

10.4 INTRODUCTION TO SELECTED COMPUTER MODELS

HEC-1 Flood Hydrograph Package

Capabilities. The source of this model, the Hydrologic Engineering Center, describes the program as follows (U.S. Army Corps of Engineers, 1986a):

> All ordinary flood hydrograph computations associated with a single recorded or hypothetical storm can be accomplished with this package. Capabilities include rainfall–snowfall–snowmelt determinations; computations of basin precipitation, unit hydrographs, kinematic wave transforms and hydrographs; routing by reservoir, storage-lag, multiple-storage, straddle-stagger, Tatum, Muskingum, and kinematic wave methods; and complete stream system hydrograph combining and routing. Best-fit unit hydrograph, loss-rate, snowmelt, base freezing temperatures and routing coefficients can be derived automatically. Automatic printer plot routines are also provided. HEC-1 may also be used to simulate flow over and through breached dams. Expected annual flood damage can also be computed for any location in a river basin.

The primary function of HEC-1 is to compute the volume and timing of stormwater runoff from the land surface to the conveyance and storage system and to route it downstream through the system, yielding a series of discharge values at predetermined locations. Two factors included in HEC-1 determine the amount and rate of runoff from the land to the conveyance and storage system: precipitation events, which establish the quantity of water available, and the natural and man-made features of the land surface, which define the amount and rate runoff. Discharges in swales, channels, storm sewers, detention/retention facilities, and other conveyance and storage works are determined by two factors: the amount and rate of flow from the land surface and the geometry and roughness of the conveyance works.

Data Requirements. Examples of data required to utilize HEC-1 if the Soil Conservation Service method is used are rainfall hyetographs, subbasin area, runoff curve number, and lag time. If open channel reaches are to be simulated, reach data requirements include length, longitudinal slope, Manning roughness coefficient, and channel geometry. Simulation of detention/retention facilities require elevation–discharge–volume relationships as well as the starting water surface elevation.

Model Availability and Support. The program is written in Fortran and is available from the Hydrologic Engineering Center (HEC) of the U.S. Army Corps of Engineers on magnetic tape in source code for mainframe or minicomputers. HEC-1 is also available for microcomputer users on the disks for execution on MS-DOS-compatible systems (512K RAM for DOS 2.1; 640K RAM for DOS 3.1). A large program, HEC-1 contains about 18,500 lines of code with about 13,500 lines of code being without comments.

The cost of the program ($200 for one copy to operate on mainframe, minicomputer, or PC as of May 1986) includes a user's manual, programmer's manual, and test data with output results. Also available from HEC at cost are related technical papers (U.S. Army Corps of Engineers, 1986a, b). For many years, HEC provided a high level of support for HEC-1 and other HEC computer programs. For example, program purchasers automatically received enhancements or corrections to the source code, and users obtained assistance directly from the HEC. As of late 1988, HEC support services were markedly reduced.

HEC-1, variations on it, and supplements to it are also available from various private organizations. Examples are Dodson and Associates, Inc. (7015 W. Tidwell Road, Suite 107, Houston, TX 77092, 713-895-8322, and Haestad Methods (37 Brookside Road, Waterbury, CT 06708, 800-422-6555).

HEC-2 Water Surface Profiles

Capabilities. The Hydrologic Engineering Center, the originator of this model, describes the program as follows (U.S. Army Corps of Engineers, 1986a):

> The program computes water surface profiles for one-dimensional steady, gradually varied flow in rivers of any cross-section. Flow may be subcritical or supercritical. Various routines are available for modifying input cross-section data; for example, for locating encroachments or inserting a trapezoidal excavation on cross-sections. The water surface profile through structures such as bridges, culverts and weirs can be computed. Variable channel roughness and variable reach length between adjacent cross-sections can be accommodated. Printer plots can be made of the river cross-sections and computed profiles. Input may be in either English or metric units.

The standard step computational procedure is used in the model for channel or channel and floodplain reaches. For a given discharge and stage at a starting cross section, a trial stage is estimated for the next upstream cross section. Using the Manning equation, the mechanical energy loss between the two cross sections is calculated on the premise that the upstream stage is correct. Then the program checks to determine if the conservation of energy principle is satisfied. Assuming that conservation of energy is violated, a second upstream stage is established and tested. The process is repeated until the upstream stage is determined which satisfies the conservation of energy. This iterative computational procedure is repeated for successive upstream reaches. The process results in calculated flood stages at each cross section. Refer to Chapter 4 for a detailed description of the standard step technique.

The hydraulic effect of hydraulic structures such as bridges, culverts, and weirs is also calculated using the conservation of energy principle. The stage upstream of hydraulic structure is determined as a function of discharge, the physical characteristics of the structure, and the known downstream stage. Starting with the downstream stage, mechanical energy losses attributed to phenomena such as expansion of flow, flow through or over the structure, and contraction of flow into the structure are calculated. Flow through or over a bridge, culvert, or weir may include various combinations of open channel, pressure, and weir flow.

Data Requirements. Examples of the extensive amounts of data required to utilize HEC-2 are flows, channel–floodplain cross sections and distances between them, and Manning roughness coefficients for the channel and each floodplain. When bridges, culverts, and other hydraulic structures are simulated, additional required data include waterway opening sizes, pier location and shape, channel bottom elevations, profiles of approach roads and similar related structures, and dam or weir crest elevation and shape. Refer to Section 4.6 for an example of contingency checks that can be used to discipline the review of voluminous output produced by HEC-2 and similar water surface profile programs.

Model Availability and Support. HEC-2 is written in Fortran and is available from the Hydrologic Engineering Center (HEC) of the U.S. Army Corps of Engineers on magnetic tape in source code for mainframe or minicomputers. It is also available for microcomputer users on disks ready to be executed on MS-DOS-compatible systems (512K RAM for DOS 2.1). The program is large, containing over 12,000 lines of code, 8300 of which are lines of code without comments.

There is a nominal charge (e.g., $200 for one copy to operate on mainframe, minicomputer, or PC as of May 1986) for the program, a user's manual, and test data with output results. Additional support documentation, such as a programmer's manual and related technical papers, are available at cost from the HEC (U.S. Army Corps of Engineers, 1986a, b). As with HEC-1, the HEC reduced support services for HEC-2 in late 1988.

HEC-2, variations on it, and supplements to it are also available from various private organizations. Examples are the previously noted firms Dodson and Associates, Inc., and Haestad Methods, and other firms, such as Plus III Software, Inc. (1 Dunwoody Park, Suite 130, Atlanta, GA 30338, 800-235-4972).

ILUDRAIN

Capabilities. This program is described by its authors as follows (Terstriep and Noel, 1985):

> ILUDRAIN is an enhanced version of the ILLUDAS model offered for many years by the Illinois State Water Survey. Since the Water Survey has decided not to support a micro-computer version of ILLUDAS outside Illinois, ILUDRAIN will fill the needs of many users who are switching from mainframe to micro applications.
>
> ILUDRAIN is a hydrologic model that generates runoff hydrographs by applying user supplied storm rainfall intensities to the paved and grassed areas of urban watersheds. The hydrographs are generated from as many (subbasins) as the user desires, using physical watershed parameters such as the slope and distribution of contributing paved and grassed areas.
>
> The hydrographs are routed downstream through the drainage system, described by the user as circular or rectangular pipes or trapezoidal open sections. Deten-

tion storage may be specified in any (subbasin), and the downstream impacts of this storage on flow and pipe size are shown.

Both tabular and graphical output are produced at the users option. Multiple hydrographs such as inflow and outflow from a detention basin may be shown on the same plot.

ILUDRAIN, and its predecessor ILLUDAS (Terstriep and Stall, 1974), are unique relative to most computer models used in urban SWM because they have some explicit design capability. Whereas, like most models, ILUDRAIN and ILLUDAS analyze a user-specified hydrologic-hydraulic system, ILUDRAIN and ILLUDAS can also size storm sewers to carry computed peak discharges.

Data Requirements. Required rainfall information includes designation of a standard or user provided rainfall hyetograph, rainfall duration, total rainfall volume, and antecedent moisture condition. Subbasin data requirements include subbasin identification, area, directly connected paved area, supplemental paved area, contributing pervious area, time of concentration, and hydrologic soil group. Examples of data to be provided for reaches include reach identification, length, slope, roughness, and geometry.

Figure 10.4 shows a watershed with much of the data required to do hydrologic–hydraulic simulation with ILUDRAIN. The watershed has six subbasins and six sewer or channel reaches, referred to as branch reaches. The main purpose of the example simulation is to determine the required concrete sewer sizes with $n = 0.013$ for branch reaches 1–0, 1–1, 1–2, 2–0, and 3–0. In addition to the data presented in Appendix B, the following conditions apply:

1. Hydrologic soil group C
2. Antecedent moisture condition 3, which is designated as "rather wet" and represents 0.5 to 1.0 total rainfall during the five days preceding the rainfall event to be simulated (Terstriep and Stall, 1974)
3. 5-year 1-hour design storm of 1.55 in. using the Huff distribution for north central Illinois (Terstriep and Stall, 1974)

ILUDRAIN input data for the example in the form of coded data sheets is presented in Appendix B. Data forms, such as these for ILUDRAIN, are typically provided for computer models. ILUDRAIN output corresponding to the Appendix B input is presented in Appendix C.

Model Availability and Support. ILUDRAIN is available from CE Software (P.O. Box 2474, Station A, Champaign, IL 61820, 217-359-5602) on disks ready to be executed on MS-DOS (128K RAM storage is required). The cost of the program ($375 as of late 1986) includes the program disk, a utility disk,

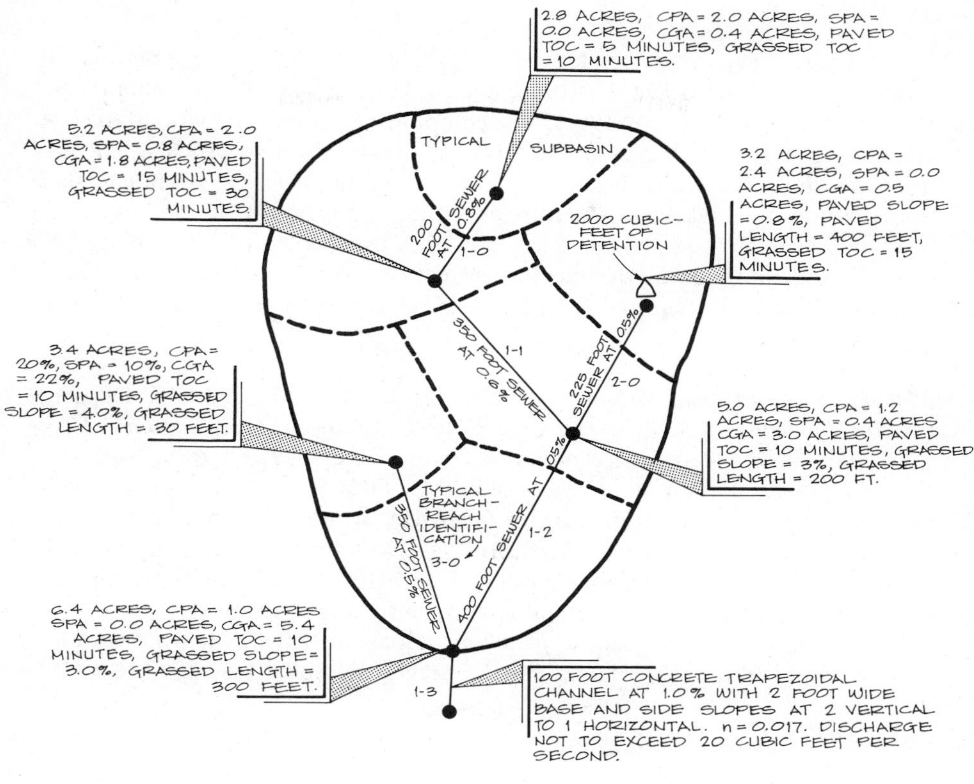

NOTE: TOC : TIME OF CONCENTRATION
CPA : CONNECTED PAVED AREA
SPA : SUPPLEMENTAL PAVED AREA
CGA : CONTRIBUTING GRASSED AREA

FIGURE 10.4 Watershed used for example of ILUDRAIN input and output (*Source:* Adapted from Terstriep and Noel, 1985)

user's manual, sample problem, coding forms, updates for one year, a license agreement, and 1 hour of consulting time. Additional support is available for a fee (Terstriep and Noel, 1985). The similar ILLUDAS program (Terstriep and Stall, 1984) is available to Illinois residents from the Illinois State Water Survey, which supports the program. Additional support documentation relevant to both ILUDRAIN and ILLUDAS is available, including reports of sensitivity studies (Donohue & Associates, 1979; Wenzel and Terstriep, 1976). Variations on ILLUDAS are available from private firms such as Dodson and Associates, Inc.

TR55

Capabilities. This program has all the capabilities of the TR55 method as introduced in Chapter 3 and as presented in detail in the TR55 manual (U.S. Department of Agriculture, 1986). The computer program is structured to correspond to major chapters in the TR55 manual. More specifically, menu-driven modules are available for runoff curve number determination, calculation of time of concentration and travel time, application of the graphical peak discharge method presented in Chapter 3, use of the tabular hydrograph method, and computation of detention/retention storage volume, as also illustrated in Chapter 3. Required data are identical to that for TR55 as introduced in Chapter 3.

Model Availability and Support. TR55 was developed by the U.S. Department of Agriculture, Soil Conservation Service. The program is written in BASIC and available for microcomputer users on two disks ready to be executed on MS-DOS-compatible systems (256K RAM). TR55 is available from the National Technical Information Service (U.S. Department of Commerce, 5285 Port Royal Road, Springfield, VA 22161, 703-487-4650). There is a nominal charge (e.g., $90 for one copy to operate on a PC as of early 1987) for the program. A separate user's manual is apparently not available.

TR55, variations on it, and supplements to it are also available from various private organizations. An example is the previously noted Hastead Methods.

Other Computer Models

As noted earlier in this chapter, numerous and varied computer models are available. Four models—HEC-1, HEC-2, ILUDRAIN, and TR55—are introduced in the preceding section of this chapter. Brief introductions to other models, such as

DR3M-QUAL (Distributed Routing Rainfall Model with Quality)
HSPF (Hydrocomp Simulation Program - Fortran)
HVM-QQS (HVM-Quantity-Quality-Simulation)
MITCAT (MIT Catchment Model)
PSRM (Penn State Runoff Model)
RUNQUAL (Runoff Quality)
STORM (Storage, Treatment, Overflow, Runoff Model)
SWWM (Storm Water Management Model)
TR20

are provided by others. For example, the PSRM, SWWM, and TR20 models are introduced by the American Public Works Association (1981); the HSPF, HVM-QQS, MITCAT, RUNQUAL, STORM, SWMM, and TR20 models are presented in summary form by Whipple et. al. (1983); and Bedient and Huber (1988) introduce HSPF, PSRM, STORM, and SWWM.

10.5 USE OF TWO OR MORE COMPUTER MODELS ON ONE PROJECT

Selected individual computer models are described in Section 10.4. In practice, two or more computer models are often used in a single project, with some of the output of one being used, usually by manual intervention, as the input to one or more other models. Individual computer programs comprising these computer model packages are selected to meet specific project needs, all of which cannot be satisfied by any one model. The useful practice of using two or more computer models may be illustrated by describing computer model packages used in two actual projects.

HEC-1 and HEC-2 Package

HEC-1 and HEC-2 were used to perform diagnostic investigations and to test alternative flood control solutions on a 2.96-mi^2 area in and near Valparaiso, Indiana (Donohue & Associates, 1984b; Walesh et al., 1987). Input to, output from, and the interrelationship between the two models are shown in Figure 10.5. As shown by the arrows in Figure 10.5, the two programs were interrelated through manual intervention for purposes of this project. For each of

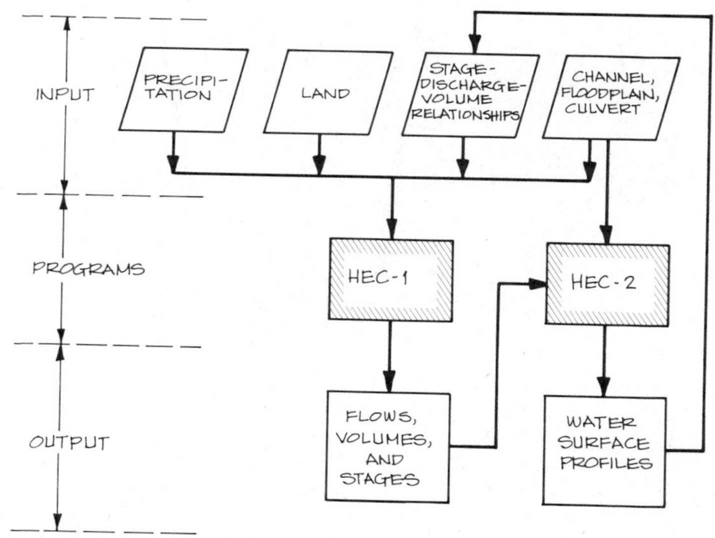

FIGURE 10.5 Hydrologic–hydraulic model composed of two HEC programs

the two computer programs, output was a final computed result or was used as input to the other program.

As used in this investigation, the function of HEC-1 was to determine the volume and timing of runoff from the land surface to swale, channel, storm sewer, or detention/retention facilities and route it downstream through the system of conveyance and storage facilities, producing a series of discharge values at predetermined locations. HEC-2 was used to calculate flood stages for selected reaches in the open channel system.

The relationship between the computer programs and the important components of the watershed system is illustrated in Figure 10.6. The entire land surface, most of the conveyance system, and all detention/retention facilities were simulated with HEC-1. Subsequently, HEC-2 was applied to selected channel reaches where detailed hydraulic analyses, such as the generation of flood stage profiles for various recurrence intervals, were required. The total of 19 simulations that were conducted included existing and future land use, a historic rainfall event, and rainfall events over a range of recurrence intervals.

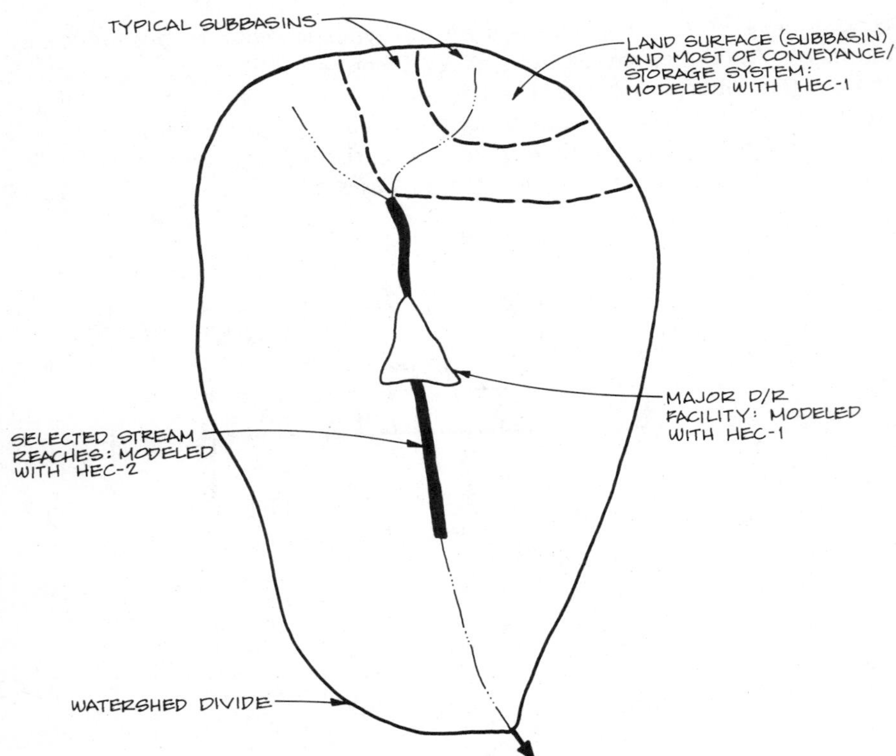

FIGURE 10.6 Watershed components and portions of the HEC-1 and HEC-2 model used to simulate them (*Source:* Adapted from Donohue & Associates, 1984b)

The watershed diagnosis accomplished with the computer model package revealed that existing detention facilities were undersized. In addition, major channels and conduits through built-up areas were found to have insufficient capacity and future urbanization would aggravate flooding throughout the watershed. Numerous alternatives were evaluated using the computer model package. The resulting plan recommended the construction of two new detention facilities, modification of two existing detention facilities, and extensive channel cleaning and maintenance.

Package of Proprietary and Public Discrete Event and Continuous Models

As noted earlier in this chapter, the Southeastern Wisconsin Regional Planning Commission was, in the mid-1970s, participating in or planning to undertake water resources studies throughout most of its 2700-mi^2 planning area. No single computer model at that time satisfied the modeling criteria. Accordingly, a package of largely existing computer programs was assembled. Figure 10.7 illustrates the overall package, identifies five computer programs or submodels comprising the package, shows interrelationships between computer programs, identifies input and output of each computer program, and indicates the use of the various outputs in comprehensive water resources planning.

The three computer programs identified as Hydrologic Submodel, Hydraulic Submodel 1, and Water Quality Submodel are three computer programs contained within a package then called the "Hydrocomp Simulation Programming" (Hydrocomp, 1972, 1976). These first three computer programs were all continuous process simulation models. Hydraulic Submodel 2 was the then current version of HEC-2, a discrete event, steady-state program. The Flood Economics Submodel was developed by the commission staff.

A major effort over a period of about one year was required to obtain and integrate the proprietary, public domain, and new software into the desired model configuration. Development of the necessary data base, refinement to the model, and the use of the model continued, on an as-needed basis, for over a decade.

10.6 CALIBRATION: KEY TO CREDIBILITY IN MODELING

Most of the algorithms contained in a hydrologic–hydraulic–water quality–economics model are mathematical approximations of complex phenomena. Therefore, a model should be calibrated before it is used to simulate streamflow, flood stage, water quality conditions, and monetary flood damages under existing and hypothetical watershed development conditions. That is, results obtained with the model should be compared to historic field data for

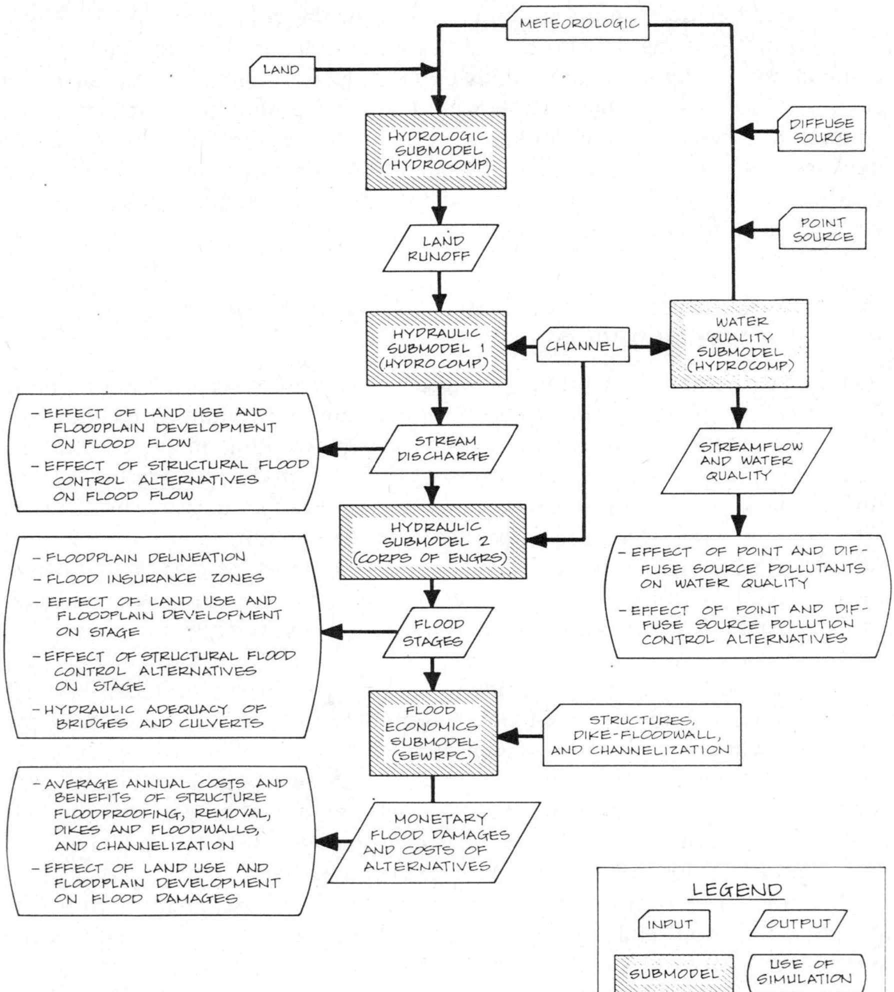

FIGURE 10.7 Hydrologic–hydraulic–water quality–flood economics model composed of Hydrocomp, HEC and specially written programs (*Source:* Adapted from Walesh, 1976)

the watershed, and if significant differences are found, adjustments should be made to the model parameters to adjust the model to the specific natural and man-made features of the watershed.

The extent to which the results of digital computer hydrologic–hydraulic–water quality–economic and other modeling enter into and influence the decision-making process and enhance the quality of those decisions is largely dependent on model credibility as perceived by the decision makers. That credibility is, in turn, determined by the thoroughness of the model calibration efforts and the effectiveness of the means used to display and communicate the results of the calibration to the decision makers.

Once a watershed model is calibrated using data for a given watershed, the simulation proceeds on the premise that the model will respond accurately to a variety of model inputs representing hypothetical watershed conditions such as land use changes and channel modifications. Therefore, the model provides a powerful analytic tool in SWM.

As noted, a model "should" be calibrated before it is used to simulate existing and hypothetical conditions. However, the practitioner will often encounter situations in which adequate data are not available for calibration and financial resources and time constraints prohibit establishing a monitoring program. Inability to calibrate does not necessarily mean that modeling should not be used. In these situations, the engineer must determine if an uncalibrated model, based on its previous use in similar situations, constitutes the most appropriate analysis and design technique relative to other approaches such as manual methods.

It is interesting to note that calibration is often expected when more sophisticated tools, such as digital computer models, electric analog models, and physical models are used, but is not expected when manual methods, such as the rational method, the Road Research Laboratory method, or the universal soil loss equation method, are employed. Linsley (1973) notes that contrary to common practice, "any hydrologic procedure will yield better results if tested against observed data and any constants appropriately fixed by data from the project area."

This section of the chapter emphasizes the importance of calibration and presents ideas and techniques for calibration of hydrologic–hydraulic–water quality–economic models. In addition to defining and stressing the necessity of calibration, the following topics pertinent to model calibration are discussed and illustrated with examples drawn from actual projects: the calibration-verification process, types and sources of calibration data, techniques for storage and display of calibration data, accuracy of data and its effect on calibration, sensitivity analyses, and calibration techniques and related considerations. Much of the material in this chapter is based on a paper by Walesh and Raasch (1978).

Calibration

According to Marsalek (1977), the two important functions of calibration are (1) to provide estimates of model input parameters that are difficult to measure directly—such as infiltration rates into or pollutant accumulation rates on the land surface—and (2) to compensate to some extent for minor deficiencies in the structure of the model. In addition to these functions, the calibration process also serves to indicate the sensitivity of the model to changes in parameters and therefore suggest the sensitivity of the actual system to changes in the system. That is, much can be learned about the system through the calibration process. Sensitivity analysis is discussed in detail later in this chapter.

Although formal parameter optimization procedures are possible, the most common approach to model calibration is a trial-and-error process concentrating on physically meaningful changes to those model parameters having the greatest impact on model output. Calibration may be viewed as an iterative process in which preliminary values of model parameters are set, the model is operated, the simulated output compared to measured output, and if necessary, changes are made in input parameters, thus initiating the second cycle of the process. The process is repeated until a satisfactory correlation is obtained between simulated and measured output.

The generalized model calibration process is illustrated in Figure 10.8. The actual process used to calibrate a package of hydrologic–hydraulic–water quality models to the 137-mi^2 urbanizing Menomonee River watershed in southeastern Wisconsin is shown in Figure 10.9.

The Ideal Calibration–Verification Process

Ideally, model calibration should be followed by verification. Model verification consists of comparison of measured response to computed output, with the latter based, in part, on empirically derived input parameters obtained

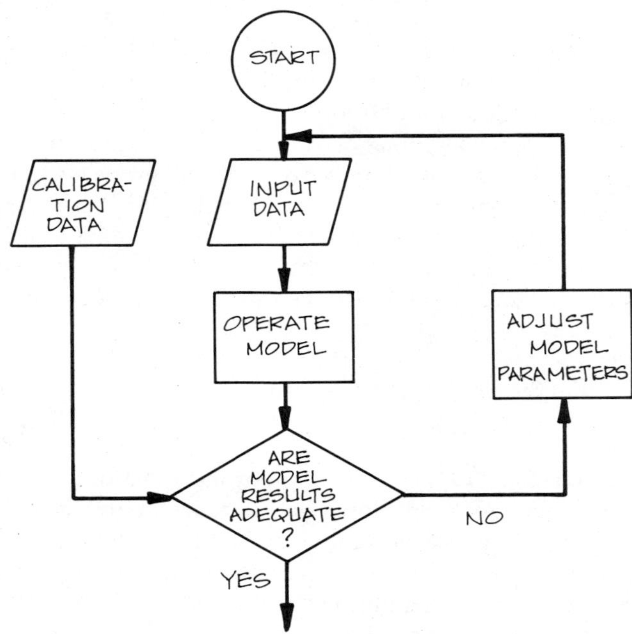

FIGURE 10.8 Generalized model calibration process

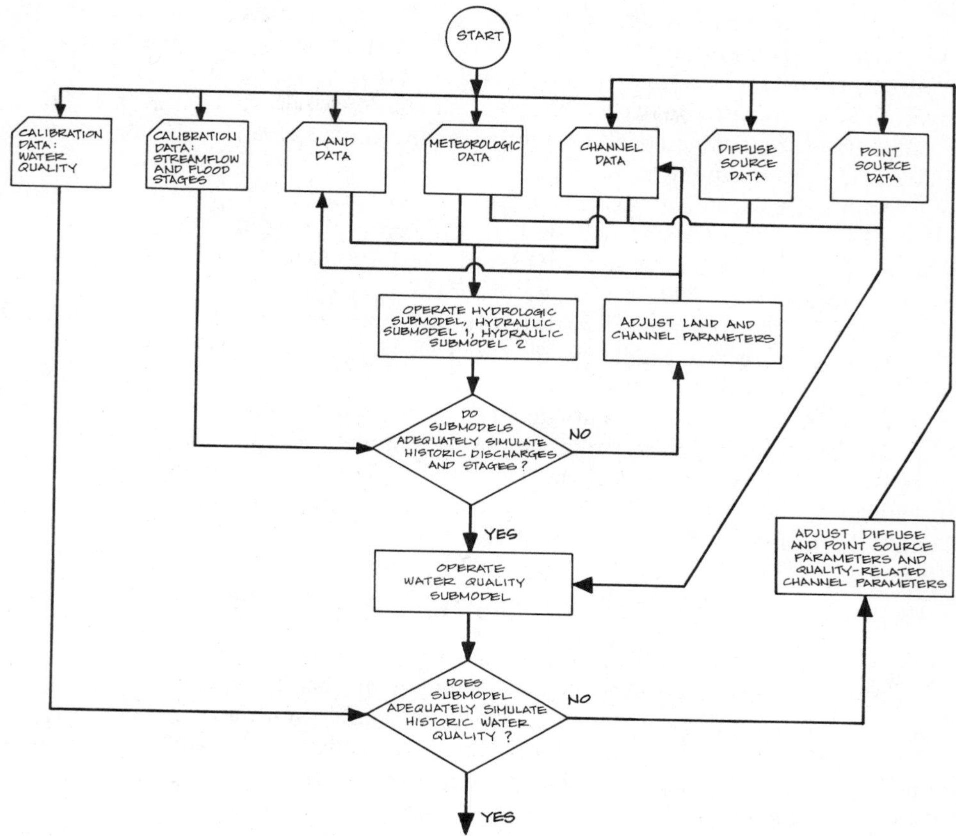

FIGURE 10.9 Process used to calibrate hydrologic–hydraulic–water quality model to the Menomonee River watershed in southeastern Wisconsin (*Source:* Adapted from SEWRPC, 1976b)

from calibration. In the calibration–verification process, the available field data are split, with part being used to calibrate the model and the remainder used for an independent verification of the calibrated model. In actual practice, data, if available at all, are often sufficient only for calibration, so that it is not usually feasible to split the data set in order to perform a calibration–verification sequence.

The literature contains a few examples of the use of the desirable calibration–verification process. One example is Larsen's (1977) study, in which a hydrologic–hydraulic model was applied to a 383-acre urban catchment in the Toronto area. Calibration consisted of modeling three summer rainfall events and comparing recorded and simulated runoff. The overall purpose was to establish numerical values for those input parameters that could not be determined accurately by direct measurements. Percent imperviousness is an example of such an input parameter, in that it could not be determined accurately because of errors in delineating the areas on aerial photographs and because of "loss" of some runoff from impervious areas to pervious areas. Model verification consisted of the subsequent simulation of 16 rainfall events and comparison of simulated and recorded hydrographs, runoff volumes, peak discharges, and time to peak.

Another example of the classic calibration–verification process is documented by Abbott (1978), who applied four continuous models and three discrete event models to a 5.5-mi^2 urban watershed near Oakland, California. The continuous models were calibrated using the first 40 percent of a 42-month continuous record of rainfall and runoff and validated using the remaining 60 percent of the record. Single-event models were calibrated using three discrete rainfall-runoff events extracted from the first part of the 42-month data set and validated using four events taken from the remainder of the record.

The remainder of this chapter emphasizes model calibration and makes no additional reference to model verification as defined above. Much of the subsequent discussion is applicable to both calibration and verification, although only the former is mentioned explicitly.

Types and Sources of Calibration Data

Many types of data are available for calibration of hydrologic–hydraulic–water quality–economic models, and such data may be obtained from a variety of sources. Examples of major categories of calibration data are: precipitation, stream discharge, river or lake stage, area of inundation, and water quality and flood damage. These data may be obtained from a variety of public and private sources. Table 10.3 is a data type–data source matrix, indicating probable sources of various types of calibration data.

When beginning to search for calibration data, there may be a tendency to think strictly in terms of quantitative data such as peak flow and maximum stage. But qualitative or semiquantitative data can also be useful. For example, knowing that a roadway was overtopped by floodwaters in the vicinity of a culvert, that extensive ponding occurred on a street, that many basements flooded in a particular area, the sewer surcharge occurred as evidenced by blown manhole covers, or that flow occurred on a floodplain can be useful in model calibration even if the peak stage, area of inundation, and other definitive data are not available.

Photographs of flooding, erosion, sedimentation, algae blooms, and other surface water problems can be useful in calibration. For example, flood stages can be deduced from photographs. Furthermore, photographs remind elected officials and citizens of the severity of the community's surface water problems. Also, to the extent that the model reproduces the historic phenomena shown on the photographs, elected officials and citizens are more likely to accept the credibility of the modeling technique and of the resulting recommendations.

Storage and Display of Calibration Data

Once calibration data have been obtained from the various sources and have been collated, the data should be managed in such a manner as to save time, reduce costs, and minimize handling errors. In some cases it may be technically and economically feasible to input the data to the model, or a special supplementary computer program, for automatic comparison of recorded data to simulated values. For example, the Hydrocomp Simulation Programming (Hydrocomp, 1976) has a feature whereby recorded daily flows may be input to the model for subsequent graphical comparison to simulated flows on a printer, thus eliminating the need for manual handling of the recorded and simulated daily flows in the calibration process.

A simple but effective means of handling certain types of historic data such as high water marks or detailed flood flow hydrographs or pollutographs, is to plot the data in a reproducible form to facilitate making prints on which simulation results can be readily plotted by hand for ease of comparison to the record data. Computer graphics software can also be used to accomplish this approach.

Some of the data will more readily be handled as "hard copy." Examples include photographs of flood inundation and information such as high water marks or water quality measurements that are widely separated in space and time.

Accuracy of Data and Its Effect on Calibration

Factors Affecting "Goodness of Fit." The "goodness of fit" or correlation between observed behavior and model output may be influenced by the following four factors (Alley, 1977):

1. The criteria used to judge the success of the calibration process, that is, the allowable tolerances or differences between model output and recorded data
2. The structure of the model, that is, the degree to which the mathematical expressions comprising the model algorithms represent actual physical process occurring in the physical system
3. The range of situations represented in the data set used for calibration, that is, the degree to which the calibration period, or periods, are representative of the totality of hydrologic–hydraulic–water quality and other situations that may occur in the system and may be of interest in SWM
4. The quality of the input and calibration data, that is, the degree to which input such as precipitation amounts, pollutant land surface accumulation rates, and channel–floodplain cross sections represent actual conditions in the simulated system

The emphasis here is on the last factor, that is, the quality of the input and calibration data and its effect on the goodness of fit. Discussions of the

TABLE 10.3 Types and Sources of Calibration Data

Types of Data		Source[a]							
Category	Subcategory	USGS and Other Federal Agencies	State DNR and Similar Agencies	Drainage, Sewerage, Flood Control District, or Similar Entity	City, Village, Town, County, or Regional Planning Commission	News Media-Newspaper, Radio, TV	Museums, Historical Societies	Residents of the Area	Engineering/ Planning Firm Having Done Water Resources Work in Area
Stream discharge	Continuous record[b]	×							
	Partial record[c]	×							
	Spec. measurements	×	×						
River or lake stage	Continuous record[b]	×							
	Partial record[c]	×		×	×				
	Spec. measurements	×	×	×	×			×	×
	Ground photos[d]				×	×	×	×	×
Area of inundation	Aerial photos		×			×	×		
	Ground Photos[d]		×	×	×	×	×	×	×
	Discrete ground observations[e]				×			×	
Water quality	Dry weather, low-flow conditions[f]	×	×	×					×
	Wet weather, high-flow conditions[f]	×	×	×					×
	Extended periods[g]								×
	Semiquantitative observations[h]		×	×	×	×	×	×	×

Flood damage	Number and type of structures damaged[i]			×	×		×	×
	Extent of agricultural land inundation	×	×	×	×		×	×
	Dollar damage[j]	×		×	×	×	×	×
	Disaster relief funds provided	×						

[a]× indicates that data are likely to be available from indicated source.

[b]Continuous record (obtained at intervals ranging from a few minutes to a day) by a float or bubble recorder or by manual observations.

[c]Partial record (obtained during extreme high or low flow or other special conditions).

[d]Ground photographs showing flood stages relative to objects of known or readily determined vertical position.

[e]When these discrete observations are applied to a topographic map, the elevation contours on the map can be used to interpolate between the observations and obtain a more complete representation of the inundated area.

[f]Physical, chemical, and biological indicators.

[g]Monitoring for a period of a week to one or more months with the intent of encountering both dry and wet weather conditions.

[h]Information about the occurrence of events such as fish kills, algae blooms, offensive odor, and high turbidity.

[i]Categorized by structure type (e.g., single-family residential, commercial) and with depth of flooding relative to structure.

[j]*Caution:* Wide variations may be expected, depending on which direct and indirect losses or costs are included and whether public- and private-sector losses are included.

In addition, the agency or consulting firm responsible for the modeling could be a source of all categories of calibration data to the extent that special studies are carried out in support of the modeling effort.

significance of the last factor, the quality of input data, are present in the literature (Alley, 1977; Marsalek, 1977; Shelley, 1975), and a summary of that information is presented here.

Types of Errors in Data. The three types of errors that may be present in either input or calibration data, and therefore may influence the success of the calibration effort, are random errors, systematic errors, and blunders. Random errors are dispersed or scattered round a central tendency and are normally as likely to be positive as they are negative and therefore will not have a significant effect on the mean, median, or mode of a set of measurements. A characteristic of random errors is that although they are definitely occurring, they are generally attributable to multiple, overlapping causes, and specific origins are very difficult to identify. In addition, random errors occur in combination with systematic errors, the second of the three types of errors possibly present in calibration and input data.

In contrast to random errors, systematic errors tend to occur with the same sign, that is, they are positive or negative relative to the true value and frequently exhibit a fixed offset from the true value or exhibit a consistent relationship between the measured value and the true value. In a statistical sense, systematic errors may be visualized as producing a distribution having a mean, median, or mode displaced from the true value of the measured parameter. Examples are: under-recording of rainfall amounts due to positioning a rain gauge in close proximity to a structure, and overmeasurement of historic high water marks due to an upward shift in the benchmark from which water levels were run.

The third type of error commonly entering into model input data or calibration data and the easiest to reduce is the blunder. Such errors are usually the easiest to discover during the calibration process because they often yield values extremely different than simulated values and because, unlike random and systematic errors, they show no consistent pattern. An example of a blunder in use of input data is incorrectly coding a coordinate on a channel–floodplain cross section, thus misrepresenting channel–floodplain conveyance. Examples of errors in calibration data are: misreading scales on water quality monitoring devices, such as dissolved oxygen or specific conductance meters, and plotting historic high water data one or more contour intervals above or below the reported value, thus grossly overstating the lateral extent of the historic floodplain. A check list should be prepared and used as a guide to examine model output for the effects of blunders in input or calibration data (SEWPRC, 1977a).

Expected Accuracy of Data. Without even considering other factors influencing the correlation between model output and measured phenomena, it is important to note that the expected accuracy of the input and calibration data in effect establish the maximum possible correlation between model output and historic fact. Alley (1977) states that "it is unrealistic to expect the calibrated model to generate outputs within plus or minus ten percent, if the data used to calibrate the model were only within plus or minus twenty percent."

In a discussion of the target accuracies for measuring discharge in storm sewers, Linsley (1973) implies that flow measurements within plus or minus 5 percent of actual flows are difficult to achieve. The U.S. Geological Survey (1985) defines "excellent" flow records as those in which about 95 percent of the daily discharges are within 5 percent; "good" is defined as having about 95 percent of the daily discharges within 10 percent, and "fair" is defined as within 15 percent. In summary, whereas it is generally agreed that flow measurements within plus or minus 5 percent of true value are desirable, such accuracy is apparently not yet often achieved, and this should be considered in comparing simulated discharges and runoff volumes to recorded values.

Precipitation data have a direct influence on calibration inasmuch as such data provide input to the hydrologic portion of models. Alley (1977) indicates that the accuracy of precipitation measuring instruments may be expected to be less than that of streamflow measurements, whereas Marsalek (1977) suggests that a desirable and achievable precipitation accuracy is about plus or minus 10 percent.

There seems to be a consensus, for example Alley (1977) and Marsalek (1977), that water quality monitoring yields accuracies significantly less than that obtainable in discharge and precipitation measurements. It is generally estimated that water quality and monitoring will, at best, yield water quality values within plus or minus 25 percent of actual in-stream conditions. However, Alley (1977) notes that such accuracies are probably not grossly incompatible with the accuracy of water quality simulation relative to water quantity simulation.

Impact of Data Errors on Calibration

As noted above, it is unrealistic to expect a model to generate output that is more accurate and precise than the input to the model and the data used to calibrate the model. The nonlinearity of hydrologic, hydraulic, and water quality processes precludes a determination, on theoretical grounds, of the impact of the input data errors on model output (Alley, 1977). Even if such a determination could be made, the effect of errors in input data would be masked or inextricably related to errors in model parameters, errors caused by model deficiencies, and errors in calibration data, all of which influence model output and therefore influence the deviation between model output and the true value of the phenomena in question.

One generalization is that of the three types of errors, systematic errors and blunders are more likely to affect the goodness of fit between model output and recorded values than are random errors, since random errors tend to be characterized by being positive and negative in an offsetting or equalizing manner.

Calibration Techniques

A variety of techniques are available to expedite the calibration process and establish the credibility of the results, and some are discussed here. The specific set of techniques applicable to a given modeling study is dependent on the particular model or models being used in the study. For example, some of the techniques presented here are applicable only to continuous process, as opposed to discrete event, models. Although all the specific calibration techniques presented here may not be applicable to a given modeling situation, they do serve as a check list from which the engineer can select or develop procedures likely to be applicable and useful in a given situation. Calibration guides or aides are available for some models.

Benefit from Local Modeling Experience. An economy of effort can occur when a given hydrology–hydraulic–water quality model is repeatedly applied in a given physiographic and climatic area. Once experience is gained using a model on watersheds having a variety of combinations or land slope, soil type, land use and cover, and channel resistance and geometry within such an area, subsequent applications of the model in the same physiographic and climatic area can benefit immensely with respect to the selection of numerical values of input parameters from the earlier studies. Whereas model parameters reflecting the climatic conditions and land surface and channel systems may be expected to vary significantly from one part of the United States to another, they may be expected to exhibit a strong similarity within climatically and physiographically homogeneous areas.

Iterative Data Collection-Model Calibration Process. It is highly desirable that an iterative process be established between data collection and model calibration. As soon as a monitoring program begins and the first set of data are available and have been subjected to quality control checks, the data should be made available for initial model calibration work. Immediate use of data in calibration will not only serve to expedite the overall modeling effort but may feed useful information back into the monitoring program, thus giving rise to an iterative data collection-model calibration process. For example, immediate analysis of streamflow monitoring data and its use in modeling have revealed significant errors in delineation of tributary areas, rating curve deficiencies, and instrumentation malfunctions.

If the monitoring is being conducted primarily to provide for model calibration, immediate use of the data in calibration may identify a point in time in which the monitoring program may be temporarily or permanently terminated, thus resulting in a cost saving to the project. Although data collection is often completed prior to use of the data in model calibration, thus eliminating the possibility of the iterative process above, there is little doubt that the iterative process should be the goal.

Professionals responsible for model setup, calibration, and application should be involved in the planning for and design of the monitoring program because of their intimate familiarity with the model capability and calibration needs. Assume, for example, that the purpose of the monitoring is to provide data for calibration of continuous hydrologic–hydraulic–water quality models. Monitoring should be conducted in each of several seasons of the year over a continuous period of time—perhaps a month—during which, hopefully, dry weather and wet weather conditions will occur so as to permit testing and calibration of all algorithms in the model. In such a monitoring program, monitoring would be conducted relatively infrequently during dry weather conditions—perhaps daily—and at increased frequency during rainfall–snowmelt events. If, in contrast, a discrete event model is to be used, the monitoring program should be designed to capture rainfall–runoff events similar in magnitude to the selected design event.

Sensitivity Analyses. An important early step in the calibration process, particularly if the modeling staff is not well acquainted with the model or models used, is the conduct of sensitivity analyses. This step in the process recognizes that although model output is influenced by a large number of input parameters, the effects of changes in those parameters may be expected to vary widely from parameter to parameter. A sensitivity analysis is a systematic examination of model response to changes in the values of the parameters and is intended primarily to provide more information about the workings of the model, thus aiding the calibration process by identifying "critical" parameters. Assuming that the model provides a reasonable reproduction of system behavior, a sensitivity analysis can have the added benefit of providing the project team with information about the behavior and response of the specific system being modeled.

Sensitivity analyses should be conducted in a systematic, parameter-by-parameter fashion. In practice, much of what is learned about sensitivity of the model to changes in the model parameters is deduced by observation during the calibration process. This does yield useful information but is not likely to be as comprehensive as a systematic sensitivity study.

The types of insights provided by sensitivity studies are illustrated by reviewing the results of two independent sensitivity studies conducted on the model Illinois Urban Drainage Area Simulator (ILLUDAS). Wenzel and Terstriep (1976) applied ILLUDAS to a 288-acre primarily residential watershed in Jackson, Mississippi. The watershed surface water system consisted of open channels and storm sewers. Hydrologic soil group C was dominant and yard and street slopes were in the range 1 to 6 percent. About 25 percent of the watershed was impervious, with almost two-thirds of the impervious area being directly connected to the watershed outlet. The watershed was partitioned into 25 subbasins for the sensitivity analysis.

Input parameters individually varied during the sensitivity analysis were antecedent moisture condition, hydrologic soil group, magnitude and frequency of storm event, percent imperviousness, time increment used for calculations, and number, and therefore size, of subbasins. Output parameters observed were peak flow, pervious area runoff volume, and pipe sizes chosen by the model when operated in the design mode. Some of the results were as follows:

1. Peak flow increased as percent impervious increased, as would be expected, but the rate of increase in peak flow diminished markedly as imperviousness increased.
2. A change in soil group from one category to an adjacent category had a 10 percent or less impact on peak flows.
3. Runoff volume, in contrast to peak flow, was very sensitive to changes in hydrologic soil group and antecedent moisture condition.
4. The most reliable results were obtained by using a computational time increment equal to or less than average paved area entry times.
5. Aggregating subbasins, which reduced the number by about one-half, had an insignificant effect on peak discharge from the watershed and on arrival time of the peak at the watershed outlet.

Donohue & Associates (1979) applied ILLUDAS to a 441-acre primarily agricultural watershed with some residential and forested areas in Mt. Pleasant, Wisconsin. This sensitivity study was conducted with full knowledge of and to complement the preceding sensitivity study. In the second sensitivity study, different independent parameters were examined and a different type of watershed was used.

The surface water system consisted largely of natural drainage features. Hydrologic soil group C was dominant and slopes were generally in the 1 to 3 percent range. About 9 percent of the watershed was impervious with less than one-third of this directly connected to the watershed outlet. The watershed was partitioned into seven subbasins for the sensitivity analysis.

Input parameters varied in the sensitivity analysis were paved and grassed area entry times, percent contributing grassed area, percent directly connected paved area, and percent supplemental paved area. In addition, results obtained with local historic rainfall events were compared to results obtained with the standard temporal distributions included in the program. Output parameters observed were peak discharge on a subbasin and watershed basis and runoff volume for the watershed. Some of the results were as follows:

1. Peak flow from the watershed was very sensitive to variations in paved and grassed area entry times. For example, when entry times were increased and decreased by 50 percent, peak discharge from the watershed was, respectively, decreased and increased by 30 percent.
2. Peak flow and total runoff volume were very sensitive to changes in directly connected paved area and supplemental paved area emphasizing the need to define carefully the means by which paved or impervious areas are hydraulically connected to the watershed outlet.

3. For short storms—approximately 1 hour or less in duration—the peak discharge was about the same when the historic temporal distribution was used as when the standard distribution available in the model was used. For longer durations, up to 15 hours, wide variations occurred in peak flow for the standard distribution versus the historic distribution.

Huber et al. (1975) describe a systematic sensitivity study conducted with both the runoff quantity and runoff quality portions of the Storm Water Management Model. The model was applied to a 135-acre urban basin and a specific storm event was simulated. The investigation revealed that runoff quantity was particularly sensitive to surface roughness in impervious areas, detention depth for impervious areas, and maximum and minimum values of infiltration. Runoff quality was highly sensitive to land use since nonpoint-source pollution land surface loading rates are a function of land use.

Physically Meaningful Parameter Adjustment. There are probably many parameter adjustments or combinations of parameter adjustments that could lead to a satisfactory calibration in a given situation, for example, reproduction of a recorded discharge, stage, or water quality condition. It is vitally important, however, that the numerical values assigned to parameters be physically meaningful.

Assume, for example, that the initial simulated stage for a specific flood event is high in a given reach relative to the flood stage profile actually recorded during the historic event. During the subsequent calibration process, the Manning roughness coefficients for the channel and floodplains are decreased, and although this produces a lower flood stage, it is still high compared to the recorded event. A further decrease in the Manning roughness coefficients could reproduce the required stage but would not be reasonable, since it would involve use of numerical values well below those expected for the physical condition of the channel and the floodplain. The calibration effort should turn at this point to other avenues, such as the possibility that the channel–floodplain geometry or bridge–culvert geometry are not adequately represented in this reach; the possibility that historic data are in error; or the possibility that the simulated flows are too high because of failure to adequately represent storage in the system due to formation of temporary impoundments upstream of constrictive culverts or bridges (Malcom, 1978).

Transition from the Typical Homogeneous Subbasin to the Typical Heterogeneous Watershed

Most gauged watersheds tend to be heterogeneous with respect to hydrologic–hydraulic–water quality factors such as soil type, ground slope, land use-cover, channel profiles, and channel–floodplain cross sections. Therefore, the calibration process should not be initiated on the entire subject watershed. Instead, the initial calibration of a model should be conducted on small subbasins that are essentially homogeneous with respect to those characteristics that are the primary determinants of the hydrologic–hydraulic–water quality response. By following this approach, only one or perhaps two sets of hydrologic–hydraulic–water quality parameters need be dealt with during each calibration run. Parameter values determined by calibration runs on small homogeneous units may then be applied to the heterogeneous watershed during the latter stage of the calibration process.

The procedure of beginning with homogeneous subbasins or subwatersheds and then doing heterogeneous subwatersheds or watersheds is illustrated in schematic form in Figure 10.10 and may be further explained by briefly describing its application to hydrologic–hydraulic simulation of the 135-mi^2 Menomonee River watershed in southeastern Wisconsin using a continuous hydrologic–hydraulic model (SEWRPC, 1976b).

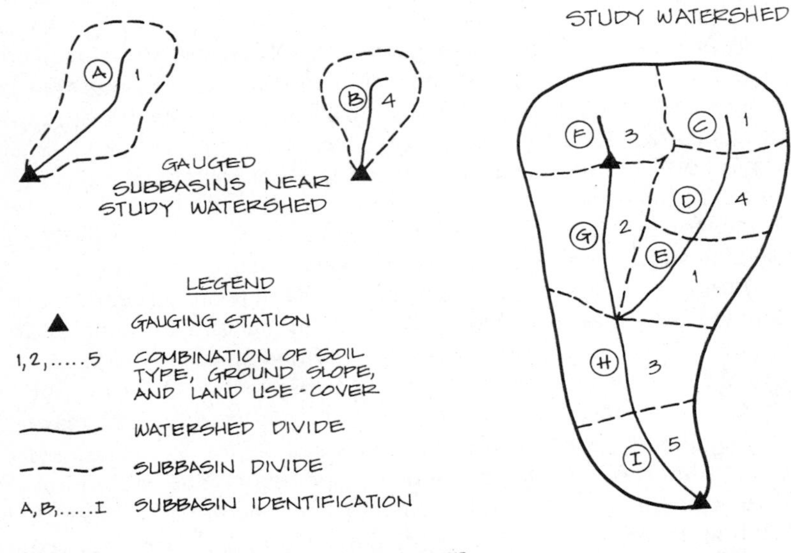

SUGGESTED CALIBRATION SEQUENCE:

1. CALIBRATE SUBBASIN A - USE RESULTS TO ESTABLISH PARAMETERS FOR SUBBASINS C AND E IN THE STUDY WATERSHED.
2. CALIBRATE SUBBASIN B - USE RESULTS TO ESTABLISH PARAMETERS FOR SUBBASIN D IN THE STUDY WATERSHED.
3. CALIBRATE SUBBASIN F IN THE STUDY WATERSHED - USE RESULTS TO ESTABLISH PARAMETERS FOR SUBBASIN H IN THE STUDY WATERSHED.
4. CALIBRATE THE ENTIRE STUDY WATERSHED - DEDUCE PARAMETERS FOR SUBBASINS G AND I.

FIGURE 10.10 Transition from small homogeneous subbasins or subwatersheds to a large heterogeneous watershed during model calibration

The Menomonee River watershed contains five major combinations of soil type and land use-cover and a wide variety of channel slopes and channel–floodplain geometry. Inasmuch as only one continuous recording stream gauge was available in the watershed—supplemented with three partial record stations—it was preferable to follow the procedure of first achieving successful calibration on relatively homogeneous subbasins within or outside the Menomonee River watershed before approaching calibration of the entire heterogeneous watershed. Accordingly, the following calibration sequence was followed:

1. Calibration of the hydrologic–hydraulic portions of the model on the gauged 25-mi^2 Oak Creek subwatershed located outside of but near the Menomonee River watershed and containing two of the five soil type and land use-cover combinations present in the study watershed
2. Calibration of the hydrologic–hydraulic portion of the model on the gauged 60-mi^2 Root River Canal subwatershed located outside of but near the Menomonee River watershed and containing two of the five soil type and land use-cover combinations present in the study watershed
3. Calibration of the hydrologic portion of the model on the gauged 50-mi^2 East Branch of the Milwaukee River subwatershed located outside of but near the Menomonee River watershed and containing one soil type and land use-cover combination present in the study watershed
4. Calibration of the hydrologic–hydraulic portions of the model on a gauged—partial record—3.3-mi^2 urban subwatershed in the Menomonee River watershed and containing two of the soil type and land use-cover combinations present in the study watershed
5. Calibration of the hydrologic–hydraulic portions of the model on a gauged—partial record—8-mi^2 rural subwatershed in the Menomonee River watershed and containing one of the soil type and land use-cover combinations present in the study watershed
6. Finally, calibration of the hydrologic–hydraulic portion of the model against streamflow records obtained near the outlet of the entire study watershed, which, in effect, reflected all five soil type and land use-cover combinations present in the study watershed

In summary, the calibration process should proceed from calibration of small, gauged, relatively homogeneous subbasins or subwatersheds to the calibration of the large, gauged heterogeneous watershed that is the subject of the study. Although the preceding specific example emphasizes hydrologic–hydraulic simulation, the process of the transition from small homogeneous areas to larger heterogeneous areas can also be used in the calibration of water quality models.

Transition from Long-Term to Short-Term Phenomena

Hydrologic–hydraulic–water quality phenomena embrace a wide range of responses in terms of duration and temporal variation. For example, at one end of the spectrum is average annual runoff or average annual transport of a pollutant from a watershed which generally exhibits relatively small year-to-year changes. At the other end of the spectrum is the flood event hydrograph or pollutograph for a watershed characterized by extremely rapid changes all occurring within a very short period, perhaps measured in hours or a fraction of a day.

The process by which hydrologic–hydraulic–water quality models are calibrated should embrace this full spectrum of long-term and short-term phenomena in order to assure and demonstrate that the model has the capability of providing a satisfactory reproduction of all types of conditions encountered within the system. Stated differently, concentration of calibration efforts on only a small segment of this spectrum is not likely to produce a model that is calibrated so as to be capable of simulating other portions of the spectrum.

Incorporation of the full spectrum of long-term to short-term phenomena in the calibration process might, in the case of a hydrologic–hydraulic model, be achieved by attaining satisfactory correlation between simulated and recorded: annual runoff volumes, seasonal runoff volumes, monthly runoff volumes, flow-duration relationships, event hydrographs, annual peaks, and discharge-frequency relationships. While the process is sequential, moving from the gross response—annual runoff—to the detailed response—event hydrographs—there is an iterative element in that achieving a satisfactory calibration at any point in the sequence may involve redoing one or more of the earlier steps.

The transition from long-term to short-term phenomena in calibrating hydrologic–hydraulic models may be specifically illustrated by reviewing the procedures used in the hydrologic–hydraulic modeling of the Menomonee River and Kinnickinnic River watersheds in southeastern Wisconsin (SEWRPC, 1976b, 1977b, 1978). Calibration on the basis of annual runoff volumes is illustrated in Figure 10.11, which compares recorded and simulated annual runoff from the Menomonee River watershed for an 11-year period for which historic data were available. Calibration on the basis of flow duration is illustrated in Figure 10.12, which compares recorded and simulated flow-duration relationships flows for about an 8-month period in the Kinnickinnic River watershed. Calibration based on monthly runoff volumes is illustrated in Figure 10.13, which compares recorded and simulated monthly runoff volumes in the Menomonee River watershed for all 129 months in the period of record.

Calibration based on runoff events is illustrated in Figure 10.14, which compares recorded and simulated hydrographs for four runoff events in the Menomonee River watershed. Calibration based on annual instantaneous peak discharges is illustrated in Figure 10.15, which compares recorded and simulated annual instantaneous peak discharges for the Menomonee River watershed for

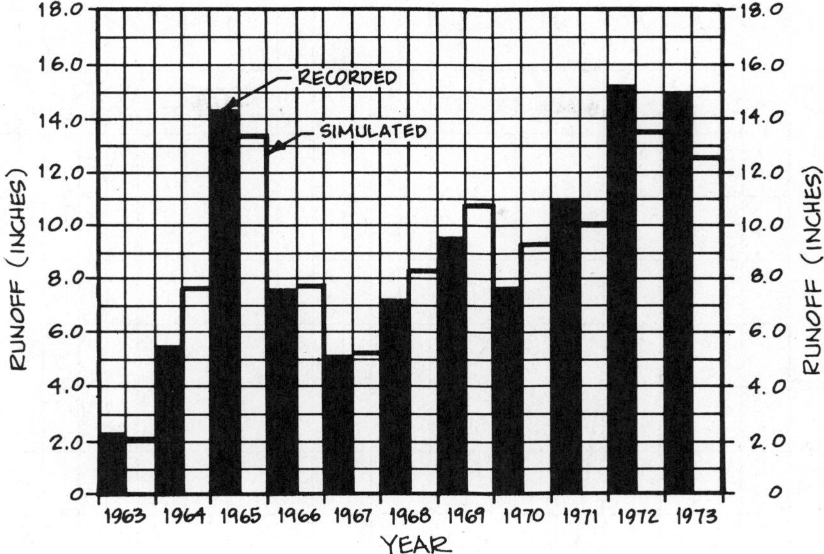

FIGURE 10.11 Recorded and simulated annual runoff volumes for the Menomonee River watershed at the Wauwatosa gauge for the period January 1963 through September 1973 (*Source:* Adapted from SEWRPC, 1976b)

a 12-year period of historic record. Finally, calibration based on discharge–frequency relationships is illustrated in Figure 10.16, which compares recorded and simulated discharge–frequency relationships for the Menomonee River watershed for the 12-year period of record.

The preceding process could be applied to calibration of a water quality model. For example, the process could begin with a comparison of annual recorded and simulated transport of selected pollutants and conclude with the comparison of recorded and simulated pollutographs for selected runoff events. Using a transition from long-term to short-term phenomena is more likely to be feasible when a continuous process, as opposed to discrete event, model is used.

Comparison to Accepted Methodology. Digital computer models may be used to generate a wide variety of outputs—for example, flows, stages, and pollutant concentrations and transport—for large and small rural and urban areas. Some of the areas or conditions modeled within a water resource system, such as a watershed, may be similar to that normally analyzed by more familiar manual methods. For example, a hydrologic–hydraulic digital computer model applied to a large urban area may include simulation of stormwater discharges for small headwater catchments that might otherwise be analyzed by the rational method. Techniques such as the rational method are

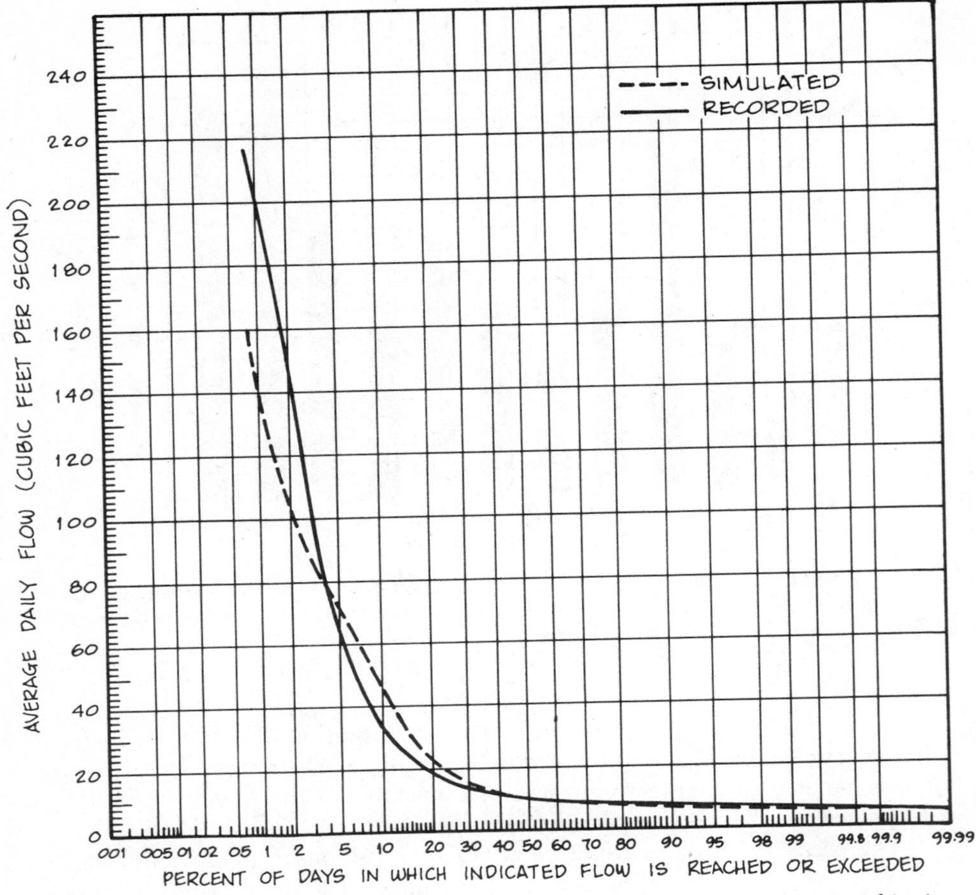

FIGURE 10.12 Recorded and simulated flow-duration relationships for the Kinnickinnic River at the 7th Street gauge for the period September 14, 1976 to April 30, 1977 (*Source:* Adapted from SEWRPC, 1978)

widely used and accepted and have some merit and credibility. Accordingly, peer acceptance of a computer model may be enhanced by comparison, where appropriate, of model results to results obtained with manual methods.

Furthermore, comparison of model results and results obtained with more traditional manual methods is inevitable. The engineer using computer modeling is advised to recognize this and incorporate the comparison into the analysis, including an explanation of differences that may arise, rather than reacting to the inevitable questions concerning such comparison after the modeling is completed.

Examples of comparison of model output to results of accepted manual methodologies include comparison of flood flows of the specified recurrence

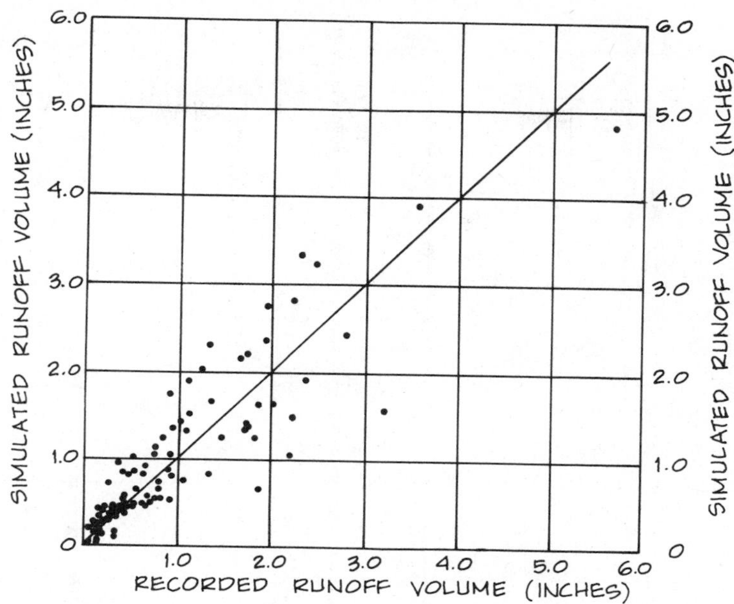

FIGURE 10.13 Recorded and simulated monthly runoff volumes for the Menomonee River watershed at the Wauwatosa gauge for the period January 1963 through September 1973 (*Source:* Adapted from SEWRPC, 1976b)

interval simulated for certain rural areas to flood flows of specified recurrence interval as obtained with empirical regional methods such as those developed by the U.S. Geological Survey; comparison of stormwater flows of specified recurrence interval as obtained with the model for small urban basins to stormwater flows of specified recurrence interval as obtained with the traditional rational method; and comparison of average annual nonpoint-source pollution loads from urban catchments as obtained with a model to loads obtained with empirical methods, such as the U.S. Environmental Protection Agency preliminary screening procedure (Heaney et al., 1976).

Calibration of Flood Damage Models. The calibration of flood damage models is different because historic monetary and nonmonetary flood damages are not usually systematically and uniformly monitored. However, data such as emergency relief grants or loans or enumeration of number of structures affected and the depth of inundation may be available in some areas for some flood events. Although such information may be incomplete and offer only a gross representation of actual direct and indirect flood damages, the information can be useful in assessing the reasonableness of simulated monetary flood damages (SEWRPC, 1976a).

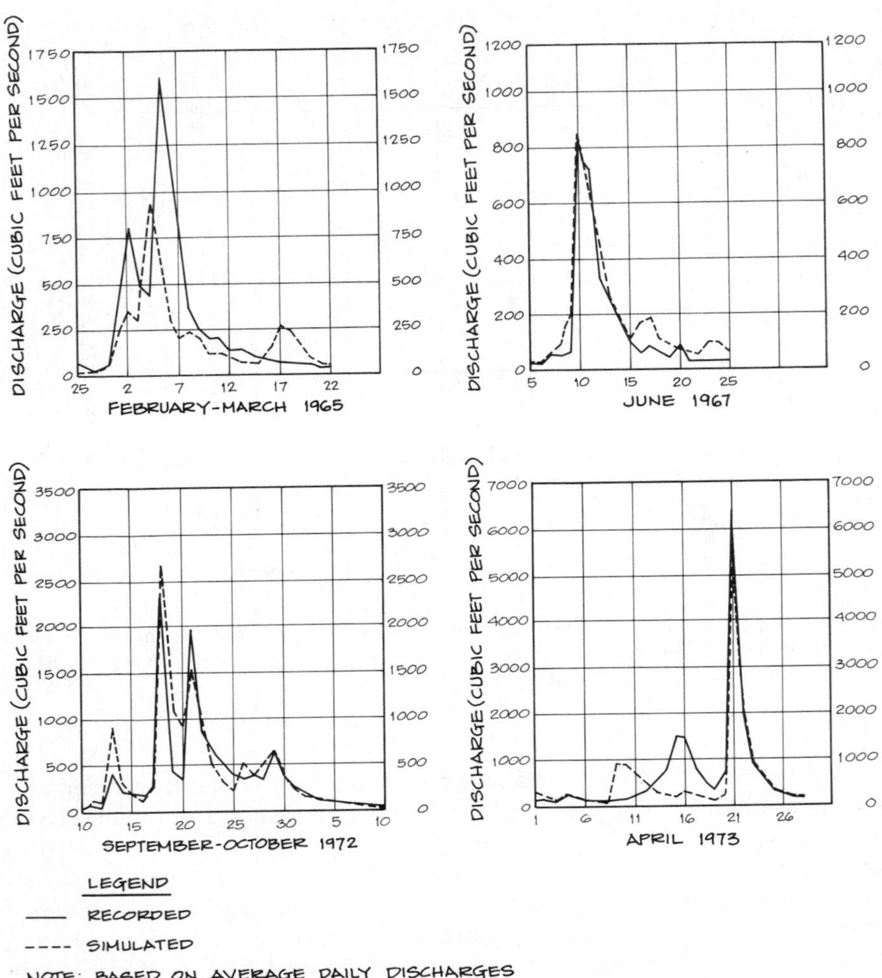

FIGURE 10.14 Recorded and simulated hydrographs for the Menomonee River watershed at the Wauwatosa gauge for four events (*Source:* Adapted from SEWRPC, 1976b)

Display of Typical Calibration Results. The portion of an engineering report describing the modeling aspects of a water resources project should include typical calibration results in graphic or tabular form, showing a comparison between recorded and simulated output. The events or situations presented in the report should be carefully selected to give a balanced representation of the model capability—that is, the examples should include both "good" and "poor" calibrations. The known or hypothesized cause of poor calibration results should be briefly discussed in the text. The display of typical calibration results will serve to establish a proper perspective on model capability and also

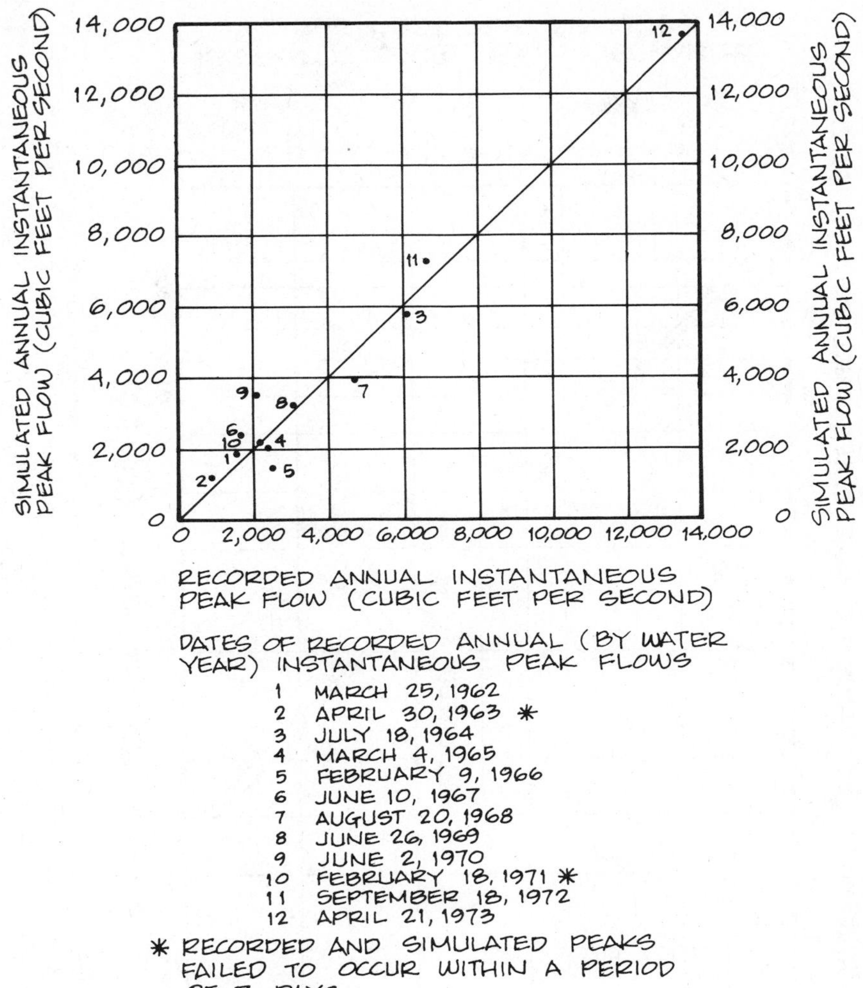

FIGURE 10.15 Recorded and simulated annual instantaneous peak flows for the Menomonee River watershed at the Wauwatosa gauge for water years 1962–1973 (*Source:* Adapted from SEWRPC, 1976b)

contribute to the credibility of the modeling particularly when viewed by peer groups.

Advantage of Continuous Models: Maximum Use of Available Data. Continuous hydrologic–hydraulic–water quality models continuously and sequentially transform a long—several days to several decades—series of historic or synthetic meteorologic data into a corresponding long series of simulated hy-

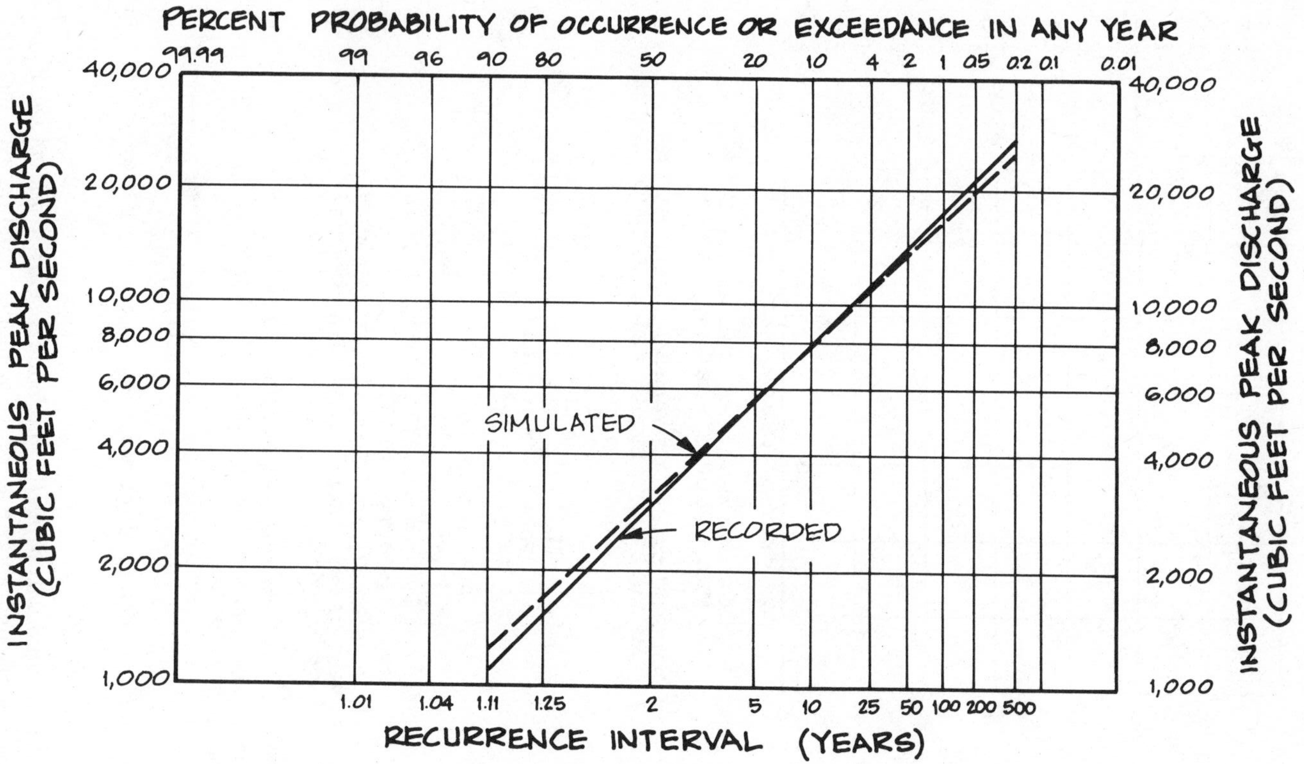

FIGURE 10.16 Recorded and simulated discharge–frequency relationships for the Menomonee River watershed at the Wauwatosa gauge for water years 1962–1973 (*Source:* Adapted from SEWRPC, 1976b)

drologic, hydraulic, and water quality information. In contrast, discrete event hydrologic–hydraulic–water quality models simulate the response of a watershed to a major rainfall or rainfall–snowmelt event and are not intended for simulation of conditions between events.

Although the advantages and disadvantages of these two fundamentally different approaches to modeling have been discussed in Section 10.2, it is important to emphasize the one advantage of continuous models over discrete event models that has a direct bearing on model calibration. Because the continuous model simulates the full spectrum of hydrologic–hydraulic–water quality responses ranging from dry weather, low-flow conditions to wet weather, high-flow conditions, the continuous model can make more extensive use of all available historic streamflow, stage, and water quality data during calibration. In contrast, the discrete event model is limited to use of historic streamflow, stage, and water quality data obtained for runoff event conditions.

Alley (1977) and Marsalek (1977) caution against focusing attention during calibration on peak flow or peak concentrations, noting that this may result in apparent reliability of calibration but does not assure model reliability during other conditions. Not only are a number of events needed for calibration, but they should encompass a wide range of storm event and antecedent conditions. This is more likely to be technically feasible with a continuous event model than with a discrete event model.

10.7 APPLICATIONS OF COMPUTER MODELS: EXAMPLES

Computer models, as noted earlier in this chapter, can be used to help address the two fundamental issues typically raised in all but the most trivial urban SWM projects. The issues are system diagnosis, or problem definition, and evaluation of solutions, or problem solving.

The two broad categories of model applications are illustrated in this section of the chapter by means of specific examples drawn from actual projects. Although numerous and varied, the examples are intended merely to suggest the wide range of useful applications of computer modeling in diagnosing urban surface water systems and in evaluating alternative solutions to problems.

Critical Storm Duration

When a discrete event model is being used, storm duration must be selected for each recurrence interval of interest early in the modeling process. Sensitivity of peak flows, volumes, and other hydrologic parameters of interest can be determined by repeatedly operating the model with different design storm durations. The duration that tends to cause the largest peak discharge, volume, or other hydrologic parameter of interest around the watershed is selected as being the most critical.

This procedure was applied to the 2.96-mi^2 Smith Ditch watershed in and near Valparaiso, Indiana (Donohue & Associates, 1984b). The previously discussed HEC-1 and HEC-2 computer program package illustrated in Figure 10.5 was used. Design storm durations of 1, 6, 12, and 24 hours were simulated for both 10- and 100-year-recurrence-interval events.

Modeling results for 11 locations around the watershed, as shown in Table 10.4, indicate that the 6-hour duration was the most critical for both the 10- and 100-year events because the 6-hour duration yielded the largest peak flows at the most locations in the watershed. Accordingly, the 6-hour duration was used for most subsequent diagnostic model runs and for the evaluation of alternative control measures.

Impact of Urban Development

The urbanization process often includes installation of large amounts of impervious surface, construction of efficient conveyance systems, and filling of floodplains. As discussed in detail in Chapter 2, urbanization without offsetting mitigating measures usually has a significant adverse impact on the quantity and quality of surface water runoff.

Computer modeling is an effective way to quantify the impact of various urbanization scenarios. For example, refer to Chapter 2 for a discussion of how the computer model package shown in Figure 10.7 was used to determine the impact of various urbanization scenarios on a 136-mi^2 urbanizing watershed in southeastern Wisconsin (Walesh and Videkovich, 1978).

Numerous other examples of the use of computer modeling to determine the impact of various aspects of urbanization can be found in the literature and in engineering reports. For example, consider an early modeling investigation of the 100-mi^2 watershed of the North Branch of the Chicago River near Chicago, Illinois (Hydrocomp International, 1969). The Hydrocomp model was used to determine the impact of watershed-wide floodplain filling on peak discharge and stage of the North Branch near the watershed outlet. Figure 10.17 shows the dramatic impact that such fill would have if a historic storm of moderate severity were to occur again. Figure 10.17 and this example are of historic interest because the North Branch of the Chicago River project, which was completed in 1969, was one of the first uses of watershed-wide computer modeling for policymaking and planning purposes.

The possibility of watershed-wide floodplain filling may seem extreme but it is not, particularly for smaller urbanizing watersheds, and should be a cause for concern. This is evidenced by the many urban watersheds which have undergone extensive if not complete floodplain filling and now experience severe flood problems. Computer modeling can be used to demonstrate the adverse consequences of this generally undesirable form of urban development.

Delineation of Floodplains

Floodplain delineations are used for many purposes, including development of floodplain land use and other regulations, establishing insurance premiums

TABLE 10.4 Simulated Peak Discharges in Cubic Feet per Second at Selected Locations in Smith Ditch Watershed for 10- and 100-Year Events of Varying Durations

Location (Points of Interest along the Length of Smith Ditch)	Simulation Description							
	10-Year				100-Year			
	1-Hour	6-Hour	12-Hour	24-Hour	1-Hour	6-Hour	12-Hour	24-Hour
North side of detention facility	190	220[a]	175	140	430	530[a]	390	250
East side of detention facility	75[a]	70	55	35	170[a]	140	95	60
Inflow to detention facility	155	230[a]	190	150	420	530[a]	400	280
Outflow from detention facility	140	230[a]	190	150	420	530[a]	400	280
West side of detention facility	270[a]	230	170	110	460[a]	410	280	180
North side of railroad	280	310[a]	260	210	490	710[a]	570	400
South side of railroad	210	220[a]	210	190	260	340[a]	300	270
Roosevelt Road	270	280[a]	230	210	380	390[a]	340	300
Marks Road	350	375[a]	310	240	610	620[a]	490	360
Silhavy Road	660	680[a]	550	370	1170	1200[a]	920	630
USH 30	810	860[a]	710	490	1530	1590[a]	1210	820

Soruce: Donohue & Associates (1984b).

[a]Largest flow for a given location and recurrence interval.

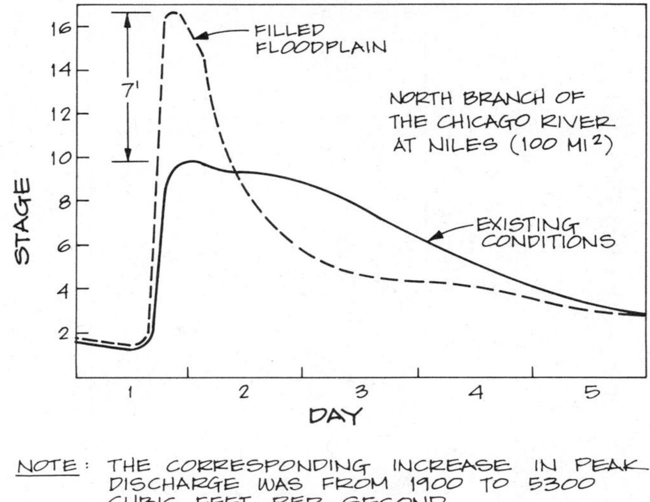

FIGURE 10.17 Effect of floodplain fill on peak discharge and stage for the North Branch of the Chicago River assuming a repeat of the June 10–14, 1967 storm which produced a 25-year-recurrence-interval discharge (*Source:* Adapted from Hydrocomp International, 1969)

under the National Flood Insurance Program, identifying riverine areas to be acquired for public park and open space use, and quantifying the impact of proposed upstream urbanization or proposed downstream hydraulic restrictions. Floodplain hydraulics is discussed in detail in Chapter 4.

A hydrologic model, such as HEC-1, is typically used to develop discharge–probability relationships at many preselected locations on the watershed stream system. Discharges corresponding to selected probabilities or recurrence intervals are then input to a hydraulic model, such as HEC-2, and the corresponding flood stage profiles are calculated. The profiles and the floodplains are typically plotted.

The preceding, now common technique of sequentially using hydrologic and hydraulic models was used to delineate floodplains along 72 mi of stream in the urbanizing Menomonee River watershed in southeastern Wisconsin (SEWRPC, 1976b). Figures 10.18 and 10.19 show, respectively, flood stage profiles and a floodplain delineation for a 4.1-mi portion of the stream system. The computer model package shown in Figure 10.7 was used to calculate flood discharges and profiles. In particular, discharges were calculated with the Hydrologic Submodel and Hydraulic Submodel 1, and flood stage profiles were calculated with Hydraulic Submodel 2.

Floodway Determination

Hydraulic and nonhydraulic factors that should be considered in establishing the limits of the floodway, an important floodplain regulatory zone, are de-

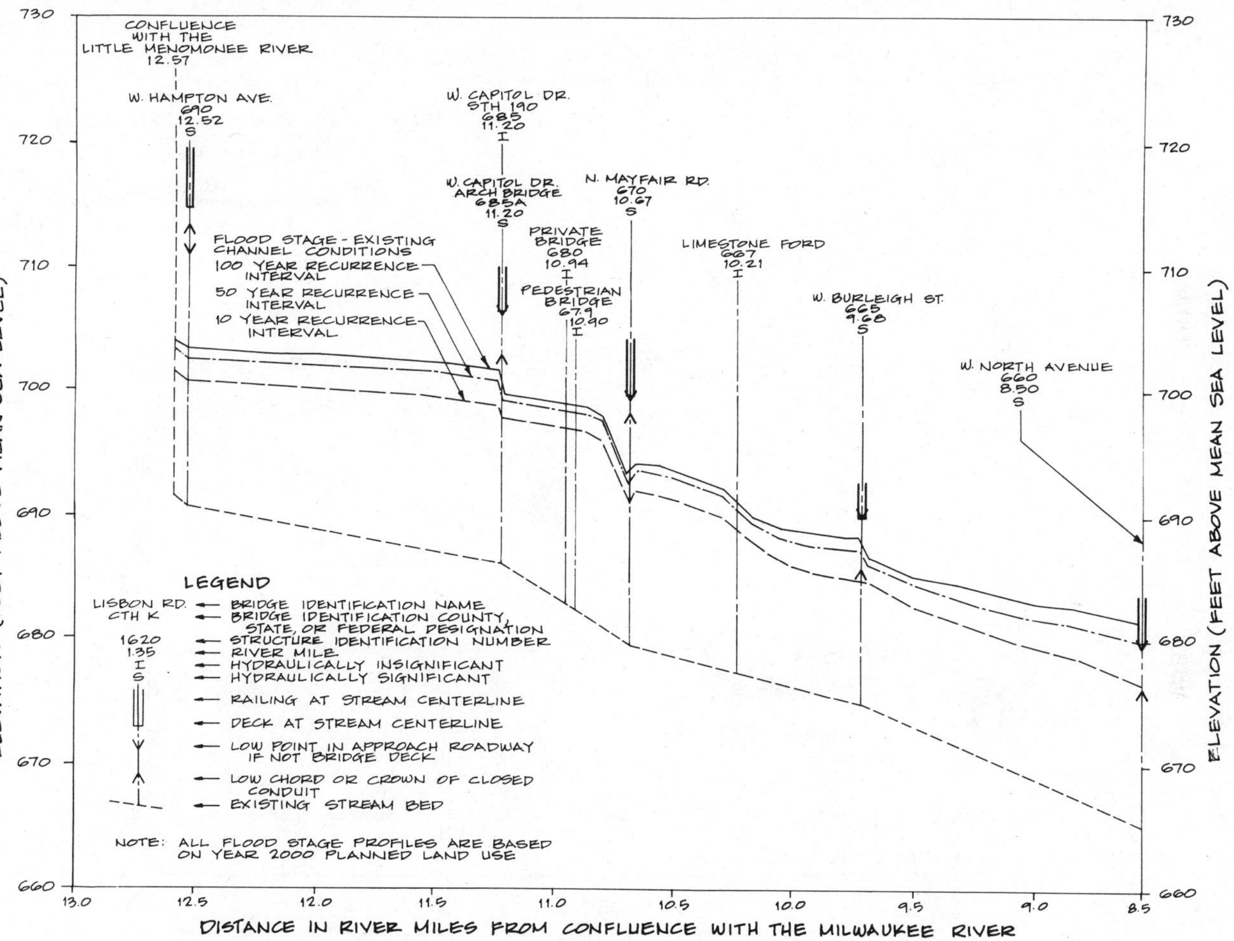

FIGURE 10.18 Flood stage and stream bed profile for a portion of the Menomonee River in southeastern Wisconsin (*Source:* Adapted from SEWRPC, 1976b)

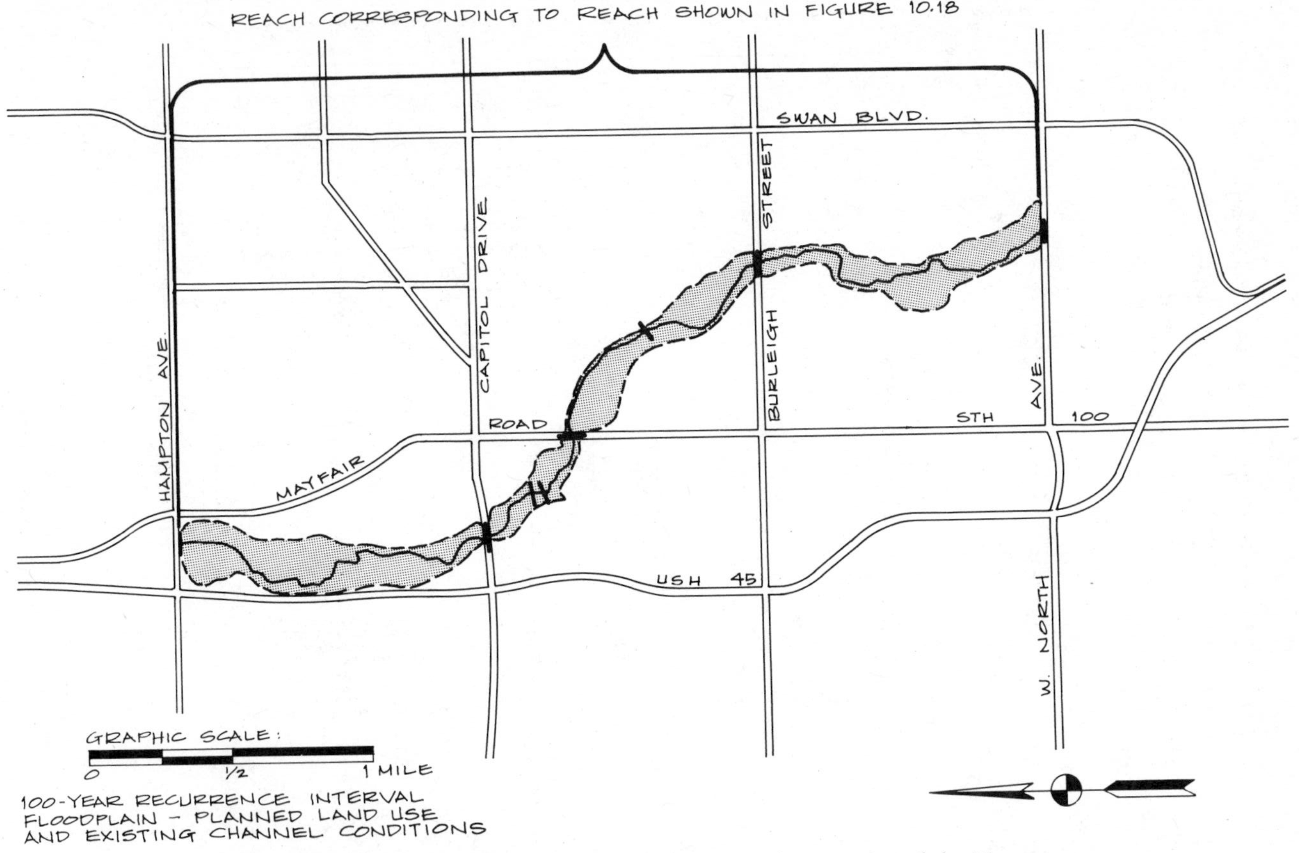

FIGURE 10.19 100–year-recurrence-interval floodplain for a portion of the Menomonee River in southeastern Wisconsin (*Source:* Adapted from SEWRPC, 1976b)

scribed in Chapter 4. Incorporating the hydraulic factors in a floodway determination usually requires the use of a hydraulic computer model such as HEC-2.

Effect of Minor Channel Modification on Flood Stages

Sometimes computer models are used to convince the public and their elected representatives that apparently promising flood control solutions are not feasible. A frequently suggested approach for resolving flood problems—but one that is usually ineffective for providing relief during major floods—is minor channel clearing, deepening, widening, and shaping. People often perceive that removal of "obstructions" from the channel, which can be accomplished at relatively little cost, will mitigate flooding. They fail to realize that during flood flow conditions, most of the flow is conveyed on the adjacent floodplains, and therefore channel clearing and modification will have little or no impact.

The Wisconsin communities of Elm Grove and Brookfield were included in a comprehensive watershed planning program (SEWRPC, 1976b) because they experienced serious flooding from Underwood Creek. Several costly, but potentially effective flood control solutions were examined, including structure floodproofing and removal, major channel modifications, and dikes and floodwalls.

There was persistent local interest in using a much less costly program of minor channel clearing and modification. Although engineering judgment clearly indicated that such a program would be ineffective, the alternative was simulated anyway in the hope of convincing residents and officials of the futility of this approach. For purposes of modeling, it was assumed that a 1.9-mi reach of Underwood Creek would be cleared of obstructions, deepened by 0.5 ft, have the bottom width increased by 20 percent and shaped to increase hydraulic efficiency. Results of HEC-1 analysis, as summarized in Table 10.5, indicate that 5-, 50-, and 100-year-recurrence-interval flood stages would be essentially unaffected. Therefore, flooding and flood damages would not be reduced with this alternative. These modeling results convinced decision makers to omit the channel clearing and modification alternative from further consideration and to recommend a detention storage–channelization–flood proofing solution, which was developed with the assistance of the computer model package shown in Figure 10.7.

Determine Hydraulic Adequacy of Existing Bridges and Culverts and Design the Hydraulic Features of New Bridges and Culverts

Computer models are often used to evaluate the hydraulic capacity of existing bridges and culverts and to design the hydraulic features of new bridges and culverts. A hydraulic model such as HEC-2 identifies those existing bridges and culverts which are overtopped, cause large backwater effects, or are other-

TABLE 10.5 Effect of Minor Channel Modifications on 5-, 50-, and 100-Year-Recurrence-Interval Flood Stages along Underwood Creek in Elm Grove and Brookfield, Wisconsin

			5-Year Recurrence Interval Conditions			
Community	River Mile Station[a]	Bridge or Culvert	Discharge (ft^3/sec)	Stage without Minor Channel-ization (feet above mean sea level)	Stage with Minor Channel-ization (feet above mean sea level)	Stage Decrease (ft)
Elm Grove	3.66		650	741.2	741.2	0.0
		Juneau Blvd.				
	3.68		600	743.0	743.0	0.0
	3.75		600	743.0	743.0	0.0
		Village Hall				
	3.77		600	743.7	743.5	0.2
	4.00		600	743.8	743.5	0.3
	4.26		600	744.5	744.2	0.3
	4.47		600	747.0	746.6	0.4
		Marcella Avenue				
	4.49		600	748.2	747.9	0.3
	4.74		600	749.0	748.7	0.3
	4.81		600	749.1	748.8	0.3
		North Avenue				
Brookfield	4.83		600	751.2	751.0	0.2
	5.02		600	751.2	751.0	0.2
	5.23		600	751.4	751.3	0.1
	5.51		600	752.6	752.4	0.2
	5.58		600	754.3	753.5	0.8
		Clearwater Road				
	5.60		600	756.2	756.2	0.0

Source: Adapted from Southeastern Wisconsin Regional Planning Commission (1976b).
[a]Upstream and downstream of bridges and culverts and at approximately 0.25-mi intervals.
[b]Stage increase.

wise unable to accommodate the specified discharge. For example, the flood stage profile in Figure 10.18 identifies bridges and culverts that have relatively small hydraulic capacities. In the case of designing new bridges and culverts, hydraulic models can be used in a trial-and-error fashion to establish or confirm the size of culverts or the size and configuration of bridge waterway openings.

Evaluate Hydraulic Adequacy of Conveyance System

As a result of additional residential development being proposed for an 85-acre watershed in Valparaiso, Indiana, the city commissioned an independent diagnosis (Walesh, 1988) of the capacity of the existing stormwater system plus

TABLE 10.5 (Continued)

50-Year Recurrence Interval Conditions				100-Year Recurrence Interval Conditions			
Discharge (ft³/sec)	Stage without Minor Channel-ization (feet above mean sea level)	Stage with Minor Channel-ization (feet above mean sea level)	Stage Decrease (ft)	Discharge (ft³/sec)	Stage without Minor Channel-ization (feet above mean sea level)	Stage with Minor Channel-ization (feet above mean sea level)	Stage Decrease (ft)
1500	745.3	745.3	0.0	1900	745.7	745.7	0.0
1500	745.4	745.4	0.0	1900	745.8	745.8	0.0
1500	745.4	745.4	0.0	1900	745.8	745.8	0.0
1500	745.5	745.5	0.0	1900	745.9	745.9	0.0
1500	745.5	745.5	0.0	1900	746.0	746.0	0.0
1500	746.1	745.9	0.2	1900	746.6	746.6	0.2
1500	748.4	748.2	0.2	1900	748.9	748.9	0.2
1500	749.0	749.2	0.2[b]	1900	749.4	749.6	0.2[b]
1500	750.9	750.9	0.0	1900	751.7	751.7	0.0
1500	751.0	751.0	0.0	1900	751.7	751.7	0.0
1500	752.1	752.1	0.0	1900	752.3	752.3	0.0
1500	752.1	752.1	0.0	1900	752.4	752.4	0.0
1500	752.6	752.5	0.1	1900	752.9	752.9	0.0
1500	753.6	753.5	0.1	1900	754.0	753.8	0.2
1500	756.3	756.3	0.0	1900	756.5	756.7	0.2[b]
1500	757.5	757.5	0.0	1900	757.9	757.9	0.0

proposed additions to the system to serve the new development. The existing and proposed stormwater systems was represented as consisting of 20 sewer reaches and two channel reaches.

Sensitivity studies conducted with ILUDRAIN, the model used to simulate the project area, indicated that the critical storm duration was 1 hour. Using ILUDRAIN operating in the design mode, progressively more severe storms—2-, 10-, 25-, 50-, and 100-year recurrence interval—were applied to the watershed. For each recurrence interval, sewer inadequacies were indicated, as shown in Table 10.6 in terms of standard pipe sizes. That is, the degree of inadequacy of a given storm sewer was measured by the number of additional standard pipe sizes that would be required to carry the flow.

As indicated by Table 10.6, 11 sewers were inadequate for the 2-year recur-

TABLE 10.6 Conveyance System Inadequacies as a Function of Storm Severity

Simulation Number	Design Storm Recurrence Interval (years)	Undersized Sewers or Channels[a]																					
		Sewer Reaches																				Channel Reaches	
		A	B	C	D	E	F	G	H	I	J	K	L	M	N	O	P	Q	R	S	T	A	B
1	2			4	1	1	2	3			1	1				1	2	1	1				
2	10			6	2	3	4	4			2	2	2			2	2	2	2		1		
3	25	2	1	7	2	3	4	5		1	2	3	2	1		3	3	2	2	1	1		
4	50	2	1	7	2	4	5	5		1	3	3	3	1	1	3	4	2	3	1	2		
5	100	2	1	7	2	4	5	5	1	1	3	3	3	1	1	4	4	3	3	1	2	O	

Source: Adapted from Walesh (1988).

[a]A numerical entry denotes an inadequate sewer and indicates relative inadequacy in terms of standard pipe sizes assuming replacement by reinforced concrete pipe. Standard pipe diameters, in inches, are: 12, 15, 18, 21, 24, 27, 30, 36, 42, 48, 54, 60, 66, 72, 78, 84, 90, 96,. . . . An "O" means bank overflow for reaches.

rence interval. Similarly, 13, 18, 19 and 20 sewers were inadequate for the 10-, 25-, 50-, and 100-year-recurrence-interval conditions, respectively. The channel reaches were adequate for all but the 100-year condition. Based on the negative results of the diagnosis of the existing and proposed stormwater system, including identification of the most inadequate sewers, several alternatives were evaluated and a system of new and parallel relief sewers was recommended. This example illustrates how a computer model can easily be used to apply increasingly larger hydrologic loads to an urban SWM system with the goal of identifying and quantifying inadequate links in the system.

Screening Analysis of Potential Detention Facility

Figure 10.20 shows the headwater portion of the Menomonee River watershed in southeastern Wisconsin. Because of a flooding problem in the indicated community and the attractive potential detention facility site immediately upstream, preliminary consideration was given to the construction of a detention facility. Using the computer model illustrated in Figure 10.7, a historic storm known to have caused flooding was simulated with and without the potential detention facility.

As shown in Figure 10.21, for this and presumably many other flood-causing events, the hydrograph is bimodal without the detention facility, indicating that the community experiences two approximately equal flood peaks. The first, sharply rising and falling peak represents the early arrival of local runoff. The second, gradually rising and falling peak is caused by the later arrival of runoff from the headwaters of the watershed. Modeling indicated that the hydrograph would be bimodal with the detention facility, but the second, up-

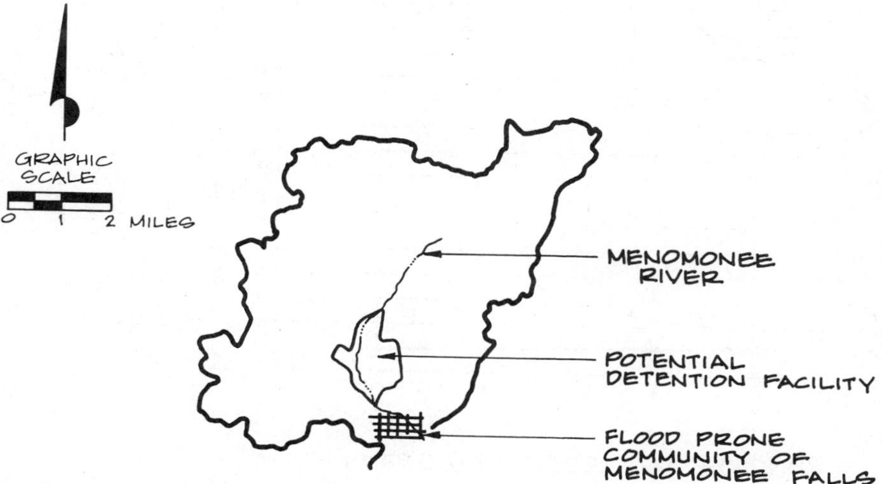

FIGURE 10.20 Headwaters of the Menomonee River watershed in southeastern Wisconsin

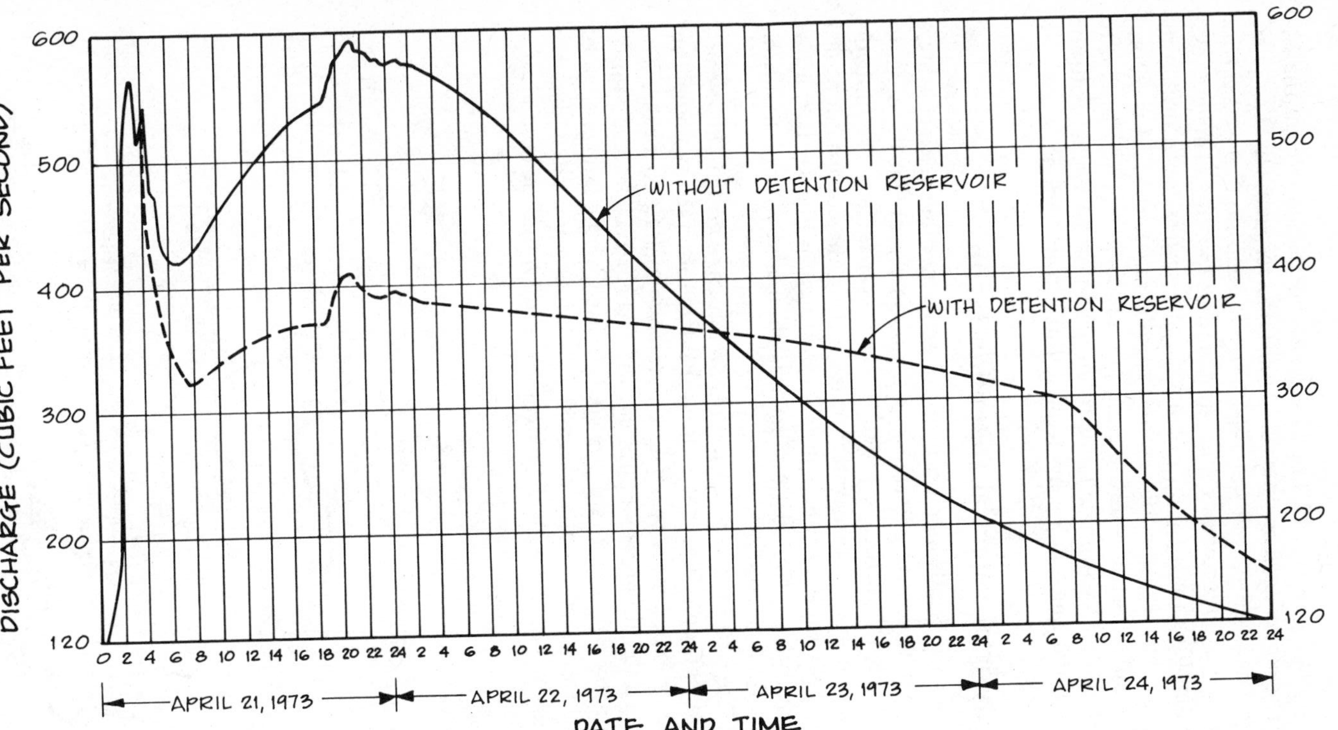

NOTE:
HYDROGRAPHS ARE FOR THE METEOROLOGICAL EVENTS RESPONSIBLE FOR THE APRIL 21, 1973 FLOOD SUPERIMPOSED ON YEAR 2000 PLAN CONDITIONS.

FIGURE 10.21 Bimodal flood flow hydrographs for the Menomonee River in Menomonee Falls, Wisconsin, with and without an upstream detention facility (*Source:* Adapted from SEWRPC, 1976b) stream originating peak would be substantially reduced. However, the detention facility would have no effect on the first, local originating peak.

Because of the negative finding of this screening diagnosis, further consideration of the detention facility was not warranted. Further hydrologic–hydraulic modeling led to the recommendation to use channel modifications as a means of flood control. This example clearly illustrates how modeling can provide insight into how a watershed system behaves, what factors it is sensitive to, and what kinds of solutions to water resources problems are likely to be technically feasible.

Detention Facility Sizing Using Pseudo-Continuous Simulation

In some urban SWM analysis and design situations, an event model with its reliance on a design storm is not technically satisfactory, while a continuous model may not be available or the cost of developing the necessary data base, particularly the long-term meteorologic portion, may be prohibitive. Pseudo-continuous simulation (Walesh et al., 1979) using a series of selective historic storms as shown schematically in Figure 10.22 may be appropriate. The pseudo-continuous process consists of the following three steps:

1. Assemble hyetographs and associated antecedent conditions for major rainfall events from a long historic record.
2. Input the selected hyetographs to the event model and obtain corresponding direct runoff hydrographs.
3. Select annual peak discharges or annual maximum volumes from the model output and perform statistical analysis resulting in a discharge–probability or volume–probability relationship.

As stated in Walesh et al. (1979), pseudo-continuous simulation differs from the conventional use of an event model in two ways:

> First, a relatively large number of actual hyetographs are used as input to the model rather than one or a few design storm hyetographs. Second, rather than trying to predetermine the recurrence interval of a peak discharge or volume of a direct runoff hydrograph obtained from the model, the output hydrographs are treated as a series of major hydrographs. A statistical analysis, similar to that which would be performed on a long historic gauging record, is then used to obtain the desired discharge–probability or volume–probability relationships.

Five applications of pseudo-continuous simulation in the Milwaukee, Wisconsin, area are described elsewhere (Walesh et al., 1979). One of the five examples is presented here to illustrate the methodology further.

Each of the five referenced applications of pseudo-continuous simulation utilized historic rainfall events as continuously recorded by the first-order National Weather Service station in Milwaukee during the period 1940–1976. As

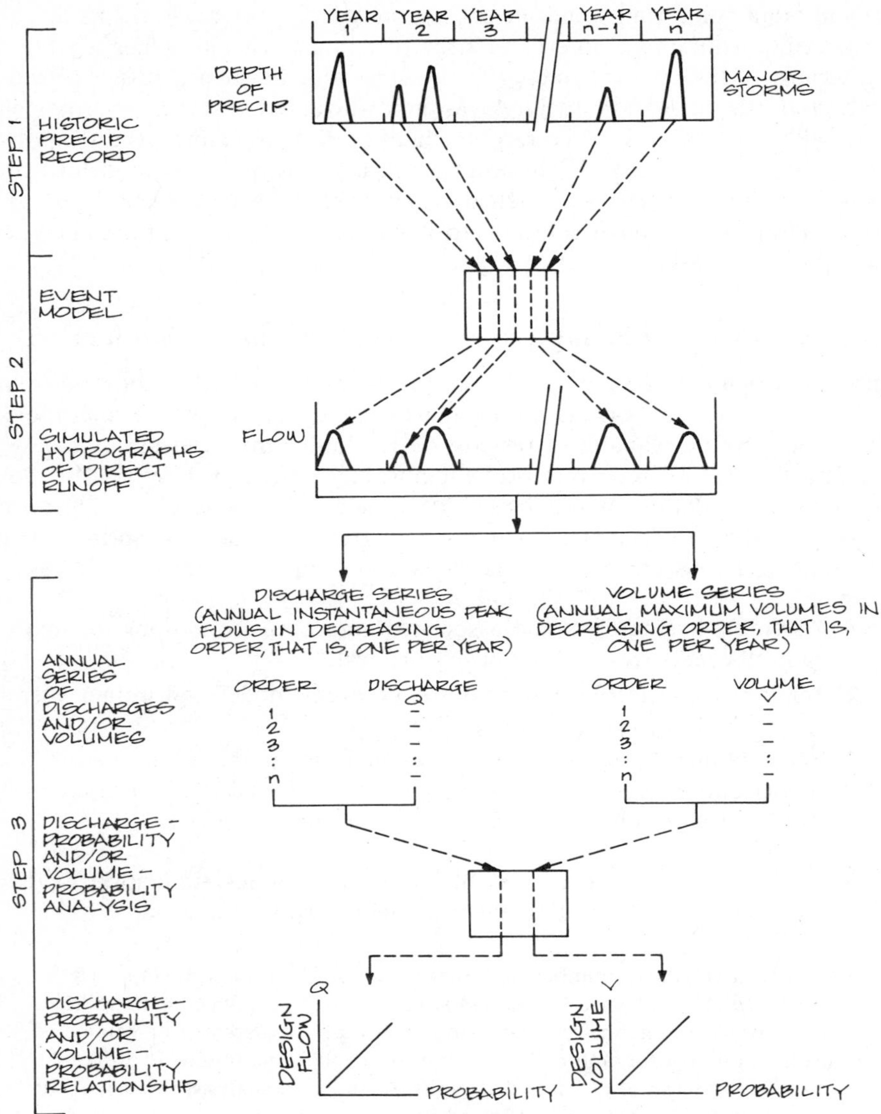

FIGURE 10.22 Pseudo-continuous simulation technique (*Source:* Adapted from Walesh et al., 1979)

described by Walesh et al. (1979), the following procedure was used to select major events from the 27-year historic rainfall record:

> A major rainfall event was defined as one in which one or more hours of continuous rainfall occurred in which the volume of rainfall was equal to or greater than that associated with a two year recurrence interval volume for any one hour or

> longer portion of the hyetograph. If an entire event or any part of an event exceeded a two year frequency, the entire event was selected for the historic record. . . . Hourly data were used for identification of major events since continuous hourly data for the period of 1940–1946 were available in tabular form. The idea was to use hourly data to scan the entire historic precipitation record and identify the major events. For purposes of scanning the historic hourly record, a precipitation event was defined as a period of precipitation preceded and followed by two or more hours of no or trace precipitation. . . . A manual search of the 37 years of hourly precipitation data identified a total of 42 rainfall events satisfying the two year recurrence interval criteria.

As shown in Table 10.7, the 42 events ranged in duration from 0.42 to 28.4 hours. The range in rainfall volumes was 0.92 to 3.85 in., and average intensities varied from 0.12 to 2.21 in./hr. Note that Table 10.7 also includes the antecedent moisture condition for each of the 42 events because that parameter is required for most event models.

For each of the 42 events, 5-minute-interval rainfall hyetographs were developed for input to the event model. The 5-minute interval was used not necessarily because it would be appropriate for all model runs but rather, to provide the flexibility to use either the small time increment or to aggregate the data to a larger more appropriate time increment. The storm selection and analysis process was tedious and time consuming, but needed to be done only once for use in all subsequent Milwaukee area projects.

One of the five applications of the pseudo-continuous simulation method was development of a volume–probability relationship for a potential detention facility which in turn could be used to determine required storage volume as a function of design recurrence interval. The potential detention site had a 3.06-mi^2 catchment. The tributary area was very flat and hydrologic soil groups C and D covered over 90 percent of the area. The analysis assumed planned land use conditions under which essentially all of the tributary area would be urbanized.

The TR-20 computer program was used, with the application of TR-20 differing from the usual approach only in that many more storms were simulated. Specifically, rainfall hyetographs and antecedent conditions were simulated for each of the 42 storm events, and annual peak storage volumes were selected from the model output. The volumes were ranked and plotting positions were calculated. The resulting volume–probability relationship is shown in Figure 10.23. The volume–probability relationship was used to determine the size and cost of various degrees of protection.

A discharge–probability relationship was also developed for the detention site. An earlier hydrologic analysis in the area using continuous simulation also produced a discharge–probability relationship at the site. The discharge–probability relationship developed from continuous simulation was compared to the discharge–probability relationship developed with the pseudo-continuous simulation technique. The 10- to 100-year recurrence interval discharges generated with the pseudo-continuous simulation approach were within 10 percent of those obtained with continuous simulation.

TABLE 10.7 Selected Characteristics of Major Historic Storms at Milwaukee, Wisconsin: 1940–1976

Event Number	Start				Stop				Duration (hr)	Precipitation (in.)	Average Intensity (in/hr)	Antecedent Moisture Condition[a] (in.)
	Year	Month	Day	Hour	Year	Month	Day	Hour				
1	1940	June	22	2:20 AM	1940	June	22	2:35 PM	12.25	3.65	0.30	0.41
2	1942	July	28	7:00 PM	1942	July	28	9:30 PM	2.50	1.67	0.67	0.48
3	1942	August	1	10:20 PM	1942	August	2	1:55 AM	3.92	2.10	0.54	2.83
4	1942	August	6	4:00 PM	1942	August	6	6:00 PM	2.00	1.66	0.83	1.71
5	1943	November	6	5:45 PM	1943	November	7	5:25 AM	11.67	2.08	0.16	0.24
6	1945	September	19	7:45 PM	1945	September	19	11:05 PM	3.33	1.21	0.36	0.00
7	1945	September	27	3:25 PM	1945	September	27	11:25 PM	7.00	1.80	0.26	0.86
8	1947	September	5	3:00 AM	1947	September	5	6:05 AM	3.08	1.52	0.49	1.47
9	1947	September	21	2:15 AM	1947	September	21	11:30 AM	9.25	2.71	0.29	0.48
10	1948	June	28	4:30 PM	1948	June	29	2:30 AM	10.00	1.89	0.19	0.00
11	1949	July	26	4:45 PM	1949	July	26	9:30 PM	4.75	1.63	0.34	0.00
12	1950	June	12	10:15 PM	1950	June	13	4:45 AM	6.42	1.80	0.28	0.00
13	1950	July	19	5:50 AM	1950	July	20	7:50 PM	26.00	3.11	0.12	2.18
14	1952	June	13	12:55 PM	1952	June	13	1:35 PM	0.67	1.46	2.19	0.58
15	1952	July	17	11:45 PM	1952	July	18	9:25 AM	9.67	2.33	0.24	1.44
16	1952	August	3	5:50 PM	1952	August	4	12:30 AM	6.75	1.89	0.28	0.00
17	1953	July	31	11:15 PM	1953	August	1	4:10 AM	4.92	1.28	0.26	0.19
18	1954	June	2	8:50 PM	1954	June	3	9:30 PM	24.67	2.86	0.12	1.45
19	1954	July	6	1:15 PM	1954	July	6	2:35 PM	1.33	1.72	1.29	0.91
20	1955	May	24	[b]	1955	May	24	[b]	4.17	1.77	0.42	0.82

21	1955	August	16	2:10 PM	1955	August	16	3:15 PM	1.08	1.63	1.51	0.00
22	1956	April	27	1:30 AM	1956	April	27	8:25 AM	6.92	1.67	0.24	0.10
23	1956	June	20	8:30 PM	1956	June	20	10:50 PM	2.33	1.36	0.58	1.72
24	1956	July	12	8:55 PM	1956	July	13	1:00 AM	4.08	1.55	0.38	1.60
25	1956	August	30	3:05 PM	1956	August	30	4:00 PM	0.92	1.58	1.72	0.22
26	1958	October	8	11:35 PM	1958	October	9	5:35 AM	6.00	2.17	0.36	0.80
27	1959	July	17	10:00 PM	1959	July	18	9:25 AM	11.42	3.85	0.34	0.31
28	1959	October	22	6:30 PM	1959	October	23	5:50 AM	11.33	2.33	0.21	0.00
29	1960	August	2	10:05 PM	1960	August	3	12:45 AM	2.67	2.16	0.81	0.10
30	1961	September	22	12:45 PM	1961	September	22	1:45 PM	1.00	2.18	2.18	0.33
31	1964	July	1	5:50 PM	1964	July	1	6:15 PM	0.42	0.92	2.21	0.00
32	1964	July	18	4:00 AM	1964	July	18	12:25 PM	6.42	2.43	0.38	2.07
33	1964	August	20	8:55 PM	1964	August	21	4:25 AM	7.67	1.66	0.22	0.19
34	1968	June	26	1:50 AM	1968	June	26	11:15 AM	9.42	2.02	0.21	3.04
35	1969	June	29	5:40 PM	1969	June	29	10:55 PM	5.25	1.94	0.37	2.74
36	1971	July	8	3:45 AM	1971	July	8	7:20 AM	3.58	1.80	0.50	0.04
37	1972	September	18	12:30 AM	1972	September	18	5:25 AM	4.83	1.73	0.36	2.09
38	1973	April	20	10:30 PM	1973	April	21	5:05 AM	6.58	2.51	0.38	0.51
39	1975	April	28	12:55 PM	1975	April	29	7:30 AM	6.58	1.72	0.26	0.57
40	1976	March	4	7:55 AM	1976	March	4	11:00 PM	15.08	2.30	0.15	1.83
41	1976	March	26	7:20 PM	1976	March	27	3:35 AM	8.25	1.96	0.24	0.00
42	1976	April	24	6:40 PM	1976	April	25	11:05 AM	28.42	3.54	0.12	0.64

Source: Walesh et al. (1979).

[a]Inches of precipitation during a 5-day period prior to the start of storms.
[b]Not available.

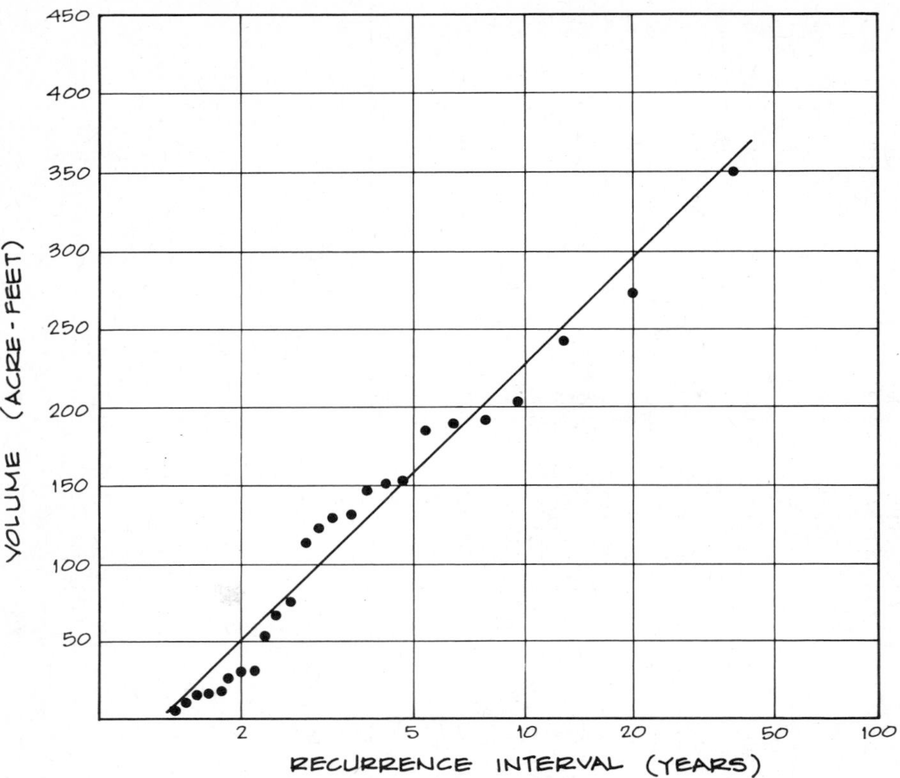

FIGURE 10.23 Detention facility volume–probability relationship developed with pseudo-continuous simulation technique

Effectiveness of a Detention Facility When Subjected to Storms of Widely Varying Severity

Some government jurisdictions have regulations which require that detention/retention facilities or other control measures be provided in new developments such that for a single specified recurrence interval, the peak discharge from the site after development will not exceed the peak discharge from the site before development. Such regulations are often silent on expected flood control performance, if any, under widely varying recurrence interval events.

Figure 10.24 shows hydrographs developed for runoff from a 240-acre watershed in Madison, Wisconsin. At the time of the project (Donohue & Associates, 1984a) the watershed was being used almost entirely for agricultural purposes but being considered for commercial and light industrial development.

Watershed analysis using HEC-1 clearly indicated, as shown in Figure 10.24, the large impact of development on peak 1-, 10-, and 100-year recur-

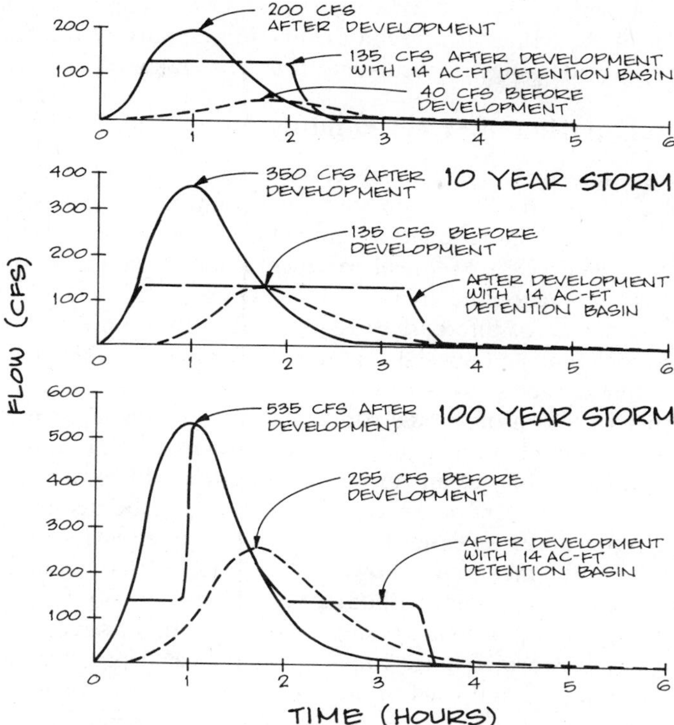

FIGURE 10.24 Performance of a detention basin for a range of recurrence intervals (*Source:* Adapted from Donohue & Associates, 1984a)

rence interval discharges in the absence of any mitigative measures such as detention/retention. For the 2-hour duration used in the analysis, 1-, 10-, and 100-year peak discharges after development would be 5.0, 2.6, and 2.1 times the peak discharges before development, respectively.

Of special interest is the expected effect of a detention facility designed to satisfy the local stormwater control ordinance. The ordinance specifies that the 10-year recurrence peak flow after development should not exceed the 10-year recurrence interval peak flow before development. A conventional 14-acre-ft detention facility would satisfy this criterion as shown in Figure 10.24 for the 10-year-recurrence-interval event. However, as also shown in Figure 10.24, the 14-acre-ft detention facility would have no effect on the after-development peak flow generated by a 100-year-recurrence-interval event. Similarly, the facility would have relatively little mitigative effect on the after-development peak flow of the 1-year event.

Refer to Chapter 8 for additional discussion of the shortcomings of the single-recurrence-interval criterion. Also discussed in Chapter 8 is the alternative of using a continuous recurrence interval criterion in designing detention/retention facilities so that such facilities function effectively for many storm events having a wide range of recurrence intervals.

Depth and Duration of Street Ponding

A 1200-acre, primarily residential combined sewer service area in Skokie, Illinois, was the subject of a flood control planning and design project culminating in the completed installation of a unique flood control system (Walesh and Schoeffmann, 1984, 1985). Frequent basement flooding caused by surcharging combined sewers plagued the area.

System evaluation of alternatives required the development and use of a computer model that had special features, including the ability to simulate conveyance and storage of surface water on street surfaces and adjacent residential properties. With the model, flow in streets and on adjacent properties was routed using the Manning equation, plus conservation of mass to account for the conveyance ability of any given street section. Stormwater ponding on street surfaces and adjacent parkways and lawns was accounted for by a reservoir routing procedure. Release of flow from the street surface to the sewer system was set at the maximum allowable release rates that could be handled by the combined sewer system without surcharging.

One required type of output was depth and duration of street ponding as a function of design storm. Figure 10.25 shows typical depth–duration output provided by the model and used in the evaluation of various alternatives.

The project resulted in the construction of a retrofit detention system consisting of flow regulators installed in catch basins to restrict flow into the combined sewer system, low berms located across streets at the selected locations to direct and temporarily pond stormwater, a system of surface and subsurface detention facilities to provide additional storage, and some relief sewers.

Impact of Water Quality Management Measures

Computer models are being increasingly used to evaluate the effect of point- and nonpoint-source pollution controls on surface water quality. In general, water quality modeling cannot yet be done with the same degree of confidence as water quantity, that is, hydrologic–hydraulic modeling. Nevertheless, many situations arise in which water quality modeling results can be a useful guide to professional judgment.

The comprehensive plan for the 136-mi^2 urbanizing Menomonee River watershed in southeastern Wisconsin (SEWRPC, 1976b) included evaluation of the impact on stream water quality of additional urban development, elimination of four remaining municipal wastewater treatment plants by intercepting the wastewater and conveying it to a regional system, and application of nonpoint-source pollution control measures.

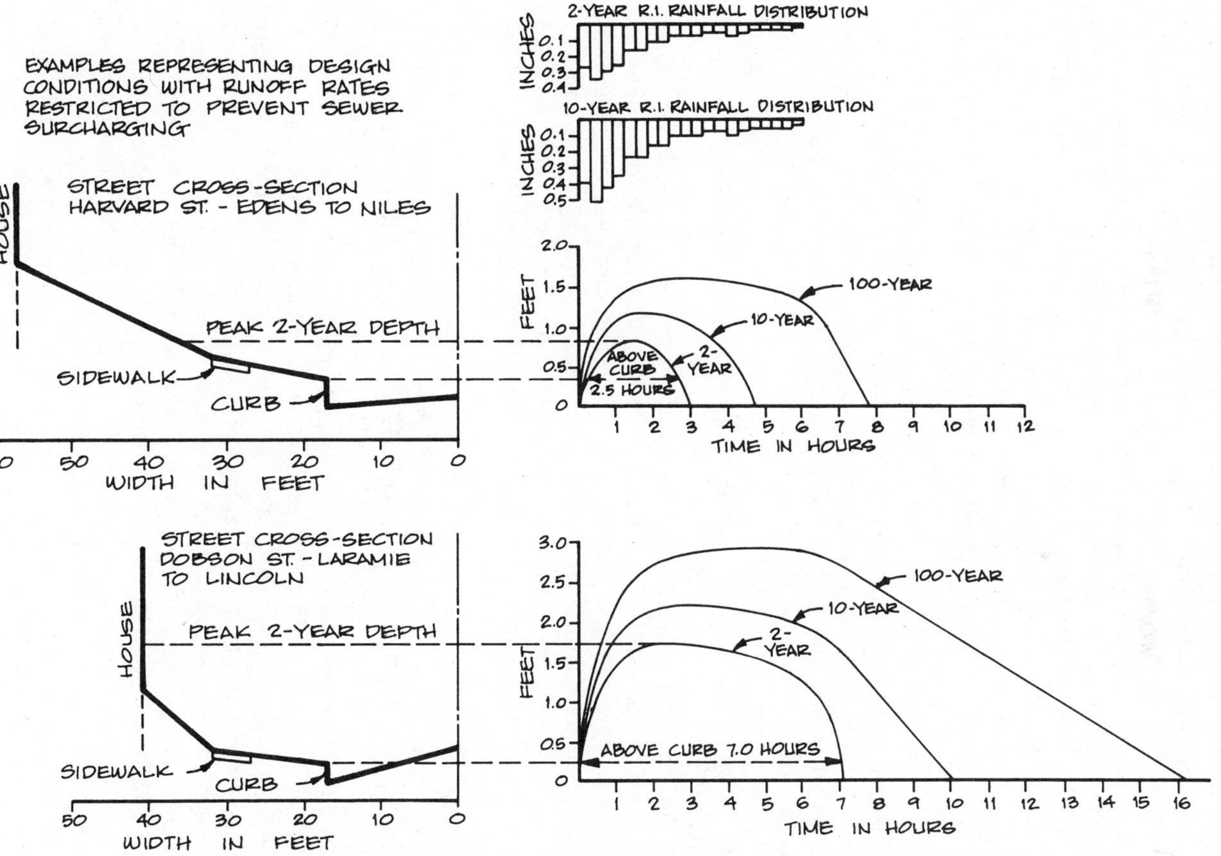

FIGURE 10.25 Depth and duration of street ponding as a function of recurrence interval (*Source:* Walesh and Shoeffmann, 1984)

The continuous simulation portion (Hydrologic Submodel, Hydraulic Submodel 1, and Water Quality Submodel) of the computer model package shown in Figure 10.7 was used to make four 10-year simulation runs of the entire watershed. The four runs represented the following four water quality–related conditions: existing land use and water quality management measures, including the four municipal wastewater treatment plants; existing land use and water quality management measures except the wastewater treatment plants are eliminated; future land use without the wastewater treatment plants and without any other water quality management measures; and future land use without the wastewater treatment plants and with nonpoint-source pollution controls.

The simulation runs without the wastewater treatment plants accounted for the effect of reducing pollutant loads as well as the effect of reducing flow. Nonpoint-source pollution management measures were simulated within the model by adjusting model input and parameters to reflect reductions in land surface accumulation rates of various pollutants, reducing phosphorus release and oxygen demand associated with stream bottom deposits, and reducing the concentration of some substances in the groundwater component of stream base flow.

Each of the four simulation runs produced a continuous series of hourly stream flow and a continuous series of hourly values for the following nine water quality parameters: dissolved oxygen, ultimate carbonaceous biochemical oxygen demand, total dissolved solids, fecal coliform count, nitrate-nitrogen, nitrite-nitrogen, ammonia nitrogen, phosphate-phosphorous, and water temperature. Almost 10 million water quality values were generated for each of the four runs.

Statistical analysis of the streamflow and water quality data series were performed by routines in the computer model and used to develop water quality–duration relationships like that shown in Figure 10.26 for dissolved oxygen at a station on the Menomonee River in the lower third of the watershed. Water quality–duration relationships were developed for each water quality perimeter for each of the four conditions at each of the 12 locations. Similar to a flow-duration curve, a water quality–duration curve is a graph of percent of time on the horizontal axis that a concentration or other water quality level shown on the vertical axis is reached and exceeded or not reached and exceeded. Like flow-duration relationships, water quality–duration relationships are an effective way to summarize and draw conclusions from a large volume of data.

Water quality–duration relationships developed for the Menomonee River watershed helped to determine the relative effect of various management measures. Using Figure 10.26 as an example, under the then-existing conditions, dissolved oxygen levels within the stream would drop below the then-mandated minimum concentration for maintenance of fish and aquatic life of 5.0 milligrams per liter (mg/L) about 7 percent of the time. Assuming additional urban development combined with the offsetting effect of removing the four municipal wastewater treatment plants and application of nonpoint-

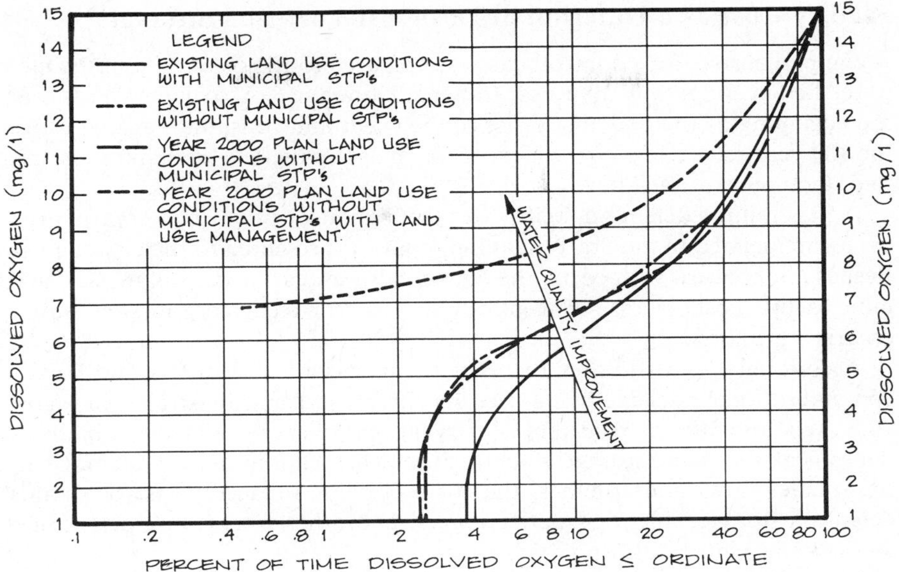

FIGURE 10.26 Dissolved oxygen–duration relationships for the Menomonee River (*Source:* Adapted from SEWRPC, 1976a)

source pollution management measures, the water quality–duration curve indicates a great improvement in dissolved oxygen levels, including remaining above the critical 5.0-mg/L level essentially all the time. Furthermore, the water quality–duration curve suggests the relatively large positive impact of nonpoint-source pollution controls.

Based in part on the water quality modeling results, the comprehensive plan for the Menomonee River watershed included recommendations to abandon the four municipal wastewater treatment plants and connect them via intercommunity trunk sewers to the regional system, provision of sanitary sewer service to most existing unsewered urban development and to most future urban development, and the implementation of nonpoint-source pollution controls in the urban and rural areas of the watershed.

Computation of Average Annual Monetary Flood Damage

The procedure for calculating average annual monetary flood damage is described in Chapter 6. As noted in that chapter, the necessary computations are voluminous, tedious, and error-prone. Therefore, computer models are usually used to develop the reach stage–damage relationship and the reach stage–probability relationship and combine them to form the reach damage–probability relationship, which yields average annual monetary flood damage. Computer programs are available from the Corps of Engineers (U.S. Army Corps of Engineers, 1986b) and private sources.

Facility Cost as a Function of Design Recurrence Interval

A comprehensive flood control plan was prepared for the 2.96-mi^2 Smith Ditch watershed in and near Valparaiso, Indiana (Donohue & Associates, 1984b). At the beginning of the planning process, the client and consultant agreed to use the 100-year-recurrence-interval event for the planning-level design of alternative flood control facilities.

A $2.9-million detention facility was the highest-priority flood control project recommended in the draft planning report as presented to the city council. Because of concern with costs, the council also asked the consultant to determine facility costs if less stringent 50- and 10-year-recurrence-interval design criteria were used.

Accordingly, two additional runs were made, with the HEC-1 and HEC-2 computer model package being used on the project and illustrated in Figure 10.5. Ten- and 50-year-recurrence-interval events were simulated to obtain reduced peak flows and storage volumes at critical locations in the system. Using the smaller flows and volumes, the corresponding reduced costs for smaller detention storage, inlet and outlet sewers, and other works were determined as shown in Figure 10.27.

Only a 9 percent cost reduction would be achieved if the design event were reduced from a 100-year recurrence interval to a 50-year recurrence interval.

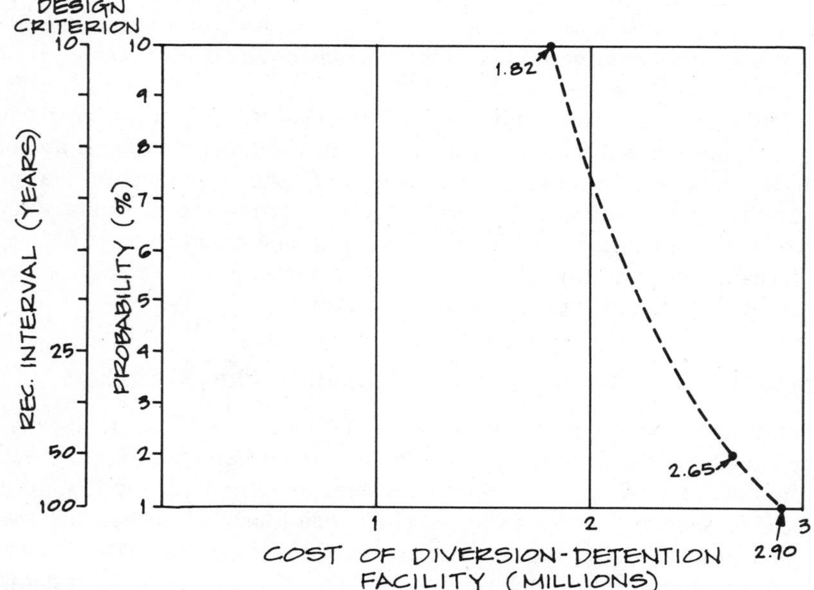

FIGURE 10.27 Cost of a flood control facility as a function of design recurrence interval (*Source:* Adapted from Donohue & Associates, 1984b)

Similarly, only a 37 percent cost reduction would occur if the design event were reduced from a 100-year recurrence interval to a 10-year recurrence interval.

Based on the relatively small sensitivity of cost to design recurrence interval, and probably based on vivid memories of a recent 100-year-recurrence-interval flood event, the council confirmed the original 100-year recurrence interval design criterion. The facility was eventually designed and built as a multipurpose flood control and recreation facility providing 100-year-recurrence-interval protection (Walesh et al., 1987).

Elected officials and other client representatives often want to change or consider changing a previously agreed upon planning and design criterion. If the project in question utilizes computer modeling, the engineer responsible for the project can respond with relatively little additional effort because of the ease with which computer models can be operated for different conditions.

10.8 COSTS AND BENEFITS OF COMPUTER MODELING

One argument given for not using available computer modeling tools in urban SWM is the "high" cost of modeling. Digital computer modeling is unquestionably much more costly than a wide spectrum of readily available alternative analytic approaches, ranging from strictly judgment to the techniques relying primarily on manual computation.

The cost of applying state-of-the-art modeling tools to a problem-prone watershed in an urbanizing area is likely to be measured in terms of thousands to tens of thousands of dollars when all modeling costs are accounted for. Cost components include assembling existing input and calibration data, conducting field inventories to acquire additional data, calibrating the model, using the model in a production mode to diagnose the existing system and evaluate alternative quantity and quality management measures, and preparing the necessary documentation.

Inasmuch as the principal reason for undertaking a study of an urbanizing watershed is to develop recommendations for resolving existing and preventing future surface water problems, it is important to consider the monetary magnitude of the minimal benefits that may result from the use of a computer model that simulates complex cause-and-effect hydrologic–hydraulic–water quality relationships. For example, major channelization projects which are one common solution to flood problems may have a unit capital cost ranging from $100 to $1000 per lineal foot. Assume $500 per lineal foot for channelization and assume that the total modeling cost for a comprehensive urban SWM plan is $100,000. The entire modeling cost would be offset if the better understanding of the system resulting from the use of the model effects a savings of only 200 ft, about 0.04 mi, of new flood control channel throughout the entire watershed. Similar arguments may be advanced for the design of other aspects of the urban SWM system, such as storm sewers, multiple-purpose reservoirs, and combined sewer overflow control devices.

The point is this: Urban SWM planning and engineering projects normally result in recommendations for the expenditure of large amounts of public funds. Additional incremental investment in engineering costs to permit the use of state-of-the-art modeling techniques will usually improve the understanding of cause-and-effect relationships. In turn, improved understanding will probably improve the quality of recommendations and is likely to reap large monetary benefits in terms of reduced construction costs or more effective allocation of construction funds.

10.9 MODELING: MAINTAINING PERSPECTIVE

During the past two decades there has been a dramatic increase in the development and use of digital computer models in urban SWM. Engineers have found these models to be very useful in diagnostic studies and in developing solutions to urban water problems for several reasons. First, model algorithms and the manner in which they approximate real processes encourage the user to seek improved understanding of the processes and, in turn, lead to model improvements and better understanding of watersheds. Second, model input requirements encourage the conduct of systematic, in-depth inventories of the natural resources and cultural features of watershed. Third, model calibration requirements can lead to optimum utilization and extrapolation of historic water quantity and quality data. Finally, model computation capabilities facilitate examination of a full spectrum of hypothetical conditions and management measures that were previously impossible because of time and budget limitations.

However, even the most current computer models have limitations. It is important to emphasize that models do not provide exact duplication of system processes. It is also important to note that most of the computer models in current use do not "make decisions" but instead provide qualified, quantified information to engineers and other decision makers to serve as one input to the decision-making process.

There is a tendency for some staff members, project managers, administrators, and decision makers in both government and private practice who are not familiar with the nature and limitations of models to have unwarranted confidence in the ability of models to reproduce the behavior of watershed systems and to place undue emphasis on model output in decision making. The anticipated or planned use of a model in an urban SWM project can actually interfere with the sound conceptualization and planning of the project by leading to insufficient consideration of a systematic approach whereby data are to be collected and analyzed, problems are to be defined, and alternatives are to be synthesized, tested, and compared. Again, this may happen because of a misunderstanding of capability.

The engineer who uses the model is much more important than the model in determining the credibility of the modeling phase of the SWM project. The degree to which the system is adequately represented in the model algorithms and by the model input is heavily influenced by the knowledge and judgment of the engineer or engineers having responsibility for the modeling. Similarly, the manner in which model output are evaluated and used in the decision-making process depends on the knowledge and judgment of the user. Caution, based on the understanding of the capabilities and limitations of models, is the key in the intelligent use of modeling in urban SWM.

It has been said that "the purpose of the computing is insight, not numbers" (Hamming, 1962). This philosophy may be applied to modeling of urban watersheds by emphasizing that the principal purpose of modeling is insight, not numbers—insight into how the system behaves, what factors it is likely to be sensitive to, how human beings may favorably or unfavorably influence those factors, and what kinds of solutions to surface water problems are likely to be technically feasible.

REFERENCES

Abbott, J., *Testing of Several Runoff Models on an Urban Watershed,* Tech. Memo. No 34. ASCE Urban Water Resources Research Program, New York, October 1978.

Alley, W. M., *Guide for Collection, Analysis and Use of Urban Stormwater Data,* Chapter 3, pp. 16–28. Report of a Conference Sponsored by the Engineering Foundation, U.S. Geological Survey and ASCE Urban Water Resources Research Council, 1977.

American Public Works Association, *Urban Stormwater Management,* Spec. Rep. No. 49. APWA, Chicago, IL, 1981.

Bedient, P. B., and W. C. Huber, *Hydrology and Floodplain Analysis,* Addison-Wesley, New York, 1988.

Brandstetter, A. B., *Assessment of Mathematical Models for Storm and Combined Sewer Management,* EPA-600/2-76-175a. Environ. Prot. Agency, Cincinnati, OH, August 1977.

Crawford, N. H., and R. K. Linsley, *Digital Simulation in Hydrology: Stanford Watershed Model IV,* Tech. Rep. No. 39. Dep. Civ. Eng., Stanford University, Palo Alto, CA, July 1966.

Donohue & Associates, Inc. *ILLUDAS Sensitivity Studies, Water Resour. Tech. Memo. No. 14,* December 1977.

Donohue & Associates, Inc., *Hydrologic Impact of Development of the Zeier Plant, Madison, Wisconsin,* April 1984a.

Donohue & Associates, Inc., *Smith Ditch Lagoon No. 1 and Hotter Lagoon Investigation, Valparaiso, Indiana,* June 1984b.

Franzini, J. B., "Flood Control—Average Annual Benefits." *Consult. Eng.* 107–109, (1961).

Geiger, W. F., S. A. Labella, and G. C. McDonald, "Overflow Abatement Alternatives Selected by Combining Continuous and Single Event Simulations." *Proc. Nat. Symp. Urban Hydrol., Hydraul., Sediment Control, 1976* pp. 71–70 (1976).

Hamming, R. W., *Numerical Methods for Scientists and Engineers,* p. vii. McGraw-Hill, New York, 1962.

Heaney, J. P., W. C. Huber, and S. J. Nix, *Storm Water Management Model. Level I. Preliminary Screening Procedures,* EPA-600/2-76-275. U.S. Environ. Prot. Agency, Cincinnati, OH, October 1976.

Huber, W. C., J. P. Heaney, M. A. Medina, W. A. Peltz, H. Sheilch, and G. F. Smith, *Storm Water Management Model, Users Manual, Version II,* EPA-670/2-75-017. U.S. Environ. Prot. Agency, Cincinnati, OH, March 1975.

Hydrocomp, Inc., *Hydrocomp Simulation Programming—Mathematical Model of Water Quality Indices in Rivers and Impoundments.* Palo Alto, CA, 1972.

Hydrocomp, Inc., *Hydrocomp Simulation Programming Operations Manual,* 4th ed. Palo Alto, CA, January 1976.

Hydrocomp International, *Simulation of Continuous Discharge in the North Branch of the Chicago River.* Palo Alto, CA, December 1969.

Larsen, E., "Recorded and Simulated Runoff From an Urban Catchment Area in Metropolitan Toronto." *Proc. Int. Symp. Urban Hydrol., Hydraul., Sediment Control, 1977.*

Linsley, R. K., *A Manual on Collection of Hydrologic Data for Urban Drainage Design.* Hydrocomp, Inc., Palo Alto, CA, March 1973.

Linsley, R. K., and N. H. Crawford, "Continuous Simulation Models in Urban Hydrology." *Geophys. Res. Lett.* **1**(1), 59–62 (1974).

Malcolm, H. R., "Culvert Design and Channel Erosion." *Proc. Res. Conf., Water Probl. Urban. Areas, 1978.* pp. 298–301.

Marsalek, J., "Data Collection, Instrumentation and Verification of Models," *Proc. Conf. Mod. Concepts Urban Drainage,* Paper No. 8 (1977).

Marsalek, J., *Research on the Design Storm Concept,* Tech. Memo. No. 33. ASCE Urban Water Resources Research Program, New York, September 1978.

McPherson, M.B., *Some Notes on the Rational Method of Storm Drain Design*, Tech. Memo. No. 6. ASCE Urban Water Resources Research Program, New York, January 1969.

Orenstein, G.S., "The True Cost of Microcomputers," *Civ. Eng.* November, pp. 57–59 (1986).

Shelley, P. E., *Collection of Field Data for Stormwater Model Calibration.* Presented at a Short Course on Applications of Stormwater Management Models, University of Massachusetts, Amherst, July 28–August 1, 1975.

Sherman, L. K., "Stream-Flow From Rainfall by the Unit-Graph Method," *Eng. News Rec.* pp. 501–505, April, (1932).

Southeastern Wisconsin Regional Planning Commission, *Flood Damage Computation Procedures in the Menomonee River Watershed.* Memorandum to the Menomonee River Watershed Committee, February 18, 1976a.

Southeastern Wisconsin Regional Planning Commission, *A Comprehensive Plan for the Menomonee River Watershed,* Plann. Rep. No. 26, Vols. 1 and 2. SEWRPC, Waukesha, WI, October 1976b.

Southeastern Wisconsin Regional Planning Commission, *Validation of Flood Discharges and Flood Stages,* Memorandum to File, April 18, 1977a.

Southeastern Wisconsin Regional Planning Commission, *Calibration of Hydrologic-Hydraulic Submodels on the Noyes Creek and Honey Creek Subwatersheds of the Menomonee River Watershed,* Mem. to File. SEWRPC, Waukesha, WI, October 1977b,

Southeastern Wisconsin Regional Planning Commission, *A Comprehensive Plan for the Kinnickinnic River Watershed,* Plann. Rep. No. 32. SEWRPC, Waukesha, WI, December 1978.

Task Committee on Quantifying Land-Use Change Effects, "Evaluation of Hydrologic Models Used to Quantify Major Land-Use Change Effects. *J. Irrig. Drain. Div., Am. Soc. Civ. Eng.* **111**(1), 1–17. (1985).

Terstriep, M. L., and D. Noel, *ILLUDRAIN Version 2.1 - Users Manual.* CE Software, Champaign, IL, May 1985.

Terstriep, M. L., and J. B. Stall, *"The Illinois Urban Drainage Area Simulator, ILLUDAS." Bull.—Ill. State Water Surv.,* **58** (1974).

U.S. Army Corps of Engineers, Hydrologic Engineering Center, *Publications Catalog.* USACE, Davis, CA, October 1986a.

U.S. Army Corps of Engineers, Hydrologic Engineering Center, *Computer Program Catalog.* USACE, Davis, CA, November 1986b.

U.S. Department of Agriculture, Soil Conservation Service, *Urban Hydrology for Small Watersheds,* 2nd ed., Tech. Release No. 55, USDA, Washington, DC, June 1986.

U.S. Geological Survey, *"Water Resources Data for Indiana, Water Year, 1984." U.S. Geol. Surv. Water-Data Rep.* **IN-84-1**, 1–292 (1985).

Walesh, S. G., *Models: Practical Tools in Urban Water Resources Planning,* Tech. Memo. No. 31. ASCE Urban Water Resources Research Program, New York, July 1976.

Walesh, S. G., *Urban Hydrology,* Chapter 5, Spec. Rep. No. 49. Am. Public Works Assoc., Chicago, IL, 1981.

Walesh, S. G., *Implementation of Microcomputers in Universities.* Presented at the Fourth International Conference on Urban Storm Drainage in Lausanne, Switzerland, August 31–September 4, 1987.

Walesh, S. G., "Analysis of the Stormwater System for the Proposed Ganz Addition to Concord Meadows in the City of Valparaiso, Indiana." 1988a.

Walesh, S. G., "Microcomputer Capability: Practitioners' Perspective," In *Proc. Nat. Conf. Hydrau. Eng.,* Colorado Springs, CO, ASCE, New York, 1988b.

Walesh, S. G., "Microcomputers in Undergraduate Education," In *Computing in Civil Engineering: Microcomputers to Supercomputers—Proc. Fifth Conf.* Alexandria, VA, Edited by K. M. Will, ASCE, New York 1988c, pp. 471–477.

Walesh, S. G., and G. E. Raasch, *Calibration: Key to Credibility in Modeling.* Presented at the ASCE Hydraulics Division Conference, University of Maryland, August 1978.

Walesh, S. G., and M. L. Schoeffmann, *Surface and Subsurface Detention in Developed Urban Areas: A Case Study.* Presented at the ASCE Urban Water '84 Conference, Baltimore, MD, May 1984.

Walesh, S. G., and M. L. Schoeffmann, "One Alternative to Flooded Basements," *Am. Public Works Assoc, Rep.* **52**(3), 6–7, March (1985).

Walesh, S. G., and D. F. Snyder, "Reducing the Cost of Continuous Simulation." *Water Resourc. Bull.* **15**(3), 644–659, June (1979).

Walesh, S. G., and R. M. Videkovich, "Urbanization: Hydrologic-Hydraulic-Damage Effects." *J. Hydraul. Div., Am. Soc. Civ. Eng.,* **104**(HY2), February, pp. 141–155. (1978).

Walesh, S. G., D. H. Lau, and M. D. Liebman, "Statistically Based Use of Event Models." *Proc. Int. Symp. Urban Storm Runoff, 1979.* pp. 75–81.

Walesh, S. G., J. A. Hardwick, and D. H. Lau, *Cost Savings Through Multi-Purpose Flood Control—Recreation Facility.* Presented at the ASCE Water Resources Specialty Conference, Kansas City, MO, March 16–18, 1987.

Wenzel, H. G., and M. L. Terstriep, "Sensitivity of Selected ILLUDAS Parameters." *Ill. State Water Surv.* August, pp. 1–25 (1976).

Wenzel, H. G., and M. L. Voorhees, *Evaluation of the Design Storm Concept.* Presented at the Fall Meeting of the American Geophysical Union, San Francisco, CA, December 1978.

Wenzel, H. G., and M. L. Voorhees, *Sensitivity of Design Storm Frequency.* Presented at the Spring Meeting of the American Geophysical Union, Washington, DC, May 28–June 1, 1979.

Whipple, W., N. S. Grigg, T. Grizzard, C. W. Randall, R. P. Shubinski, and L. S. Tucker, *Stormwater Management in Urbanizing Areas,* Chapter 5. Prentice-Hall, Englewood Cliffs, NJ, 1983.

Williams, G. P., "Manning Formula—A Misnomer." *J. Hydraul. Div., Am. Soc. Civ. Eng.* **96** (HY1), 193 (1970).

11

MANAGEMENT MEASURES

In keeping with the definition of urban SWM presented in Chapter 1, surface water management measures are defined as those structural entities and non-structural measures intended to reconcile the water conveyance and storage function of depressions, lakes, swales, channels, and floodplains with the space and related needs of an expanding urban population. Management measures are the works that are built and the actions that are taken.

Many management measures are available for solving existing surface water quantity and quality problems and preventing the development of new problems. The challenge in urban SWM planning and design is to find, for a particular watershed system, the best combination of structural and nonstructural management measures and integrate them into an effective whole.

This chapter has a threefold purpose. First, a comprehensive summary of structural and nonstructural measures available for the control of the quantity and quality of urban surface water is presented. This "laundry list" of available management measures and their intended function should be of value to the engineer or other professional by providing a point of departure in beginning to search for management measures applicable to a particular situation. The second purpose of the chapter is to introduce the salient planning and design characteristics of each measure and to identify some of the positive and negative features of each measure to aid the practitioner in determining whether or not the management measure warrants further consideration for a particular situation. The final purpose of this chapter is to provide references to papers, reports, textbooks, reference books, and other literature which are likely to be of value to the engineer or other professional who desires further information to use in the planning for or design of a particular management measure or combination of management measures.

This chapter is not a design manual. Instead, it is a practical introduction to the many, diverse structural and nonstructural management measures available for the control of the quantity and quality of urban surface water.

11.1 DEFINITION OF STRUCTURAL AND NONSTRUCTURAL MANAGEMENT MEASURES

Available management measures may be broadly categorized as structural or nonstructural. Structural measures are typically major public works projects and as such require moderate to major planning and design efforts, formal approval by one or more government units and agencies, letting of construction contracts, and moderate-to-large capital investments and operation and maintenance commitments. Examples of structural measures used primarily to control the quantity of urban surface water are detention/retention facilities, upstream storage and diversion works, channel modifications or enclosures, dikes and floodwalls, and bridge and culvert alteration or replacement. Similarly, examples of structural measures used mainly for controlling the quality or urban surface water are sedimentation basins, artificial or restored wetlands, and infiltration systems.

In contrast, nonstructural measures usually involve little or no construction. Nonstructural measures can often be implemented quickly, sometimes unilaterally by individuals and business and other private entities, and typically require small-to-moderate capital investments. Examples of nonstructural measures intended primarily for flood control are flood insurance, floodproofing of structures and facilities, relocation or demolition of structures and facilities, and emergency action plans invoked under conditions of expected, impending, or actual flooding. Similarly, an example of a nonstructural measure used mainly for enhancing the quality of urban surface water runoff is slope and swale treatment primarily to control erosion.

Some management measures are effective in managing both the quantity and quality of surface water. Examples of dual-purpose structural measures are channel modification or enclosure and sedimentation basin–detention/retention systems. Examples of dual-purpose nonstructural measures are reservation of land, control of land use, and inspection and maintenance manuals.

In general, structural measures tend to be more applicable to already developed areas and nonstructural measures tend to be more appropriate for undeveloped areas. Stated differently, structural measures tend to be remedial, whereas nonstructural measures tend to be preventive.

Some surface water quantity and quality management measures are not clearly categorized as structural or nonstructural. For example, floodproofing could be considered a structural measure because it typically includes structural modification of an existing building or facility or special structural features incorporated into a new building or facility. However, floodproofing is typically implemented essentially unilaterally by the private sector and does not usually qualify as a public works project. Therefore, floodproofing is categorized as a nonstructural management measure in this book. Also categorized as nonstructural measures, for similar reasons, are relocation or demolition of structures and facilities and slope and swale treatment.

Structural and nonstructural quantity and quality management measures described in this chapter are summarized in Table 11.1. For each management measure, Table 11.1 indicates whether the measure is intended for quantity control, quality control, or both. The order in which management measures

TABLE 11.1 Structural and Nonstructural Measures Available for Controlling the Quantity and Quality of Urban Surface Water

Management Measure	Purpose	
	Quantity Control	Quality Control
Structural		
Channel modification or enclosure	×	×
Storage or diversion upstream of flood-prone area	×	
Dikes and floodwalls	×	
Bridge and culvert alteration or replacement	×	
Sedimentation basin		×
Sedimentation basin–wetland system		×
Sedimentation basin–detention/retention system	×	×
Infiltration system		×
Sedimentation basin–wetland system		×
Sedimentation basin–wetland–detention/retention system	×	×
Nonstructural		
Reservation of land for recreation and other compatible open space use	×	×
Control of land use outside flood-prone areas	×	×
Flood insurance	×	
Floodproofing of existing or proposed structures and facilities	×	
Relocation or demolition of structures and facilities	×	
Slope and swale treatment		×
Community utility and service policies	×	×
Inspection and maintenance program	×	×
Emergency action program	×	
Surface water management manual	×	×
Education	×	×

are presented in Table 11.1 corresponds to the order in which they are discussed in this chapter.

11.2 STRUCTURAL MEASURES

Channel Modification or Enclosure

Description and Function. Modification of a channel or enclosure of a channel in a large underground conduit are often technically feasible ways to lower the flood stage of a creek or river where it passes through or near urban development. Because uncontrolled urbanization often results in dense urban development accompanied by expansion onto floodplains, channel modification or enclosure is frequently one of the remaining few, although very expensive, feasible flood control alternatives.

The usual function of a channel modification or enclosure is to develop a lower, hydraulically more efficient conveyance facility that can carry flood discharges at a much lower stage relative to that which would exist under natural or existing conditions. One effect of channel modifications—or channelization as it is sometimes called— is to reduce greatly the width of the floodplain in the area contiguous with the channel modification, as illustrated in Figure 11.1. Channel modifications may include one or more of the following major

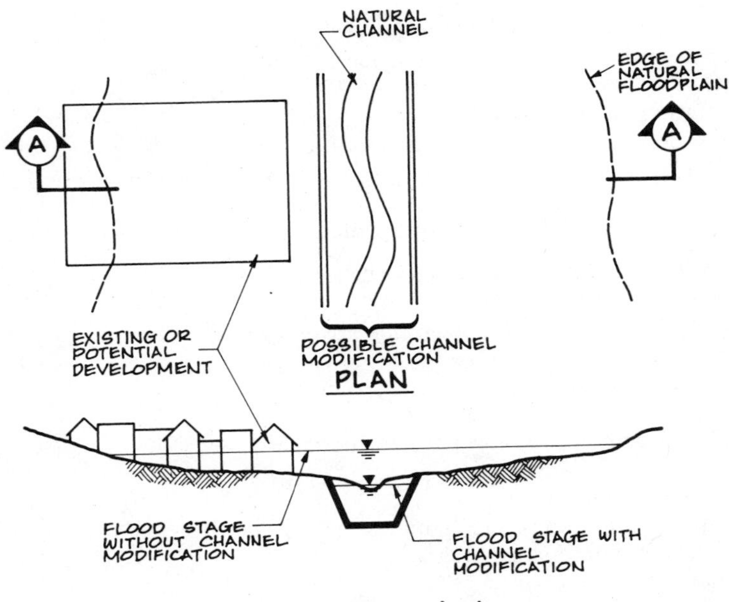

FIGURE 11.1 Channel modification and resulting reduction in flood stage

changes to a natural channel: straightening, deepening, and widening; placement of concrete, gabion, or other armoring material on the invert and sidewalls; and reconstruction of selective bridges and culverts as needed.

As illustrated in Figure 11.2, enclosure in a conduit has an effect similar to channelization in that it reduces flood stages and effective floodplain width. In addition, routing the waterway through an underground conduit also makes the space immediately above the conduit available for other use. Channel enclosure can be achieved with one or more parallel standard corrugated metal or reinforced-concrete pipe sections. If necessary, a special cast-in-place reinforced concrete conduit can be used.

Sometimes channel modification or enclosure is a means of controlling erosion of relatively narrow, steep-sided channels in urban areas. Such erosion may be threatening structures and facilities along the channels or may cause objectionable downstream sedimentation. In these situations, reducing flood stage may be a secondary consideration. An example is presented here to illustrate the erosion control motivation for some channel modification and enclosure projects. Another purpose of the example is to suggest consideration of a full range of armoring systems when planning and designing channel modifications and enclosures, regardless of whether the purpose is flood control, stabilization, or erosion control.

Figure 11.3 shows four alternative means considered for stabilizing an 800-ft channel in Fond du Lac, Wisconsin (Donohue & Associates, 1979b).

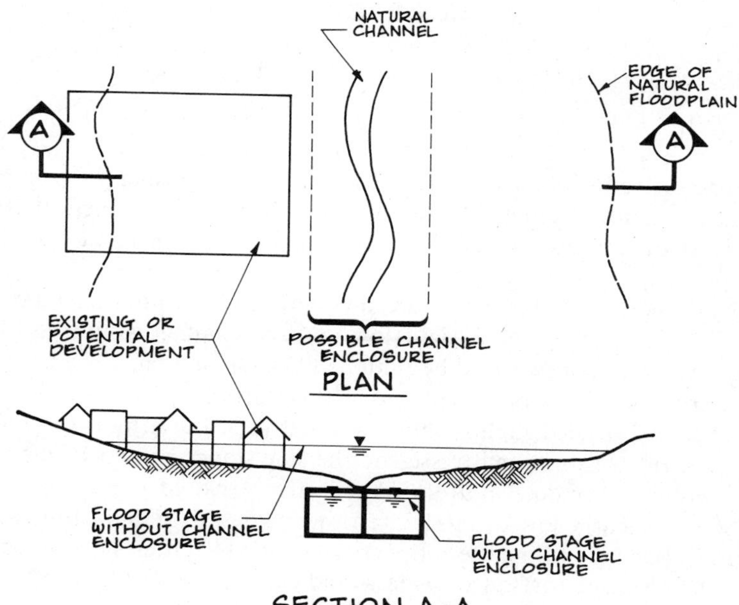

FIGURE 11.2 Channel enclosure and resulting reduction in flood stage

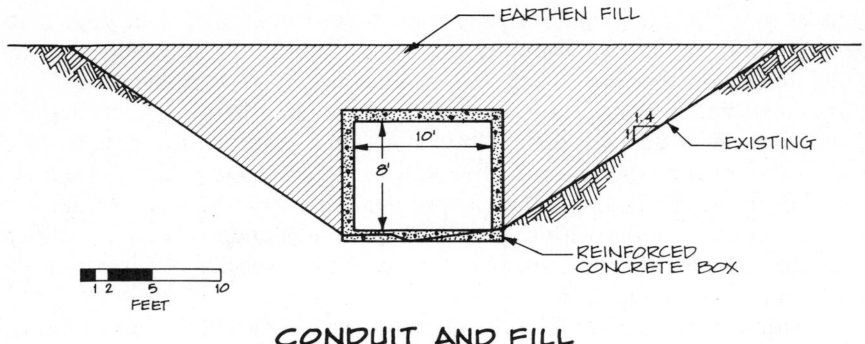

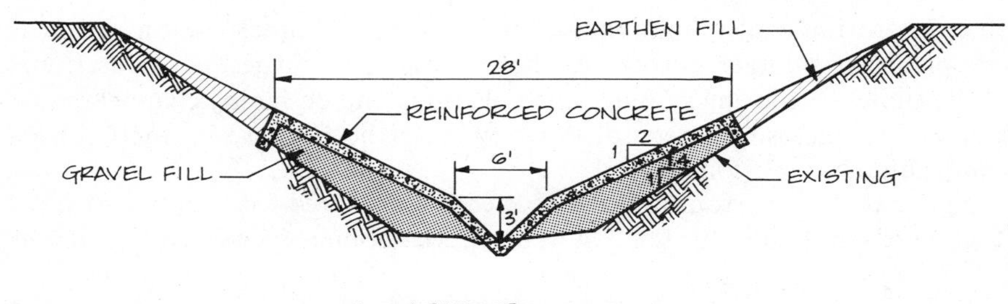

FIGURE 11.3 Alternative means of stabilizing Dutch Gap Creek in Fond du Lac, Wisconsin *(Source:* Adapted from Donohue & Associates, 1979b)

Worsening lateral erosion of the existing channel was posing a growing threat to adjacent homes. The city wanted a cost-effective and aesthetically acceptable means of halting erosion for flows up to 100-year-recurrence-interval event.

With alternative 1, a large reinforced-concrete box conduit would be placed in the channel to carry the design flow and the channel would be filled. In alternative 2, side slopes would be flattened slightly and more than half of the channel cross section would be lined with reinforced concrete. Side slopes would also be flattened slightly under alternative 3, but armoring would consist of rock-filled gabion mattresses on the side slopes and rock-filled gabion baskets along the channel bottom. Alternative 4 would use a medium-sized storm sewer to carry low-to-moderate flows. Sidewall stabilization would be achieved by partially filling the entire cross section to flatten side slopes significantly. The resulting turf-lined swale would carry flows in excess of the sewer capacity.

The 1979 estimated construction costs of alternatives 1, 2, 3, and 4, were

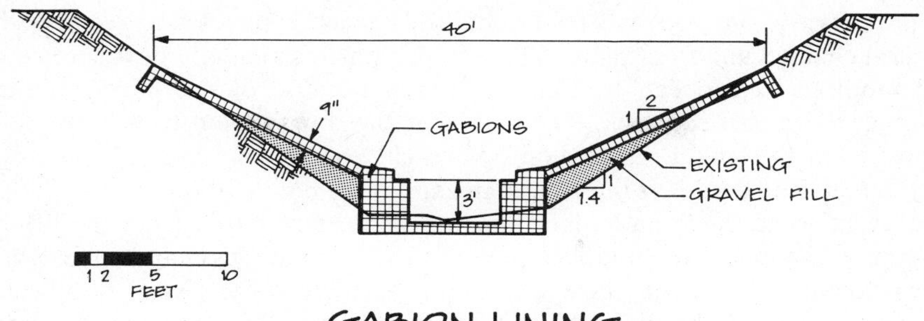

GABION LINING

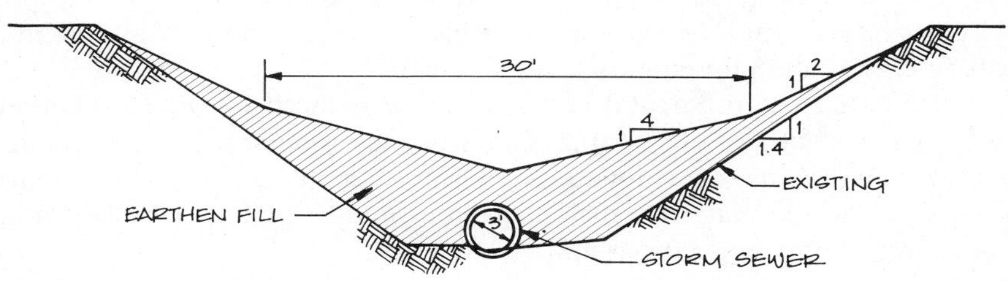

LOW FLOW STORM SEWER AND TURF OVERFLOW CHANNEL

$645,000, $335,000, $215,000, and $125,000, respectively. Alternative 3 was selected and constructed.

Planning and Design Considerations. The modified channel or underground conduit is usually sized to convey safely a major flood flow such as the 100-year recurrence interval discharge. Relative to the cross-sectional area of the natural channel, this criterion typically requires that the modified channel or underground conduit have a very large cross-sectional area.

Channel modifications usually result in significant increases in flow velocity, particularly during flood conditions. Therefore, the modified channel is typically protected with a concrete liner or other form of armoring. Another important design consideration in channel modification or enclosure is that the natural channel and floodplain upstream and downstream of the modified reach must be altered to provide smooth hydraulic transitions. One transition is required at the upstream end, where the natural channel and floodplain converge on and connect to the modified reach, and another transition is needed at the downstream end, where the modified reach expands and reenters the natural channel and floodplain. These transitions, particularly downstream of the modified reach, can be long, primarily because of the lowered grade through the modified reach. The length of the downstream transition could even exceed the length of the modified reach.

Planners and designers must be cognizant of the potential flood aggravation and negative aesthetic and environmental impacts resulting from modifying a natural channel and floodplain. Downstream discharges may increase and downstream flooding may be aggravated as a result of the loss of floodwater storage capacity in the channelized or enclosed reach. Erosion and sedimentation may occur upstream and downstream of the modified reach as described in Chapter 1. Because of the potential for varied and significant off-site—mainly downstream—effects, channel modifications and enclosures are very likely to require local, state, and other permits. Such projects should not proceed past the conceptual stage without a clear understanding of all potential negative effects and all regulatory requirements.

Aesthetic and environmental impacts can be reduced by careful planning and design. Potentially useful approaches include retaining as much natural vegetation as feasible, providing a curvilinear alignment, using more natural appearing armoring materials such as gabions and riprap, and implementing an adequate maintenance program.

Chapter 5 includes suggested analysis and design guidelines and techniques for the hydraulic, safety, aesthetic, erosion, and sedimentation control and other aspects of channel modifications and enclosures. Additional planning and design criteria and procedures are provided by the Corps of Engineers (U.S. Department of the Army, 1970). The subject is also treated in text and reference books such as Chow (1959), French (1985), Kinori and Mevorach (1984), Linsley and Franzini (1979), and Petersen (1986).

Storage or Diversion Upstream of a Flood-Prone Area

Description and Function. Upstream storage and upstream diversion may be an effective flood mitigation approach. Although storage and diversion function differently, they have similar effects on a flood-prone area, as shown in Figure 11.4, in that both are intended to reduce peak flows and therefore reduce the peak stage and width of the floodplain. Upstream storage temporarily impounds water, thus permitting a slow, controlled release and reduced downstream peak flow. With upstream diversion, some flow is directed out of the watershed with the same downstream results.

Rather than divert flow out of the watershed, a variation on upstream diversion is to divert excess flows around the flood-prone area, with the diverted flow returned to the principal stream downstream of the flood-prone area. Another variation on upstream diversion is diversion from the channel to a temporary storage facility from which water is returned by gravity or pumping

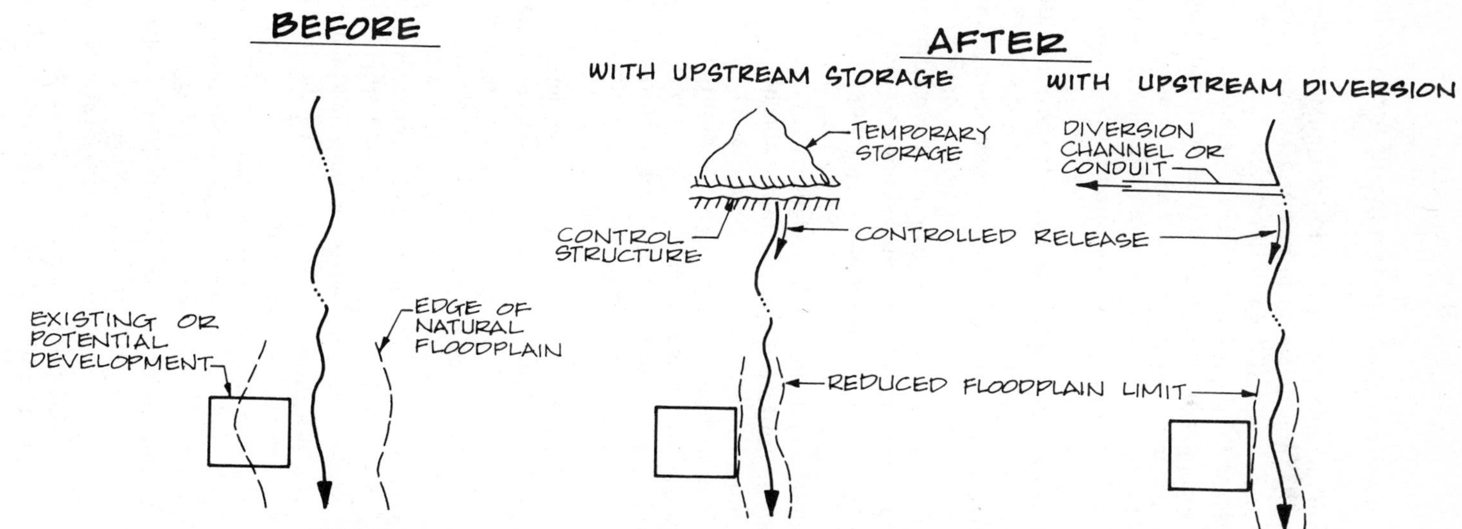

FIGURE 11.4 Upstream storage or diversion and resulting reduction in floodplain width

to the main channel. This form of diversion is the off-line detention/retention discussed in Chapter 8.

Upstream diversion, as shown in Figure 11.4, is fundamentally different in function and magnitude than diversion of lateral drainage, discussed later in this chapter in conjunction with dikes and floodwalls. Lateral drainage diversion typically requires a relatively small installation intended to control runoff enroute to a major stream by diverting it around an existing or potential flood-prone area. In contrast, upstream diversion works are usually large facilities constructed on a major stream and intended to reduce peak flood flow, flood stage, and floodplain width throughout the flood-prone area.

Planning and Design Considerations. Upstream storage should be viewed as a broad concept to prompt consideration of the full range of possibilities. For example, while a traditional large upstream surface storage facility as suggested by Figure 11.4 may be an option, other storage options may exist, depending on the nature of the flood problem and the existing surface water system. These options might include rooftop detention storage, as used in Denver (American Public Works Association, 1974). Another idea is detention storage on streets accomplished with small berms and flow regulators as installed in Skokie, Illinois, and illustrated in Figures 11.5 and 11.6 respectively, and detention storage in reinforced-concrete tanks beneath streets as also constructed in Skokie, Illinois, and illustrated in Figures 11.7 and 11.8 (Walesh, 1986; Walesh and Schoeffmann, 1984, 1985). These nontraditional storage approaches tend to consist of several or many sites distributed throughout a problem area. Furthermore, nontraditional means of storage can often be retrofitted into urban areas that are already developed as illustrated by the Skokie system.

Diversion should also be viewed broadly when considering options. For example, diversion could include reworking the existing local surface and subsurface drainage system to reroute flows to elements having greater capacities. Some of the street berms installed in Skokie, Illinois, are designed to reroute surface flow patterns.

Planning and design criteria, procedures, and techniques for detention/retention facilities are presented in Chapter 8. Included in that chapter are references to numerous papers, reports, and other materials. Chapter 5 presents additional analysis and design guidelines relevant to the hydraulics, safety, aesthetic, and other aspects of storage facilities.

The preceding section of this chapter cites reports and texts and reference books that present planning and design criteria for channel modification or enclosure. The references cited are also useful for the design of diversion channels and conduits. Additional ideas on diversions are presented by Kinori and Mevorach (1984) and Linsley and Franzini (1979).

Upstream storage or diversion facilities must typically be sized to store or convey runoff from a major flood such as the 100-year-recurrence-interval event. The target floodplain width and stage and corresponding peak flow at

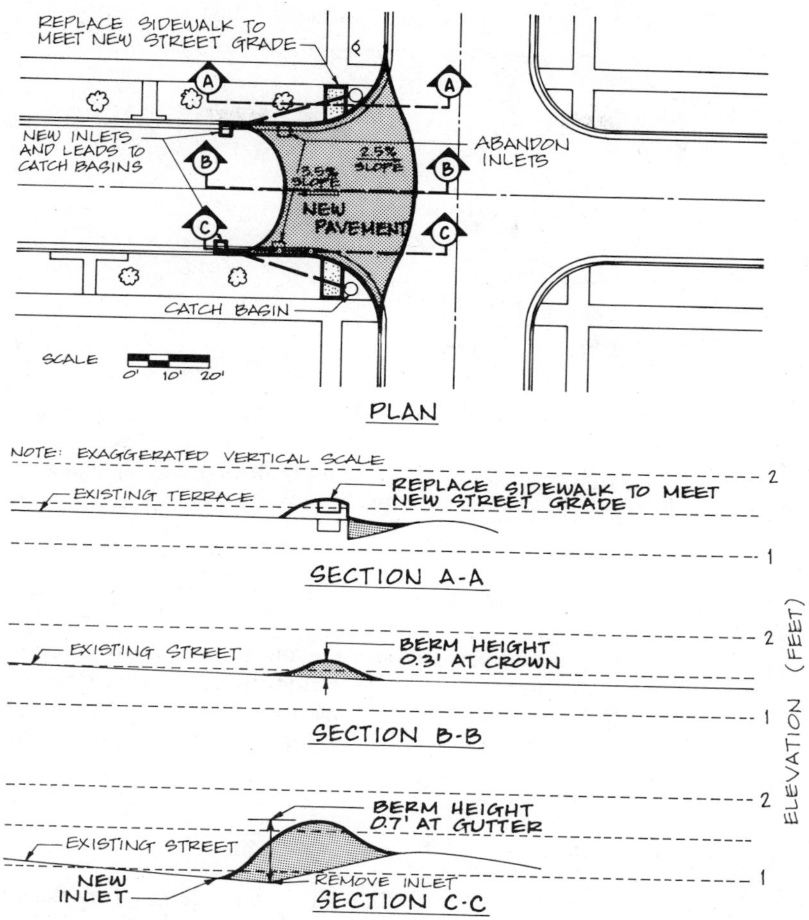

FIGURE 11.5 Typical street berm in Skokie, Illinois *(Source:* Walesh, 1986)

the existing or potential flood-prone site must be determined. Then the upstream storage or diversion facilities are sized to accommodate the excess runoff volume or discharge. Previously presented factors to consider in the design of channel modification and enclosure apply to diversion channels or conduits.

In addition to hydrologic and hydraulic considerations, upstream storage or diversion typically must address unique and sensitive environmental and legal issues. For example, there may be considerable resistance to construction of either a wet or a dry upstream water storage facility because of aesthetic impacts and because of effects on flora and fauna. Similarly, a potential upstream diversion project may encounter legal prohibitions against interwatershed transfers. Regulatory and other legal aspects of upstream storage and

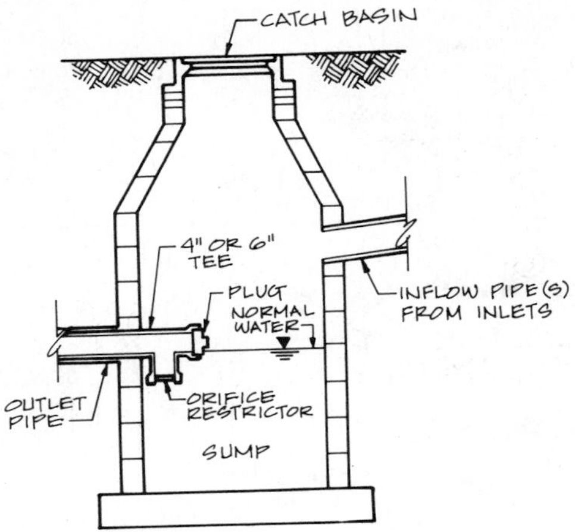

FIGURE 11.6 Typical flow regulator in catch basin installation used in Skokie, Illinois *(Source:* Adapted from Walesh and Schoeffmann, 1984; Walesh, 1986)

diversion should be researched and addressed before such projects proceed beyond the concept stage.

A positive feature of both upstream storage and upstream diversion is that a single facility has the potential to mitigate flooding in several downstream reaches or communities. A negative aspect of upstream storage and upstream diversion is the false sense of security that may develop, leading to further encroachment into flood-prone areas.

Dikes and Floodwalls

Description and Function. Earthen dikes and concrete or steel floodwalls may be an effective means of providing flood protection. Cross sections of a typical earthen dike and a typical concrete floodwall are shown in Figure 11.9. Dikes and floodwalls are intended to prevent overland flow laterally from a

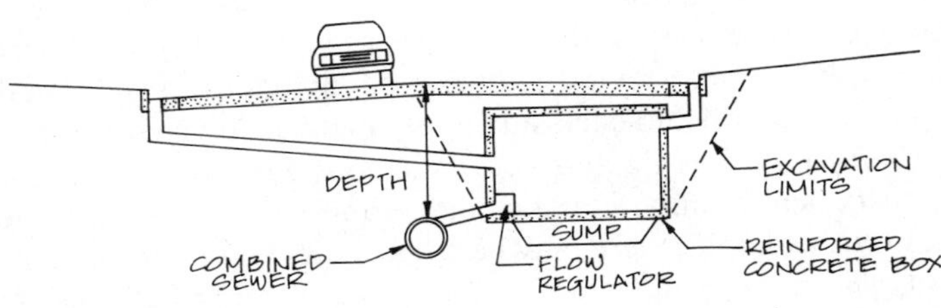

FIGURE 11.7 Schematic of subsurface detention facility used to prevent surcharging of combined sewers in Skokie, Illinois *(Source:* Adapted from Walesh and Schoeffmann, 1984; Walesh, 1986)

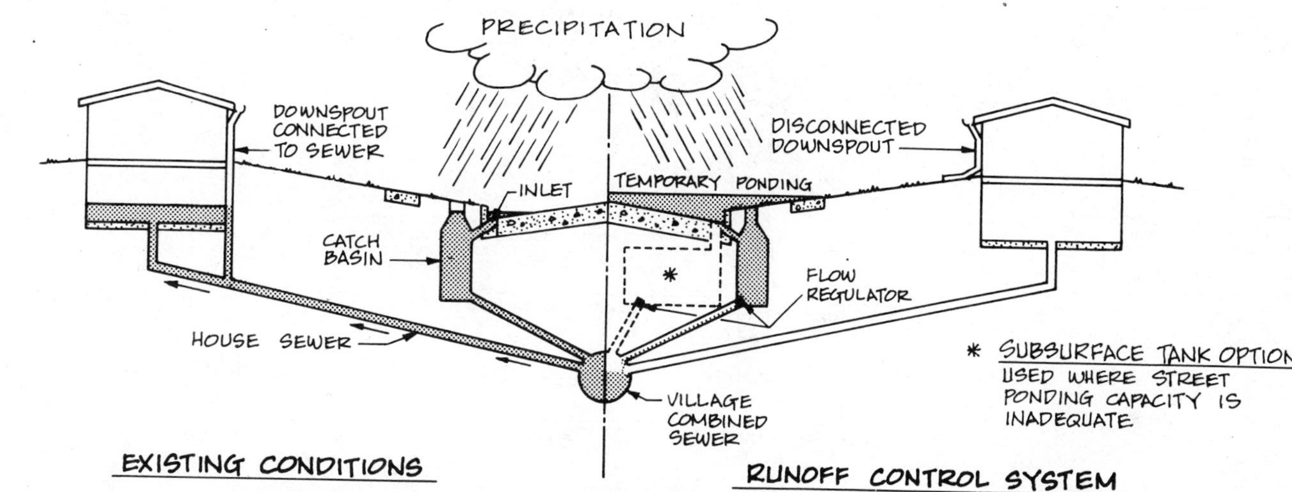

FIGURE 11.8 Schematic of the runoff control system for Skokie, Illinois, which includes flow regulators, on-street detention, and subsurface detention *(Source:* Adapted from Walesh and Schoeffmann, 1984; Walesh, 1986)

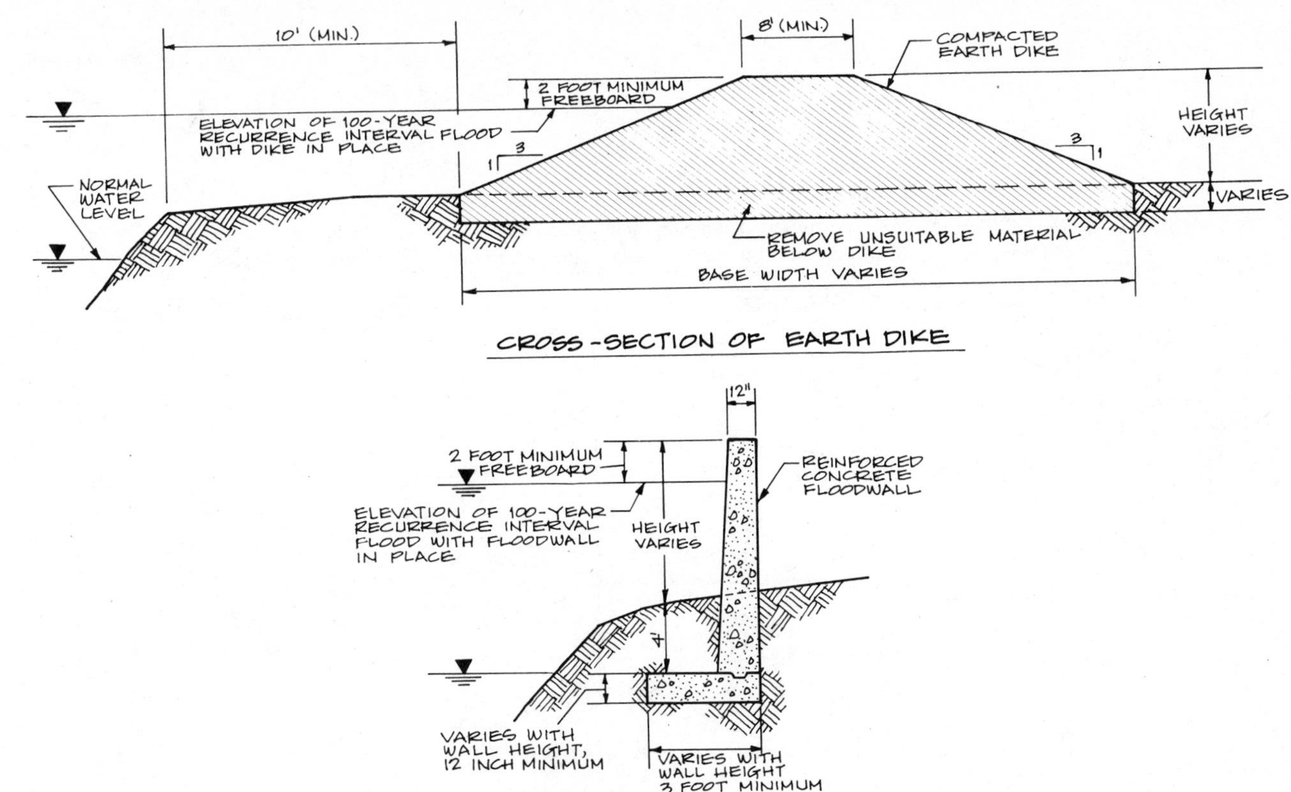

FIGURE 11.9 Typical earthen dike and concrete floodwall (*Source:* Adapted from Southeastern Wisconsin Regional Planning Commission, 1978)

creek or river into adjacent flood-prone areas. In coastal areas, dikes and floodwalls are more properly called seawalls and must also resist wave forces. Some river or coastal sites may be best suited to combination structures consisting of a floodwall constructed on top of a dike.

Some floodwalls and all dikes are gravity structures; that is, their structural stability is achieved primarily by the weight of the structure. In contrast, some floodwalls are cantilevers and are structurally stable because of the passive resistance of the surrounding earth.

Planning and Design Considerations. An important design factor is the construction space available between the stream channel and the edge of the area to be protected. When space is at a premium, floodwalls are usually used because they require a narrower strip of land than dikes. However, the space advantage of floodwalls may be offset by their typically greater cost per lineal foot.

The design of dikes and floodwalls must include a hydraulic evaluation of possible upstream increases in flood stage as a result of floodplain constrictions. Upstream stage increases may increase the size of flood-prone areas, thereby increasing flood risks and enlarging the area that should be included in floodplain regulations. The possible need for super elevation on the outside of bends should be investigated.

Other important dike and floodwall planning and design considerations are provision of adequate freeboard; structural stability; need for and means of achieving control of seepage under or through the structure; establishing an acceptable maximum side slope for a dike to assure structural stability and to facilitate mowing and other maintenance; and provision of a means of access for maintenance and repair, which in the case of a dike is usually a road along the crest. In the case of steel piling, site soils should be studied to determine if excessive corrosion will occur.

Dikes along creeks and streams require turf, riprap, gabions, or other armoring material to provide erosion protection on the side of the structure toward the creek or river. Similarly, floodwalls along creeks or rivers may be protected from erosion by placing rock or similar material at the base of the wall on the river or creek side. Coastal area dikes, floodwalls, or seawalls may be protected against wave forces and effects by armoring all or part of the seaward face or by means of an offshore breakwater.

Planning and design criteria and procedures are provided by the Corps of Engineers (U.S. Dept. of the Army, 1948, 1978) and the U.S. Nuclear Regulatory Commission (1976). Dikes and floodwalls are also discussed in text and reference books such as Kinori and Mevorach (1984), Linsley and Franzini (1979), and Petersen (1986). Some of the hydraulic, safety, aesthetic, and other guidelines presented in Chapter 5 are applicable to dikes and floodwalls.

An area protected by dikes and floodwalls may require supplemental drainage works to convey surface water runoff generated in the protected area to some discharge point off site. Special facilities may also be required to divert lateral flow, that is, flow moving from higher ground to the creek or river, around the protected area. Finally, supplemental drainage works may be needed to prevent backup of creek or river water into the protected area during flood stages. These facilities constitute what is known as the interior drainage system.

Depending on the expected volume of surface water runoff generated within the protected area, temporary on-site storage and pumping facilities may be needed. These facilities will temporarily store stormwater, which, after flooding has subsided on the adjacent creek or river, may be pumped over the dike or floodwall or discharged by gravity.

A dike or floodwall system may be intentionally penetrated by one or more outfalls. These outfalls provide gravity drainage of surface water from within the protected site to the creek or river. During major flood events, high creek or river levels may reverse the flow and result in movement of floodwaters from the creek or river into the protected area. Flow reversal may be prevented using flap gates such as illustrated in Figure 11.10. These gates function as valves that normally pass water from the protected area to the creek or river but close when the hydraulic head on the creek or river side exceeds the head in the protected area. When the flap gates or similar controls are closed, pumps may be required to provide adequate interior drainage.

To the extent that it is technically feasible, lateral drainage should be intercepted and diverted around the protected area to reduce the volume of surface water that must be accommodated within the protected area. This may be accomplished by swales or berms, as illustrated in Figure 11.11.

Interior drainage planning and design criteria and procedures are provided by the Corps of Engineers (U.S. Department of the Army, 1965) and treated in textbooks and reference books, such as Kinori and Mevorach (1984), Linsley and Franzini (1979), and Petersen (1986).

A positive aspect of dike and floodwall systems is that communities can often act unilaterally to protect existing or proposed development within their own corporate limits provided that potential upstream effects are considered.

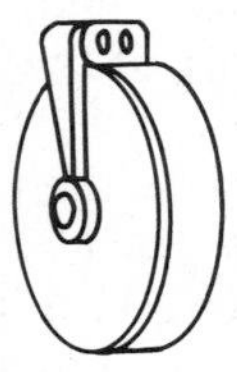

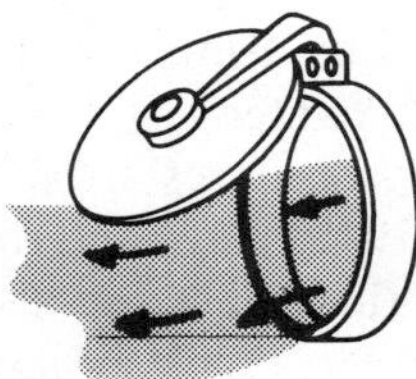

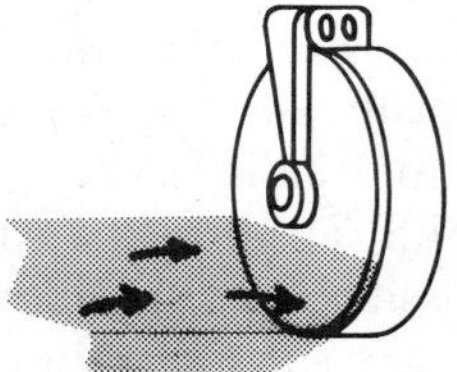

FIGURE 11.10 Flap gate installed on storm sewer outfall *(Source:* Adapted from Federal Emergency Management Agency, 1986b)

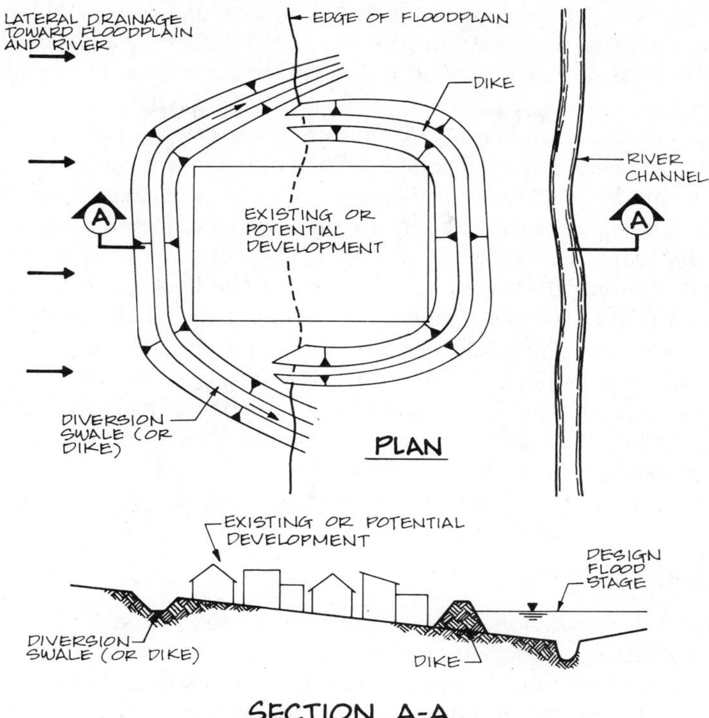

FIGURE 11.11 Intercepting swale or berm intended to convey lateral drainage around an area protected with a dike

On the negative side, dikes and floodwalls may be aesthetically objectionable, particularly from the perspective of those residents who purchased property near a creek or river because they valued the view. Finally, dikes and floodwalls may impart a false sense of security, which could in turn lead to inadequate inspection and maintenance, thus increasing the likelihood of catastrophic flooding of apparently well protected areas.

Bridge and Culvert Alteration or Replacement

Description and Function. The purpose of bridge and culvert alteration or replacement is to reduce or eliminate backwater effects that significantly contribute to upstream flooding. Potential alterations include modifying inlet condition hydraulics at a bridge or culvert, enlarging the waterway opening of a bridge, or installing one or more parallel conduits at a culvert.

Planning and Design Considerations. The technical feasibility of bridge or culvert modifications or replacement requires a careful hydraulic analysis to predict the impact on upstream stages for a range of flows usually up to and including events as severe as the 100-year-recurrence-interval event. The basis for such analysis are channel and floodplain hydraulic techniques introduced in Chapter 5.

French (1985), Normann (1975), and the U.S. Department of Transportation (1965) indicate that inlet geometry, including features such as side-tapered wingwalls and beveled edges at the entrance to conduits, can significantly increase inlet capacity for a given upstream stage or reduce upstream stage for a given discharge. Chow (1959), French (1985), and the U.S. Department of Transportation (1970) discuss sensitivity of bridge backwater to factors such as the constriction of the channel and floodplain caused by the approach roads, pier shape, the angle between piers and the stream flow, and pier length–width ratios. These studies and reports suggest that some poorly designed bridges and culverts, particularly culverts operating under inlet control, could be modified to reduce upstream backwater effects significantly. The cost of such modifications or even replacement of problematic bridges and culverts might be warranted if upstream flood damages are significantly reduced.

Sedimentation Basin

Description and Function. As defined in Chapter 9, a sedimentation basin (SB) is a facility intended to trap suspended and buoyant debris carried by stormwater runoff and the potential pollutants adsorbed onto or absorbed into the solids. A sedimentation basin schematic is included in Figure 9.1.

SBs range from crude, temporary facilities at construction or development sites, intended to retain suspended solids and other materials on the site during construction, to permanent facilities designed and operated for the long-range management of surface water runoff quality.

Although the state of the art of SB design and operation is rudimentary as it relates to control of the full spectrum of NPS pollutants in urban areas, the prognosis is positive. Results of a few field and laboratory studies indicate the SBs have suspended solids trap efficiencies in excess of 90 percent. Furthermore, SBs may significantly reduce the concentration of pollutants other than the suspended solids. Examples of these potential pollutants are phosphorus, pesticides, heavy metals, and bacteria (McCuen et al., 1983; Urbonas and Ruzzo, 1986; Walesh, 1980).

Planning and Design Considerations. SBs should be viewed broadly to prompt consideration of a wide spectrum of potential applications. Although a traditional surface sedimentation basin, like that shown in Figure 9.1, is often a feasible option, other possibilities may exist, depending on the nature of the sedimentation and related pollution problems and on the characteristics of the surface water system.

For example, in Skokie, Illinois, eight large subsurface stormwater detention facilities, each of which includes a separate sedimentation section or sump, as shown schematically in Figure 11.7, were constructed beneath the streets (Walesh, 1986; Walesh and Schoeffmann, 1984, 1985). The primary purpose of the sump was to confine suspended solids in a relatively small area near the entrance of the subsurface tanks rather than to distribute the suspended solids throughout the tanks. Similar opportunities may arise elsewhere for removing suspended solids and associated pollutants from surface water runoff.

Planning and design criteria, procedures, and techniques for SBs are presented in Chapter 9. Additional analysis and design guidelines relevant to the hydraulic, safety, aesthetic, and other aspects of SBs are presented in Chapter 5.

Sedimentation Basin–Wetland System

Description and Function. If circumstances are appropriate, an artificial or restored wetland in series with and preceded by an SB may be an effective means of removing a significant portion of the suspended solids, nutrients, and other NPS pollutants from surface water runoff. McCuen et al. (1983) present a summary of field studies which suggests the management potential of natural and artificial wetlands. Reports of in-depth monitoring studies of wetland–SB systems are available (Martin and Smoot, 1986).

The wetland SB idea may be illustrated by considering a facility constructed to protect the quality of McCarron Lake in Roseville, Minnesota (Walesh, 1986). The 72-acre lake receives runoff from a completely urbanized 1.6-mi watershed. The lake, which is heavily used for fishing, swimming, and boating and provides a setting for residential neighborhoods and a county park, was experiencing deteriorating water quality. Problems included weed and algae growth and localized sediment accumulation. A one-year monitoring study confirmed that the lake was eutrophic. Surface water runoff into the lake was determined to be the principal contributor of nutrients and suspended solids.

Figure 11.12 shows the system that was designed and constructed to alleviate most of the lake's nutrient and sediment problems. The two-component system consists of a SB followed by a restored wetland along the drainageway that receives 80 percent of the flow into the lake. Crosswise low berms and native vegetation were used to restore the remnant wetland and increase its size to 8 acres. The sedimentation basin–wetland system is expected to remove 75 percent or more of the total phosphorus and solids.

Planning and Design Considerations. With the possible exception of the SB component, proved and accepted planning and design methodologies are not available for SB–wetland systems. Accordingly, such projects require considerable judgment, should be approached with caution, and should utilize a multidiscipline team providing expertise in hydrology, hydraulics, chemistry, and biology.

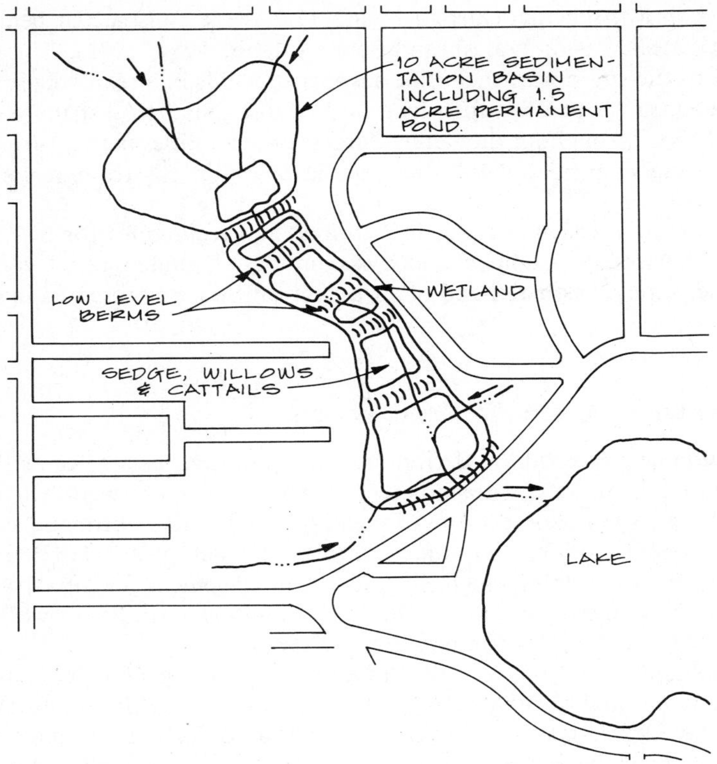

FIGURE 11.12 Sedimentation basin–wetland facility in Roseville, Minnesota *(Source:* Adapted from Walesh, 1986)

Sedimentation Basin–Detention/Retention System

Description and Function. A sedimentation basin–detention/retention system is intended to provide both quantity and quality control. The SB basin component removes suspended solids and absorbed and adsorbed potential pollutants plus buoyant debris for small runoff events, typically up to and including the 1- or 2-year-recurrence-interval event. In contrast, the D/R facility component provides flood control for moderate to major events up to and including the 100-year-recurrence-interval event.

Figure 11.13 is a plan view of an existing sedimentation basin–detention system in Madison, Wisconsin (Raasch, 1982; Walesh, 1986). A review of the circumstances leading to the construction of the Madison facility illustrates some of the planning and design considerations. Prior to construction of the project, runoff from a 640-acre urbanized watershed flowed through the site from south to north, causing flooding problems in the residential area to the north and depositing debris and sediment in the natural area south of the resi-

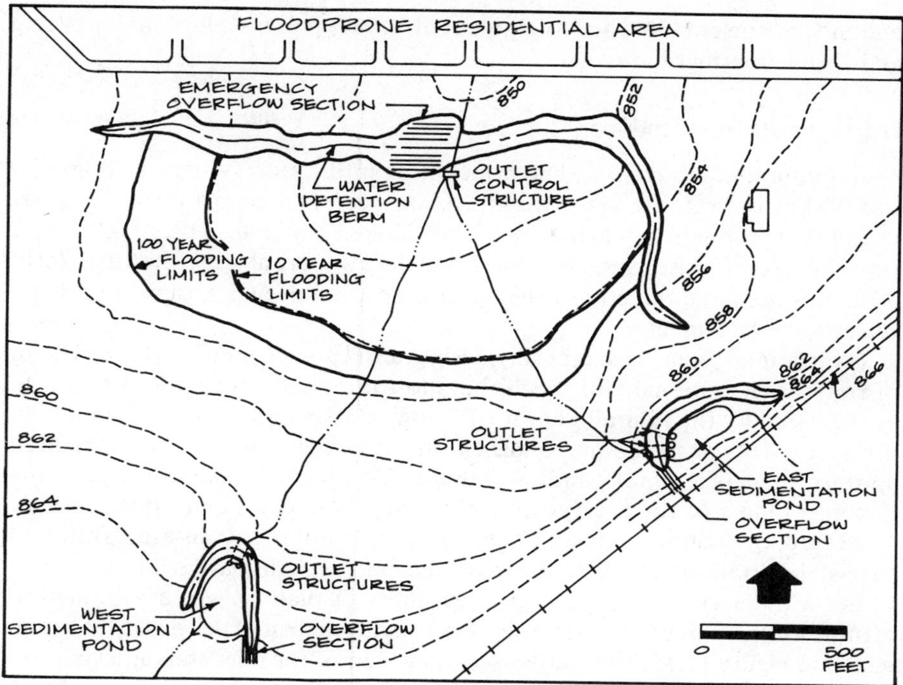

FIGURE 11.13 Sedimentation basin–detention facility system at Madison, Wisconsin (*Source:* Adapted from Raasch, 1982; Walesh, 1986)

dential area. All interested parties agreed on the sedimentation basin–detention system because it had the potential to mitigate both the flooding problem and the debris and sediment problem.

As shown in Figure 11.13, the two SBs that receive all inflow to the site are located unobtrusively on the site's perimeter. The SBs are designed to trap sediment and associated pollutants, floatables, and other debris. Potentially damaging flood flows pass through the SBs and are controlled by the downstream detention area. The 2,200-ft-long berm that forms the detention area was designed and constructed with an irregular cross section and curvilinear alignment to minimize its visual impact on the aesthetically sensitive natural area.

Planning and Design Considerations. Planning and design criteria are available for sedimentation basin–detention/retention facilities in that criteria are available for SBs and for D/R facilities. The planning and design of D/R facilities is discussed in Chapter 8 and the planning and design of SBs is presented in Chapter 9. Chapter 5 presents various analysis and design guidelines relevant to the hydrologic, safety, aesthetic, and other aspects of D/R facilities and SBs. Schueler (1987) and Whipple et al. (1987) present additional planning and design considerations.

Infiltration System

Description and Function. The principle of infiltration systems as it applies to SWM is to divert the first wash-off of surface water runoff through a filter media with subsequent discharge of the filtered water to either the surface water or groundwater system. That is, surface water, at least the first flush, does not discharge directly to the surface or groundwater system until it receives some treatment.

Infiltration systems are typically designed to capture the first 0.5 to 1.0 in. of runoff. The rationale for this is that most rainfall events are less than 1 in. as suggested by the rainfall data for Boston, Massachusetts, presented in Figure 11.14. This observation, coupled with the recognition that not all rainfall appears as surface runoff, indicates that infiltration systems have a potential to control most of the average annual runoff (Wanielista and Yousef, 1986).

According to Shaver (1986), examples of infiltration systems are infiltration basins, infiltration trenches, drywells, porous asphalt pavement, vegetated swales with check dams, and vegetative filters. A plan view of a hypothetical off-line infiltration basin system serving a small residential development is presented in Figure 11.15, and a cross-sectional view of the inlet and outlet control structure is shown in Figure 11.16. Key components of the infiltration basin

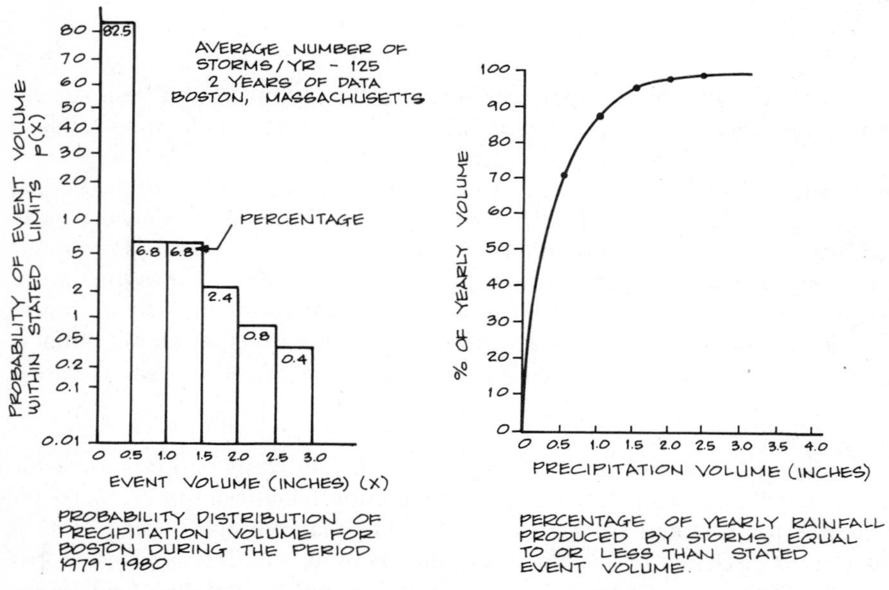

FIGURE 11.14 Summary of rainfall event volumes for Boston, Massachusetts (*Source:* Wanielista and Yousef, 1986)

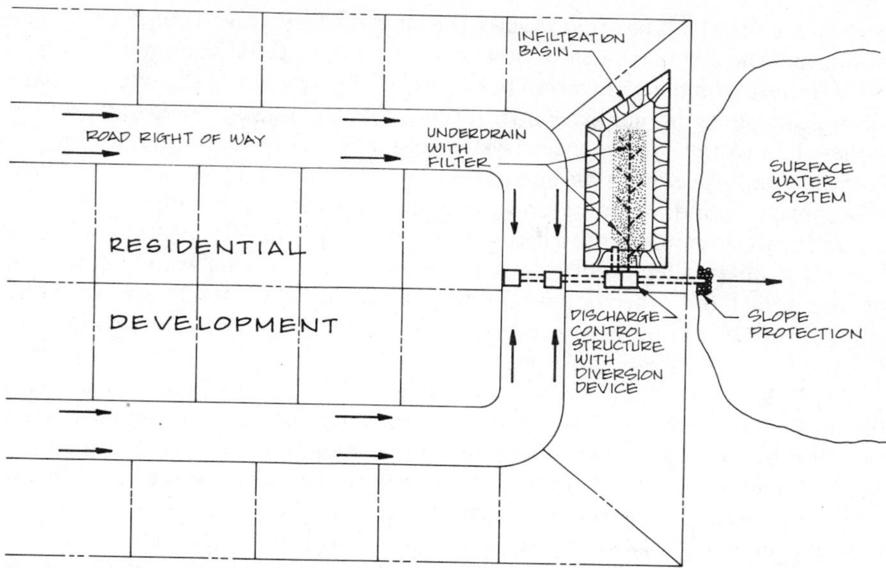

FIGURE 11.15 Off-line infiltration system serving a residential development *(Source:* Adapted from Pinellas Park Water Management District, 1982)

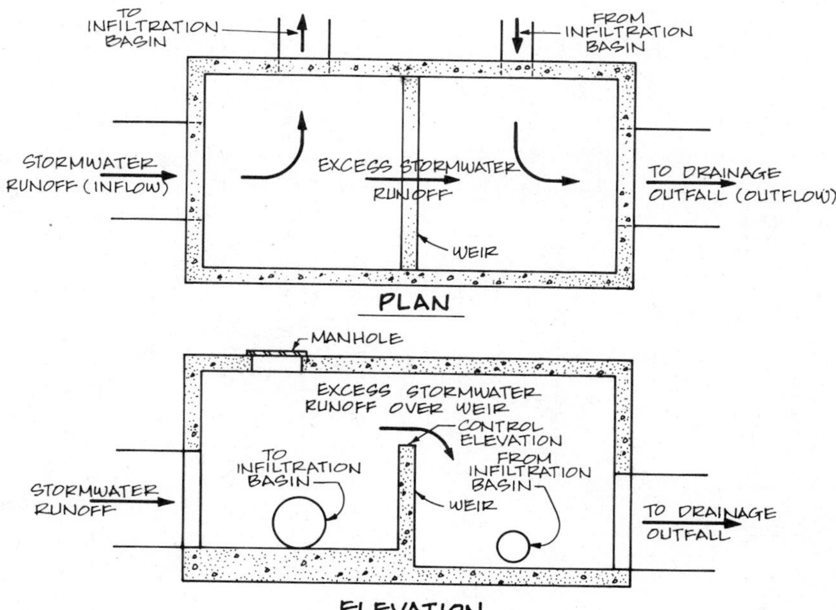

FIGURE 11.16 Hydraulic control structure for off-line infiltration system *(Source:* Adapted from Pinellas Park Water Management District, 1982)

system are the structure that diverts the first flush of runoff into the off-line basin, and the off-line basin underlaid with filter material and an underdrain.

A variation on the infiltration basin system illustrated is discharge to a basin with ultimate discharge, after infiltration, directly to the groundwater system rather than to the surface water system. Another variation on the infiltration basin system is use of an on-line basin as shown in plan in Figure 11.17. With this approach, and assuming that the basin has sufficient volume, water quality and water quantity control could be achieved. That is, infiltration would occur for all smaller events, thus protecting the receiving waters from suspended solids and other potential pollutants, and flow attenuation would occur for moderate to major events, thus helping to mitigate downstream flooding.

Urbonas and Ruzzo (1986) describe and present design criteria for a system intended to remove phosphorus from stormwater runoff by using infiltration or filtration preceded by sedimentation. As shown in Figure 11.18, the system is essentially a sedimentation basin followed by an infiltration basin. The SB can either contain a permanent pool or be dry, that is, contain water only during or immediately after a runoff event. According to Urbonas and Ruzzo (1986), the wet sedimentation is preferred because of factors such as longer detention time, less likelihood of short circuiting, and enhanced biological action.

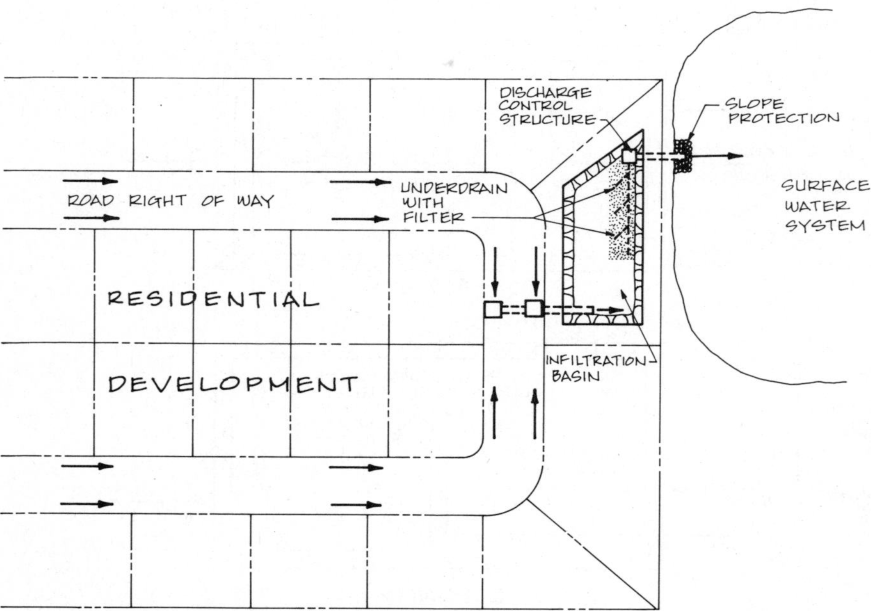

FIGURE 11.17 On-line infiltration system serving a residential development *(Source:* Adapted from Pinellas Park Water Management District, 1982)

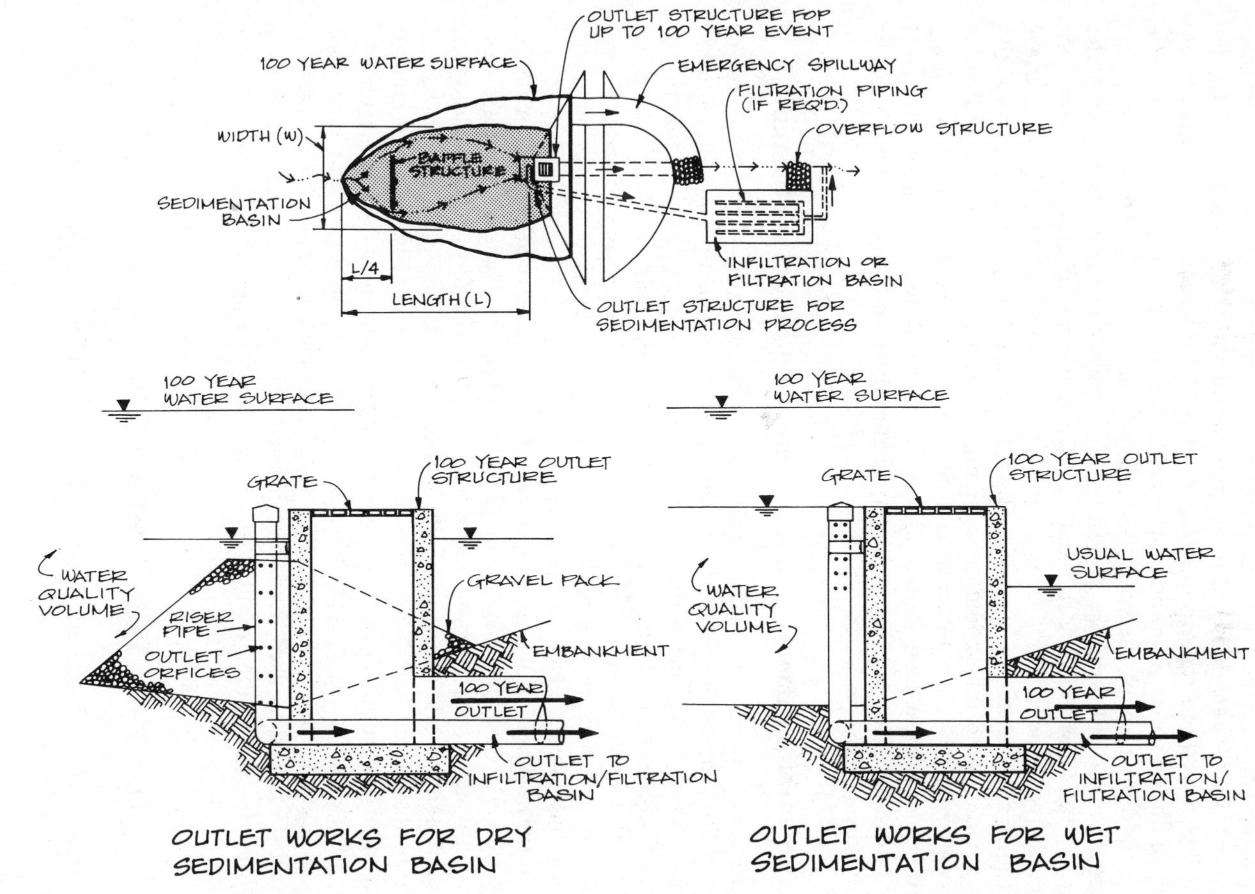

FIGURE 11.18 Sedimentation basin–infiltration system *(Source:* Adapted from Urbonas and Ruzzo, 1986)

Planning and Design Considerations. Shaver (1986) suggests factors to consider in planning and designing infiltration systems. On-site soil characteristics such as infiltration rates are important, particularly if effluent is to be discharged directly to the groundwater system. Another factor is the maximum period during which ponding is acceptable. The bottom of the infiltration system should be above the seasonal high-groundwater table and bedrock. Site topographic conditions are also important, including the proximity of water supply wells and building foundations. Design of an infiltration system must include recommended maintenance procedures and frequency. If the tributary area has a high fraction of impervious surface or extensive development activity, frequent and major maintenance may possibly be required, including replacement of the filter material and underdrain system. Finally, the planning and design professional should address the question of potential pollution of groundwater as a result of leaching of materials from the filter media into the groundwater system.

Sedimentation Basin–Wetland–Detention/Retention System

The recent emergence of the hybrid sedimentation basin–wetland system for controlling the quality of surface water runoff and the hybrid sedimentation basin–detention/retention system for controlling the quality and quantity of runoff suggests further hybridization which would yield a sedimentation basin–wetland–detention/retention system. Figure 11.19 shows conceptual presentations of a sedimentation basin–wetland–retention system and a sedimentation basin–wetland–detention system.

In the sedimentation basin–wetland–D/R sequence, the SB would have the function of removing suspended solids, absorbed and adsorbed substances, and buoyant debris. The wetland would function as a physical and biological filter to remove additional suspended solids, colloidal material, nutrients, and other substances. The D/R component would be primarily for flood control but might also result in further water quality enhancement. If it is a retention facility, as shown in the upper part of Figure 11.19, that is, a permanent pool is maintained, then further settling out of suspended solids may occur. If a detention facility is used, as shown in the lower part of Figure 11.19, it may include an infiltration or filtration system to provide further water quality enhancement.

11.3 NONSTRUCTURAL MEASURES

Reservation of Land for Recreation and Other Compatible Open Space Use

Description and Function. The principal purpose of reserving flood-prone lands for recreation and other open space uses is to prevent the construction of flood-prone buildings and facilities. Examples of open space uses, besides

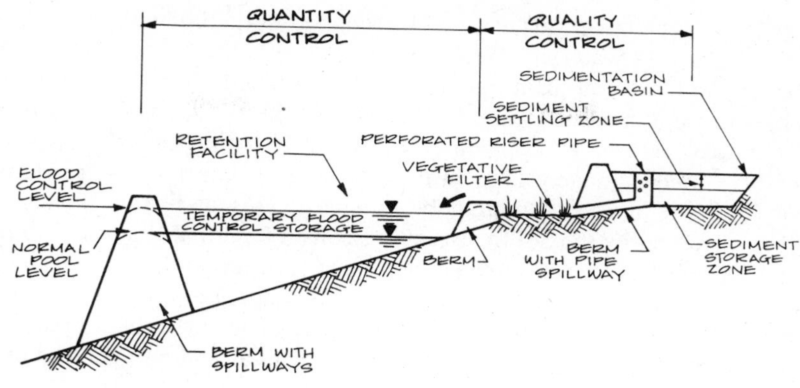

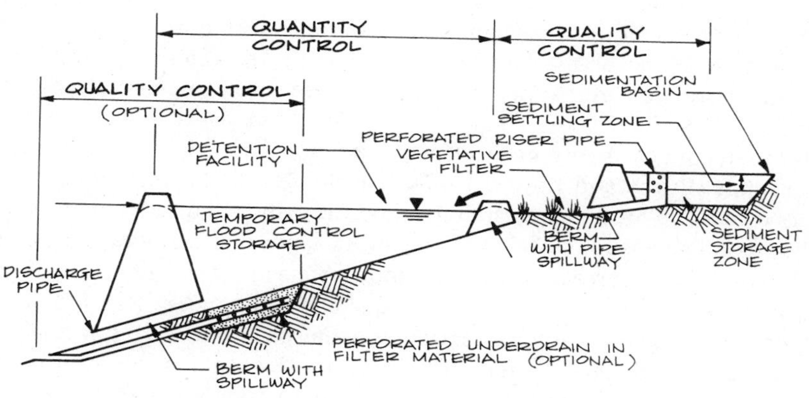

FIGURE 11.19 Sedimentation basin–wetland–detention/retention facilities

recreation, that are compatible with flooding include agriculture, parking, and storage of some materials. Besides flood mitigation, other potential benefits of reserving flood-prone lands include retaining aesthetic features often linked with lands contiguous to surface waters, protecting the flora and fauna unique to surface water environments, and preserving historic cultural features, which tend to be concentrated along surface water systems (Walesh, 1973, 1976).

Water quality enhancement is another possible benefit of reserving flood-prone and other lands in an essentially undeveloped state. The resulting strip or corridor of land and vegetation along a creek or river or around a lake will serve as a buffer between the urban development on one side and the surface waters on the other side. Some suspended solids, nutrients, organic material, and other nonpoint-source pollutants transported by surface runoff and enroute to the creek, river, or lake will be retained in the buffer zone, thus protecting the surface waters.

Reservation of flood-prone lands for recreation and other flood-compatible uses may result in a reduction in the potential property tax base for the directly affected areas. However, depending on the manner in which the open space land is used and managed, land value and therefore the tax base for lands in proximity to the reserved land may be enhanced. This land value enhancement effect was studied by the Southeastern Wisconsin Regional Planning Commission (1977). The investigation concluded that most public open space land had a favorable effect on the value of a contiguous residential property or residential property offering a view of the open space. The favorable impacts were found to increase with the size of the open space areas and number and quality of the visual and other amenities available.

Planning and Design Considerations. Flood-prone lands must be delineated before they can be reserved. Principles and procedures for a delineation of flood-prone areas along creeks and streams are presented in Chapter 4. Once delineated, four major mechanisms are available for reserving flood-prone lands for recreation and other compatible open space uses (Walesh, 1973, 1976).

The first, and probably most effective method is outright purchase using public funds (Burby and Kaiser, 1987). Advantages of this technique include definitiveness and legal incontestability, and the principal disadvantage is cost.

The second means for reserving flood-prone lands is to encourage or require developers of large parcels to dedicate flood-prone areas to a government unit or agency. Community zoning and land division ordinances often require developers to provide minimum recreation or open space area. Flood-prone areas may meet these dedication requirements. A variation on dedication of land to a public unit or agency applicable to large planned developments is giving title to and responsibility for flood-prone land to a property owners' association. The principal advantage of the dedication approach is reduced public cost, at least for land acquisition, recognizing that the cost of recreation and other open space development and maintenance of those facilities will be incurred in perpetuity by the government unit or agency. A disadvantage of dedication is that considerable negotiation may be required to accomplish the objective.

The third mechanism used for reservation of flood-prone lands for recreation and other compatible open space uses is acquisition of an easement rather than the property. A governmental unit or agency could acquire a restrictive easement which would prohibit the landowner from developing the land in a manner incompatible with its flood-prone status. Limited public access to and use of the land might be included in the easement. The easement might also include a reduction in the property tax rate paid by the private landowner consistent with the reduced use options. An advantage to the easement approach is that both public and private entities may be able to achieve their land use objectives and, relative to land acquisition, there is a smaller expenditure of public funds. As with the dedication mechanism, a negative aspect of easement acquisition is that considerable negotiation may be required.

The fourth and final principal means of reserving flood-prone lands for recreation and other compatible open space uses is application of regulations. As described in Chapter 1, such regulations may include one or more of the following regulatory measures: a zoning ordinance, a land division ordinance, a sanitary or health ordinance, and a building ordinance.

Control of Land Use Outside Flood-prone Areas

Description and Function. Chapter 2 demonstrates how land use can have a significant negative impact on the quantity, as well as the quality, of surface water runoff. Accordingly, it may be desirable to control the manner in which land is developed and used outside flood-prone areas, as well as within flood-prone areas, as discussed in the preceding section.

Planning and Design Considerations. Recommendations to implement land use controls should be supported by investigations which quantify the impact of various development scenarios on surface water runoff rates and volumes and on nonpoint-source pollution loads. Hydrologic techniques useful for performing such analysis are described in Chapter 3. Methods for estimating nonpoint-source pollution yields as a function of land use are presented in Chapter 7. Impact of urbanization analysis are usually implemented on a production basis with computer models like those described in Chapter 10.

One mechanism for implementing land use controls is preparation and adoption of a land use plan. Supporting mechanisms, as discussed in Chapter 1, include zoning, land division, sanitary and health, and building ordinances. These ordinances can include control of the amount of impervious surface, requirements that discharge–probability relationships after development be no more severe than discharge probability relationships before development, and stipulations that erosion and sedimentation must be controlled during construction.

Flood Insurance

As explained in Chapter 1, one national purpose of the National Flood Insurance Program is to spread out federal emergency funding by subsidizing some insurance premiums. The second federal purpose is to encourage the implementation of floodplain regulations throughout the United States.

From the individual property owner's or renter's perspective, the principal value of flood insurance is protection against large monetary losses resulting from flooding. Property owners and tenants should be informed about and urged to consider purchasing flood insurance. Even if a community is committed to the construction of flood mitigation works, such as a channel modification or a detention/retention facility, property owners and tenants may want to acquire flood insurance on an interim basis until such facilities are completed and operable.

Floodproofing of Existing and Proposed Structures and Facilities

Description and Purpose. Floodproofing consists of structural changes and certain procedures undertaken to mitigate flood damage to a single residential, commercial, or industrial building or other facility or to a small group of buildings or facilities. Even with successful application of floodproofing, overland flooding and the related inconvenience will continue to occur in flood-prone areas.

From a technical perspective, most floodproofing techniques are applicable to existing and new buildings and facilities. For example, if a new house is to be placed in a flood-prone area, it could be floodproofed by building it on columns, walls, or fill. An existing house in a flood-prone area could be floodproofed by raising it and constructing a new column, wall, or fill support system.

Although necessary, technical feasibility is not a sufficient reason for floodproofing or any other management measure. Although from a technical perspective floodproofing can be applied to new buildings and facilities assuming that extra construction and operation and maintenance costs are acceptable, new construction should be prohibited in flood-prone areas because of environmental, aesthetic, and other considerations, unless there is some compelling reason to do otherwise.

Floodproofing can be categorized as dry and wet and as permanent and temporary. The two-way categorization was used to develop the summary of floodproofing techniques presented in Table 11.2. The two categories are not mutually exclusive. For example, and as illustrated by the format used in Table 11.2, both dry and wet floodproofing techniques can be partitioned into permanent measures and temporary actions.

Dry floodproofing means that structural facilities and complementary measures are implemented to prevent floodwaters from entering an otherwise flood-prone building or facility. Examples of permanent dry floodproofing techniques are elevating buildings on columns or other supports (Figures 11.20 and 11.21), reinforcing walls (Figure 11.22), waterproofing construction joints (Figure 11.23) and sealing cracks and openings Figure 11.24), anchoring a structure to its foundation (Figure 11.25), preventing backup of sewage and water via floor drains (Figure 11.26) and control of hydrostatic pressure beneath floors (Figure 11.27). Temporary dry floodproofing techniques include flood shields for doors, windows, and other openings (Figure 11.28), applying impermeable sheet material to exterior walls (Figure 11.29), encircling a building or facility with sandbags (Figure 11.30) and inserting a standpipe in the basement floor drain (Figure 11.26).

Under the wet floodproofing approach, flood and other water is allowed

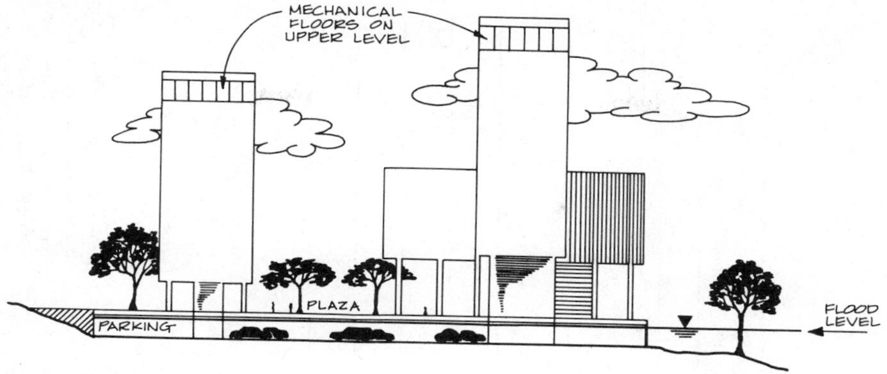

FIGURE 11.20 Constructing a new building on columns or other supports to elevate flood-prone areas above flood stage *(Source:* Adapted from Owen, 1981)

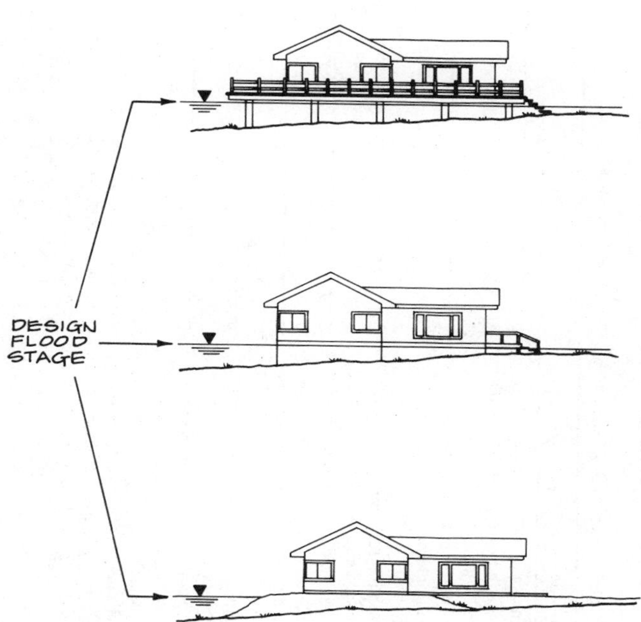

FIGURE 11.21 Means of elevating an existing or new building above flood stage *(Source:* Adapted from Illinois Department of Transportation, 1983)

TABLE 11.2 Floodproofing Techniques

Dry Floodproofing (Keep Water Out of Building or Facility)		Wet Floodproofing (Let Water Into Building or Facility but Minimize Damage)	
Permanent Measures	Temporary Actions	Permanent Measures	Temporary Actions
Document floodproofing procedures in writing indicating who is to do what and when.	→	→	→
Organize and train a floodproofing team to implement measures as needed.	→	→	→
Advise building residents or users of the floodproofing provisions.	→	→	→
Advise local officials of the floodproofing plan and give them a copy.	→	→	→
Inspect and maintain floodproofing measures.	→	→	→
Develop and maintain an emergency warning system.	→	→	→
Evacuate people and selected equipment, records, and supplies (as an added precaution).	→	→	→
Construct new building above or raise existing building above flood stage (Figures 11.20 and 11.21).	Install shields or bulkheads on doors, windows, and other openings (Figures 11.24 and 11.28).	Relocate electric and other vulnerable equipment above flood levels (Figure 11.31).	Move floodprone materials to areas above flood stage.
Surround building with an earthern berm, concrete wall, or other barrier.	Place polyethylene material or other seal on exterior walls (Figures 11.24 and 11.29).	Provide openings to release air trapped beneath floors or ceiling by rising floodwater.	Turn off electric, gas, and water service.
Construct wall or berm around all or a portion (e.g., stairwell to lower level) of structure.	Reinforce walls at windows and other critical areas (Figure 11.22).	Anchor buoyant objects such as tanks (Figure 11.24).	Disconnect and move water vulnerable electrical and mechanical equipment (e.g., pumps, blower motor on furnace) (Figure 11.24).
	Utilize standby pump system as needed.	Construct low walls around equipment (Figure 11.32).	

Reinforce wall and floors (Figure 11.22). Install waterproof seals at structural joints (e.g., between basement walls and floors) (Figure 11.23). Replace basement and other windows with glass block or otherwise completely close (Figure 11.23). Seal cracks and openings in exterior surfaces (Figure 11.24). Install drains and sump pumps to evacuate water (Figures 11.24). Attach superstructure to its foundation (Figure 11.25). Install automatic backwater valve and manually operated gate or similar valve on sanitary lateral or install overhead sewer and pump (Figures 11.24 and 11.26). Install drains and pump to relieve pressure below floor (Figure 11.27).	Surround building with sandbags, earthern dike, earth fill crib, or other barrier (Figure 11.30). Install standpipe in basement floor drain (Figure 11.26).	Install pressure relief valves and blow out plugs in basement floor (Figures 11.33 and 11.34).	Cover water vulnerable equipment and materials with impermeable sheet material (Figure 11.24). Flood basement with clean water. Fill buried tanks used to store fuel or other liquids (Figure 11.24).

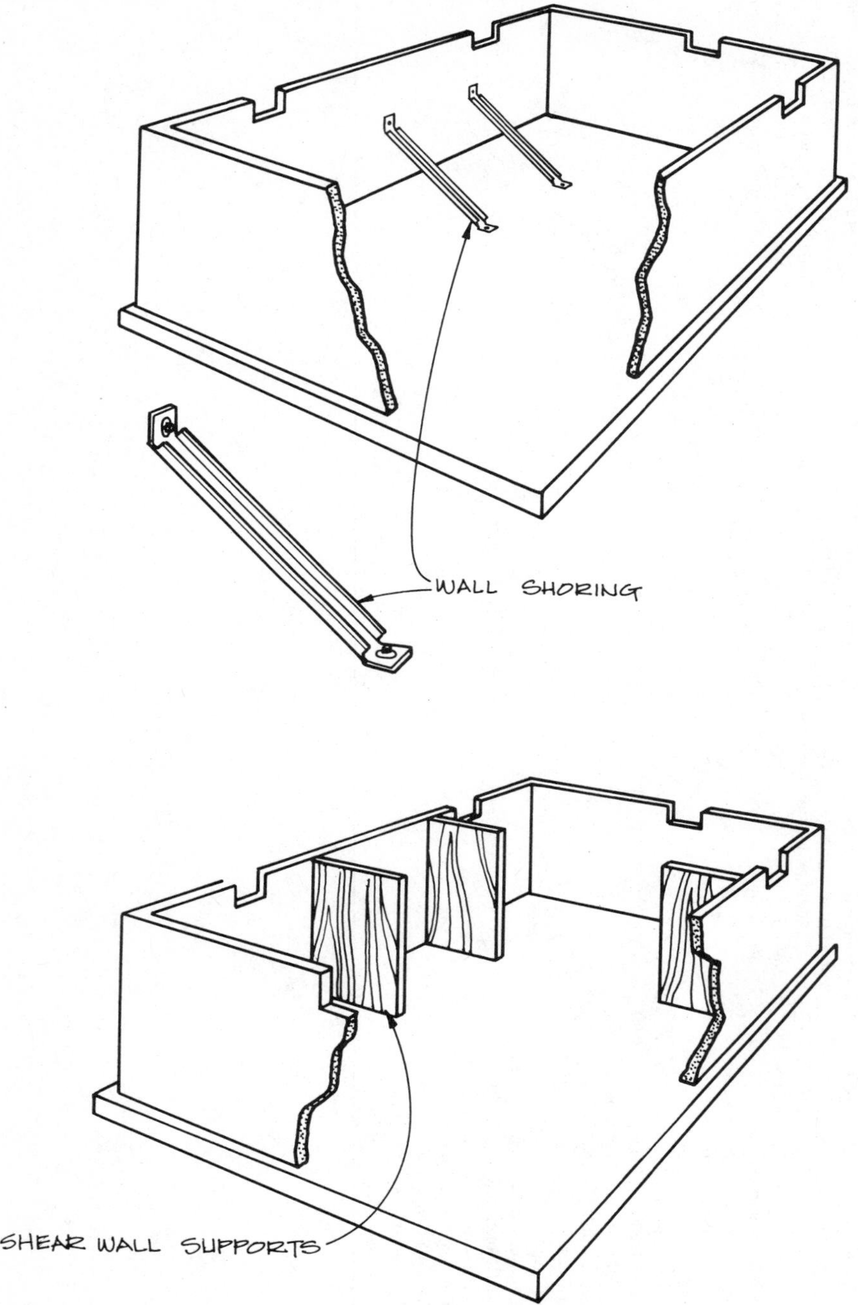

FIGURE 11.22 Methods of reinforcing basement walls to resist flood–induced lateral hydrostatic pressure *(Source:* Adapted from Federal Emergency Management Agency, 1986a)

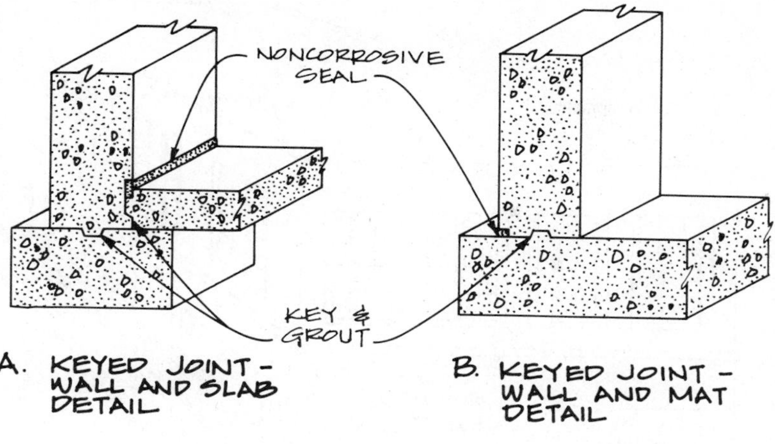

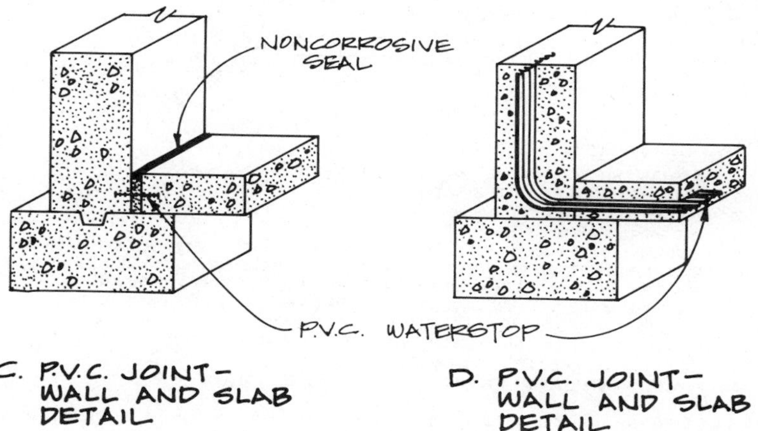

FIGURE 11.23 Means of waterproofing the wall–floor connection *(Source:* Adapted from Federal Emergency Management Agency, 1986b)

to enter a building or a facility, but structural changes are made and other measures are instituted to reduce damage significantly to the otherwise protected building or facility and its contents. Examples of permanent wet floodproofing methods are relocating electrical and other vulnerable equipment above flood levels (Figure 11.31), anchoring buoyant objects such as tanks (Figure 11.24), constructing low walls around water vulnerable equipment (Figure 11.32) and installing pressure relief valves (Figure 11.33) or blow out plugs (Figure 11.34) in basement floors. Temporary wet floodproofing tech-

EXPLANATION

1. PERMANENT CLOSURE OF OPENING WITH MASONRY
2. COATING TO REDUCE SEEPAGE
3. VALVE ON SEWER LINE
4. UNDERPINNING
5. INSTRUMENT PANEL RAISED ABOVE EXPECTED FLOOD LEVEL
6. MACHINERY PROTECTED WITH POLYETHYLENE COVERING
7. STRIPS OF POLYETHYLENE BETWEEN LAYERS OF CARTONS
8. UNDERGROUND STORAGE TANK PROPERLY ANCHORED
9. CRACKS SEALED WITH HYDRAULIC CEMENT
10. RESCHEDULING HAS EMPTIED THE LOADING DOCK
11. STEEL BULKHEADS FOR DOORWAYS
12. SUMP PUMP AND DRAIN TO EJECT SEEPAGE

FIGURE 11.24 Combination of permanent and temporary floodproofing measures *(Source:* Adapted from Sheaffer et al., 1967)

niques include turning off utilities, moving floodprone materials above expected flood stage (Figure 11.24), covering vulnerable equipment and materials with impermeable sheet material (Figure 11.24), flooding basements with clean water, and filling buried tanks used to store fuel or other liquids (Figure 11.24).

Permanent floodproofing techniques are always in place and, assuming proper maintenance, they usually function regardless of human intervention. In contrast, temporary floodproofing requires human intervention and at least a small degree of pre-flood preparation.

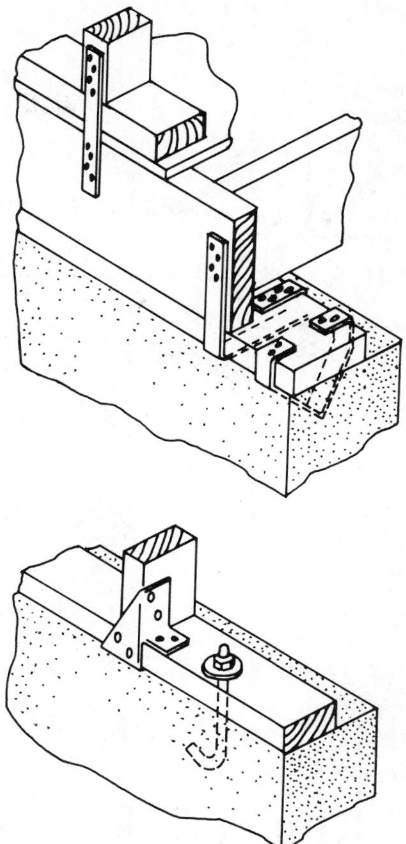

FIGURE 11.25 Anchoring a structure to its foundation to prevent lateral movement or flotation *(Source:* Adapted from Owen, 1981)

Need for Professional Services. Many floodproofing techniques appear deceptively simple in function and in installation to owners or tenants of flood-prone buildings and facilities. Accordingly, floodproofing may be inadvertently applied in a manner that actually increases the threat to the structural integrity of a building or facility and further endangers its inhabitants or users. For example, sealing of basement windows with brick or block and installation of a standpipe in the basement floor drain may appear sufficient to a homeowner who does not understand the potential catastrophic consequences of upward hydrostatic forces on a basement floor and horizontal hydrostatic forces on basement walls. Furthermore, closing off windows and installing a standpipe are within the capability of many owners and tenants. If they proceed, which they can easily do without the knowledge of public officials or

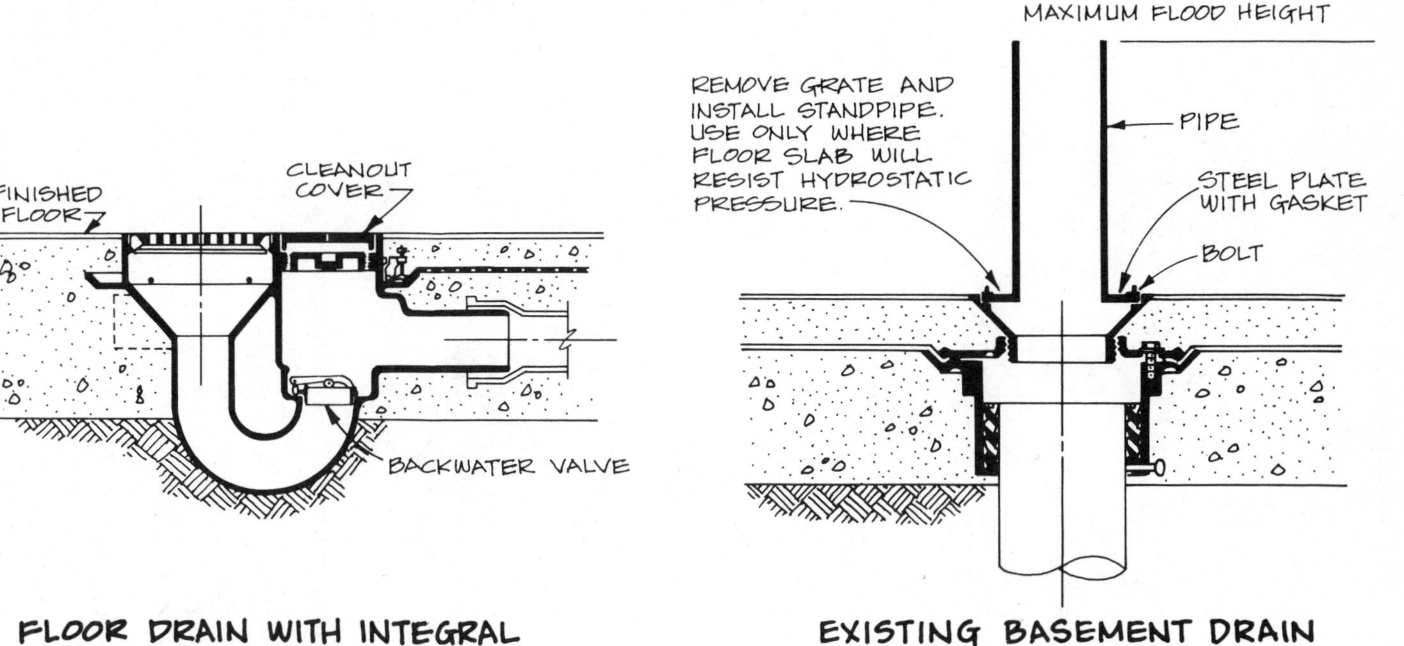

FIGURE 11.26 Methods of preventing back up of sewage and floodwater into building basements via floor drains *(Source:* Adapted from Federal Emergency Management Agency, 1986b)

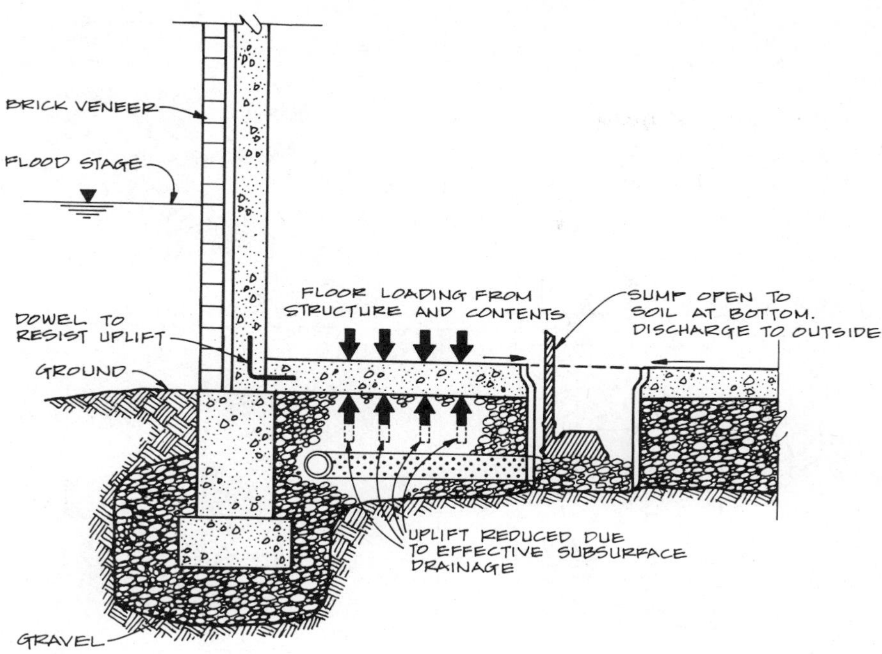

FIGURE 11.27 Drain and pump to relieve flood-induced pressure beneath floor *(Source:* Adapted from Federal Emergency Management Agency, 1986b)

knowledgeable professionals, the building may incur damage during the next flood far in excess of that which would have occurred had the basement simply filled with floodwater and wastewater.

As another example, a building may be pushed or lifted off its foundation by dynamic or buoyant forces during a flood as a result of failure to connect the superstructure structurally to its foundation. The above-grade wall of any building may collapse as a result of temporarily floodproofing the building by covering the external surfaces with impermeable sheet material but failing to apply adequate internal wall bracing at windows and other critical points to offset externally applied hydrostatic forces. Finally, the interior of a wet floodproofed building may be damaged as a result of an unsecured fuel oil storage tank tipping and spilling its contents.

With few exceptions, building and facility floodproofing should be done under the guidance of a qualified professional engineer. Because the cost of professional services may be high relative to the capital and other costs of the floodproofing measures, a community or other governmental unit may consider retaining a qualified engineer to provide the services, thus capitalizing on the economy of scale in the engineer's efforts. Such an arrangement also provides an opportunity for the governmental unit to pay all or part of the engineering fee with the understanding that the owners of floodproofed buildings

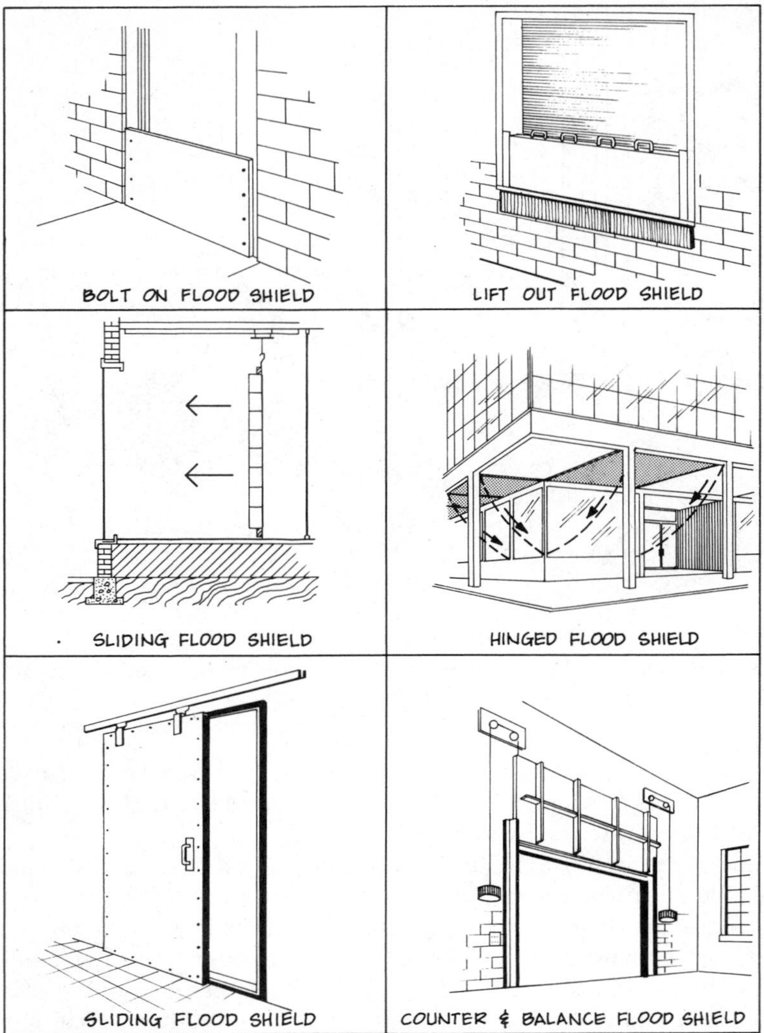

FIGURE 11.28 Typical flood shields *(Source:* Adapted from Federal Emergency Management Agency, 1986b)

and facilities will be responsible for the cost of installing and maintaining floodproofing measures.

Planning and Design Factors. A fundamental decision to make in planning and designing a floodproofing system is to choose between dry and wet floodproofing. The issue of permanent versus temporary techniques and physical versus procedural techniques then needs to be addressed.

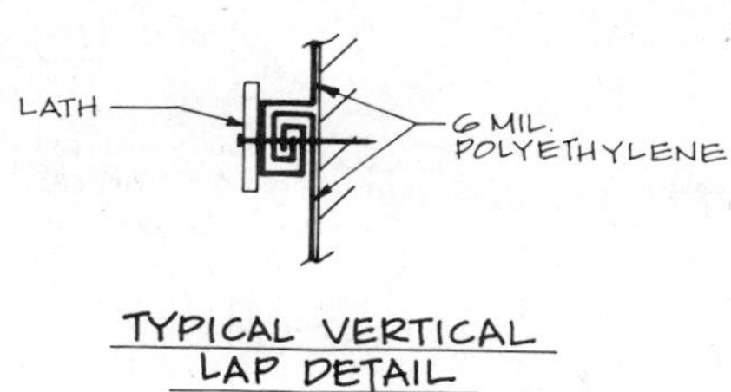

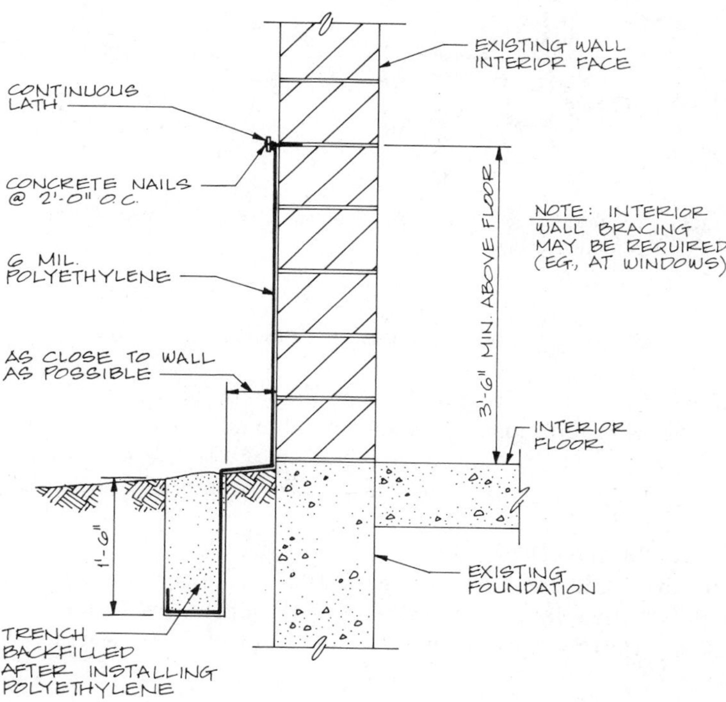

FIGURE 11.29 Temporary impermeable membrane applied to an exterior wall *(Source:* Adapted from Donohue & Associates, 1979a)

The decision to use temporary as opposed to permanent floodproofing measures must include an analysis of the likely available warning time. In urban areas, creeks and streams tend to be flashy in that runoff occurs quickly and water rises rapidly in swales, creeks, and streams. Sufficient time may not be available to implement temporary floodproofing measures.

Regardless of the types of floodproofing measures to be implemented, they should be planned and designed to withstand hydrostatic forces, dynamic forces, and potential erosion effects typically associated with major flood events, usually up to and including the 100-year-recurrence-interval flood. A

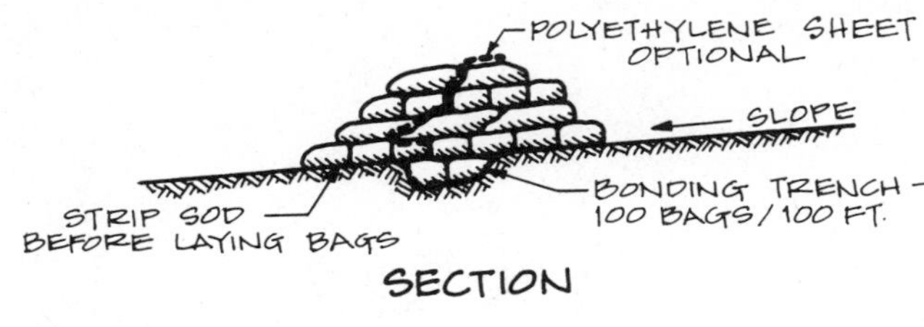

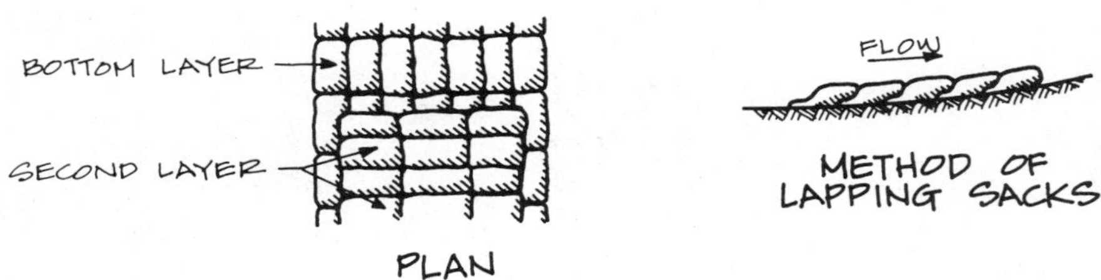

FIGURE 11.30 Technique for constructing sandbag dike *(Source:* Adapted from Federal Emergency Management Agency, 1986b)

freeboard of at least 1 ft above the design flood should be incorporated into the design.

Floodproofing facilities and equipment and materials intended for use during flood events require inspection and maintenance. The design of a floodproofing system should include an enumeration of inspection and maintenance tasks and the frequency with which those tasks should be conducted.

Planning and Design Aides. During the last decade, numerous reports, papers, and other materials have become available to assist engineers and other professionals in the planning and design of floodproofing systems. Information on floodproofing existing single-family residential structures is provided

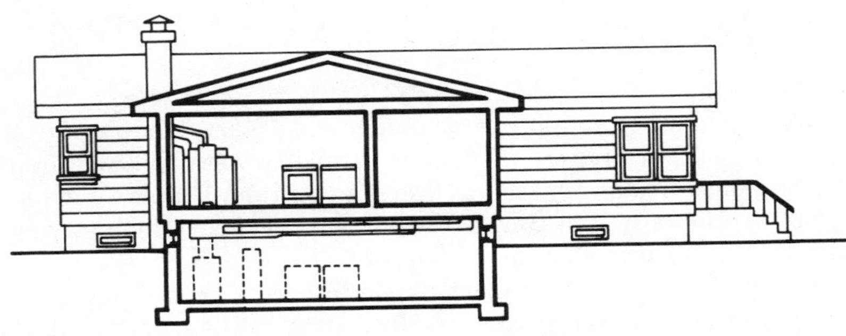

FIGURE 11.31 Permanently relocating utilities and equipment to higher floors *(Source:* Adapted from Illinois Department of Transportation, 1983)

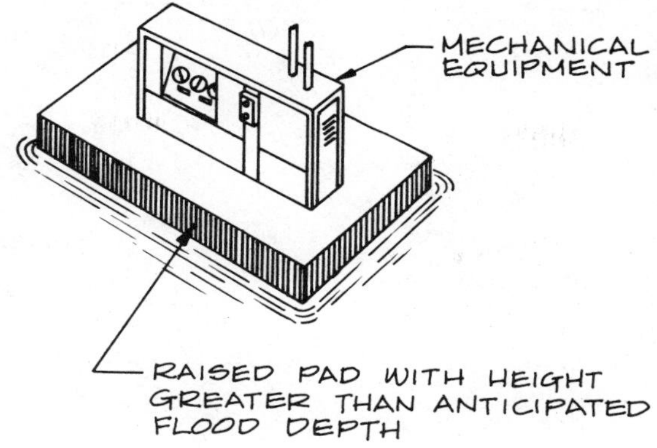

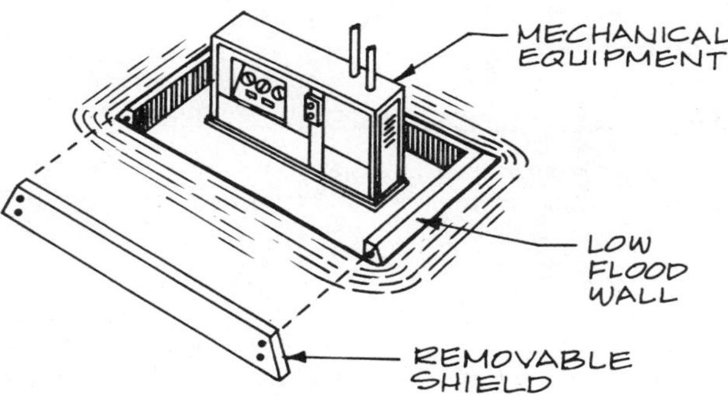

FIGURE 11.32 Means of protecting equipment and facilities subject to shallow flooding (*Source:* Owen, 1981)

by the Federal Emergency Management Agency, (FEMA, 1986a), Flanagan (1984), and the Illinois Department of Transportation (1982). Flanagan's report includes photographs of many and varied floodproofing techniques.

Detailed discussions of floodproofing existing residential structures by elevating them are presented by FEMA (1986c), Illinois Department of Transportation (1983), and U.S. Army Corps of Engineers (1977). FEMA (1985) provides detailed procedures for the installation of manufactured homes in flood-prone areas. For information on floodproofing of primarily new residential structures by designing them to be elevated above flood stage, refer to the National Association of Home Builders (1977) and U.S. Department of Housing and Urban Development (1976).

Floodproofing of commercial, public, and other nonresidential buildings and facilities is covered in Bresenhan (1979), Donohue & Associates (1979a),

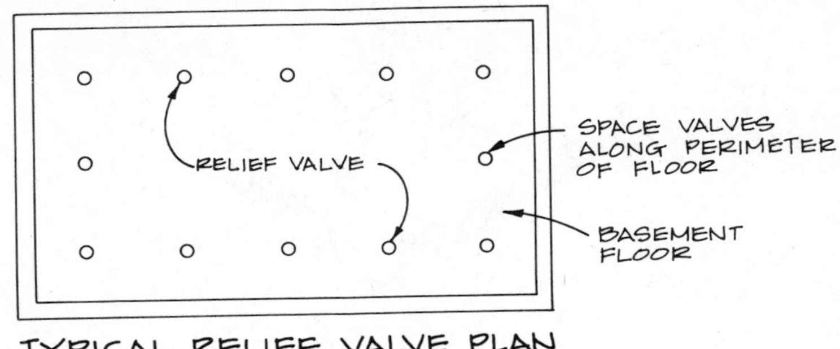

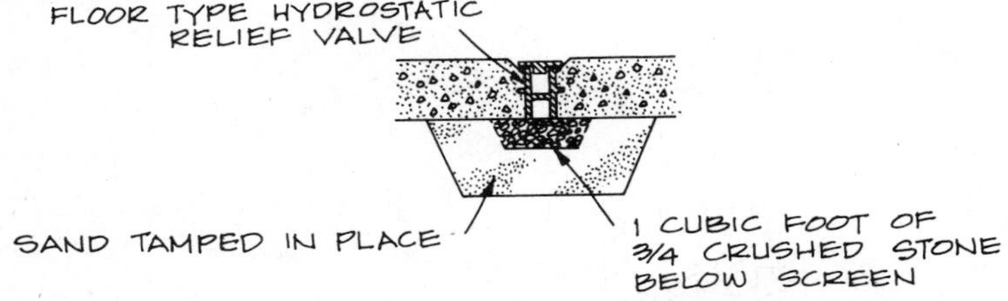

FIGURE 11.33 Pressure relief values for installation in basement floor to relieve flood-induced pressure *(Source:* Adapted from Federal Emergency Management Agency 1986b)

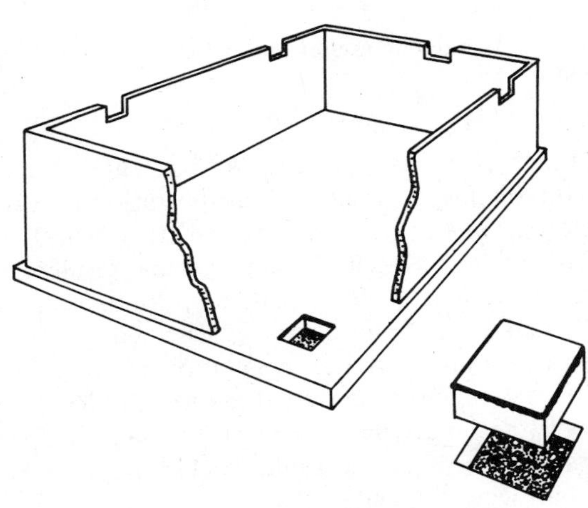

FIGURE 11.34 Floor blow-out plug to relieve flood-induced pressure beneath floor *(Source:* Adapted from Federal Emergency Management Agency, 1986a)

FEMA (1986b), Flanagan (1984), Flynn et al. (1984), Owen (1981), and Walesh (1982a, 1983). Additional floodproofing ideas and information are set forth in FEMA (1986a), Pace (1978), Pace and Campbell (1978), Sheaffer et al. (1967), U.S. Army Corps of Engineers (1972, 1987), and U.S. Nuclear Regulatory Commission (1976).

Floodproofing cost data and benefit-cost methodology are presented by Bresenhan (1979), Carson (1975), FEMA (1986a,b), Flack (1978), Johnson (1978), McKeever (1977), U.S. Army Corps of Engineers (1977, 1981), and U.S. Department of Housing and Urban Development (1976).

Relocation or Demolition of Structures and Facilities

Description and Purpose. Sometimes floodproofing of existing structures and facilities may not be a technically practical or economically feasible approach because of factors such as poor structural condition or the design flood stage being high relative to the building or facility. Structures that cannot be floodproofed, but are structurally sound, may be candidates for relocation to higher ground in the immediate area. Other structures may have to be demolished.

Examples of successful structure relocation and demolition projects are available to use as guides. Approximately 130 families and a few businesses were evacuated from an island on the Mississippi River in Prairie du Chien, Wisconsin, as part of a program carried out in the late 1970s and early 1980s. In most cases, the relocation consisted of the city purchasing the buildings and lots using Corps of Engineers and U.S. Department of Housing and Urban Development funds. The former owners generally used the money received from the sale of their property to relocate elsewhere in Prairie du Chien, but some left the area. A small percentage of the buildings were moved to nearby, but higher locations in the city. In those instances, which were the most common, where the city purchased a building and its site from its owner, the city recouped some of its costs by allowing contractors to demolish the buildings for a fee, with the contractor having the benefit of the salvaged material. The evacuated island is being converted to tourist and recreation use, with focal points being a marina and an historic mansion (*La Crosse Tribune,* 1979; *Milwaukee Journal,* 1978; Walesh, 1982b).

During the same period as the Prairie du Chien relocation and demolition project, the business district of Soldiers Grove, Wisconsin, was relocated from the Kickapoo River floodplain to higher ground in the immediate vicinity. In this federally financed project, existing buildings were abandoned and demolished and the businesses that were originally in the flood-prone area were relocated into a cluster of new buildings on higher ground adjacent to the former business area. The new business community was designed with extensive solar heating and lighting features (League of Wisconsin Municipalities, 1983). Another report (Minnesota Department of Natural Resources, no date) describes acquisition and relocation projects in four Minnesota communities.

Planning and Design Considerations. The nature of historic flooding is an important determinant of the likelihood of success of a relocation and demolition project. For example, in both the Prairie du Chien and Soldiers Grove situations, historic flooding was frequent and deep—waist high or more and generally above first-floor elevations throughout the flooded areas. Therefore, the intensity of local interest was high and sustained. Futhermore, both projects involved heavy influx of federal funding.

Structure relocation and demolition projects can be very complex and very costly. The cost of structural removal and relocation, regardless of how it is financed, includes structure and site acquisition costs, structure demolition or moving costs, site restoration costs, the cost of acquiring and developing new sites for relocated structures, and occupant relocation costs.

In addition to flood damage mitigation, a potentially positive aspect of structure relocation or demolition is the increased opportunity to redevelop the recreation, plant and wildlife, and aesthetic features of the flood-prone areas. The Prairie du Chien and Soldiers Grove projects are realizing these benefits.

A potential negative feature of structure relocation or demolition is the tax base loss to the community by removing taxable property from within its corporate limits. Some or all of the lost revenue might be offset by the reduced costs of providing municipal services, such as police and fire protection, schools, water supply and sewerage, and the reduced costs of flood fighting and flood-related services and postflood cleanup and repair. Furthermore, many of the former residents of a flood-prone area may construct new homes or businesses within the community or relocate their structures to previously undeveloped land within the community, thereby adding to the tax base.

Even if the negative financial aspects of structure relocation or demolition are resolved or accepted, directly affected residents may suffer the intangible but nevertheless real cost of losing their neighborhood. Even if a group of contiguous buildings are relocated in a cluster and reoccupied by their former owners or tenants, the overall setting and environs will be dramatically changed. That is, whereas it may be technically and financially feasible to relocate structures, it is not feasible to move a neighborhood intact to a new location.

Slope and Swale Treatment

Description and Function. The purpose of slope and swale treatment is to reduce the likelihood of erosion by protecting soil from being dislodged by rainfall, moving water, and wind. Mild slopes can be seeded and protected by straw or other mulch. Lateral, essentially horizontal diversion ditches may be constructed along the top of steep slopes to intercept runoff and divert it around the slope. Such ditches may also carry runoff to slope drains, which in turn convey flow to the base of the slope. Other means available for protecting steeply sloped areas include riprap and gabions.

Roadside and other swales may be protected from erosion by such means as staked sod, jute netting, excelsior matting, and fiberglass netting over seed and mulch; or armoring with rock, asphalt, or concrete. Backup erosion protection for large areas can be provided by installing temporary silt fences composed of a filter fabric supported on a temporary wire fence or placing straw bale barriers along the lower edge of the site. These systems will trap suspended solids in the event that the upstream erosion protection proves to be inadequate.

In addition to controlling erosion, establishing turf in a roadside or other swale may control NPS pollutants. A comparative study in Florida suggests that grassy roadside swales may be very effective, compared to curb and gutter systems, in controlling NPS pollutants (Brevard County Water Resources Department, 1982). Adjacent 12- and 15-acre residential areas, one with a grassy swale system and the other with curb and gutter, were monitored for runoff and various NPS pollutants. Yields of biochemical oxygen demand, total nitrogen, and total suspended solids were about two orders of magnitude greater for the curb and gutter system than for the swale system. Runoff volumes and peaks were also much greater for the curb and gutter system. Finally, the present worth cost of the swale system, including maintenance, was reported to be significantly less than the present worth cost of the curb and gutter system.

Planning and Design Considerations. As indicated in Chapter 7, sediment yields from areas under construction are orders of magnitude greater than areas protected with vegetation or impermeable surfaces. Accordingly, slope and swale treatments should be applied as soon as possible after development begins. Furthermore, some treatments, such as silt fences and straw bale barriers along the perimeter of development sties, should be placed prior to the initiation of construction.

When vegetation is used for slope or swale treatment, the type of vegetation must be consistent with soil and moisture conditions and vegetation must be maintained. Some slope and swale treatments, such as silt fences and straw bale barriers, require frequent maintenance if they are to function effectively. A concise overview of erosion control measures, concentrating on slope and swale treatments, has been presented by Lord (1986). Various vegetative practices are also described by Novotny and Chesters (1981) and Schueler (1987).

Community Utility and Service Policies

Local governmental units and agencies have the option of adopting policies affecting the extension of public utilities and services so as to discourage or prevent development of flood-prone and environmentally sensitive areas. These policies may explicitly address utilities and services such as streets, water supply, sanitary sewerage, gas, and electric. Community utility policies can be formally prepared in written form and adopted by governing bodies, or they can evolve over a period of time as a consistent set of decisions rendered by governing bodies and the professional staff of governmental units and agencies.

A community's long-term interest is probably best served by planning and designing utilities and services to discourage development of problematic or sensitive areas. For example, sanitary sewers and water mains for areas contiguous with floodplains should be placed and sized on the assumption that the floodplains will not be developed. In contrast, placing a major sewer or water main close and parallel to a floodplain or terminating it at the floodplain limit may encourage proposals for residential, commercial, or industrial development in the flood-prone area.

The credibility and defensibility of community utility and service policies that discourage development of flood-prone or environmentally sensitive areas depends on supporting data. Examples of such data are delineations of flood-prone areas, characterization of soils relative to suitability for urban development, water table elevations, enumeration of plant and wildlife resources, and identification of historic and cultural resources. The credibility and defensibility of community utility and service policies will be further enhanced to the extent that they are a means of furthering the objectives of adopted land use plans and land use zoning.

Inspection and Maintenance Program

Even the simplest surface water system requires an ongoing inspection and maintenance program if the facilities are to function in accordance with plans and designs. Surface water systems are very robust, particularly with respect to structural integrity, in that the emergency system described in Chapter 1 is almost always present, whether by accident or by design, to convey excess flows or to store excess volume.

However, even if a surface water facility is not structurally at risk because of inadequate inspection and maintenance, its function may be seriously impaired. For example, a detention/retention facility may not pass low flows on the rising limb of the hydrograph because the outlet is blocked by debris. This will reduce the amount of storage available and the downstream flow attenuation achieved during the higher-flow portions of the hydrograph. The same facility will not drain adequately by gravity after the event. Similarly, a weed- or debris-obstructed channel will function, but at a greatly reduced capacity, thus increasing flood risks for contiguous properties.

Many and varied inspection and maintenance activities are possible, as indicated by Table 11.3, which enumerates inspection and maintenance activities and the intended purpose of each activity. Additional, more detailed inspection and maintenance procedures are listed in Chapter 5. The responsible government unit or agency should develop an inspection and maintenance program tailored to the particular surface water system. The program should be documented in written form.

During the 1970s, when concern with NPS pollution began to expand, mechanical and vacuum street sweepers were seen as being a potentially effective management measure in urban areas and one that could be easily engaged in by municipalities. By removing solids from street and parking lot surfaces prior to wash-off by rainfall or snowmelt runoff, the solids and the associated pollutants would be prevented from entering surface water bodies. Furthermore, existing street-sweeping practices could be improved by increasing sweeping frequency, utilizing wire brushes, decreasing the forward speed of the street sweepers, increasing brush speed, and using more vacuum-type sweepers (Amy et al., 1974).

Unfortunately, subsequent field studies (e.g., Collins and Ridgway, 1980; Terstriep et al., 1986) revealed that street sweeping was generally not an effective surface water management measure. Of all the solid material typically picked up by a street sweeper, a disproportionate amount is in the larger-particle category. However, a disproportionate amount of the potential pollutants are associated with the smaller particles, which tend to escape the sweeper and remain on the street or parking lot to be washed off with the next runoff event.

In summary, street sweeping has not proved to be as effective an NPS management measure as originally expected. However, the importance and effectiveness of the traditional function of street sweeping, removal of nuisance and aesthetically objectionable debris and material, is in no way diminished by the unsatisfactory performance of street sweepers in controlling NPS pollution.

Although the principal purpose of an inspection and maintenance program is to assure that facilities function as intended to control the quantity and quality of surface water, aesthetic and other benefits are typically achieved. Experience suggests that good inspection and maintenance of the more visible surface water facilities by the responsible governmental unit or agency tends to encourage supplemental maintenance by contiguous landowners. For example, assume that a community regularly inspects flood control channels and detention/retention facilities and performs basic maintenance, such as mowing grass, revegetating eroded areas, and removing debris. Then adjacent property owners or tenants are very likely to perform complementary maintenance, such as picking up debris and cutting weeds and brush near and perhaps in the flood control facility.

Emergency Action Program

An emergency action program (EAP) is a set of preplanned and integrated steps taken by a community or group of communities before flooding or other surface water problems occur and when such problems are expected, impending, occurring, and after they occur. An EAP plan for a community is similar to a set of procedural floodproofing measures sometimes prepared, as discussed earlier, for an individual flood-prone structure or facility. The purpose of the EAP is to mitigate damage and disruption and reduce threats to the life and health of area residents.

The preparation of an EAP for an urban area must recognize the flashy nature of urban streams. The elapsed time from the onset of a heavy rainfall

TABLE 11.3 Maintenance Program for a Surface Water System

		Purpose				
Activity	Frequency	Aesthetic Improvement	Encourage Voluntary Care by Citizens	Maintain Turf to Resist Erosion	Maintain Coveyance and Storage Capacity	Maintain Capability to Remove Solids, Nutrients, and Other Pollutants
Mowing and weed, brush, and tree control by cutting and/or herbicide application	As needed Periodically	×	×	×	×	×
Trash and debris cleanup	After major runoff events Periodically	×	×		×	
Revegetation by seeding, fertilization, and mulching	As needed	×	×	×		×
Inspecting and cleaning trash and safety racks and/or grates, sluice gates and other control devices, motors, and pumps	During and after major runoff events Periodically	×	×			×

Cleaning catch basins and/or inlets	During and after major runoff events Periodically			×	×
Inspecting and cleaning or dredging storm sewers, combined sewers, channels, and detention/retention facilities	During and after major runoff events	×	×	×	
Remove settled solids and settled and floating debris from sedimentation basins	Periodically	×	×		×

Source: Adapted from Donohue & Associates (1984); Engemoen et al. (1983); Hurlbert (1980); Hunter and Tucker (1982); Schoenbeck (1979).

until peak discharges and stages occur is usually measured in hours, not days. Accordingly, very little time is typically available to prepare for the onset of a flood.

An EAP might consist of the following four phases, the first of which should be completed before a particular flood event begins to occur:

1. *Pre-flood Preparation.* This phase may include mapping flood-prone areas; stockpiling of materials such as sandbags, sand, and impermeable membrane; identifying the location of and making arrangements for the use of necessary equipment, such as trucks, front-end loaders, portable pumps; inspecting and maintaining flood control facilities; and locating and making arrangements for the use of emergency shelters for evacuees.
2. *Monitoring and Warning.* This phase of the EAP may include monitoring of upstream gauges, like that shown in Figure 11.35, and stage sensors and alarms; patrolling low-lying areas to note the water stage relative to flood stage conditions; listening to National Weather Service flash flood watch and warning bulletins; broadcasting emergency messages over radio and public address systems to warn residents of low-lying areas; and activating a siren warning system which uses a special signal or pattern to indicate that flooding is expected or imminent.
3. *Flood Fighting.* This phase may include evacuating residents of threatened areas, closing roads according to a preplanned schedule, providing for medical care, reinforcing police protection, using portable pumps to relieve surcharging in sanitary sewers, sandbagging, building temporary earth dikes, activating floodproofing measures, and continuously inspecting flood control facilities.
4. *Postflood Cleanup.* Activities representative of this phase are removing sandbags and temporary dikes; helping evacuees return to their residences and businesses; repairing damage to public utilities and facilities and restoring service; applying for disaster aid from regional, state, and federal agencies; and critiquing the EAP.

Ideas and information for use in preparation of EAP are available from the Commonwealth of Pennsylvania (1978), FEMA (1979, 1986a), Hilton (1980), Owen (1977), Susquehanna River Basin Commission (1976), and the U.S. Army Corps of Engineers (1987).

Surface Water Management Manual

A SWM manual for a particular community or group of governmental units and agencies is an organized collection of approaches, analytic tools and techniques, and procedures. SWM manuals are usually designed to complement and support the SWM program in a community or group of communities, and, in particular, to help with the implementation of adopted master plans.

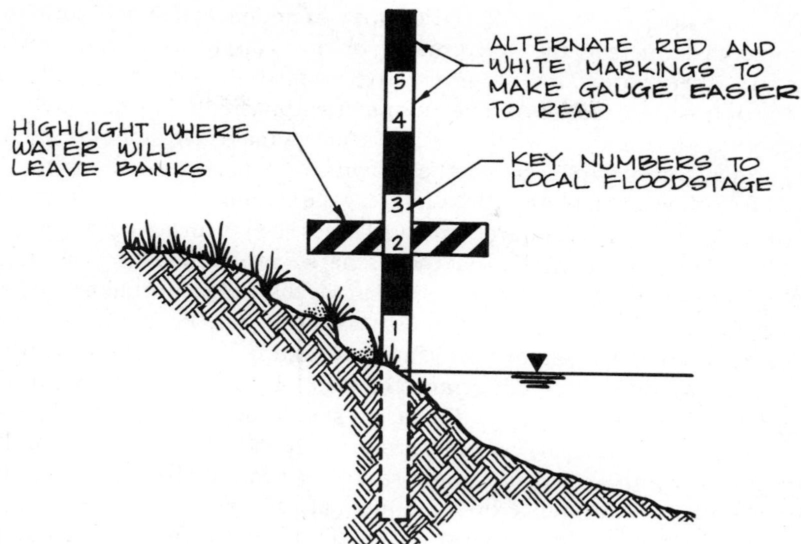

FIGURE 11.35 Simple staff gauge to warn of dangerously high flood stages *(Source:* Adapted from Illinois Department of Transportation, 1982)

Manuals are usually developed for use by engineers and other professionals in private and public practice who have responsibility for planning, designing, constructing, and operating SWM facilities and systems or for reviewing plans, designs, and recommendations prepared by others. The existence of a manual should not preclude the staff of a government unit or agency from stipulating different or additional requirements in unusual situations. Similarly, the government unit or agency should consider requests from other professionals to deviate from the applicable SWM manual.

A well-prepared manual can be an aid both to professionals employed in governmental units and agencies and to professionals employed in the private sector. For example, subjecting land developers and their professionals to similar review expectations and procedures encourages administrative consistency in the government unit or agency having review authority. A well-written and illustrated manual can also minimize the administrative effort of the professional staff of a governmental unit or agency by eliminating the need for providing detailed explanations to developers. Developers and their professional staff can benefit from a well-prepared SWM manual by having a clear a priori understanding of the expectations of the governmental unit or agency.

A SWM manual can be performance oriented or procedure oriented. With the performance-oriented approach, emphasis is placed on the expected performance of the resulting SWM system. Performance-oriented manuals typically do not prescribe analysis and design methodologies to be used by professionals, nor do they specify specific types of management measures to be recommended. In fact, the performance-oriented approach encourages engineers and other professionals to be creative in satisfying the intent of the manual. In contrast, procedure-oriented manuals emphasize use of particular analysis and design techniques and specify certain management measures.

Consider, for example, the manner in which the performance-oriented and procedure-oriented manuals would address the design of a surface water system for a small residential development. Under the performance-oriented format, the manual might specify that the discharge–probability relationship after development should be no more severe than the discharge–probability relationship before development. The design professional would be free to recommend a conveyance/storage system satisfying these and other performance criteria. Under the procedure-oriented format, the manual would probably specify that after–development discharge or discharges should not exceed before development discharge or discharges. Furthermore, the manual would probably require that the control of surface water runoff be achieved by on-site detention.

To reiterate, performance-oriented manuals are generally preferred. They encourage engineers and other professionals to be creative and resourceful, and therefore are more likely to result in overall optimum surface water management systems.

An example outline for a surface water management manual is presented in Table 11.4. Examples of SWM manuals in use include simple, short, and generic manuals, such as those jointly prepared by the Urban Land Institute, the American Society of Civil Engineers, and the National Association of Home Builders (1975). An example of a comprehensive, detailed, and area-specific manual is that which is used in Denver, Colorado (Denver Urban Drainage and Flood Control District, 1969). Special-purpose manuals have

TABLE 11.4 Example Outline for a Surface Water Management Manual[a]

Executive Summary
Objectives
- Support the master plan
- Address quantity and quality
- Encourage administrative consistency
- Minimize administrative effort
- Optimize developer's efforts
- Contribute to data files
- Enhance capability of unit or agency staff
- Encourage acceptance of modern techniques

Principles
- Surface water management as a necessary public service
- Manual is performance oriented, not procedure oriented
- Importance of watersheds
- Need to identify and mitigate off-site effects
- Emergency and convenience systems
- Preference for gravity operation

TABLE 11.4 (Continued)

Regulatory framework
- Federal authority, programs and regulations
- State authority, programs and regulations
- Regional authority, programs and regulations
- Local authority, programs and regulations

Criteria for design, operation, and maintenance
- Acceptable hydrologic techniques
- Acceptable hydraulic techniques
- Acceptable nonpoint pollution techniques
- Sewer criteria
- Channel and swale criteria
- Culvert and bridge criteria
- Detention/retention criteria
- Street and parking lot criteria
- Erosion and sedimentation control criteria
- Other nonpoint control criteria
- Floodproofing criteria

Submittal requirements
- Studies and preliminary design
  - Hydrologic–hydraulic procedures
  - Hydrologic–hydraulic data
  - Runoff rates and volumes
  - Alternatives evaluated
  - Recommended structural facilities and nonstructural measures
  - Cost estimates
  - Relationship to master plan
  - Implementation sequence
- Final Design
  - Property and related lines
  - Existing utilities and facilities
  - Grades
  - Bedding
  - Cost estimates
  - Approvals required
  - Construction schedule
  - Means of finance
  - Specifications

Finance
- Federal assistance possibilities
- State assistance possibilities
- Local financing

References

Appendices
- Abbreviations
- Glossary

[a]This example lists some typical, but not necessarily all, topics for a surface water management manual.

been prepared, such as the manual for erosion and sedimentation control used in northeastern Illinois (Northeastern Illinois Soil Erosion and Sedimentation Control Steering Committee, 1981) and a similar manual used in Georgia (Georgia State Soil and Water Conservation Committee, 1979). Comprehensive and detailed statewide manuals also exist, such as the manual used in Iowa (Rossmiller, 1982).

Education

Flooding, erosion and sedimentation, nonpoint-source pollution, and other water resources problems in urban areas are typically not attributable to a failure in technology. Sufficient technology has been available for decades to prevent flooding, pollution, and other problems that occur repeatedly in urban areas. The problem is primarily one of failing to apply existing technology and failing to learn from the mistakes of others. Education of elected officials, the business community, and the general citizenry promises to be effective by leading to more enlightened individual and corporate decisions and activities.

Elected officials can attend special-purpose external seminars, arrange for in-house seminars, and study brief guides and manuals (e.g., Urban Land Institute, 1975, 1978). Some municipality associations prepare special-purpose manuals for elected officials. For example, the Indiana Association of Cities and Towns is preparing a Guide to Municipal Services for local officials which includes a chapter on the fundamentals of stormwater management.

Leaders of the business community can use similar education approaches. In addition, the business community can arrange for informative speakers at meetings of civic and service clubs. Certain members of the business community, particularly realtors and lenders, can exert a significant influence on development in flood-prone areas if they are educated as to the risks involved or are required to take special precautions. For example, and as explained in Chapter 1, federally insured lending institutions are discouraged from granting mortgages on floodplain structures in communities not participating in the National Flood Insurance Program. To meet this requirement, the lending institutions must at least inquire about the flood-prone status of properties for which mortgages are requested. In some states, realtors are required to investigate and report the flood-prone status of property or risk revocation of their licenses.

Improved citizen understanding of surface water management can be achieved with many of the public involvement techniques presented in Chapter 12. Although master plan preparation is the focus of Chapter 12, the public interaction techniques presented there are applicable to SWM in general.

To reiterate a point made in Chapter 5, moving floodwaters are insidiously dangerous because the public, particularly children, are attracted to water and generally fail to appreciate the large dynamic and buoyant focus exerted by floodwaters. This danger should be stressed in public education programs.

A possibly unique aspect of solving urban NPS pollution problems, relative to the resolution of other urban water resources problems, is the potential significant role of the individual citizen. Direct individual action is relatively ineffective, other than as part of the political process, in the resolution of water resources problems such as flooding, water supply, and the development of water-based recreation. That is, there is little, if anything, the individual can do to control floods, to provide significant increases in water supply, or to establish water-based recreation facilities.

An early nonpoint-source pollution study (AVCO Economic Systems Corporation, 1970) reported on the development and use of an environmental index for urban areas based on housing condition, vacant lot condition, and parcel deficiencies. The housing condition component of this index was determined by categorizing the condition of each structure as being sound, deteriorating, or dilapidated, based on factors such as amount of floor space, the relative amount of window space, the presence of decay, and evidence of structural deficiencies. The vacant lot component of the index was determined by categorizing the condition of vacant lots as being good, fair, or poor, based on factors such as presence of weeds or brush, sanitary deficiencies, and the presence of refuse, trash, rubble and carrion. The parcel deficiency component of the environmental index was determined by summing, for the area in question, deficiencies such as refuse storage sites, rubble accumulations, burners, lumber, junked cars, dilapidated sheds, and the presence of livestock, poultry, dogs, and privies. It is apparent that the index is highly sensitive to the manner in which the citizens in a particular area maintain their properties and dispose of waste materials.

An analysis of the quality of runoff from a number of catchments having a wide range of environmental indices revealed that the concentration of pollutants in runoff was highly correlated with the environmental index, characteristics of the drainage system, and the degree of development. More particularly, the environmental index was found to be a good single prediction variable for bacterial pollution in surface water runoff.

The study cited suggests that the actions, or inactions, of individuals and property owners may significantly influence the quality of surface water runoff in urban areas and therefore the quality of creeks, rivers, lakes, and other surface water bodies. Many of the measures available for urban NPS pollution control can be accomplished at little or no cost, with the basic requirement being cooperative, direct participation by enlightened citizens. Examples of these low-cost or no-cost measures that may be applied by citizens in urban areas, in response to the incentives of education and specific ordinances and other regulations, are: control of littering by domestic animals; proper application of chemical and organic fertilizers and pesticides on lawns and gardens; control of litter and debris on private property; control of litter in public areas through use of trash receptacles; careful material storage; proper maintenance of septic systems; and elimination of connections between clear water sources such as sump pumps, downspouts, and foundation drains and sanitary systems to prevent overloading the sewer system and treatment plants.

REFERENCES

American Public Works Association, *Practices in Detention of Urban Stormwater Runoff,* Spec. Rep. No. 43. APWA, Chicago, IL. 1974.

Amy, G., R. Pitt, R. Singh, W. L. Bradford, and M. B. LaGraff, *Water Quality Management Planning for Urban Runoff,* EPA 440/9-75-004. U.S. Environ. Prot. Agency, Washington, DC, December 1974.

AVCO Economic Systems Corporation, *Storm Water Pollution From Urban Land Activity.* Federal Water Quality Administration, Department of Interior, Washington, DC, July 1970.

Bresenhan, T. P. (Ed.), *Industrial Flood Preparedness,* Proceedings of the Flood Warning and Flood Proofing Seminar for Industry, Williamsport, PA, April 16–17, Pennsylvania Department of Community Affairs, Harrisburg, PA, 1979.

Brevard County Water Resources Department, *Evaluation of Two Best Management Practices-A Grassy Swale and Retention/Detention Pond.* BCWRD, Merritt Island, FL, January 1982.

Burby, R. J., and E. J. Kaiser, *An Assessment of Urban Floodplain Management in the United States: The Case for Land Acquisition in Comprehensive Floodplain Management,* Tech. Rep. No. 1. Association of State Floodplain Managers, Madison, WI, June 1987.

Carson, W. D., *Estimating Costs and Benefits for Non-Structural Flood Control Measures.* U.S. Army Corps of Engineers, Hydrologic Engineering Center, Davis, CA October, 1975.

Chow, V. T., *Open Channel Hydraulics.* McGraw-Hill, New York, 1959.

Collins, P. G., and J. W. Ridgway, "Urban Storm Runoff Quality in Southeast Michigan." *J. Environ. Eng. Div. Am. Soc. Civ. Eng.* **106**(EE1), 153–162 (1980).

Commonwealth of Pennsylvania, *Flash Flood Handbook,* September 1978.

Denver Urban Drainage and Flood Control District, *Urban Storm Drainage Criteria Manual.* DUDFCD, Denver, CO, March 1969.

Donohue & Associates, Inc., *A Temporary Floodproofing Plan for Diecast Division of Tecumseh Products Company,* 1979a.

Donohue & Associates, Inc., *Dutch Gap Creek Improvements-Preliminary Engineering,* City of Fond du Lac, WI, December 1979b.

Donohue & Associates, Inc., *Smith Ditch Lagoon No. 1 and Hotter Lagoon Investigation, City of Valpraiso, IN,* June 1984.

Engemoen, M., M. Mercer, and R. E. Krempel, *Developing an Effective Storm Drainage Maintenance Program.* Presented at the Storm Water Management Course, University of Wisconsin-Extension, Madison, December 7–8, 1983.

Federal Emergency Management Agency, *Economic Feasibility of Floodproofing-Analysis of Small Commercial Building.* FEMA, Washington, DC, June 1979.

Federal Emergency Management Agency (FEMA), *Manufactured Home Installation in Flood Hazard Areas,* FEMA 85. FEMA, Washington, DC, September 1985.

Federal Emergency Management Agency (FEMA), *Flood Emergency and Residential Repair Handbook,* FIA-13. FEMA, Washington, DC, March 1986a.

Federal Emergency Management Agency (FEMA), *Floodproofing Non-Residential Structures,* FEMA 102. FEMA, Washington, DC, May 1986b.

Federal Emergency Management Agency (FEMA), *Retrofitting Flood-prone Residential Structures,* FEMA 114. FEMA, Washington, DC, September 1986c.

Flack, J. E., "Economic Analysis of Structural Floodproofing."*J. Water Resour. Plann. Manage. Div.,Am. Soc. Civ. Eng.* November (1978).

Flanagan, L. N., (Ed.), *Floodproofing Systems and Techniques-Examples of Flood Proofed Structures in the United States.* U.S. Army Corps of Engineers, Washington, DC, December 1984.

Flynn, T. J., S. G. Walesh, J. G. Titus, and M. C. Barth, *Greenhouse Effect and Sea Level Rise: A Challenge for This Generation,* Chapter 9. Van Nostrand-Reinhold, New York, 1984.

French, R. H., *Open Channel Hydraulics.* McGraw-Hill, New York, 1985.

Georgia State Soil and Water Conservation Committee, *On-Site Erosion Control,* November (1979).

Hilton, R. E., *Flood Emergency Evacuation Plans.* Jacksonville District, Corps of Engineers, FL, 1980.

Hunter, M. R., and S. L. Tucker, "Contract Maintenance for Drainage and Flood Control Facilities." *Flood Hazard News,* Vol. 12, No. 1. Urban Drainage and Flood Control District, Denver, CO, December 1982.

Hurlbert, D., "Catch Basin Cleaning Program Reduces Flooding Complaints." *Public Works* June, pp. 98–99 (1980).

Illinois Department of Transportation, Division of Water Resources, *Elevating or Relocating a House to Reduce Flood Damage,* Local Assistance Ser. 3C, IDOT, Chicago, IL, October 1983.

Illinois Department of Transportation, Division of Water Resources, *Protect Your Home from Flood Damage,* Local Assistance Ser. 3B. IDOT, Chicago, March 1982.

Johnson, W. K., *Physical and Economic Feasibility of Non-structural Floodplain Management Measures.* U.S. Army Corps of Engineers, Hydrologic Engineering Center and Institute for Water Resources, March 1978.

Kinori, B. Z., and J. Mevorach, *Manual of Surface Drainage Engineering, Vol. II.* Am. Elsevier, New York, 1984.

La Crosse Tribune, "Prairie du Chien Expects $1 Million for More Flood Relocation," September 8 (1979).

League of Wisconsin Municipalities, "The Fourth of July Flood of '78 Cinches Relocation to Higher Ground."*The Municipality* August, pp. 181–182 (1983).

Linsley, R. K., and J. B. Franzini, *Water-Resources Engineering,* 3rd ed. McGraw-Hill, New York, 1979.

Lord, B. N., "Effectiveness of Erosion Control." *In Urban Runoff Quality—Impact and Quality Enhancement Technology,* Proc. Eng. Found. Conf., edited by B. Urbonas and L. A. Roesner, pp. 281–289. New England College, Henniker, NH, 1986.

Martin, E. H., and J. L. Smoot, "Constituent-Load Changes in Urban Stormwater Runoff Routed Through A Detention Pond-Wetlands System in Central Florida," U.S. Geological Survey, Water Resources Investigations Rep. 85–4310, Washington, DC. 1986.

McCuen, R. H., S. G. Walesh, and W. J. Rawls, "Control of Urban Stormwater Runoff by Detention and Retention." *Misc. Publ. U.S., Dep. Agric.* **1428,** 1–75 (1983).

McKeever, S. R., *Floodproofing: An Example of Raising a Private Residence.* Department of the Army, Corps of Engineers, South Atlantic Division, Atlanta, GA, March 1977.

Milwaukee Journal, "Relocation Projects Spurred by Floods," February 2 (1978).

Minnesota Department of Natural Resources, *Reducing Flood Damages by Acquisition and Relocation: The Experiences of Four Minnesota Communities,* Floodplain Manage. Inf. Brochure No. 1. MDNR, St. Paul, MN. (no date).

National Association of Home Builders Research Foundation, *Manual for Construction of Residential Basements in Non-Coastal Flood Environs - A Study and Preparation of a Guide Manual for the Design and Construction of Floodproofed Residential Basements.* Prepared for the Federal Insurance Administration, Department of Housing and Urban Development, March 1977.

Normann, J. M., "Improved Design of Highway Culverts." *Civ. Eng. (N.Y.)* March, pp. 70–73 (1975).

Northeastern Illinois Soil Erosion and Sedimentation Control Steering Committee, *Procedures and Standards for Urban Soil Erosion and Sedimentation Control in Illinois,* October 1981.

Novotny, V. and Chesters, G., *Handbook of Nonpoint Source Pollution,* Chapter 11. Van Nostrand Reinhold, New York, NY, 1981.

Owen, H. J., *Cooperative Flood Loss Reduction—A Technical Manual for Communities and Industry,* U.S. Water Resources Council, Washington, DC June 1981.

Owen, H. J., *Guide for Flood and Flash Flood Preparedness Planning.* U.S. Department of Commerce, Washington, DC. May 1977.

Pace, C. E., *Tests of Brick-Veneer Walls and Closures for Resistance to Floodwaters,* Misc. Pap. C-78-16. Concrete Laboratory, U.S. Army Engineer Waterways Experiment Station, Vicksburg, MS, December 1978.

Pace, C. E., and R. L. Campbell, *Structural Integrity of Brick-Veneer Buildings,* Tech. Rep. C-78-3. Concrete Laboratory, U.S. Army Engineer Waterways Experiment Station, Vicksburg, MS, May 1978.

Petersen, M. S., *River Engineering.* Prentice-Hall, Englewood Cliffs, NJ, 1986.

Pinellas Park Water Management District, *Stormwater Management—Pinellas Park Water Management District—Stormwater Runoff Control Rules.* Pinellas, Park, FL, 1982.

Raasch, G. E., "Urban Stormwater Control Project in an Ecologically Sensitive Area." *Proc. Int. Symp. Urban Hydro., Hydraul., Sediment Control, 1982.*

Rossmiller, R. L., *Iowa Urban Stormwater Management Manual.* November 1982. Engineering Extension Service, Iowa State University, Ames, IA.

Schoenbeck, R. H., *Maintenance and Operation of Detention Systems.* Presented at the Storm Water Detention Systems Institute, University of Wisconsin-Extension, Madison, April 17–18, 1979.

Schueler, T. R., *Controlling Urban Runoff: A Practical Manual for Planning and Designing Urban BMP's.* Metropolitan Washington Council of Governments, Washington, DC, July, 1987.

Shaver, H. E., "Infiltration as a Stormwater Management Component." In *Urban Runoff Quality—Impact and Quality Enhancement Technology,* Proc. Eng. Found. Conf., edited by B. Urbonas and L. A. Roesner, New England College, Henniker, NH, 1986.

Sheaffer, J. R., W. J. Bauer, P. Gold, J. E. Hackett, R. Lewis, A. Miller, and J. Rowley, *Introduction to Flood Proofing—An Outline of Principles and Methods.* Center of Urban Studies, University of Chicago, Chicago, IL, April 1967.

Southeastern Wisconsin Regional Planning Commission, *A Regional Park and Open Space Plan for Southeastern Wisconsin,* SEWRPC, Waukesha, WI, November 1977.

Southeastern Wisconsin Regional Planning Commission, *A Comprehensive Plan for the Kinnickinnic River Watershed.* SEWRPC, Waukesha, WI, December 1978.

Susquehanna River Basin Commission, *Neighborhood Flash Flood Warning Program Manual,* October 1976.

Terstriep, M. L., D. C. Noel, and G. M. Bender, "Sources of Urban Pollutants—Do We Know Enough?" In *Urban Runoff Quality—Impact and Quality Enhancement Technology,* Pro. Eng. Found. Conf., edited by B. Urbonas and L. A. Roesner, pp. 107—121, New England College, Henniker, NH, 1986.

Urban Land Institute, American Society of Civil Engineers, and National Association of Home Builders, *Residential Storm Water Management.* ULI/ASCE/NAHB, Washington, DC, 1975.

Urban Land Institute, American Society of Civil Engineers, and the National Association of Home Builders, *Residential Erosion and Sediment Control-Objectives, Principles, and Design Considerations.* ULI/ASCE/NAHB, Washington DC 1978.

Urbonas, B., and W. P. Ruzzo, "Standardization of Detention Pond Design for Phosphorus Removal." In *Urban Runoff Pollution* (H. C. Torno, J. Marsalek, and M. Desbordes, Eds.), NATO Adv. Sci. Inst. Ser., Ser. G, Vol. 10, Chapter 8, Springer-Verlag, New York, 1986.

U.S. Army Corps of Engineers, *Flood Proofing Regulations. USACE,* Washington, DC, June 1972.

U.S. Army Corps of Engineers, Institute for Water Resources, *Cost Report on Non-Structural Flood Damage Reduction Measures for Residential Buildings Within the Baltimore District.* Kingman Building, Fort Belvoir, VA, July 1977.

U.S. Army Corps of Engineers, Chicago District, *Preliminary Cost Estimates for Non-Structural Plan Formulation.* USACE, Chicago, IL, October 1981.

U.S. Army Corps of Engineers, Omaha District, *Emergency Flood Fight Manual.* USACE, Omaha, NB, April 1987.

U.S. Department of the Army, Corps of Engineers, *Wall Design-Flood Walls,* EM-1110-2-2501. USDA, Washington, DC, January, 1948.

U.S. Department of the Army, Corps of Engineers, *Interior Drainage of Leveed Urban Areas: Hydrology,* EM-1110-2-1410. USDA, Washington, DC May 1965.

U.S. Department of the Army, Corps of Engineers, *Hydraulic Design of Flood Control Channels,* EM-1110-2-1601. USDA, Washington, DC, July 1970.

U.S. Department of the Army, Corps of Engineers, *Design and Construction of Levees,* EM 1110-2-1913. USDA, Washington, DC, July 1978.

U.S. Department of Housing and Urban Development, Federal Insurance Administration, *Elevated Residential Structures-Reducing Flood Damage Through Building Design: A Guide Manual.* USDHUD, Washington, DC, December 1976.

U.S. Department of Transportation, Federal Highway Administration, *Hydraulic*

Charts for the Selection of Highway Culverts, Hydraul. Eng. Cir. No. 5. USDT, Washington DC, December 1965.

U.S. Department of Transportation, Federal Highway Administration, *Hydraulics of Bridge Waterways,* 2nd ed., Hydraul. Des. Ser. No. 1. USDT, Washington, DC, 1970.

U.S. Nuclear Regulatory Commission, *Flood Protection for Nuclear Power Plants,* Regul. Guide No. 1.102, Rev. 1. USNRC, Washington DC, September 1976.

Walesh, S. G., "Floodland Management: The Environmental Corridor Concept." *Hydraul. Eng. Environ.-Proc. 21st Annu. Hydraul. Div. Spec. Conf., Am. Soc. Civ. Eng., 1973,* pp. 105-111.

Walesh, S. G., "Floodplain Management: The Environmental Corridor Concept." *Tech. Rec. Southeast. Wis. Reg. Plann. Comm.* **3**(6), 1–13 (1976).

Walesh, S. G., *Detention/Retention for Enhancement of the Quality of Storm Water Runoff: State of the Art.* Presented at the Fourth Annual Conference, Wisconsin Section, American Water Resources Association, Madison, WI, February 14–15, 1980.

Walesh, S. G., *Emergency Floodproofing of the Tecumseh Diecast Plant.* Presented at the Seminar on Residential and Commercial Floodproofing, St. Paul, MI, March 1–2, 1982a.

Walesh, S. G., *June 9, 1982 Visits to Soldiers Grove and Prairie du Chien, Wisconsin,* Memo., Donohue & Associates, Inc., June 14, 1982b.

Walesh, S. G., *Flood Protection and Floodproofing Status of Hazardous Waste Management Facilities Located in Coastal and Riverine Floodplains—Results of Site Visits.* Presented at the Conference on the Effects of Sea Level Rise on the Coastal Economy and Environment sponsored by U.S. EPA and ICF, Inc., Washington, DC, March 30, 1983.

Walesh, S. G., "Case Studies of Need-Based Quality—Quantity Control Projects. In *Urban Runoff Quality—Impact and Quality Enhancement Technology,* Proc. Eng. Found. Conf. edited by B. Urbonas and L. A. Roesner, pp. 423–437, New England College, Henniker, NH, 1986.

Walesh, S. G., and M. L. Schoeffmann, *Surface and Subsurface Detention in Developed Urban Areas: A Case Study.* Presented at the American Society of Civil Engineers Water Resources Planning and Management Division Conference, Baltimore, MD, May 1984.

Walesh, S. G., and M. L. Schoeffmann, "One Alternative to Flooded Basements." *Am. Public Works Assoc. Rep.* **52**(3), 6–7, March (1985).

Wanielista, M.P., and Y.A. Yousef, "Best Management Practices Overview." *In Urban Runoff Quality—Impact and Quality Enhancement Technology.* Proc. Eng. Found. Conf. edited by B. Urbonas and L. A. Roesner, pp. 314–322, New England College, Henniker, NH, 1986.

Whipple, Jr., W., R. Kropp, and S. Burke, "Implementing Dual-Purpose Stormwater Detention Programs", Water Resources Planning and Management Div., Am. Soc. Civ. Eng., **113,** 6 (1987) 779–792.

12

PREPARATION OF A MASTER PLAN

Urban surface water management (SWM) is defined simply in Chapter 1 as "everything done to remedy existing surface water problems and to prevent the occurrence of new problems." It is also defined in that chapter as consisting of planning, design, construction, and operations functions. This chapter focuses on the planning phase of urban SWM.

Ideally, an urban SWM master plan should be prepared for an urban area prior to implementing structural and nonstructural management measures. That is, planning should precede the design, construction, and operation phases. Unfortunately, and as also noted in Chapter 1, the planning phase of urban SWM often receives little or no attention. It's been said, "if we fail to plan, we plan to fail." By focusing on planning, this chapter is dedicated to helping in a small way to rectify the widespread absence of planning.

Described in this chapter is a proven process for preparing urban surface water master plans that are intended to be implemented. The process integrates essentially all of the SWM fundamentals and most of the tools and techniques presented in preceding chapters. Before presenting the master planning process, the definition of and need for master planning are presented. Then the typical motivation for master planning is discussed as are some critical aspects or features of successful master planning efforts. The chapter concludes with a presentation of ideas on the economy of scale in master planning and on the costs and benefits of master planning.

12.1 DEFINITION OF AND NEED FOR MASTER PLANNING

Surface water master planning can be defined in two ways: (1) in terms of the immediate product, and (2) in terms of the process used to produce the product. Defined in terms of the immediate product, a master plan is a document or set of documents that contains three types of recommendations:

1. Structural Management Measures: a recommended system of measures, such as sewers, channels, detention/retention facilities, sedimentation basins, and dikes and floodwalls with costs
2. Nonstructural management measures: a recommended set of measures, such as land acquisition, regulations, manuals of practice, flood insurance, structure floodproofing or removal, inspection and maintenance programs, emergency programs, and education programs, with costs, to the extent that costs can be estimated
3. Plan Implementation Program: a recommended program that indicates when plan elements are to be implemented, who has the principal responsibility for implementing each element, and how the elements are to be implemented.

In short, focusing on the product and as illustrated in Figure 12.1, a master plan answers the what, when, who, how, and other questions needed to take action. More specifically, these questions are: What has to be done to remedy existing surface water problems or prevent surface water problems from occurring? When do activities have to occur or facilities be constructed? Who is responsible for doing what has to be done? How will the project be financed?

Instead of defining master planning in terms of the immediate product, master planning may be defined as a dynamic but systematic and disciplined process. In keeping with the process orientation, Linsley and Franzini (1979) define planning as "the orderly consideration of a project from the original statement of purpose through the evaluation of alternatives to the final decision on a course of action." The master planning process, as illustrated in Figure 12.2, includes establishing objectives and standards; conducting an inventory of the natural and cultural features of the planning area; analyzing

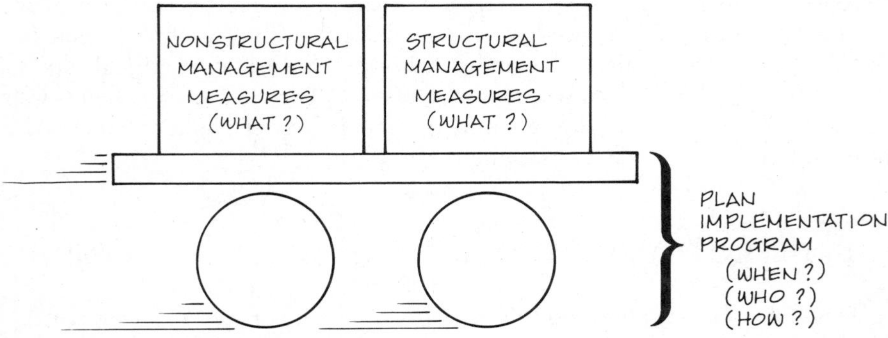

FIGURE 12.1 Master planning defined in terms of products

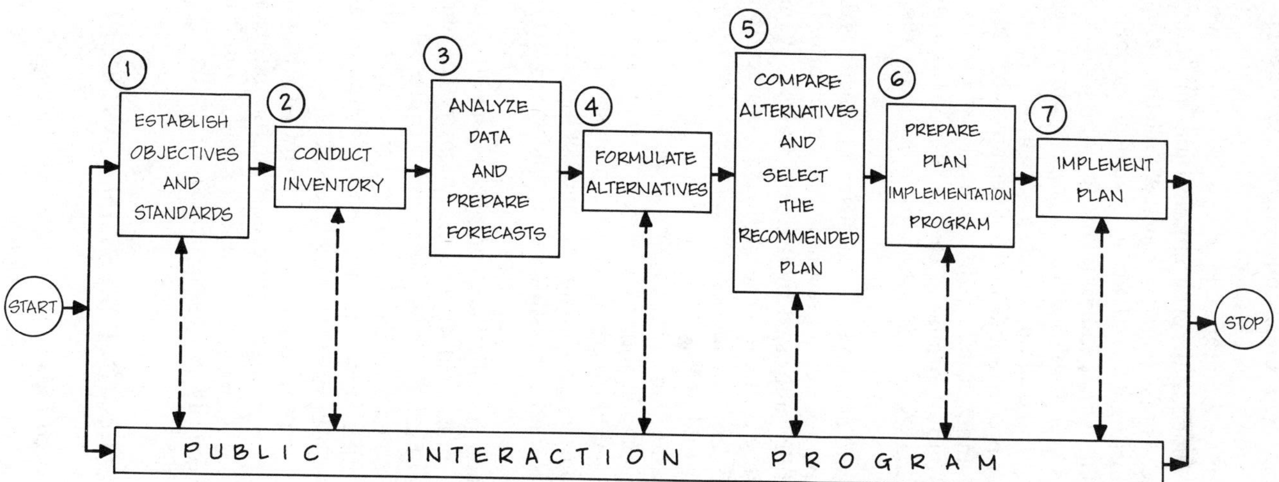

FIGURE 12.2 Master planning defined in terms of process

data, defining problems, and forecasting various future scenarios; creating potential alternatives; comparing alternatives and selecting the optimal combination of alternatives and subalternatives to constitute the recommended plan; preparing a plan implementation program; and implementing the plan.

In keeping with the process theme, Grigg (1986) defines planning as "studying what to do" and distinguishes planning from decision making or "deciding what to do." The point is that studying and deciding are two different activities or processes. And as exemplified by the preparation of surface water master plans, the studying and deciding processes are usually carried out by different groups. A team of professionals and technicians prepares the master plan, incorporating significant interaction with the public and with elected and other officials. Another group of primarily elected officials usually influenced by the public typically makes decisions hopefully based on the findings of the master plan.

Because planning and deciding are two different functions done by different groups, planners must not be so presumptuous as to think that their recommendations will be fully embraced by decision makers. Planners can greatly enhance the probability of acceptance of their recommendations if the work is of high quality and if the planners practice effective communication with all interested individuals and groups.

While defining master planning, it is important to define what master planning is not. A master plan is not an engineered design in the traditional sense. Implementation of facilities recommended in a master plan requires preparation of design and construction documents and detailed cost estimates, obtaining necessary permits, and addressing other concerns typical of public works projects.

Urban surface water problems are complex because they involve technical, economic, environmental, legal, financial, administrative, and political facets. Some of these facets of urban SWM problems have internal conflicts and there are conflicts between the facets. Furthermore, some facets are ambiguous and defy complete quantification. Nevertheless, complex urban surface water quantity and quality problems must be addressed. We cannot claim that plans will lead to the best or optimum solution or even that a best or optimum solution exists. We can be confident, however, that the master planning process will lead to a good course of action and will avoid a multitude of erroneous, and probably unnecessarily expensive, courses of action.

12.2 FLUCTUATING INTEREST IN MASTER PLANNING

There is an obvious need for surface water master plans for urban areas. However, master planning is still the exception, rather than the rule, throughout the United States. Furthermore, when surface water master planning is undertaken, it is usually done in reaction to serious flooding, pollution, or other problems.

Low Priority of Surface Water Systems

However rational master planning for surface water management may be and may sound to the professional community, it is difficult to "sell" community officials and citizens on the concept, let alone convince them to commit the necessary resources. There are simply too many other competing services and facilities that need attention and appear to be more important to the community than the surface water system.

The public and many elected officials take the surface water system—channels, sewers, detention/retention facilities, pumping stations, dikes, floodwalls—for granted. This is understandable because, as discussed in Chapter 1, many components of the surface water system are not visible or readily apparent to the casual observer. Some of the components are underground and some are small and easily overlooked. Others, if carefully designed, tend to conform to natural topography and are not readily apparent. The surface water system tends to be visible or noticed only when it malfunctions or is alleged to do so.

In addition, and as also elaborated on in Chapter 1, the urban surface water system functions infrequently, that is, during and immediately after rainfall or snowmelt events. Most other municipal services are not only visible but function on an essentially continuous basis. Examples are the sanitary sewerage, water supply, electrical, telephone, television, natural gas, streets and highways, and police and fire protection systems.

Being Prepared to Seize the Opportunity

Accordingly, public interest in and willingness to pay for planning, designing, constructing, and operating surface water facilities tends to rise and fall figuratively, if not literally, with the floodwaters as illustrated in Figure 12.3. This is particularly true for the planning function, which generally seems to enjoy the least support from elected officials and the public. During and immediately after a flood, political leaders and citizens are willing to fund remedial efforts, sometimes even master planning projects, as illustrated in Figure 12.3 and also in Figure 12.4. However, months later, when the master plan is done or the design is completed and costly recommendations are made, public interest wanes, little or nothing is done, and the cycle is repeated, as suggested by Figures 12.3 and 12.4.

Engineers and other professionals in public service or private practice having urban SWM responsibilities should prepare themselves by being familiar with master planning—the process, the product, and the long-term benefits. Inevitably, opportunities will arise, usually in response to serious flooding, pollution, or other events, in which a master planning effort could be initiated. Those who are prepared will be able to seize that opportunity and proceed with the planning effort.

Rushing into a master planning effort after and as a result of a disastrous flood or other event is not as desirable as undertaking planning before disaster

FIGURE 12.3 Fluctuating interest in urban surface water management *(Source:* Adapted from Debo, 1979)

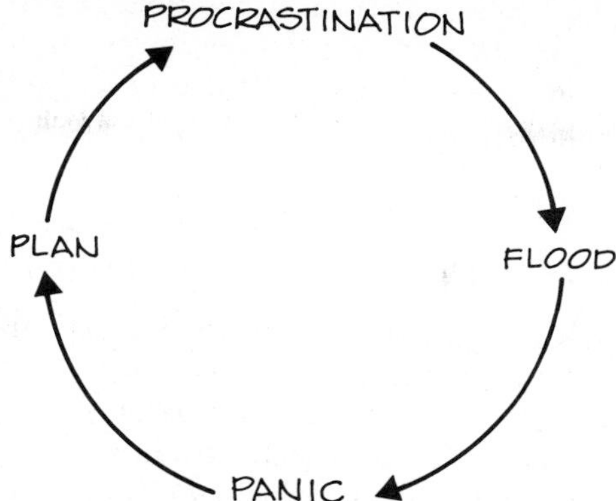

FIGURE 12.4 The "hydroillogical" cycle (*Source:* Adapted from Cyre and Shearer, 1987)

occurs. However, even this reactionary approach to master planning is preferable to designing and constructing hastily selected and costly facilities without the benefit of a plan.

12.3 MASTER PLANNING PRINCIPLES

Master plans must be tailored to the communities and watersheds for which they are being prepared. However, experience indicates that respecting certain overriding planning principles will greatly enhance the likelihood of success.

Focus on Land Use

Existing and future surface water quantity and quality problems are inextricably tied to existing and future land use patterns. Master planning for urban SWM must respect the principle of the interdependence of land and water resources.

As demonstrated repeatedly in Chapter 2, land use significantly influences the volume and rate of runoff from the land surface to the surface water system; therefore, land use is a primary determinant of the location and severity of urban flooding problems. The general location of public and private wastewater treatment facility discharges to the surface water system will be influenced by the overall land use pattern, as will the nature and location of solid waste disposal and toxic and hazardous waste sites. The nature and density of land use will also affect the quantity and quality of pollutants, solid wastes, and toxic and hazardous waste materials received by and passed through such facilities. Similarly, and as also demonstrated in Chapter 2, the type and quantity of nonpoint-source pollution and the points of entry of those pollutants into the surface water system and groundwater system are largely determined by existing and future land use patterns.

Ideally, master plans directed to the resolution of existing surface water quantity and quality problems and the prevention of future problems should be done within the framework of an agreed-upon area-wide land use plan. In lieu of a plan, an enlightened forecast of likely future land use is absolutely necessary.

Phase by Prioritized Watersheds

Watersheds are the basic foundation for the preparation of surface water master plans (Brown, 1972). As shown in Figure 12.5, watershed divides are not respecters of city, village, county, state, and other civil division boundaries. However, watersheds do define the system within which the hydrologic cycle functions, and therefore must be the principal physical basis for master planning.

Occasionally, as illustrated by case 1 in Figure 12.5, a watershed of interest will be wholly contained within one civil division. In this situation, a civil division can unilaterally prepare a master plan for the watershed except for possible downstream effects outside the watershed and the civil division. The more common situation is shown in case 2, in which at least minimal inter-civil division cooperation is required to prepare a master plan.

Concept of and Need for Phasing. Individual communities or groups of contiguous communities usually do not have the financial and professional personnel resources required to prepare master plans for all the pertinent water-

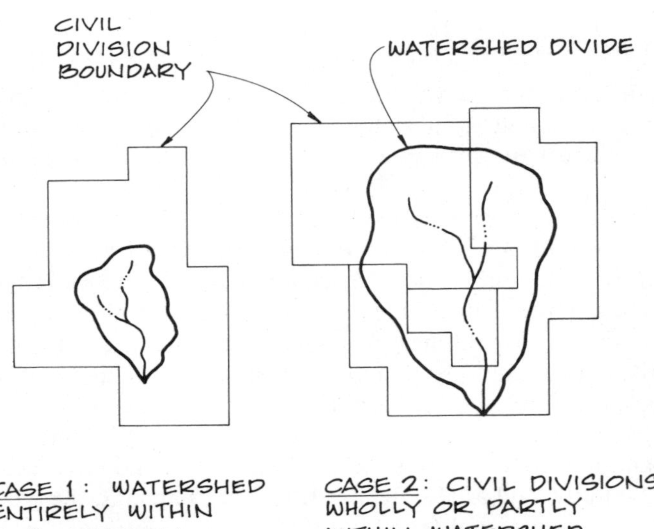

FIGURE 12.5 Relationships between watershed divides and civil division boundaries

sheds simultaneously, that is, for all the watersheds in the community plus portions of those watersheds outside the community. A phased master planning approach spread over years or even decades is likely to be much more feasible. Major watersheds are delineated, each is characterized as to the relative seriousness of the existing and potential surface water problems, and the prioritized sequence is established.

The prioritization concept may be illustrated with the assistance of Figure 12.6, which shows three different hypothetical watersheds. Watershed A is completely developed and already has serious flooding problems. The lower or downstream portion of watershed B is developed, although only minor flooding problems occur. Expected development in the upstream portion of watershed B will, in the absence of preventive measures, cause serious flooding in the lower reaches. Although watershed C is in a natural condition, urban development is expected to occur. The logical priority within which to prepare and begin implementing master plans for the hypothetical situation is: watershed A, watershed B, and then watershed C.

In actual situations, many other factors would be used to characterize the seriousness of existing and potential surface water problems and to otherwise establish the planning priority. Examples of these other factors are the extent and seriousness of point- and nonpoint-source pollution problems, the need for park and open space, the expected rate of urbanization in undeveloped or

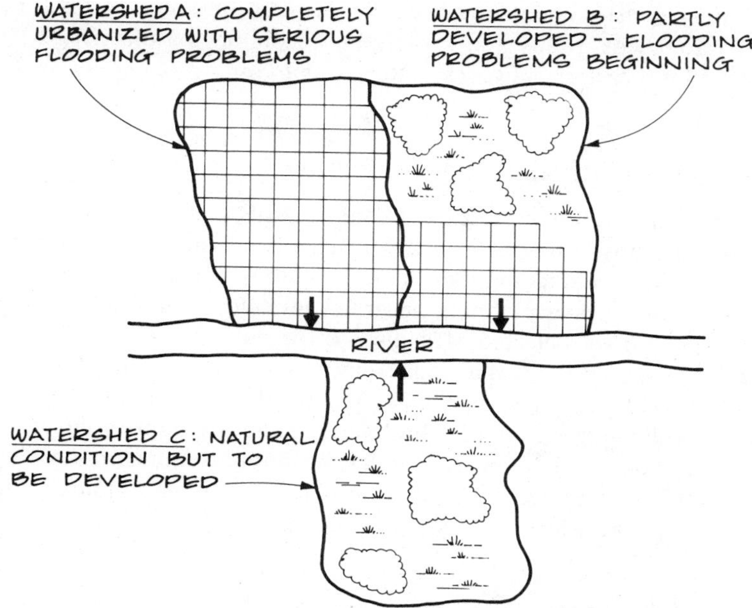

FIGURE 12.6 Categorization of urban area watersheds to help establish priorities for preparing master plans

partially developed areas, and the demonstrated interest of residents and political leaders in undertaking a master planning project.

Examples of Phasing. The areal scale of and time frame for prioritized master planning can vary widely depending on circumstances. For example, faced with growing flooding, erosion, and sedimentation problems, Bettendorf, Iowa, initiated a surface water master planning effort in October, 1983. The total planning area consisted of 11 watersheds covering about 11 mi^2, which included most of the city plus contiguous tributary areas. Master plans were prepared on a priority basis for eight watersheds over a period of three years. The watershed sizes ranged from 0.25 to 3.3 square miles. In some cases, master plan completion was followed immediately by design and construction of recommended facilities. The master planning and subsequent design and contruction program was accepted by elected officials partly because program phasing on a priority basis spread planning, design, and construction costs in a logical fashion over several fiscal years (Lau, 1988).

On a much larger area scale and over a much longer time scale, a major watershed master planning effort is being carried out in southeastern Wisconsin. The 2700-mi^2 planning region encompasses only 5 percent of the state but contains about half of the state's population and therefore about half of its urban development. A total of 11 watersheds are located wholly or partly within the region. Faced with existing and potential water resources problems resulting from urban development, a watershed master planning effort was begun in 1964. Master plans have been completed or are in progress for seven of the 11 watersheds. Many and varied plan recommendations have been implemented [Southeastern Wisconsin Regional Planning Commission (SEWRPC), 1986].

Involve Elected Officials and the Public

As noted previously, the interest of elected officials and the public in surface water facilities tends to fluctuate widely, usually being most intense during and immediately after a damaging and disrupting flood event. Primarily for this reason, successful master planning programs include meaningful involvement of and interaction with elected officials and the public.

A master planning effort that fails to include a public interaction program plans to fail. Although said over a century ago and in an entirely different context, the following words of Abraham Lincoln are appropriate to urban surface water master planning today: "With public sentiment, nothing can fail; without it nothing can succeed. Consequently, he who molds sentiment goes deeper than he who enacts statutes or pronounces decisions" (Helweg, 1985).

Objectives of the Public Interaction Program. Interaction programs have three objectives. The first objective is to demonstrate to the public and their elected representatives that engineers and other professionals having responsibility for the urban surface water system are aware of the problems, at least in a general sense; want to learn more about them; and want to seek solutions. Sometimes the more vociferous citizens and officials need an opportunity to vent their frustration with surface water problems and the apparent inability of responsible parties to solve those problems. Public information meetings held early in the master planning process can provide this opportunity and, hopefully, enable these frustrated individuals to become positive participants in the master planning process.

The second objective of public involvement efforts is to gather supplemental data and information pertinent to the master planning effort. Interested citizens, if informed about a master planning program, are likely to contribute photographs of flooding, information on high water marks, and other useful data and information. Similarly, but on a larger scale and in a more formal manner, various government units and agencies are likely to offer potentially useful data, reports, and other information if they are informed about a master planning effort and invited to contribute.

The third and final objective of public interaction efforts is to build support for rapid plan implementation. Enlightened citizens and officials who have been informed about a master planning program and have been given an opportunity to participate in preparation of a master plan are likely to become supporters of the plan, help to interpret it for others, and otherwise help to implement it.

Public Involvement Techniques. The first involvement of elected officials and the public should coincide with the beginning of the master planning effort and should continue throughout the planning program. Numerous and varied involvement techniques are available and can be used to design a program tailored to a particular situation. Specific ideas on how to conduct effective public meetings are offered by Connor (1977) and Helweg (1985). Costello (1974) discusses public involvement techniques and Hillegass et al. (1970) describe the strategy of involving influential members of the community in the planning process. Use of structured face-to-face negotiations between environmentalists, developers, and government units to resolve conflicts is illustrated by Priscoli (1988). Sargent (1972) introduces the term "fishbowl planning" to denote frequent public involvement throughout the planning process and gives examples.

Some form of initial information meeting or meetings should be held. The planning staff should use the occasion to explain the overall purpose of and time schedule for the master planning effort. Citizens and officials should be invited to express concerns, offer suggestions, and contribute data and information. The initial information meeting may be a formal hearing or a casual conversation among interested parties. Such meetings are often best held in the neighborhoods, areas, or communities having the more serious flood problems. Potential sites include municipal buildings, schools, churches, and private homes.

Members of the planning staff typically perform a field reconnaissance of the planning area by walking and driving along and across the surface water system and making notes and taking photographs. This reconnaissance provides another opportunity for informal, one-on-one interaction with inquisitive members of the public.

One or more public meetings should also be held when alternative structural or nonstructural solutions have been developed or a preliminary recommended plan prepared. Elected officials and the public should have an opportunity to learn about and comment on the alternatives.

Meetings should be supplemented with other appropriate information dissemination and interaction techniques. Press releases or interviews with the press can be effective. Some communities have newsletters in which they include progress reports on the master planning program. Special brochures have been prepared by communities, as have videotapes and slides for use on local television or for presentation to service clubs and civic groups.

Advisory committees can be effective. For example, over the past two decades the Southeastern Wisconsin Regional Planning Commission used advisory committees to guide its watershed master planning efforts. During the 4-year period ending in October 1976, the Commission prepared a master plan for the 135-mi^2 Menomonee River watershed, which contains all or parts of four counties, seven cities, six villages, and four towns. The 19-person advisory committee included two elected officials, 10 engineers and other professionals holding senior positions in local communities and agencies, two local business people, three state officials, a representative of an environmental group, and a representative of the academic community. The advisory committee conducted six meetings for officials and the public, helped provide data and information for use in preparing the plan, met regularly for detailed reviews of the evolving planning report, and helped to interpret the plan to their sectors of the watershed community (SEWRPC, 1976).

On a much smaller scale, the village of Flossmoor, Illinois, used an advisory committee to help guide the preparation of a master plan for a 4.7-mi^2 area encompassing most of the community. The committee included a consulting engineer, a contractor, a realtor, and a retired Army Corps of Engineers official (Lau, 1988).

The city of Bettendorf, Iowa, in preparing its surface water master plan, used the city council as an ad hoc advisory committee. The planning team, consisting of city officials and their consultants, met regularly with the council throughout the planning program to present progress reports and to obtain input from council members (Lau, 1988).

Explore Creative Options

As explained in Chapter 1, floodplain management in the United States evolved over the past three decades from a single purpose–single means approach to a multiple purpose–multiple means approach. Similarly, as described in Chapter 1, the use of detention/retention facilities has evolved over the past approximately two decades from the single quantity control phase to the current multiple-purpose quantity–quality–recreation phase.

The master planning process should capitalize on these specific evolutionary developments by giving due consideration to a multiple purpose–multiple means approach to urban SWM. Furthermore, master planning, carried by the momentum of recent developments, should move beyond the current of the state of the art and seek to conceive and develop even better approaches to urban SWM.

Many and varied structural and nonstructural techniques have been used successfully in recent years to supplement the proven structural approaches, such as storm sewers, channel modifications, and dikes and floodwalls. Examples of newer ideas planned, designed, and constructed or implemented in recent years, as discussed primarily in Chapter 11, include permanent and temporary floodproofing of individual buildings, acquisition and removal of flood-prone buildings, acquisition or regulation of flood-prone lands, rooftop and subsurface storage, street detention, infiltration of surface water into the groundwater system, and user fees.

Use Computer Modeling

With the possible exception of simple watershed systems, master planning should be done with the aid of digital computer models. Chapter 10 provides both a comprehensive and an in-depth treatment of computer modeling in urban SWM. Stressed in that chapter is the idea that digital computer models usually provide the most technically and economically sound engineering means of achieving the necessary understanding of the spatial and temporal fluctuations in surface water quantity and quality under existing and hypothetical future conditions.

Computer models enable the planning team to utilize extensive hydrologic, hydraulic, water quality, and economic data and knowledge by providing a cost-effective means of performing voluminous and repetitive data management, mathematical calculations, and logic operations. Prior to the advent of digital computer models, effective hydrologic, hydraulic, water quality, and economic simulation was not feasible. Computer modeling provides a means to answer, at least in an approximate fashion, the numerous ‘what if?’’ questions that should arise and be answered during master planning.

12.4 THE PLANNING PROCESS: AN OVERVIEW

A pragmatic approach to surface water master planning is to view planning as a seven-step process with a parallel public interaction program, as illustrated in Figure 12.2. The process and variations on it has been successfully used on many projects. (Costello, 1974; McPherson, 1977; Ringenoldus, 1970; Sheaffer et al., 1982). The seven-step process is described in detail in subsequent sections of the chapter. In a fundamental sense, effective surface water master planning is application of the engineering approach at its best— thorough, systematic, and imaginative. Effective master planning is hard work, combining rigorous technical challenge with a demanding communication effort.

The topic of planning the planning project is not treated explicitly in this chapter. Implicit in this chapter is the need for careful planning of a planning project. Included should be consideration of organizing the project team; establishing a communication, coordination, and control process for members of the planning team and other participants; identifying tasks and estimating personnel and other resources for each task; and scheduling tasks. Helweg (1985) provides a very useful discussion of planning the planning project, and Grigg (1985) discusses organizing for water planning.

12.5 STEP 1: ESTABLISH OBJECTIVES AND STANDARDS

Definitions and Use

An objective is a goal or end toward which a plan is directed. A standard is a criterion, preferably quantitative, associated with an objective and used as a basis of comparison to determine the adequacy of an alternative intended to meet the objective. Stated differently, the objective is the "target" and the standard is used to measure the degree of success in "hitting the target."

The adopted or agreed-upon objectives and standards should guide the planning process. Technical, economic, environmental, legal, financial, administrative, political, and other constraints may prevent complete achievement of the established objectives or, more particularly, the supporting standards. Nevertheless, the objectives and standards should guide the planning process and should be used to determine the expected success of the recommended plan. Refer to American Society of Civil Engineers (1984) for a synopsis of many aspects of water resources planning objectives, with emphasis on the role of the engineer in establishing objectives and using them in preparing plans.

Timing

It is tempting at the beginning of an apparently complex master planning effort to avoid systematic establishment of objectives and standards because of the press of other activities and in order to get on with the inventory, analysis, and other phases of the project. Similarly, establishing objectives and standards can be put off indefinitely during the master planning effort.

Objectives and standards must be established at the beginning of the master planning process for two reasons. First, the development of objectives and standards provides an opportunity for the planning staff, which will typically be the in-house staff of a government unit or agency or an in-house staff working with the staff of a consulting firm, to firm up and agree upon their understanding of what is to be accomplished. Second, the objectives and standards provide the planning staff with focus and perspective during the preparation of the plan. In essence, guidelines provide the basis for and give direction to the planning process.

Origin of Objectives and Standards

Drafting and refining meaningful objectives and standards requires input from the various individuals or groups that have responsibility for or interest in the planning area. Examples are engineers and other professional employees of governmental units and agencies, elected officials, individual citizens and citizen environmental and other groups, and consultants participating on the planning team.

The critical public interaction program mentioned earlier should help to identify the overriding concerns and values of the various interest groups, which in turn can be translated into operational objectives and standards. For example, if the public values and is willing to pay for natural, water-oriented public open space, the objectives and standards could prescribe preservation of natural areas along streams and waterways.

Some objectives and standards, particularly the latter, will be prescribed by rules and regulations of governmental units and agencies having jurisdiction in the planning area. For example, some state agencies specify a minimum freeboard for dikes and floodwalls which should be included in the supporting standards. Other objectives and standards will be based in part on the cumulative master planning experience of engineers and other professionals serving on the planning team.

Examples of Objectives and Standards

In most cases, objectives and standards are stated in relatively brief form and integrated; that is, objectives and the supporting standards are not stated separately. An example of the less formal and briefer approach is provided by the objectives and standards developed by the client and consultant near the beginning of a master planning program for a 6.5-mi^2 urbanizing area in and contiguous with the city of Valparaiso, Indiana. The following objectives and standards, quoted directly from the planning report (Donohue & Associates, 1984), were used in the development and evaluation of alternatives:

1. The major stormwater system should be configured and sized so as to store or convey runoff from the 100-year recurrence interval–6-hour-duration rainfall occurring under future land use conditions. System design must also include evaluation of interaction between components.
2. Flooding problems should be resolved as close to their point of origin

as possible, and the problem should not be shifted from one area to another.

3. The first step toward resolution of sanitary and combined sewer backup problems is resolution of the extensive and serious surface flooding problem. Much of the sewer backup problem is probably caused by the heavy load imposed by surface flooding, not necessarily by hydraulic constrictions in the sewer system.
4. In areas undergoing development, the stormwater system should be planned and designed so as to generally conform with natural drainage patterns.
5. To the extent feasible, gravity inflow and outflow should be used for D/R facilities in the interest of minimizing the cost and simplifying operation and maintenance. If a completely gravity-driven facility is not feasible, the next best alternative is gravity inflow with pumped outflow.
6. An integrated system of a few large, publicly owned and maintained D/R facilities is preferable to many, small, privately owned facilities.
7. The aesthetic and recreational aspects of stormwater management alternatives should be evaluated because of the proximity of potential facilities to residential neighborhoods, because of the land-intensive nature of D/R facilities, and because of the potential for cost-effective, multipurpose use—for example, stormwater control and recreation—of some facilities.

Much less often, objectives and standards are developed in great detail and presented in a more formal format. For example, over a period of two decades the Southeastern Wisconsin Regional Planning Commission (1976) developed objectives for water quantity and quality control facilities, land use, sanitary sewer systems, and other physical elements of the planning region. One of the water control facility objectives is "an integrated system of drainage and flood control facilities and floodland management programs which will effectively reduce flood damage under the existing land use pattern of the watershed and promote the implementation of the watershed land use plan, meeting the anticipated runoff loadings generated by the existing and proposed land uses." Thirteen specific standards support this single objective. Some of these standards, in greatly abbreviated form, are:

1. Bridges and culverts carrying freeways and expressways over perennial waterways should accommodate the 100-year-recurrence-interval discharge without overtopping the roadway.
2. New and replacement bridges and culverts should accommodate the 100-year-recurrence-interval discharge with no more than a 0.5-ft stage increase.
3. Dikes and floodwalls should convey the 100-year discharge with a freeboard of at least 2 ft.

4. Upstream or downstream increases in flood discharges and stages caused by new channel modifications, dikes, floodwalls, and other facilities should be offset by complementary facilities for the storage and movement of floodwaters.
5. Land acquisition, easements, floodplain regulations, and other nonstructural flood mitigation measures should encompass at least the 100-year-recurrence-interval floodplain.

12.6 STEP 2: CONDUCT INVENTORY

Planning for the future of a watershed requires an appreciation of the past and an understanding of the present. Data and information obtained and organized during the inventory provide the factual basis for the surface water master plan.

Categories of Data and Information

Chapter 3 introduces the idea of organizing all hydrologic and hydraulic data in one of the following categories: (1) completed or ongoing studies, (2) natural resources data, and (3) infrastructure data. These data categories can also be used to organize the larger mass of data and information typically collected during the inventory phase of a master planning program. Furthermore, most of the hydrologic–hydraulic data types, sources, and uses discussed in Chapter 3 and presented in Table 3.1 are directly applicable to the master planning effort and should be included in the inventory phase of the master plan. The calibration section of Chapter 10 includes, in Table 10.3, an additional enumeration of data types and sources. A similar set of data types and sources is presented in Table 4.1.

However, the inventory effort in the master plan requires even more data and information. Examples of natural resource data and information not included in the previously noted tables, but often pertinent to master planning are groundwater quality and identification of areas having unique or special natural, historical, recreational or aesthetic values. An example of infrastructure data and information not included in the other listings in this book, but usually useful in master planning programs, is enumeration of outdoor recreation facilities and needs.

Management of the Inventory Phase

The inventory phase of a master planning program is usually one of the most costly in terms of staff time and, therefore, dollars. Even when existing sources of data and information are plentiful, well organized, and are put to optimum use, labor-intensive, time-consuming, and otherwise costly field work and activities are usually required. For example, the field survey and related office work needed to obtain the physical dimensions of one bridge for subsequent hydraulic analysis could easily exceed several engineer and technician hours and cost in excess of $300. As another example, labor and equipment required for the installation and first-year operation of a stream gauging station, exclusive of water quality monitoring, could easily exceed $10,000.

An axiom of master planning is that whereas it is difficult to obtain the personnel and monetary resources needed to conduct the inventory, it is equally difficult to manage the inventory and, in particular, to know when to stop the effort. The test that should be applied to each task in the inventory is whether or not the resulting data and information are needed for the planning program.

Inventory work can be monotonous and tedious, and there may be a tendency to compromise consistency and quality. Because the inventory provides the factual basis for all subsequent efforts, the inventory must be carefully managed to assure maintenance of an acceptable level of quality throughout.

Use of a digital computer-based land data management system like that described by Walesh et al. (1977) might be warranted in a master planning project. Such systems are designed to store, retrieve, analyze, and display land data. In this context, land data are defined as all types of natural resource and cultural feature data having an areal characteristic. Walesh et al. (1977) describes one project in which a 136-mi^2 watershed was partitioned into about 35,000 cells each 2.5 acres in size. Up to a dozen types of land data, such as civil division identification, subbasin identification, hydrologic soil type, wildlife habitat with value rating, and land use, were loaded into each cell comprising the land data management system. The computerized land data base was subsequently used to assemble tabular data, delineate environmentally sensitive corridors as defined by various combinations of land data types, identify lands most likely to require erosion control measures, prepare input for hydrologic–hydraulic computer models, and draw artwork for the project report. Ideas and examples presented by Hancock and Heaney (1987) suggest the possibility of a land data management system being programmed into spreadsheet software.

One way to optimize the inventory phase is to begin the subsequent analysis and forecast phase as soon as possible after starting the inventory. By so doing, an iterative process can be established involving the collection of data and information and the use of these data and information. As soon as the inventory begins and initial data are available, analysis of these data should begin. Immediate use of data will not only serve to expedite the overall planning effort but is likely to direct useful feedback into the inventory program, thus providing an opportunity for favorable adjustments.

Documentation

The inventory effort, with emphasis on findings, should be presented in at least summary form in the planning report. All sources should be carefully and completely cited. Documentation of the inventory effort in the planning report serves two functions. First, documentation helps to establish the credibility of the overall planning effort by demonstrating that the master plan is built on a foundation of substantiated data and information. Second, documentation in a public or at least readily accessible report makes the data and information and the sources of that data and information available to other professionals for subsequent planning and design efforts.

12.7 STEP 3: ANALYZE DATA AND PREPARE FORECASTS

The analysis and forecast phase of the master planning process uses the data and information obtained and organized under the inventory phase. The planning team must resist the temptation to tabulate but not evaluate, that is, collect and present data but fail to analyze the data for understanding of past, current, and hypothetical future conditions.

The two purposes of this phase are to understand the present state of surface water resources in the watershed and to formulate future conditions or scenarios. Examples of issues raised and questions asked during the analysis and forecast phase are:

1. In the absence of any additional special controls, how will land development progress in the watershed? What other scenarios are possible?
2. What are the location, type, severity, and cause of historic flooding? Assuming that no mitigation efforts are instituted, to what extent will flooding problems be aggravated as a result of future land development scenarios in the watershed?
3. What are the location, type, severity, and cause of surface water pollution? What are the relative influences of point and nonpoint sources of pollution? What water uses are inhibited or prevented? Assuming that no corrective measures are implemented, to what extent will surface water pollution problems be aggravated as a result of various future development scenarios?

Tools and Techniques

Numerous and varied tools and techniques are used to analyze data and prepare forecasts. Many of the tools and techniques associated with the water resources field are described in or are the principal subject of earlier chapters in this book. Examples are impact of urbanization studies (Chapter 2), hydrologic analysis (Chapter 3), floodplain and floodway delineation (Chapter 4), hydraulic analysis (Chapter 5), computation of average annual monetary flood damage (Chapter 6), estimation of nonpoint-source pollution loads (Chapter 7), detention/retention analysis (Chapter 8), and use of computer modeling (Chapter 10).

Other tools and techniques, originating outside the water resources field but having application to urban surface water master planning, are described elsewhere. For example, Howe (1971) presents various methods for analyzing the relative value of recreation facilities and natural areas. Included are use of surrogates such as travel costs to a recreational facility as a measure of recreational benefits and application of numerical uniqueness indices for evaluating the overall aesthetic value of a natural area.

In addition to requiring the use of many and varied sophisticated tools and techniques, the analysis and forecast effort is usually complex because it requires integration of many diverse factors. For example, and as stated elsewhere (SEWRPC, 1976), delineation of floodplains in an urbanizing area, one of the usual analysis and forecast efforts, requires consideration and integration of the following factors: "precipitation characteristics; relationship between basin morphology and runoff; effect of urbanization and soil properties on runoff volume and timing; effect of hydraulic characteristics of the stream network on streamflow; relationships between streamflow, flood stage, and frequency of flood occurrence; seasonal influence and influence of floodland storage and conveyance."

In selecting analytic tools and techniques for the analysis and forecast, the planning team should resist the tendency to either "overkill" or to "oversimplify." That is, the tools and techniques to be used should be matched to the nature of the planning area. For example, the surface water system, the infrastructure, and existing and potential surface water related problems in most watersheds are usually of sufficient complexity to require the use of hydrologic, hydraulic, and other computer modeling techniques. However, occasionally, plans are prepared for small watersheds having relatively simple surface water systems and an uncomplicated infrastructure. In such cases, manual methods, such as the rational method or TR55, may be adequate for hydrologic and hydraulic analyses and forecasts.

Planning Period

By definition, a plan must be designed to meet conditions up to some future point in time. Several factors should be considered in selecting the planning period. The first factor is the expected economic life of major public works and other facilities typically recommended in surface water master plans. The planning period should be of sufficient duration to assure that essentially full benefit will be derived from the facilities throughout their useful life. That is, the need for the facilities or the ability of the facilities to meet demands should not terminate too early in their useful life. Linsley and Franzini (1979) provide guidance on the average useful or physical life of various hydrologic–hydraulic works. However, as noted by James and Lee (1971), Linsley and Franzini (1979), and others, the economic life of a facility may be less than its physical life. The expected economic life should govern in planning.

The second factor is the period over which forecasts will be reasonably accurate. The accuracy of forecasts is likely to decrease as the planning period increases. As noted by Howe (1971) and others, inaccuracy of long-term forecasts coupled with the small present values of benefits and costs far in the future, tends to favor shorter planning periods.

A third factor is the target date or design year for relevant existing plans. For example, a flood control master plan prepared in 1984 for a 6.5 mi^2 area in and near Valparaiso, Indiana (Donohue & Associates, 1984), used the year 2000 as the target data. This year was selected because the previously adopted land use plan, to which the flood control plan was closely linked, used the year 2000 as the design year.

12.8 STEP 4: FORMULATE ALTERNATIVES

Formulation of alternatives is the most important part of the master planning process because the substance of the recommended plan originates in this step. Alternative formulation is a partly creative, partly systematic effort where alternatives are conceptualized, screened, and if promising, further developed. The essential conceptual, technical, economic, environmental, financial, legal, administrative, political, and other features of each alternative are examined as illustrated in Figure 12.7.

Chapter 11 provides a "shopping list" of structural and nonstructural measures available for management of the quantity and quality of surface water. Measures presented in Chapter 11 and in other similar listings may serve as a starting point for formulating alternatives. But the alternative formulation step should not be constrained by Chapter 11 measures or other enumerations of management measures. All such lists should be point of departures for a creative effort to find potentially workable remedial and preventive measures.

Analyze Data and Prepare Forecasts

Numerous and varied tools and techniques are available to develop, qualify, and quantify alternatives. Many of these tools and techniques were described in earlier chapters of this book. Examples are hydrologic analysis (Chapter 3), floodplain and floodway delineation (Chapter 4), hydraulic analysis (Chapter 5), computation of average annual monetary flood damage and flood mitigation benefits (Chapter 6), detention/retention facility sizing (Chapter 8), and use of computer modeling (Chapter 10).

Additional tools and techniques, many of which originate outside the water resources field but have application to urban surface water master planning, are available. Examples of available tools and techniques include use of basic engineering economics (e.g., James and Lee, 1971) and application of linear and dynamic programming (Aguilar, 1973; Helweg, 1985; Loucks et al., 1981). Howe (1971) and Linsley and Franzini (1979) discuss using the difference in cost of the next-best feasible alternative and the alternative under consider-

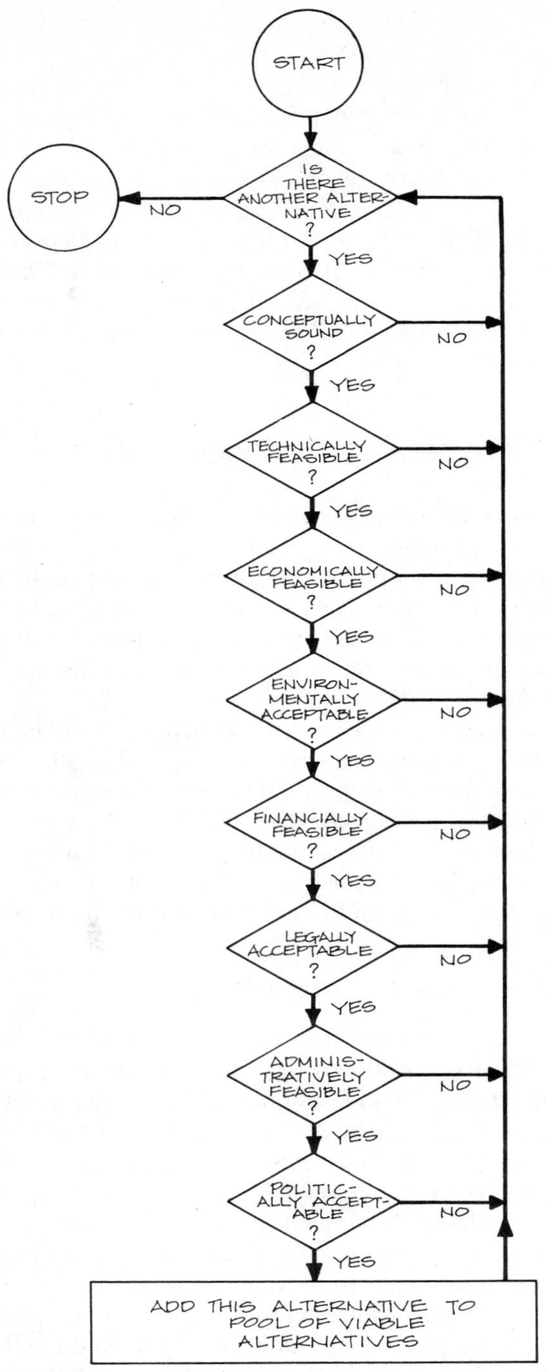

FIGURE 12.7 Development of an alternative *(Source:* Adapted from Southeastern Wisconsin Regional Planning Commission, 1976)

ation as a measure of benefits of the alternative under consideration. As noted earlier, Howe (1971) also mentions use of surrogates to quantify benefits and use of uniqueness indices to quantify aesthetic values of natural areas.

Conceptual Feasibility

In conceptualizing and beginning to develop alternatives, techniques such as brainstorming, "back-of-the envelope calculations," reference to fundamentals such as conveyance and storage concepts, and combining features of other alternatives can be effective. Premature attention, at the concept level, to details, potential constraints, and proven measures can stifle the creativity needed to explore all possibilities fully.

Although the conceptual evaluation indicates otherwise, sometimes it is necessary for a planning team to develop further an alternative that is not conceptually sound. Certain officials or the public at large may misperceive the feasibility of a particular alternative, even though professional judgment indicates that the alternative is technically, economically, or otherwise infeasible. For an example, refer to the section titled "Effect of Minor Channel Modification on Flood Stages" in Chapter 10.

Technical Feasibility

As a potential structural management alternative moves from the concept stage, its location must be determined, the facility must be sized in accordance with the established objectives and standards, and technical feasibility must be determined. In some cases, computer modeling will be needed to determine the overall size and principal physical features of alternatives as illustrated by examples presented in Chapter 10. Occasionally, simple manual computations can be used to size and determine the technical feasibility of a potential alternative.

If the master planning effort was motivated by a recent flood, and many are, elected officials and the public will usually want to know how the recommended plan would function if the flood would occur again. Development and testing of each alternative and of the recommended plan should include the necessary analysis to answer the inevitable question about recent major floods. A description of expected performance of the recommended plan, assuming a repeat of the historic flood or floods, should be included in the planning report.

Economic Feasibility

The next logical test is economic feasibility. At one end of the spectrum, this might be a value judgment as to whether or not the benefits that would be derived from a particular facility or approach would warrant the cost. At the other end of the spectrum, a formal benefit-cost analysis might be conducted using the methodology presented in Chapter 6.

Determination of economic feasibility requires a preliminary estimate of costs and, to the extent possible, benefits. As a particular alternative successfully passes subsequent tests of environmental, financial, legal, administrative, and political feasibility and acceptability, cost estimates are likely to be refined. However, estimating costs for an alternative in a master plan cannot and need not be done with the same accuracy as estimating costs for a designed facility for which plans and specifications have been prepared. In planning, cost estimates need to be of sufficient accuracy to guide decision making.

If, subsequent to completion of a master plan, an alternative goes through the design process, additional site data acquired as part of the design and additional facility detail resulting from the design are likely to affect costs. Accordingly, a significant contingency—15 percent or more—should be applied to planning-level cost estimates. Provisions should also be made for engineering and administrative cost associated with designing and constructing structural management measures and implementing nonstructural measures.

Costs estimates appearing in a planning report for various alternatives or recommended facilities should be carefully qualified. Factors that should be addressed include the year of the cost estimate, the magnitude of the contingency, allowances for engineering and administrative costs, and other factors, such as the passage of time, which are likely to affect costs. The following statement taken from a planning report (Donohue & Associates, 1985) illustrates one way to qualify the cost estimates: "Costs presented in this report reflect 1984 conditions. Total costs are the sum of land acquisition and construction costs, and engineering and administrative costs calculated as 10 percent of the sum of all preceding costs. Construction costs were estimated based on unit costs and quantities for each project component. . . . Actual construction costs will be established at the time of bidding and be determined by the size and location of a facility or group of facilities to be constructed under one contract, the construction market, and the availability of materials. With the passage of time, inflation and other construction market pressures may affect costs."

An example of the cost estimate for a detention facility component of a recommended watershed plan (Donohue & Associates, 1984) is presented in Table 12.1. In addition to the contingency costs, estimates for plan components should tend toward conservatism, that is, should tend toward the high side. For example, in Table 12.1, earthwork was estimated at $4.00 per cubic yard based on the then-prevailing rates for excavation with a short haul off the site. This recommended plan element was designed, and construction began in 1985 and was completed in 1987. Actual excavation was accomplished at about $1.00 per cubic yard because of the coincidental and very favorable interrelationship between the excavation needs at the detention project and fill needs at a nearby highway project (Walesh et al., 1987, 1988).

The inability to perform a complete quantitative economic evaluation of an

TABLE 12.1 Costs for an Off-Channel Detention Facility Used to Illustrate a Format for Presenting Costs

Item	Unit	Quantity	Unit Cost (dollars)	Total Cost (thousands of dollars)
Land acquisition	acre	11	25,000	275
Site clearing	square yard	53,000	0.50	27
Earthwork	cubic yard	210,000	4[a]	840
Cunette (1300 ft long, 10 ft wide, 6 in. thick)	square yard	1,444	30	43
Place topsoil, seed, fertilizer, and mulch	square yard	53,000	1.50	80
Diversion from Smith Ditch				
Diversion structure	each	1	25,000	25
54-in. reinforced concrete pipe	lineal foot	650	190	124
Inlet structure	each	1	25,000	25
Diversion from county market				
Diversion structure	each	1	25,000	25
60-in. reinforced concrete pipe	lineal foot	800	210	168
Diversion from combined sewer area				
54-in. reinforced concrete pipe	lineal foot	1350	190	257
78-in. reinforced concrete pipe	lineal foot	1300	260	338
Inlet/outlet structure	each	1	25,000	25
Outlet sewer 18-in. reinforced concrete pipe	lineal foot	650	60	39
		Subtotal		2291
		Contingencies (15%)		344
		Subtotal		2635
		Engineering and administration (10%)		264
		Total		$2899

Source: Adapted from Donohue & Associates (1984).

[a]Excavate with short haul off site.

alternative should not lead to rejection of the alternative. Whereas the costs of alternative structural or nonstructural management measures can usually be estimated with sufficient accuracy for planning purposes, the benefits often defy monetary quantification. Therefore, a quantitative benefit-cost analysis cannot be completed. Benefits may be intangible, that is, not measurable in monetary terms, but nevertheless be real and significant.

Assume, for example, that one alternative under consideration in an urban setting is a multipurpose flood control and recreation facility consisting of a linear system of detention facilities interspersed with attractively landscaped active and passive recreation facilities and features. Annualized costs of land acquisition, construction, and operations can readily be estimated. However, total annual benefits would be virtually impossible to quantify in monetary terms even if flood control benefits are quantified. The aesthetic, recreational, and environmental benefits of the project would probably elude monetary quantification, but nevertheless could be very significant.

Environmental Acceptability

Positive and negative environmental effects should be determined. The word "environment" is used here in a broad sense to include water, land, flora, and fauna as well as the overall urban scene. For example, preventing floodplain development by acquiring floodplains and designating them for a public open space use may have several positive environmental impacts, such as preserving habitat for plant and wildlife, protecting a fishery, and preserving acoustical and visual buffers. Construction of a detention/retention facility by excavating a flat, barren, unattractive site might provide an opportunity to enhance the urban environment by providing relief, landscaping, and provision of outdoor recreation facilities.

Financial Feasibility

Each alternative should next be tested for financial feasibility. Economic feasibility does not necessarily mean financial feasibility. That is, a potential alternative may promise to provide benefits in excess of costs, however benefits and costs are determined, but there may be a very small likelihood of being able to finance the alternative.

For example, assume that a large regional detention/retention facility intended to accommodate future runoff from the watershed, consisting of all or parts of several communities, is being considered. Assume further that the facility is technically and economically feasible and promises to be environmentally acceptable. If there is a low likelihood that the communities involved can work together to finance the facility, it may have to be dropped from further consideration.

Legal and Administrative Feasibility

Each potential alternative should be tested for legal and administrative feasibility. Promising alternatives may have to be dropped from further consideration because of legal or administrative barriers. For example, a technically, economically, and environmentally feasible but very costly system of conveyance works may be a possible solution to a serious flooding problem for several contiguous communities located wholly or partly within a particular watershed. Assume that success of the system depends on it being financed by establishing a surface water utility and by charging a user fee to all properties. If state law prohibits establishment of a district and charging a user fee, the alternative may have to be abandoned. As a variation, assume that state laws permit the establishment of a surface water utility and use of user fees. Studies may reveal that the project is administratively infeasible because there is no means of setting up and operating an intercommunity utility.

Political Acceptability

Political acceptability, the last test of an alternative, is difficult to assess. Assuming an ongoing, parallel public interaction program, as illustrated in Figure 12.2, the planning team should have a good sense of the likelihood of overall community acceptance of a particular alternative. The team must realize that alternatives which prove to be rational based on technical, economic, environmental, financial, legal, and administrative feasibility and acceptability may not be politically acceptable.

For example, in the early 1980s the city of Fond du Lac, Wisconsin, which is located on the shores of Lake Winnebago, the largest lake within the state, filled in a decorative park pond because of the near drowning of an unattended child. Several flood control planning and design efforts were under way in and near the city at that time. Because of the antipond political climate, detention/retention was probably not a feasible alternative.

Importance of Documentation

Regardless of which alternatives are ultimately selected to form the recommended plan, all possible alternatives should be fully documented, with costs and positive and negative features, in the planning report. Other potential alternatives omitted at the conceptual stage should be discussed in a general fashion.

Documentation of the preceding steps in the planning process, with emphasis on the alternatives considered, gives credence to the recommendations. Furthermore, plan implementation may take many years, during which unanticipated changes may occur in the planning area. Accordingly, previously eliminated alternatives may become attractive. The degree to which such alternatives are reconsidered may be dependent on the extent to which they are documented in the planning report and available in the public domain.

12.9 STEP 5: COMPARE ALTERNATIVES AND SELECT THE RECOMMENDED PLAN

Having formulated a set of possible alternative solutions to flooding, pollution, and other surface water–related problems, the master planning process is now directed at selecting a subset of alternatives to comprise a recommended plan. The essential features of each alternative must be presented to and compared in summary form for members of the planning team and eventually to other decision makers and the public at large.

Various graphic and tabular formats may be used. For example, Figures 12.8 and 12.9 are examples of simple graphic presentations in plan of a channel–levee alternative and a detention alternative (Donohue & Associates, 1985). Figure 12.10 is a cross-sectional representation of two channel-modification alternatives (SEWRPC, 1976). Note that the cross-sectional view uses the same vertical and horizontal scale and includes typical common objects to aid in understanding the relative height and size of key elements.

Graphic presentation should be supplemented with tabular summaries of key features of each alternative, such as size, costs, benefits, and positive and negative attributes. Table 12.2 is an example of a tabular summary of features for management measures considered for a river reach in Elm Grove, Wisconsin (SEWRPC, 1976). An example of a more modest approach is shown in Table 12.3, which presents the principal features and costs of four channel-modification alternatives intended to control streambank erosion along an 800-ft-long reach of channel (Donohue & Associates, 1979).

Alternatives should be compared on the basis of how well they achieve established objectives. Figure 12.11 illustrates an objectives versus alternatives matrix which uses symbols in the body of the matrix to show the degree to which each alternative satisfies the established objectives.

As the positive and negative features of each alternative are considered and compared with features of other alternatives, the possibility of composite alternatives often arises. For example, alternatives 8 and 9 listed in Table 12.2 are composite alternatives because they combine channelization, storage, and floodproofing.

Composite alternatives might achieve a technically improved solution or might be a means of obtaining support among decision makers and the public. As an example, major channelization may be reduced to a level acceptable to antichannelization groups if supplemented with upstream detention or retention. Or an extensive problematic floodproofing project may be reduced to manageable proportions if combined with the acquisition and removal of those buildings which are least amenable to floodproofing or least likely to be floodproofed. In the Madison, Wisconsin area a proposed single-purpose detention

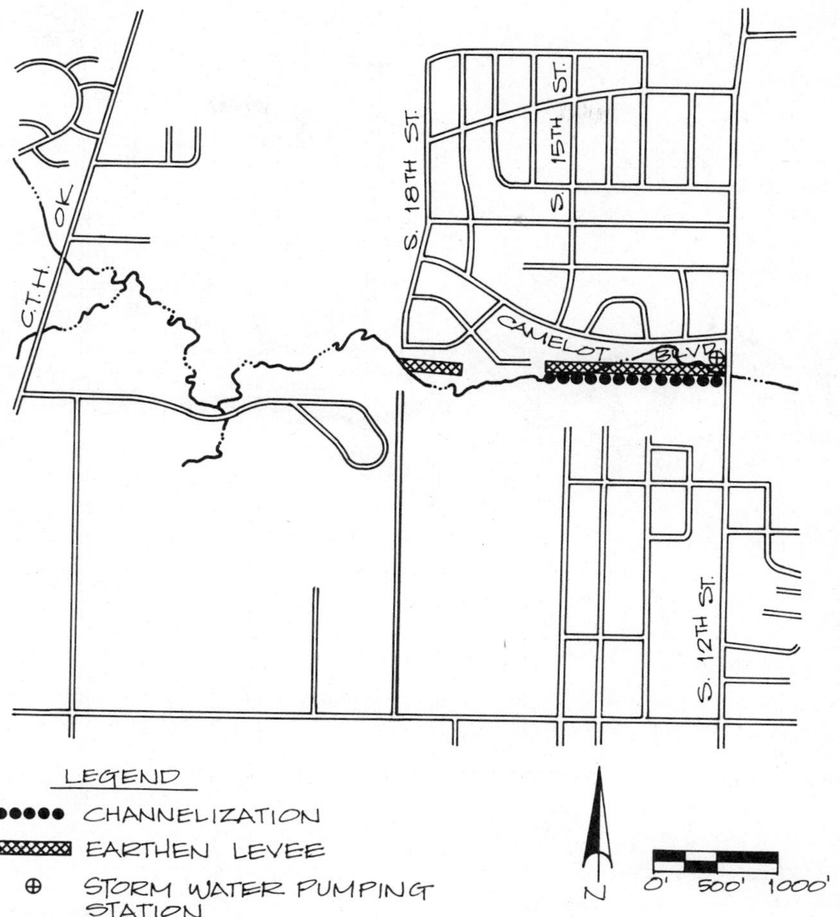

FIGURE 12.8 Plan view of a channelization and levee alternative (*Source:* Adapted from Donohue & Associates, 1985)

facility was initially unacceptable because of concern with sediment and debris accumulation at the site. The facility was rendered acceptable by combining it with two sedimentation basins at the point of entry of local streams (Raasch, 1982).

12.10 STEP 6: PREPARE PLAN IMPLEMENTATION PROGRAM

As indicated by the planning process set forth in Figure 12.2, upon completion of step 5, the master planning process has presumably determined what is to

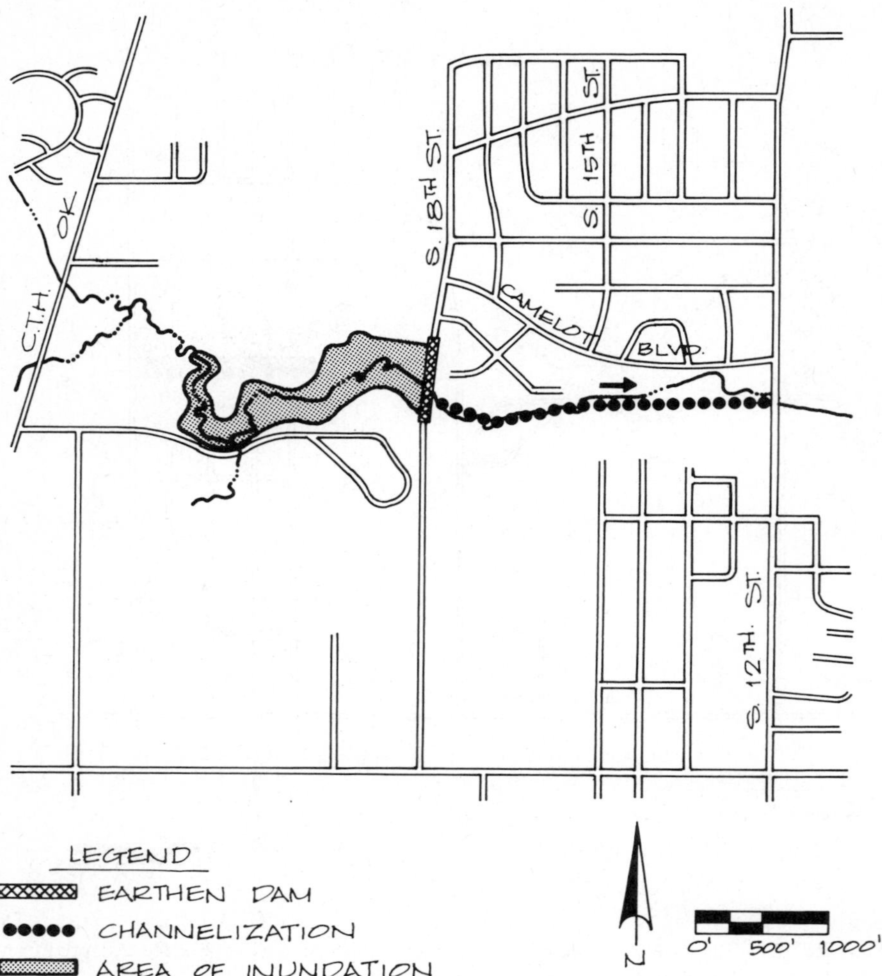

FIGURE 12.9 Plan view of a detention facility and channelization alternative *(Source:* Adapted from Donohue & Associates, 1985)

be done. As suggested by Figure 12.1, the questions of when the plan elements are to be implemented, who has primary responsibility for implementing them, and how the implementation is to be carried out, including the financing, remain to be answered. Step 6 addresses preparation of an implementation program. As stated by Machiavelli: "Any plan that does not carry with it its own plan for implementation, is worthless as a plan and should, therefore, be abandoned" (Kurz, 1973).

The master planning process contains a paradox. As emphasized throughout this and other chapters, urban SWM must be carried out on a watershed

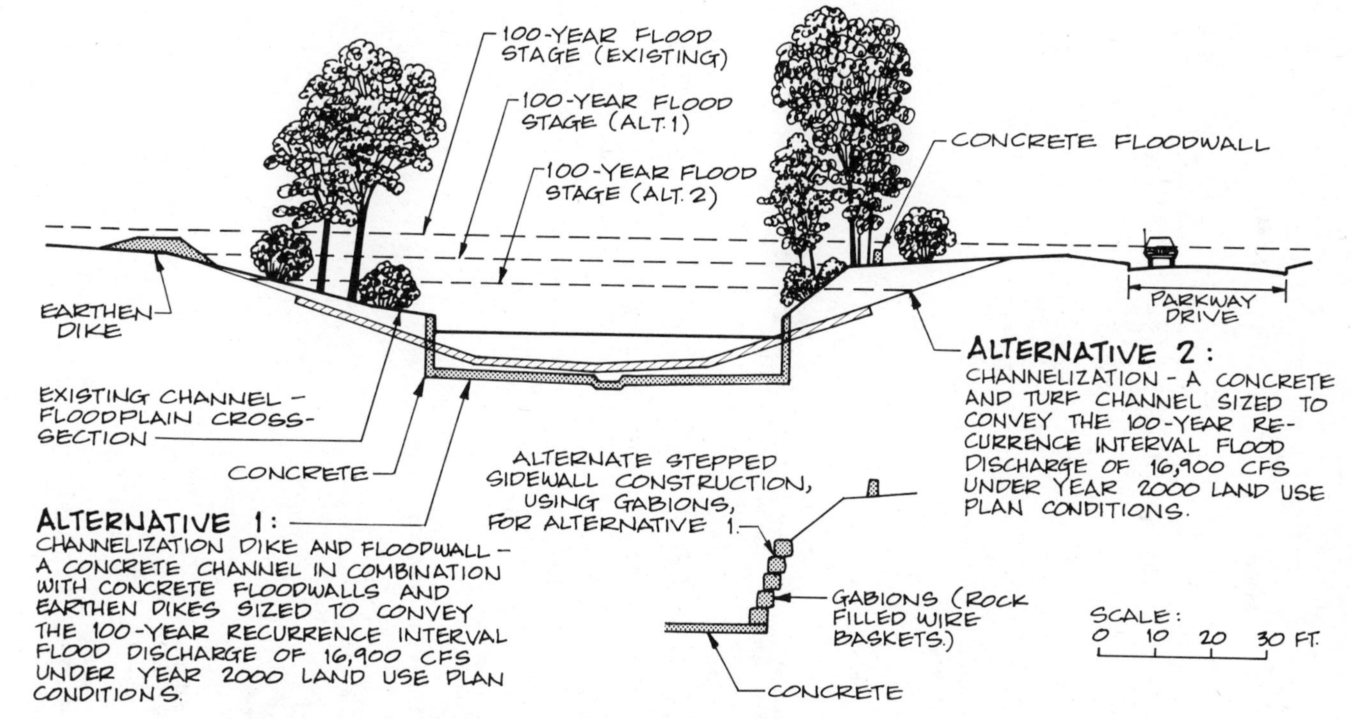

FIGURE 12.10 Cross sections illustrating existing conditions, a channelization–dike–floodwall flood control alternative, and a channelization alternative (*Source:* Adapted from Southeastern Wisconsin Regional Planning Commission, 1976)

TABLE 12.2 Principal Features and Costs and Benefits of Floodplain Management Alternatives for the Village of Elm Grove, Wisconsin

Alternative				Economic Analysis[a,b]								
				Capital Cost								
No.	Name	Description	Technically Feasible	Item	(thousands of dollars)	Annual Amortization of Capital Cost (thousands of dollars)	Annual Operation and Maintenance Cost (thousands of dollars)	Total Annual Cost (thousands of dollars)	Annual Benefit (thousands of dollars)	Annual Benefit Minus Annual Cost (thousands of dollars)	Benefit Cost Ratio	Economically Feasible
1	No action		Yes							−362.8	0.00	No
2	Detention storage	215-acre-ft detention reservoir on Dousman Ditch in the City of Brookfield.	Yes	Reservoir and associated works.	514.2[d]	32.6[d]	5.1[d]	37.7[d]	160.0[d]	122.3	4.24	Yes
3	Structure floodproofing and removal	Floodproof up to 156 residential and 24 commercial and other structures. Remove up to 19 residential structures.	Yes	Floodproofing.	580.0	118.8	0.0	118.8	362.8	244.0	3.05	Yes
				Removal.	1290.0							
				Subtotal	1870.0							
4	Major channel modification	2.00 mi of major channelization. Replace seven stream crossings.	Yes	Channeliz.	1865.1	232.3	1.0	233.3	362.8	129.5	1.56	Yes
				Bridges.	1796.1							
				Subtotal	3661.2							
5	Minor channel modification	1.92 mi of minor channel clean-out.	No									
6	Dikes and floodwalls	2.42 mi of dikes. 1.60 mi of floodwall. Replace 13 stream crossings. Install six storm water pumping stations.	Yes	Dikes.	692.0	303.3	11.2	314.5	362.8	43.3	1.15	Yes
				Floodwalls.	1574.5							
				Bridges.	2123.6							
				Pumping stations.	390.0[e]							
				Subtotal	4780.1							
7	Bridge and culvert alteration or replacement	Replace the four stream crossings causing the greatest backwater.	No									

TABLE 12.2 (Continued)

Nontechnical and Noneconomic Considerations[c]		Recommended?
Positive	Negative	
No expenditures.	Total flooding problem remains.	No
Potential to retain public space along Dousman Ditch primary environmental corridor.	Resolve only about one-half of the flood problem. May encourage new flood-prone development. Need for village of Elm Grove to coordinate design, construction, and financing with the city of Brookfield.	No
Immediate partial flood relief at discretion of property owners. Most of the costs could be borne by beneficiaries.	Complete, voluntary implementation unlikely and therefore left with significant residual flood problem. Overland flooding and some attendant problems remain. Some floodproofing is likely to be applied without adequate professional advice and, as a result, structure damage may occur.	No
Opportunity to develop an urban-oriented parkway through the business–commercial area.	Aesthetic impact in residential areas.	No
		No
	Aesthetic impact of visual barrier. Pumping station operation and maintenance are critical to effective functioning of the system.	No
		No

(continued)

TABLE 12.2 Principal Features and Costs and Benefits of Floodplain Management Alternatives for the Village of Elm Grove, Wisconsin

Alternative				Economic Analysis[a,b]								
				Capital Cost								
No.	Name	Description	Technically Feasible	Item	(thousands of dollars)	Annual Amortization of Capital Cost (thousands of dollars)	Annual Operation and Maintenance Cost (thousands of dollars)	Total Annual Cost (thousands of dollars)	Annual Benefit (thousands of dollars)	Annual Benefit Minus Annual Cost (thousands of dollars)	Benefit Cost Ratio	Economically Feasible
8	Channeliz. storage composite	215-acre-ft detention reservoir. 2.00 mi of major channelization. Replace seven stream crossings.	Yes	Reservoir. Channeliz. Bridges. Subtotal	514.2 1688.6 1553.7 3756.5	238.2	6.1	244.3	362.8	118.5	1.49	Yes
9	Storage–major channeliz.–intermediate channeliz.–floodproofing composite	215-acre-ft detention reservoir. 0.91 mi of major channelization. 1.14 mi of intermediate channeliz. Replace seven stream crossings. Floodproofing up to 105 residential structures.	Yes	Reservoir. Major channelization. Intermediate channelization. Bridges. Floodproofing. Subtotal	514.2 1055.1 122.9 1553.7 26.3 3272.2	207.6	6.6	214.2	362.8	148.6	1.69	Yes

Source: Adapted from Southeastern Wisconsin Regional Planning Commission (1976).

[a]Economic analyses are based on 1975 costs, an annual interest rate of 6 percent and a 50-year amortization period and project life. Average annual flood damage is $362,800.

[b]Economic analyses were not done for technically impractical alternatives.

[c]Presented only for technically and economically feasible alternatives and alternative 1.

[d]Based on the assumption that the total average annual cost of $46,200 for the detention reservoir would be shared by the village of Elm Grove and the city of Brookfield in proportion to the flood damage mitigation benefits derived by each community.

[e]Present worth cost based on a 25-year economic life.

TABLE 12.2 (Continued)

Nontechnical and Noneconomic Considerations[c]		
Positive	Negative	Recommended?
Opportunity to use the channelization component for development of an urban oriented parkway through the business–commercial area.	Aesthetic impact of channelization component in residential areas. Need for village of Elm Grove to coordinate design, construction, and financing of detention storage with the city of Brookfield. Upstream storage may encourage new flood-prone development.	No
Potential to retain public open space along Dousman Ditch primary environmental cooridor. Floodproofing component would provide immediate partial flood relief at discretion of property owners. With floodproofing component, some costs could be borne by beneficiaries. Opportunity to use the channelization component for development of an urban oriented parkway through the business–commercial area.	Storage component may encourage new flood-prone development. Complete, voluntary implementation of floodproofing unlikely and therefore left with significant residual flood problem. Overland flooding and some attendant problems remain with floodproofing. Some floodproofing is likely to be applied without adequate professional advice and, as a result, structure damage may occur. Although less than that of major channelization, the intermediate channelization component will have an aesthetic impact in residential areas. Erosion and attendant downstream deposition and maintenance requirements are likely for the intermediate channelization component. Need for village of Elm Grove to coordinate design, construction, and financing of detention storage with the city of Brookfield.	Yes

TABLE 12.3 Principal Features and Costs of Alternative Channel Improvements for Dutch Gap Creek in Fond du Lac, Wisconsin

	Alternative			
	1	2	3	4
Description	Reinforced-concrete box conduit 8 ft by 10 ft inside dimension. Fill in channel to top of banks. Riprap or gabions at downstream end.	Concrete-lined open channel 2 horizontal to 1 vertical side-slope. 28-ft top width. Riprap or gabions at downstream end.	Gabion-lined open channel 6-ft bottom width. 2 horizontal to 1 vertical side-slope. 35- to 40-ft top width.	Storm sewer and turf overflow channel 36-in. diameter pipe. 4 horizontal to 1 vertical side slope. 30-ft top width. Gabion walls at upstream and downstream ends. Grate at inlet to pipe and inlets along channel.
Capital cost[a]	\$644,000	\$344,000	\$214,000	\$126,000
Positive features	Minimal maintenance. Allows trees and shrubs to be replaced and visual barriers reestablished.	Minimal maintenance.	Relatively natural appearance. Existing vegetation at top of banks could remain.	Area would remain aesthetically attractive.
Negative features	Most of existing vegetation would be removed for construction. Aesthetic effect of stream would be lost.	Most of existing vegetation would be removed for construction. Starkness of concrete. High velocities increase safety hazard.		Maintenance of turf and inlets is critical to proper operation. Aesthetic effect of stream would be lost.
Recommended ?	No	No	No	Yes

Source: Adapted from Donohue & Associates (1979).

[a] 1979 costs.

OBJECTIVES	ALTERNATIVES: 1 DO NOTHING	2 FLOOD PLAIN MANAGEMENT	3 LEVEES TO PROTECT RESIDENTIAL AREA	4 LEVEES EXTENDING DOWNSTREAM	5 LEVEES FROM ROUTE 101 DOWNSTREAM TO RESIDENTIAL AREA	6 BYPASS CHANNEL SOUTH OF RESIDENTIAL AREA	7 BYPASS CHANNEL TO NORTHEAST	8 CHANNEL IMPROVEMENT	9 STORAGE DAM	10 RELOCATION AND FLOOD PLAIN MANAGEMENT	11 SETBACK LEVEES BOTH BANKS	12 SETBACK LEVEE LEFT BANK AND LEVEE RIGHT BANK
MAXIMUM PREVENTION OF FLOOD DAMAGES	○	⊘	●	●	●	●	⊘	⊘	●	⊘	●	●
LEAST DISRUPTIVE TO PHYSICAL ENVIRONMENT	●	●	⊘	⊘	⊘	⊘	⊘	⊘	⊘	⊘	○	○
LEAST HARMFUL TO FISH AND WILDLIFE	●	●	⊘	⊘	○	●	⊘	⊘	⊘	●	⊘	⊘
RATIO OF BENEFITS TO COSTS	NR	NR	●	⊘	⊘	⊘	⊘	○	○	○	○	○
MINIMIZES CONSTRUCTION COSTS	NR	NR	⊘	○	○	○	⊘	⊘	○	○	○	○
MINIMIZES LOCAL COSTS	NR	⊘	●	●	⊘	⊘	⊘	⊘	⊘	○	○	○
LEAST DISRUPTIVE TO PEOPLE	●	⊘	●	●	●	●	●	●	●	○	○	⊘
GUIDES FLOOD PLAIN USE	○	●	○	○	○	○	○	○	○	●	⊘	⊘

ACHIEVEMENT LEVEL

● HIGH ⊘ INTERMEDIATE ○ LOW NR - NOT RATED

FIGURE 12.11 Objectives versus alternatives matrix, indicating the extent to which each alternative would satisfy established objectives (*Source:* Adapted from Sargent, 1972)

wide basis because the watershed, and its constituent subbasins and subwatersheds, comprise the basic hydrologic–hydraulic–water quality system. Although the technical work is carried out on a watershed-wide basis, the plan implementation program must focus on individual government units and agencies in the watershed and on key officials and leaders within those units and agencies if implementation is to be achieved. The master plan must be prepared on a watershed basis but implemented largely on a local basis.

Who Is Responsible for Taking the Lead in Implementing Various Plan Elements?

Responsibility for taking the lead in implementing elements of a recommended master plan can reside with government units and agencies at all levels, advisory committees, and even private entities. The number of implementors can vary widely depending on the size and complexity of the master plan. If a flood control plan is prepared for a small watershed within a single community, the group of lead implementors might be a few community officials. In contrast, the plan for the 136-mi^2 Menomonee River watershed in southeastern Wisconsin identified 43 implementation units and agencies, ranging from local governments units to federal agencies, each having some responsibility for at least one of 30 plan elements. The 30 plan elements were categorized within the following broad areas: land use, floodplain management, and water quality management (SEWRPC, 1976).

For relatively complex master plans having numerous elements and numerous implementors, implementation responsibility can be summarized using a matrix like that presented schematically in Figure 12.12. If the master plan is relatively simple, implementors and their responsibility can be presented in narrative form within the planning report.

When Should Plan Elements Be Implemented?

Although all the recommended elements included in a master plan are important to the overall program, some elements have priority over others. Accordingly, the plan implementation program needs to identify when each element is to be implemented.

In the context of elements of a master plan, "when" can take on one of two meanings. First, it can mean the relative order in which plan elements are to be implemented. Second, "when" can mean the preceding plus absolute dates or milestones at which each element is to be initiated or completed. The first meaning, relative order, is the most common.

The relative order and possibly the absolute time of expected implementa-

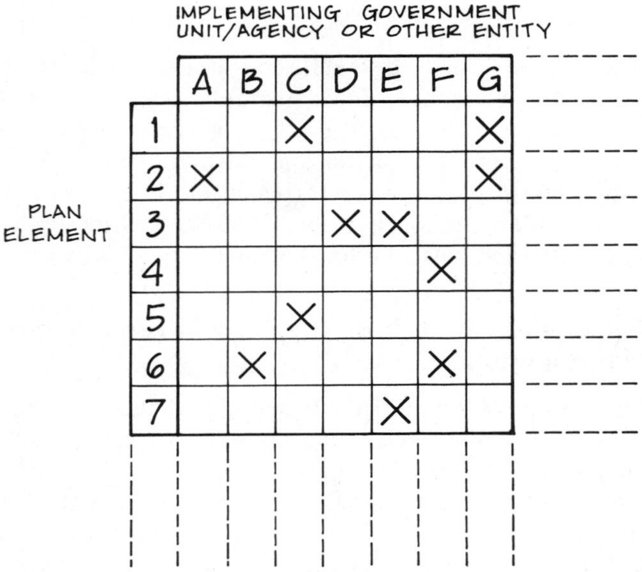

IMPLEMENTING GOVERNMENT UNIT/AGENCY OR OTHER ENTITY

PLAN ELEMENT	A	B	C	D	E	F	G
1			X				X
2	X						X
3				X	X		
4						X	
5			X				
6		X				X	
7					X		

FIGURE 12.12 Schematic of an implementation matrix for a master plan

tion are determined by consideration of three factors. The first or technical factor recognizes that operation of certain components of a surface water system is dependent on the existence of other components. For example, a detention facility cannot function as designed unless the conveyance or diversion works are in place to service the facility. Similarly, channel modifications are usually constructed in an upstream direction in order to fully utilize completed sections, that is, avoid backwater effects from downstream unimproved reaches.

The second ranking factor is cost-effectiveness. Higher priority should be given to those recommended plan elements likely to provide the greatest flood mitigation, water quality improvement, or other benefits. For example, a sedimentation basin–artificial marsh system intended to reduce the influx of suspended solids, debris, and nutrients into a heavily used urban lake would probably have a higher priority than a removal of sediment from the bottom of a smaller less used lake.

The third or opportunity factor recognizes that some recommended elements of the master plan will be implemented only if decisive action is taken immediately or at some other opportune time to capitalize on circumstances. For example, the master plan for a 6.5-mi^2 area within and contiguous with Valparaiso, Indiana (Walesh et al., 1987, 1988) recommended immediate design and construction of a particular detention facility partly because there was coincident interest in the development of an outdoor recreation facility in conjunction with the flood control facility.

If the master plan is not too complex, the implementation sequence or schedule can be presented in the report simply as a prioritized list. Another method is to place plan elements into categories of relative importance, such as first priority, second priority, and third priority. The implementation sequence may also be presented as a suggested schedule of capital and operation and maintenance costs by year. For complex master plans, a network diagram may be used, as illustrated schematically in Figure 12.13. Circles indicate plan elements, and interconnecting arrows show the sequence of and interdependence between plan elements.

How Will the Plan Be Implemented?

The "how" portion of the plan implementation program is more difficult to generalize than the "who" or "when" portions. Special opportunities that could have a bearing on the implementation of a plan should be identified. The usual emphasis is on possible sources of financing for the recommendations.

For example, the previously mentioned master plan for Valparaiso, Indiana (Donohue & Associates, 1984), recommended a second, single-purpose detention facility. As a possible means of implementing the facility, the plan recommended that the city negotiate with the Indiana Department of Highways to reach an agreement in which the material would be excavated from the site to form the detention facility and be used as needed fill for an adjacent highway

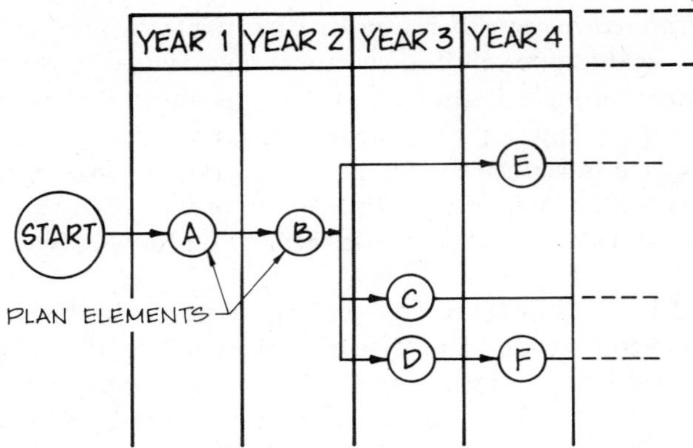

FIGURE 12.13 Schematic of a network diagram used to show the implementation schedule for a master plan

project. Within 4 years after completion of the plan, the suggested negotiations were successfully completed and the detention facility was designed and constructed.

Some communities, such as Fort Collins, Colorado, have successfully used a utility approach. User fees generate funds that are used for many aspects of urban SWM including implementing master plans (Cyre and Shearer, 1987; Krempel, 1988).

Other aspects of how a plan is to be implemented are clear identification of endorsements desired, agreements needed, and approvals and permits required. For example, implementation of a master plan for a watershed that encompasses all or parts of several communities may benefit from official endorsement by the governing bodies of each community, may require intercommunity agreements, and may require permits from state and federal agencies.

12.11 STEP 7: IMPLEMENT THE PLAN

Of all the steps in the surface water master planning process, plan implementation is the most unpredictable. The extent to which a comprehensive master plan is accepted and the enthusiasm with which public professional staff, elected officials, and the public will press for implementation will depend on two factors.

The first factor is the credibility of the plan as determined by the quality of the technical work and the thoroughness of the public involvement program carried out during the planning process. Kurz (1973) used a case study approach to evaluate five unsuccessful (i.e., nonimplemented) plans for various public facilities in the Puget Sound area of the state of Washington. Included were plans and proposals for expressways and bridges, a dam and reservoir, sewerage facilities, and an air transportation system. These plans were not implemented because of deficiencies, all of which eroded credibility of the plans. Identified deficiencies were lack of clearly stated objectives and standards, inadequate public involvement efforts, poor coordination with and between local government units and agencies, and a narrow approach to identification, development, and testing of alternatives.

The second factor influencing plan implementation is the persistence or reoccurrence of the flooding, pollution, or other surface water problems or the level of concern that past disasters will occur again.

After the master plan is completed, the first factor is history and out of control of the team that prepared the plan. Furthermore, after completion of the master plan, the team of professionals and technicians, usually consisting of governmental staff and their consultants that prepared the plan is typically disbanded. The plan must largely stand on its own unless the professional staff of the responsible government unit or agency aggressively provides plan implementation services. That is, the professional staff can work to enhance the second factor.

Some examples of plan implementation activities include presentations in response to invitations from governmental bodies, environmental organizations, service clubs, and other interested groups; response to questions concerning the master plan or requests for data and information generated during the plan preparation process; and updating of the plan either periodically or as changing conditions may require. Plan implementation also may be advanced by involvement, hopefully successful, in legal challenges to plan findings or recommendations. Another means of encouraging plan implementation is to document measures or indicators of success resulting from implementation of plan elements.

The ultimate test of a master plan is the extent to which existing surface water problems have been mitigated and the degree to which potential surface water problems have been prevented from occurring. As step-by-step implementation proceeds, each achievement should be strategically publicized. Most elements of surface water master plans are paid for with public funds, and the public should be informed about the return on their investment.

12.12 ECONOMICS OF MASTER PLANNING

Experience suggests that the benefits of master planning for urban surface water resources usually exceeds the costs. However, little literature is available in the area of the economics of planning; that is, the costs and benefits of preparing the plans are not documented. As a means of encouraging consideration of master planning, some experienced-based, intuitive ideas on the economics of planning are presented here. Helweg (1985) presents additional ideas and information under the overall theme of plan analysis.

Economy of Scale

Experience with the preparation of master plans for watersheds ranging in size from a few hundred acres to several hundred square miles suggests that the

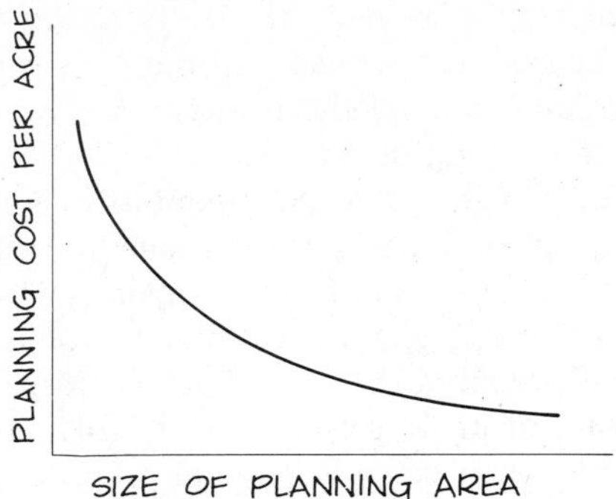

FIGURE 12.14 Conceptual relationship between the unit cost of preparing a master plan and the size of the planning area

relation between planning costs per unit area versus the size of the planning area follows the general relationship shown schematically in Figure 12.14. That is, empirical evidence indicates that there is a significant economy of scale operable in master planning in that the cost per acre of planning decreases as the size of the planning area increases. Economy of scale occurs because some aspects of the planning process need be performed only once regardless of the size of the watershed, or the amount of time required to complete certain work elements is not directly related to the watershed size.

The economy of scale suggests that rather than doing a series of master plans for small watersheds spread over an extended period, it is probably much more economical to do the master plan for the entire potential planning area or at least for groups of watersheds. If a government unit or agency or groups of government units or agencies are contemplating surface water master planning but wish to defer a final decision, a pilot study on one small watershed might be advisable to establish the overall approach. Once the approach is refined and acceptable to all concerned, master planning for the entire area or large portions of it should be undertaken to realize cost savings resulting from the economy of scale.

Apparent Benefits versus Costs

Structural and nonstructural surface water management measures can be very costly . For example, one lineal foot of major concrete channel modification may require a capital expenditure of a thousand dollars and 1 acre of flood-prone land acquisition may cost thousands of dollars. Because the potential expenditures are so large, master planning should be an important part of the process. A small investment in planning may yield a large return in terms of either reducing capital costs of management measures or improving the way the capital expenditures are made.

For example, assume that a master plan is prepared for a 4-mi^2 urban watershed at a total cost of $40,000. Using the previously cited unit cost, the total planning cost is equivalent to the capital cost of 40 ft of channel. That is, if the master plan can result in eliminating the need to construct just 40 ft of channel, the plan will have "paid for itself."

To reiterate, the costs of preparing master plans for urban surface water facilities tend to be very small compared to the capital costs of structural and nonstructural management measures typically implemented in urban areas, usually without the benefit of master planning. Therefore, the expenditure for planning is very likely to yield a large benefit in terms of either reduced costs of management measures or more effective use of funds.

REFERENCES

Aguilar, R. J., *Systems Analysis and Design in Engineering, Architecture, Construction, and Planning*. Prentice-Hall, Englewood Cliffs, NJ, 1973.

American Society of Civil Engineers, Committee on Social and Environmental Objectives of the Water Resources Planning and Management Division, *Social and Environmental Objectives in Water Resources Planning and Management*. ASCE, New York, 1984.

Brown, L. R., "Are River Basins the Best Framework for Water Resources Planning, Development and Management?" *Water Resour. Bull.* **8**(2), 401–403, April (1972).

Connor, D. M., "Transforming a Meeting from Confrontation to Cooperation." *Civ. Eng. (N.Y.)* February, **47**(2), 57–59 (1977).

Costello, L. S., "Urban Water Resources—Some Planning Fundamentals." *J. Urban Plann. Dev. Div., Am. Soc. Civ. Eng.* **100**(UP1), 57–71 (1974).

Cyre, H., and J. S. Shearer, *Stormwater Management Financing*. Presented at the 14th Annual Water Resource Planning and Management Conference-ASCE, Kansas City, MO, March 1987.

Debo, T., *Urban Drainage Programs and Ordinances,* Proc. Southeast Conf. Urban Stormwater Manage. Sponsored by the Water Institutions and Water Agencies of the Southeast States, Raleigh, NC, 1979.

Donohue & Associates, Inc., *Dutch Gap Improvements—Preliminary Engineering, City of Fond du Lac, WI,* December 1979.

Donohue & Associates, Inc., *Smith Ditch Lagoon No. 1 and Hotter Lagoon Investigation, City of Valparaiso, IN,* June 1984.

Donohue & Associates, Inc., *Flood Control Plan for Fisherman's Creek Watershed, City of Sheboygan, WI,* February 1985.

Grigg, N. S., *Water Resources Planning*. McGraw-Hill, New York, 1985.

Grigg, N. S., *Urban Water Infrastructure: Planning, Management and Operations*. Wiley, New York, 1986.

Hancock, M. C., and J. P. Heaney, "Water Resources Analysis Using Electronic Spreadsheets." *J. Water Resour. Plann. Manage.,* **113**(5), 639–658 (1987).

Helweg, O. J., *Water Resources Planning and Management*. Wiley, New York, 1985.

Hillegass, T. J., C. C. Schimpeler, and W. L. Grecco, "Community Decision Structure and Urban Planning Process." *J. Urban Plann. Dev. Div., Am. Soc. Civ. Eng.* **96** (UP1), 17–22 (1970).

Howe, C. W., *Benefit-Cost Analysis for Water System Planning,* Water Resour. Monogr. No. 2. Am. Geophys. Union, Washington, DC, 1971.

James, L. D., and R. R. Lee, *Economics of Water Resources Planning.* McGraw-Hill, New York, 1971.

Krempel, R. E., "Stormwater Management by Utility Approach," edited by S. R. Abt and J. Gessler, pp. 1234–1239. Proc. 1988 Hydraulic Engineering Conf., Colorado Springs, CO, ASCE, New York, 1988.

Kurz, J. W., "Transformation of Plans Into Actions." *J. Urban Plann. Dev. Div., Am. Soc. Civ. Eng.* **99** (UP2), 184–191 (1973).

Lau, D. H., Donohue and Associates, Inc., personal communication, February 1988, Itasca, IL.

Linsley, R. K., and J. B. Franzini, *Water Resources Engineering,* 3rd ed. McGraw-Hill, New York, 1979.

Loucks, D. P., J. S. Stedinger, and D. R. Haith, *Water Resource Systems Planning and Analysis.* Prentice-Hall, Englewood Cliffs, NJ, 1981.

McPherson, M. B., *Urban Runoff Control Planning.* Urban Water Resour. Res. Counc., Am. Soc. Civ. Eng., New York, 1977.

Priscoli, J. D., "Conflict Resolution in Water Resources: Two 404 General Permits." *J. Water Resour. Plann. Manage. Div., Am. Soc. Civ. Eng.* **114**(1), 66–77 (1988).

Raasch, G. E., "Urban Stormwater Control Project in an Ecologically Sensitive Area." *Proc. Int. Symp. Urban Hydrol., Hydraul., Sediment Control, 1982* 187–192 (1982).

Ringenoldus, J. C., "Water Resources and Regional Land-Use Plan." *J. Sanit. Eng. Div., Am. Soc. Civ. Eng.* **96**(SA6), 1311–1320 (1970).

Sargent, H. L., Jr., "Fishbowl Planning Immerses Pacific Northwest Citizens in Corps Projects." *Civ. Eng. (N.Y.)* **42**(9), September, 54–57 (1972).

Sheaffer, J. R., K. R. Wright, W. C. Taggart, and R. M. Wright, *Urban Storm Drainage Management,* Chapter 3. Dekker, New York, 1982.

Southeastern Wisconsin Regional Planning Commission (SEWRPC), *"A Comprehensive Plan for the Menomonee River Watershed,* Plann. Rep. No. 26. SEWRPC, Waukesha, WI, October 1976.

Southeastern Wisconsin Regional Planning Commission (SEWRPC), *Twenty Fifth Annual Report.* SEWRPC, Waukesha, WI, July 1986.

Walesh, S. G., J. A. Hardwick, and D. H. Lau, "Multi-purpose Flood Control-Recreation Facility: Case Study," edited by K. E. Bobay, pp. 126–137. Proc. Ninth Ann. Water Resour. Sym., Indiana Water Resources Association, Greencastle, IN, 1988.

Walesh, S. G., J. A. Hardwick, and D. H. Lau, *Cost Savings Through Multipurpose* for Resource Planning." *J. Water Resour. Plann. Manage. Div., Am. Soc. Civ. Eng.* **103**(WR2), 177–192 (1977).

Walesh, S. G., J. A. Hardwick, and D.H. Lau, *Cost Savings Through Multipurpose Flood Control-Recreation Facility: Case Study.* Presented at the 14th Annual Water Resources Planning and Management Division Conference-ASCE, Kansas City, MO, March 1987.

APPENDIX A

GLOSSARY

Antecedent moisture condition (AMC). Soil moisture at the onset of a rainfall event. The U.S. Soil Conservation Service defines AMC in terms of total rainfall during the 5 days immediately preceding the rainfall event. Dry AMC conditions mean less than 1.4 inches, average is 1.4 to 2.1 inches, and wet is greater than 2.1 inches.

Backwater. The increase in stage, or elevation of the water surface, on the upstream side of a bridge, culvert, other hydraulic structure, object, or deposit above that which would occur in the absence of the structure, object, or deposit.

Channel. Long and narrow continuous, low-lying area that, except in time of flooding, contains all of the flow of a river, stream, or other waterway.

Control structure. A structure at the discharge point of a detention/retention (D/R) facility intended to control the rate of discharge from the facility and the normal water level within the facility. With respect to flow control, the structure is usually designed to safely pass extremely high flows, minimize discharge during all other runoff events, and provide for slow discharge from the D/R facility after runoff events.

Convenience (minor) system. Those portions of the surface water control system intended to mitigate nuisance flooding during frequent rainfall events (e.g., up to about the 2-year recurrence interval). Example components of the convenience system are storm sewers, inlets and catch basins, backyard, sideyard and other swales, and building placement and grades.

Crest gate. A temporary or movable gate installed on top of a spillway crest to provide additional storage or prevent flow over the crest.

Critical depth. The flow depth at which specific energy is at a minimum for a given discharge in a particular open channel cross section through the channel. Critical depth corresponds to the apex of the specific energy curve. At critical depth, gravity and inertial forces are equal and the Froude number is 1.0.

Cunette. Longitudinal channel constructed along the center and lowest part of a channel or through a D/R facility and intended to carry low flows. Also called a *trickle channel.*

Detention facility. A surface water runoff storage facility that is normally dry but is designed to hold (detain) surface water temporarily during and immediately after a runoff event. Examples of detention facilities are: natural swales provided with crosswise earthen berms to serve as control structures, constructed or natural surface depressions, subsurface tanks or reservoirs, rooftop storage, and infiltration or filtration basins. See also *retention facility.*

Discharge probability relationship. A graph of annual instantaneous peak discharge (or other hydrologic quantity) on the vertical axis, versus probability and/or recurrence interval on the horizontal axis. The graph provides a means of estimating the flow that will be reached or exceeded in a given year at a specified probability, or a means of estimating the probability that a specified discharge will be reached or exceeded in a given year.

Divide. Imaginary line indicating the limits of a subbasin, subwatershed, or watershed.

Emergency (major or overflow) system. Those portions of the surface water control system intended to mitigate danger to human life and serious damage to buildings and other facilities in the event of a rare, major flood event. Example components of the emergency or major system are larger channels and storm sewers, street cross sections and longitudinal grades, building grades, and D/R facilities. These facilities, operating in combination with the convenience or minor system, are typically designed to convey or store up to the 100-year recurrence interval flow. The level of emergency protection to be provided may be limited by cost limitations or physical constraints.

Floodplain fringe. Portion of the floodplain outside of the floodway which is covered by floodwaters during the 100-year recurrence interval flood. It is generally associated with shallow, standing or slowly moving water rather than deep, rapidly flowing water.

Flood stage profile. A graph of flooding condition water surface elevation versus distance along a river or stream. The profile may correspond to an historic flood event or an event of a specified recurrence interval. The channel bottom, as well as bridges, culverts, and dams, are usually shown on the flood stage profile.

Flooding. Temporary inundation of all or part of the floodplain along a well-defined channel or temporary localized inundation occurring when surface water runoff moves via surface flow, swales, channels, and sewers toward well-defined channels. Flooding is not necessarily synonymous with flooding problems.

Flooding problem. Disruption to community affairs, damage to property and facilities, and danger to human life and health that occurs when land use is incompatible with the hydrologic–hydraulic system.

Floodplain. Wide, flat to gently sloping area contiguous with and usually lying on both sides of the channel. The floodplain is typically very wide relative to the channel. The floodplain is more subtle, that is, more difficult to observe than the channel, particularly in developing or developed urban areas, where the topographic discontinuity between the floodplain and adjacent higher ground is masked by urban development.

Floodproofing. Modifications to an existing or new structure or facility to minimize flood damage. Floodproofing may involve a combination of structural provisions, and changes or adjustments to properties and structures subject to flooding.

Floodway. A designated portion of the 100-year recurrence interval floodplain that will convey the 100-year flood discharge with specified small upstream and downstream stage increases. The floodway includes the channel and defines that portion of the floodplain not suited for human habitation because of the hazard to life and property. The floodway sometimes includes areas having depths and velocities in excess of established values.

Flow regulator. Device inserted in an inlet, catchbasin, or outlet control works, and intended to restrict flow into a receiving pipe or channel so as to prevent surcharge or overload of the receiving conveyance works. Temporary storage of water occurs immediately upstream of the flow regulator. The devices result in less flow passing the control section for a given head or head range. Several types of commercial flow regulators are available and other types are specially fabricated by the user. Also sometimes referred to by the more restrictive term, *inlet control device.*

Freeboard. The vertical distance from the design water surface up to the point of overtopping of a channel, dike, floodwall, dam or control structure. Freeboard is a safety factor intended to accommodate the possible effect of unpredictable obstructions, such as ice accumulation and debris blockage, that could increase stages above the design water surface.

Gabions. Rock-filled wire baskets used to protect and stabilize the bottom and side walls of channels.

Gauging station. A site on a stream where systematic observations of gauge height and discharge are obtained. The station usually has an automatic recording gauge for continuous or intermittent measurement of stages.

Hydraulics. Analysis of the physical behavior of water as it flows over the land surface; on streets and parking lots; within sewers and stream channels and on natural floodplains; under and over bridges, culverts, and dams; and through lakes and impoundments. Simply stated, hydraulics is the study of flow paths, velocities, and stages.

Hydraulic gradient. The change in head divided by the change in distance in the direction of greatest change.

Hydrograph. A graph of discharge, on the vertical axis, versus time, on the horizontal axis. A hydrograph shows the volume of the stream flow passing a point— the area beneath the curve—and the time-variation of discharge rate as it passes.

Hydrology. Study of the physical behavior of water from its occurrence as precipitation through its movement on or beneath the earth's surface; to its entry into sewers, streams, D/R facilities, lakes, reservoirs and the oceans; and its return to the atmosphere via evaporation. Simply stated, hydrology is the study of spatial and temporal changes in water volumes and discharge rates.

Hyetograph. A graph of rainfall intensity, on the vertical axis, versus time, on the horizontal axis. A hyetograph shows the volume of precipitation at a point—the area beneath the curve—and the time-variation of intensity.

Imperviousness. The portion of a subbasin, subwatershed, or watershed, expressed as a percentage, that is covered by surfaces such as roof tops, parking lots, sidewalks, driveways, streets, and highways. Impervious surfaces are important because they will not absorb rainfall and, therefore, cause almost all of the rainfall to appear as surface runoff.

Infiltration. Water, other than wastewater, that enters a sanitary sewer system (including sewer service connections) from the ground through such means as defective pipes, pipe joints, connections, or manholes. Infiltration does not include, and is distinguished from, inflow. Also, passage of water through the air-soil interface.

Inflow. Water, other than wastewater, that enters a sanitary sewer system (including sewer service connections) from sources such as roof leaders, cellar drains, yard drains, area drains, foundation drains, drains from springs and swampy areas, manhole covers, cross connections between storm sewers and sanitary sewers, catch basins, cooling towers, surface runoff, street wash waters, or drainage. Inflow does not include, and is distinguished from, infiltration.

Infiltration/inflow. The total quantity of water from both infiltration and inflow without distinguishing the source.

Inlet control device. See *Flow regulator.*

Major system. See *Emergency system.*

Manning roughness coefficient. A dimensionless coefficient used in Manning's equation to account for frictional losses in steady uniform flow.

Minor system. See *Convenience system.*

Overflow system. See *Emergency system.*

Piezometer. Nonpumping well used to measure the potentiometric surface at a point in an aquifer.

Recurrence interval. The average time interval, usually in years, between the occurrence of a flood or other hydrologic event of a given magnitude or larger. The reciprocal, or inverse, of the recurrence interval is the probability (chance) of occurrence, in any year, of a flood equalling or exceeding a specified magnitude. For example, a flood that would be equalled or exceeded on the average of once in 100 years would have a recurrence interval of 100 years and a 0.01 probability, or 1 percent chance of occurring or being exceeded in any year.

Retention facility. A stormwater storage facility that normally holds water at a controlled level to serve functions such as recreation, aesthetic, and water supply. Stormwater runoff is temporarily stored above the controlled stage. Examples of types of retention storage reservoirs are permanent ponds in residential and commercial areas and in open spaces. See also *detention facility.*

Riprap. A layer, facing, or protective mound of rock randomly placed to prevent erosion, scour, or sloughing of a structure or embankment. Also refers to the rock used to provide the protection.

Runoff curve number. A rainfall–runoff parameter commonly used in U.S. Soil Conservation Service hydrologic procedures. The larger the runoff curve number, the greater the percentage of rainfall that will appear as runoff. The runoff curve number is a function of soil type, land use, and land management practices.

Sedimentation basin (SB). A surface water runoff storage facility intended to trap suspended solids, suspended and buoyant debris, and adsorbed or absorbed potential pollutants that are carried by surface water runoff. The SB may be part of an overall multipurpose D/R facility.

Sluice gate. A gate which can be raised or lowered by sliding in vertical guides.

Spillway. A structure over or through which flood flows are discharged.

Stage. Elevation of the water surface above a given datum.

Subbasin. The smallest unit into which the land surface is subdivided for hydrologic studies. All the area within a subbasin discharges at one point.

Subwatershed. Drainage area composed of two or more subbasins.

Swale. A natural or constructed open channel usually lined with grass and having a mild longitudinal slope and a triangular or trapezoidal cross section with mild side slopes. Swales usually carry flows only during or immediately after rainfall or snowmelt events. Swales vary in size from small conveyances providing drainage along roadways and behind or between buildings to larger waterways. Constructed swales should have cross-section shapes and longitudinal and lateral slopes designed to carry the design flow without damage or unsightly erosion/sedimentation.

Thalweg. Line following the deepest part of a channel bottom.

Time of concentration. Time necessary for surface runoff to reach the outlet of the drainage area from the most remote point in the drainage area. The term "remote" is used to denote the point most remote in time, not necessarily distance, from the outlet of the drainage area.

Trash rack. A screen or bars located in the waterway at an intake to prevent the entry of floating or submerged debris.

Trickle channel. See *Cunette.*

Urban surface water management (SWM). Development and implementation of a combination of structural and nonstructural measures intended to reconcile the water conveyance and storage function of natural depressions, lakes, swales, channels, and floodplains with the space and related needs of an expanding urban population. Simply stated, urban SWM is everything done to remedy existing surface water problems and to prevent the occurrence of new problems. From a functional perspective, urban SWM consists of planning, design, construction, and operation functions.

Watershed. A drainage area composed of two or more subbasins or two or more subwatersheds.

Weir. A small dam or wall across a stream to raise the upstream water level.

Wetlands. Those areas where water is at, near, or above the land surface long enough to be capable of supporting aquatic or hydrophytic vegetation, and which have soils indicative of wet conditions.

APPENDIX B

INPUT DATA ON CODING FORM FOR ILUDRAIN EXAMPLE

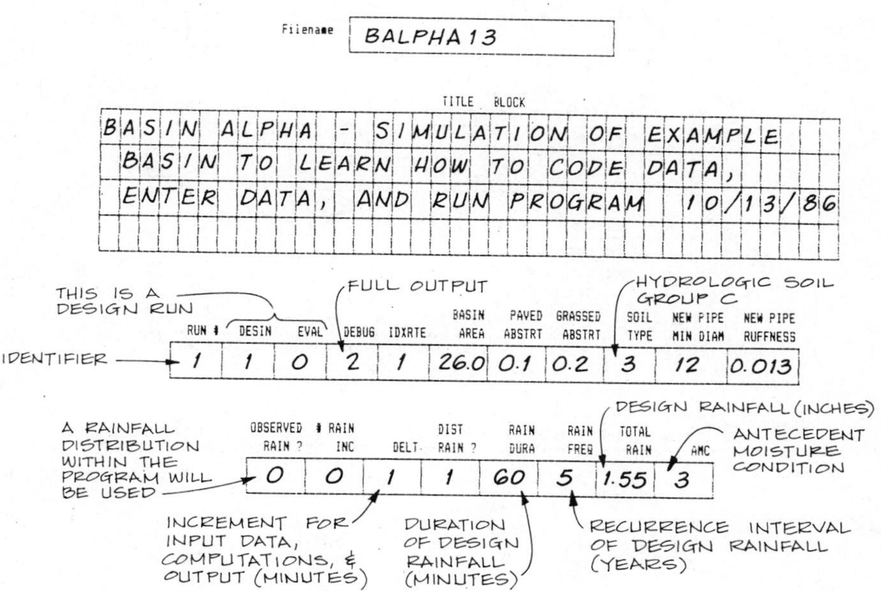

Filename: BALPHA13

TITLE BLOCK

BASIN ALPHA - SIMULATION OF EXAMPLE
BASIN TO LEARN HOW TO CODE DATA,
ENTER DATA, AND RUN PROGRAM 10/13/86

RUN #	DESIN	EVAL	DEBUG	IDXRTE	BASIN AREA	PAVED ABSTRT	GRASSED ABSTRT	SOIL TYPE	NEW PIPE MIN DIAM	NEW PIPE RUFFNESS
1	1	0	2	1	26.0	0.1	0.2	3	12	0.013

OBSERVED RAIN ?	# RAIN INC	DELT	DIST RAIN ?	RAIN DURA	RAIN FREQ	TOTAL RAIN	AMC
0	0	1	1	60	5	1.55	3

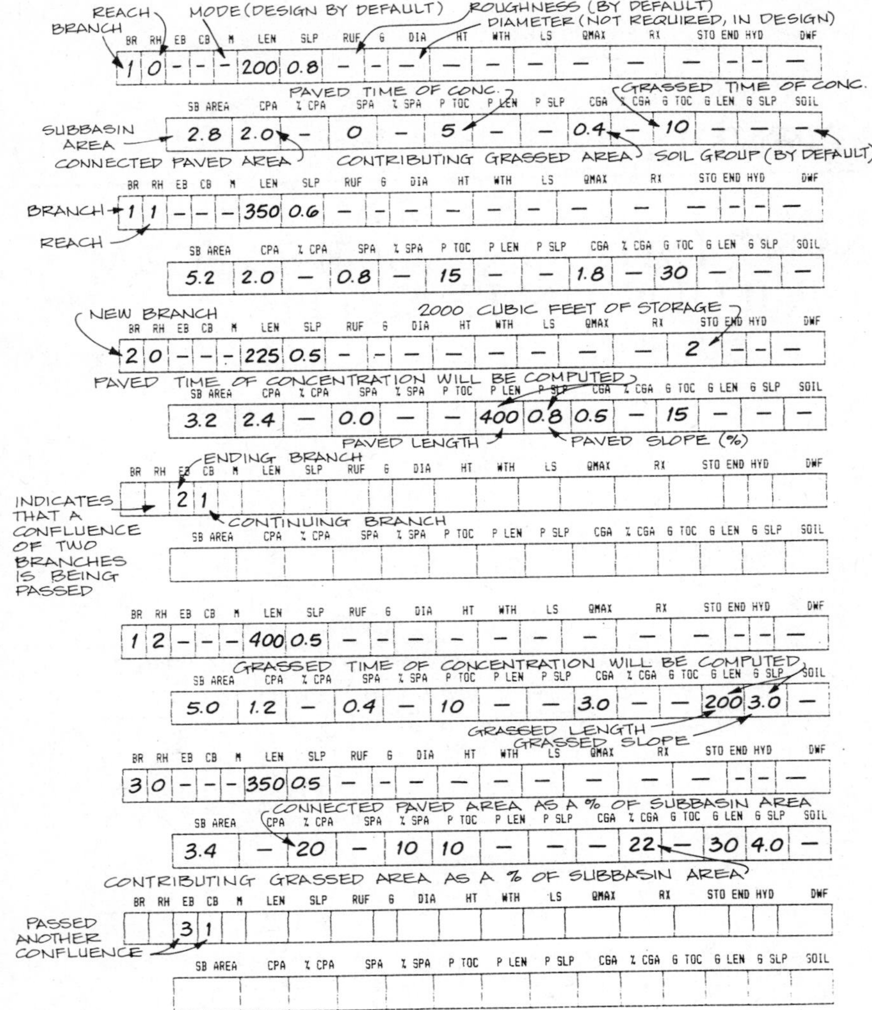

BR	RH	EB	CB	M	LEN	SLP	RUF	G	DIA	HT	WTH	LS	QMAX	RX	STO	END	HYD	DWF
1	0	–	–	–	200	0.8	–	–	–	–	–	–	–	–	–	–	–	–

SB AREA	CPA	% CPA	SPA	% SPA	P TOC	P LEN	P SLP	CGA	% CGA	G TOC	G LEN	G SLP	SOIL
2.8	2.0	–	0	–	5	–	–	0.4	–	10	–	–	–

BR	RH	EB	CB	M	LEN	SLP	RUF	G	DIA	HT	WTH	LS	QMAX	RX	STO	END	HYD	DWF
1	1	–	–	–	350	0.6	–	–	–	–	–	–	–	–	–	–	–	–

SB AREA	CPA	% CPA	SPA	% SPA	P TOC	P LEN	P SLP	CGA	% CGA	G TOC	G LEN	G SLP	SOIL
5.2	2.0	–	0.8	–	15	–	–	1.8	–	30	–	–	–

BR	RH	EB	CB	M	LEN	SLP	RUF	G	DIA	HT	WTH	LS	QMAX	RX	STO	END	HYD	DWF
2	0	–	–	–	225	0.5	–	–	–	–	–	–	–	–	2	–	–	–

SB AREA	CPA	% CPA	SPA	% SPA	P TOC	P LEN	P SLP	CGA	% CGA	G TOC	G LEN	G SLP	SOIL
3.2	2.4	–	0.0	–	–	400	0.8	0.5	–	15	–	–	–

BR	RH	EB	CB	M	LEN	SLP	RUF	G	DIA	HT	WTH	LS	QMAX	RX	STO	END	HYD	DWF
		2	1															

SB AREA	CPA	% CPA	SPA	% SPA	P TOC	P LEN	P SLP	CGA	% CGA	G TOC	G LEN	G SLP	SOIL

BR	RH	EB	CB	M	LEN	SLP	RUF	G	DIA	HT	WTH	LS	QMAX	RX	STO	END	HYD	DWF
1	2	–	–	–	400	0.5	–	–	–	–	–	–	–	–	–	–	–	–

SB AREA	CPA	% CPA	SPA	% SPA	P TOC	P LEN	P SLP	CGA	% CGA	G TOC	G LEN	G SLP	SOIL
5.0	1.2	–	0.4	–	10	–	–	3.0	–	–	200	3.0	–

BR	RH	EB	CB	M	LEN	SLP	RUF	G	DIA	HT	WTH	LS	QMAX	RX	STO	END	HYD	DWF
3	0	–	–	–	350	0.5	–	–	–	–	–	–	–	–	–	–	–	–

SB AREA	CPA	% CPA	SPA	% SPA	P TOC	P LEN	P SLP	CGA	% CGA	G TOC	G LEN	G SLP	SOIL
3.4	–	20	–	10	10	–	–	–	22	–	30	4.0	–

BR	RH	EB	CB	M	LEN	SLP	RUF	G	DIA	HT	WTH	LS	QMAX	RX	STO	END	HYD	DWF
		3	1															

SB AREA	CPA	% CPA	SPA	% SPA	P TOC	P LEN	P SLP	CGA	% CGA	G TOC	G LEN	G SLP	SOIL

TRAPEZOID → G
ARBITRARY LARGE DEPTH → HT
BOTTOM WIDTH → WTH
DOWNSTREAM MOST REACH → END
LATERAL SLOPE (V:H) → LS
MAXIMUM ALLOWABLE FLOW → QMAX

BR	RH	EB	CB	M	LEN	SLP	RUF	G	DIA	HT	WTH	LS	QMAX	RX	STO	END	HYD	DWF
1	3	—	—	1	100	1.0	—	3	—	10	2	2	20	—	—	1	—	—

SB AREA	CPA	% CPA	SPA	% SPA	P TOC	P LEN	P SLP	CGA	% CGA	G TOC	G LEN	G SLP	SOIL
6.4	1.0	—	0.0	—	10	—	—	5.4	—	—	300	3.0	

BR	RH	EB	CB	M	LEN	SLP	RUF	G	DIA	HT	WTH	LS	QMAX	RX	STO	END	HYD	DWF

SB AREA	CPA	% CPA	SPA	% SPA	P TOC	P LEN	P SLP	CGA	% CGA	G TOC	G LEN	G SLP	SOIL

BR	RH	EB	CB	M	LEN	SLP	RUF	G	DIA	HT	WTH	LS	QMAX	RX	STO	END	HYD	DWF

SB AREA	CPA	% CPA	SPA	% SPA	P TOC	P LEN	P SLP	CGA	% CGA	G TOC	G LEN	G SLP	SOIL

BR	RH	EB	CB	M	LEN	SLP	RUF	G	DIA	HT	WTH	LS	QMAX	RX	STO	END	HYD	DWF

SB AREA	CPA	% CPA	SPA	% SPA	P TOC	P LEN	P SLP	CGA	% CGA	G TOC	G LEN	G SLP	SOIL

BR	RH	EB	CB	M	LEN	SLP	RUF	G	DIA	HT	WTH	LS	QMAX	RX	STO	END	HYD	DWF

SB AREA	CPA	% CPA	SPA	% SPA	P TOC	P LEN	P SLP	CGA	% CGA	G TOC	G LEN	G SLP	SOIL

BR	RH	EB	CB	M	LEN	SLP	RUF	G	DIA	HT	WTH	LS	QMAX	RX	STO	END	HYD	DWF

SB AREA	CPA	% CPA	SPA	% SPA	P TOC	P LEN	P SLP	CGA	% CGA	G TOC	G LEN	G SLP	SOIL

BR	RH	EB	CB	M	LEN	SLP	RUF	G	DIA	HT	WTH	LS	QMAX	RX	STO	END	HYD	DWF

SB AREA	CPA	% CPA	SPA	% SPA	P TOC	P LEN	P SLP	CGA	% CGA	G TOC	G LEN	G SLP	SOIL

APPENDIX C

PARTIAL OUTPUT FOR ILUDRAIN EXAMPLE

ILLDRAIN — URBAN DRAINAGE AREA HYDROLOGIC SIMULATOR
Version 2.1 Updated May 1985

TIME-SHIFT ROUTING METHOD ACTIVATED

Basin Alpha - Simulation of Example
Basin to Learn How to Code Data/
Enter Data/Run Program 10/13/86

60 1 MIN. INCREMENTS OF RAIN IN INCHES

RAINFALL PATTERN

PEAK VOL./MIN

0.000	0.062	0.062	0.069	0.074	0.074	0.076	0.076	0.068	0.067
0.061	0.053	0.053	0.046	0.046	0.037	0.031	0.030	0.026	0.026
0.023	0.022	0.021	0.019	0.019	0.016	0.015	0.015	0.016	0.015
0.016	0.012	0.012	0.012	0.012	0.012	0.012	0.011	0.011	0.011
0.011	0.011	0.011	0.010	0.010	0.010	0.010	0.010	0.010	0.010
0.010	0.010	0.010	0.010	0.010	0.010	0.010	0.010	0.010	0.010
0.010									

HUFF DISTRIBUTION

Histogram of Distributed Rainfall

Time since start:	0	12	24	36	48	60	Mn
Accumulated rain:	0	0.794	1.139	1.305	1.434	1.550	In
Percent of storm:	0	51.20	73.50	84.20	92.50	100.0	%

TOTAL RAIN = 1.55 INCHES

RUN NUMBER	BASIN AREA ACRES	TIME INCREMENT MINUTES	SOIL TYPE 1234=ABCD
1	26.00	1.0	3

TOTAL RAIN INCHES	FREQUENCY YEARS	DURATION MINUTES	AMC PCT	IMP ABS INCHES	PER ABS INCHES
1.55	5	60	67.	0.10	0.20

— BEGIN BRANCH 1 - REACH 0 —

PAVED AREA HYDROGRAPH (FOR UPSTREAM MOST SUBBASIN)

PEAK Q

0.000	0.000	0.576	2.231	3.998	5.775	7.588	8.826	8.810	8.640
8.330	7.787	7.245	6.705	6.209	5.630	5.103	4.554	4.089	3.624
3.286	3.069	2.843	2.657	2.471	2.294	2.139	2.015	1.938	1.860
1.860	1.779	1.699	1.618	1.538	1.457	1.457	1.432	1.407	1.383
1.358	1.333	1.333	1.314	1.296	1.277	1.259	1.240	1.240	1.228
1.215	1.203	1.190	1.178	1.178	1.172	1.166	1.159	1.153	1.147
1.147	0.918	0.688	0.459	0.229					

64 MINUTES OF RUNOFF

GRASSED AREA HYDROGRAPH (FOR UPSTREAM MOST SUBBASIN)

```
0.000 0.000 0.000 0.000 0.000 0.055 0.145 0.235 0.314 0.392
0.461 0.518 0.576 0.623 0.671 0.705 0.731 0.755 0.774 0.794
0.753 0.677 0.599 0.528 0.461 0.396 0.345 0.293 0.252 0.211
0.184 0.160 0.138 0.122 0.105 0.094 0.084 0.075 0.068 0.061
0.059 0.057 0.055 0.052 0.048 0.044 0.045 0.046 0.048 0.048
0.048 0.047 0.049 0.050 0.051 0.052 0.052 0.053 0.055 0.056
0.057 0.054 0.051 0.047 0.044 0.040 0.037 0.033 0.029 0.025
0.021 0.017 0.013 0.009 0.005
```

PEAK Q AT 20 MINUTES

74 MINUTES OF RUNOFF

LOCAL SURFACE HYDROGRAPH (FOR UPSTREAM MOST SUBBASIN)

```
0.000 0.000 0.576 2.231 3.998 5.830 7.733 9.061 9.125 9.032
8.791 8.306 7.821 7.329 6.880 6.335 5.833 5.309 4.863 4.418
4.039 3.746 3.441 3.185 2.931 2.690 2.484 2.308 2.189 2.071
2.044 1.940 1.837 1.740 1.643 1.551 1.541 1.507 1.476 1.444
1.417 1.390 1.388 1.366 1.344 1.321 1.304 1.286 1.288 1.275
1.263 1.250 1.239 1.228 1.229 1.224 1.218 1.212 1.208 1.203
1.204 0.972 0.739 0.506 0.273 0.040 0.037 0.033 0.029 0.025
0.021 0.017 0.013 0.009 0.005
```

PEAK Q

SUM OF PRECEDING TWO HYDROGRAPHS

```
ROUTING CONTINUITY CHECK — INFLOW VOLUME =     11331 CU FT
ROUTING CONTINUITY CHECK — OUTFLOW VOLUME =    11331 CU FT
```

DESIGN Q

```
- B-R - 1 - 0 ------------------------------------------------
LOCAL    AREA :     CPA =  2.00      SPA =  0.00      CGA =  0.40
UPSTREAM AREA :     CPA =  2.00      SPA =  0.00      CGA =  0.40
RAINFALL FACTOR = 1.000    BASE FLOW =  0.00 CFS    SOIL  TYPE = 3

REACH :  LENGTH =    200 FT      SLOPE = 0.80 %      ROUGH = 0.0000

PEAKS(CFS) :  INLET =   9.125  DESIGN =   9.125  OUTLET =   9.089
VOLS :  ROUTED =        11331  GROSS =        11331  %PAVED = 92.13

------------------  DESIGN  CHARACTERISTICS  -------------------
PIPE :  DIAMETER =  18 INCHES                ROUGHNESS = 0.0130
DESIGN :  CAP =  9.389 CFS    VEL = 5.313 FPS    TRV =  0.551 MIN
----------------------------------------------------------------
```

SELECTED PIPE SIZE FOR 1-0 AS REQUESTED

ACTUAL PIPE HYDRAULICS FOR SELECTED PIPE SIZE

TRAVEL TIME

ROUTED DESIGN HYDROGRAPH

```
0.000 0.000 0.259 1.320 3.026 4.821 6.685 8.329 9.089 9.083
8.923 8.573 8.088 7.600 7.127 6.635 6.109 5.597 5.108 4.663
4.248 3.907 3.609 3.326 3.071 2.823 2.597 2.405 2.254 2.136
2.058 1.997 1.894 1.794 1.696 1.602 1.547 1.526 1.493 1.461
1.432 1.405 1.389 1.378 1.356 1.334 1.313 1.296 1.287 1.282
1.270 1.257 1.245 1.234 1.229 1.227 1.221 1.216 1.210 1.206
1.204 1.100 0.867 0.634 0.401 0.168 0.039 0.035 0.031 0.027
0.023 0.020 0.015 0.011 0.007 0.003
```

PEAK Q SLIGHTLY REDUCED

HYDROGRAPH FROM SUBBASIN AFTER ROUTING THROUGH DESIGNED PIPE

```
---- BEGIN BRANCH  1 - REACH  1 ----


PAVED AREA HYDROGRAPH
0.000 0.000 0.192 0.744 1.333 1.925 2.529 3.134 3.681 4.213
4.701 5.125 5.549 5.916 6.282 6.578 6.826 6.875 6.535 6.158
```

```
5.748  5.320  4.880  4.484  4.101  3.737  3.437  3.137  2.894  2.652
2.480  2.329  2.186  2.071  1.956  1.870  1.792  1.715  1.654  1.593
1.558  1.523  1.488  1.447  1.405  1.364  1.350  1.335  1.321  1.302
1.283  1.265  1.254  1.244  1.234  1.221  1.209  1.197  1.190  1.184
1.178  1.095  1.013  0.930  0.851  0.773  0.694  0.616  0.537  0.459
0.382  0.306  0.229  0.153  0.076
```

```
GRASSED AREA HYDROGRAPH
0.000  0.000  0.000  0.032  0.238  0.446  0.661  0.877  1.069  1.256
1.425  1.568  1.711  1.831  1.951  2.042  2.113  2.181  2.238  2.296
2.342  2.386  2.426  2.459  2.494  2.518  2.543  2.568  2.594  2.621
2.648  2.664  2.680  2.697  2.715  2.733  2.751  2.767  2.782  2.798
2.815  2.832  2.849  2.864  2.880  2.895  2.912  2.928  2.913  2.722
2.529  2.330  2.130  1.954  1.784  1.630  1.504  1.377  1.274  1.171
1.097  1.026  0.957  0.900  0.843  0.797  0.753  0.713  0.679  0.645
0.621  0.596  0.571  0.545  0.518  0.491  0.475  0.459  0.441  0.424
0.406  0.388  0.372  0.356  0.340  0.324  0.307  0.289  0.274  0.259
0.243  0.227  0.211  0.194  0.179  0.163  0.148  0.132  0.115  0.099
0.083  0.067  0.050  0.034  0.017
```

```
UPSTREAM ROUTED PLUS SURFACE HYDROGRAPH
 0.00   0.00   0.45   2.10   4.60   7.19   9.88  12.34  13.84  14.55
15.05  15.27  15.35  15.35  15.36  15.25  15.05  14.65  13.88  13.12
12.34  11.61  10.92  10.27   9.67   9.08   8.58   8.11   7.74   7.41
 7.19   6.99   6.76   6.56   6.37   6.20   6.09   6.01   5.93   5.85
 5.81   5.76   5.73   5.69   5.64   5.59   5.57   5.56   5.52   5.31
 5.08   4.85   4.63   4.43   4.25   4.08   3.93   3.79   3.68   3.56
 3.48   3.22   2.84   2.46   2.10   1.74   1.49   1.36   1.25   1.13
 1.03   0.92   0.82   0.71   0.60   0.49   0.48   0.46   0.44   0.42
 0.41   0.39   0.37   0.36   0.34   0.32   0.31   0.29   0.27   0.26
 0.24   0.23   0.21   0.19   0.18   0.16   0.15   0.13   0.12   0.10
 0.08   0.07   0.05   0.03   0.02
```

TOTAL HYDROGRAPH FROM UPSTREAM TWO SUBBASINS AND ENTERING 1-1.

```
ROUTING CONTINUITY CHECK — INFLOW VOLUME =    30246  CU FT
ROUTING CONTINUITY CHECK — OUTFLOW VOLUME =   30246  CU FT
```

CUMMULATIVE CONNECTED PAVED AREA — CUMMULATIVE SUPPLEMENTAL PAVED AREA — CUMMULATIVE CONTRIBUTING GRASSED AREA

```
- B-R - 1 - 1 ----------------------------------------------------
LOCAL    AREA :     CPA =   2.00      SPA =   0.80      CGA =   1.80
UPSTREAM AREA :     CPA =   4.00      SPA =   0.80      CGA =   2.20
RAINFALL FACTOR = 1.000     BASE FLOW =   0.00 CFS     SOIL  TYPE = 3

REACH :   LENGTH =   350 FT      SLOPE = 0.60 %       ROUGH = 0.0000

PEAKS(CFS) :   INLET =   9.056   DESIGN =  15.361   OUTLET =  15.353
VOLS :   ROUTED =       30246   GROSS =         30246   %PAVED = 69.03

----------------------  DESIGN  CHARACTERISTICS  -----------------
PIPE :  DIAMETER =  24 INCHES                     ROUGHNESS = 0.0130
DESIGN :   CAP =  17.512 CFS      VEL = 5.574 FPS     TRV =   0.928 MIN
------------------------------------------------------------------
```

SELECTED PIPE SIZE FOR 1-1.

```
ROUTED DESIGN HYDROGRAPH
 0.00   0.00   0.03   0.57   2.28   4.78   7.39  10.05  12.45  13.89
14.59  15.07  15.27  15.35  15.35  15.35  15.24  15.02  14.60  13.83
13.06  12.29  11.56  10.87  10.23   9.62   9.04   8.54   8.08   7.72
 7.39   7.17   6.97   6.75   6.55   6.36   6.20   6.08   6.00   5.92
 5.85   5.80   5.76   5.72   5.69   5.64   5.59   5.57   5.56   5.50
 5.29   5.07   4.84   4.62   4.42   4.23   4.07   3.92   3.78   3.67
```

```
3.55  3.46  3.19  2.81  2.44  2.07  1.72  1.48  1.36  1.24
1.12  1.02  0.91  0.81  0.70  0.59  0.49  0.47  0.46  0.44
0.42  0.40  0.39  0.37  0.36  0.34  0.32  0.31  0.29  0.27
0.26  0.24  0.23  0.21  0.19  0.18  0.16  0.15  0.13  0.11
0.10  0.08  0.07  0.05  0.03  0.02
```

```
—— BEGIN BRANCH   2 - REACH   0 ——
```

← BEGINNING OF A NEW BRANCH

```
PAVED AREA HYDROGRAPH
 0.00  0.00  0.70  2.71  4.85  7.00  9.16 10.60 10.57 10.36
 9.99  9.33  8.68  8.03  7.44  6.75  6.11  5.45  4.89  4.34
 3.94  3.68  3.41  3.18  2.96  2.75  2.56  2.42  2.32  2.23
 2.23  2.13  2.04  1.94  1.84  1.75  1.75  1.72  1.69  1.66
 1.63  1.60  1.60  1.58  1.55  1.53  1.51  1.49  1.49  1.47
 1.46  1.44  1.43  1.41  1.41  1.41  1.40  1.39  1.38  1.38
 1.38  1.10  0.82  0.54  0.26
```

```
GRASSED AREA HYDROGRAPH
 0.00  0.00  0.00  0.00  0.00  0.05  0.14  0.22  0.30  0.37
 0.43  0.49  0.54  0.59  0.63  0.66  0.69  0.71  0.73  0.75
 0.76  0.77  0.78  0.79  0.80  0.75  0.67  0.59  0.52  0.46
 0.40  0.35  0.30  0.25  0.21  0.18  0.16  0.14  0.13  0.11
 0.10  0.09  0.08  0.08  0.07  0.07  0.07  0.07  0.06  0.06
 0.06  0.06  0.06  0.06  0.06  0.06  0.06  0.06  0.07  0.07
 0.07  0.07  0.06  0.06  0.06  0.05  0.05  0.05  0.04  0.04
 0.04  0.03  0.03  0.03  0.02  0.02  0.02  0.01  0.01
```

```
LOCAL SURFACE HYDROGRAPH
 0.00  0.00  0.70  2.71  4.85  7.05  9.30 10.82 10.87 10.73
10.42  9.82  9.22  8.62  8.07  7.41  6.79  6.16  5.62  5.08
 4.70  4.45  4.19  3.97  3.76  3.50  3.23  3.01  2.84  2.69
 2.63  2.48  2.33  2.19  2.05  1.93  1.91  1.86  1.82  1.77
 1.73  1.69  1.68  1.66  1.63  1.60  1.58  1.55  1.55  1.53
 1.52  1.50  1.49  1.47  1.48  1.47  1.46  1.46  1.45  1.44
 1.45  1.16  0.88  0.60  0.32  0.05  0.05  0.05  0.04  0.04
 0.04  0.03  0.03  0.03  0.02  0.02  0.02  0.01  0.01
```

} TOTAL HYDROGRAPH FROM SUBBASIN

INDEX

Allowable release rate, 107–111, 113–114, 122–123
Alternative comparison in master planning, 480–489
Alternative formulation in master planning, 473–480
American Public Works Association (APWA), 3
American Society of Civil Engineers (ASCE), 18, 101, 327
Arid areas, 46–47
Art of hydrology, 90
Atlanta, Georgia, 294
Average annual monetary benefits (AAMB), 203, 205, 213–215, 326
Average annual monetary (flood) damage (AAMD), 64–65, 141, 203, 205
  calculation of, 207–215, 320, 324, 326, 329, 357, 383
  uses of, 64–65, 205–207, 214–215

Baltimore, Maryland, 38
Berms, on streets, 380, 400–401
Bettendorf, Iowa, 260, 462, 464
Big Thompson River, 3–5
Bimodal hydrograph, 371
Binomial expression, 13
Boston, Massachusetts, 58–60, 412
Bridge and culvert alteration or replacement, overview, 393, 407–408
British Road Research Laboratory (BRRL) Method, 124–132, 339
Brookfield, Wisconsin, 367–369

California, 65
Channel:
  modification or enclosure, overview, 393–398
  shortening, 35–37
  widening, 35, 37–39
Chicago, Illinois, 67, 102–103, 111, 362
Community utility and service policies, overview, 393, 437–438
Computer models:
  basics of, 319–320
  calibration of, 339–361
  Distributed Routing Rainfall Model with Quality (DR3M-QUAL), 334
  HEC-1, 269, 271, 323, 328–331, 334–336, 362, 364, 367, 378, 384
  HEC-2, 159–161, 328, 330–331, 334–337, 362, 364, 367, 384
  HVM-Quantity-Quality-Simulation (HVM-QQS), 334
  Hydrocomp Simulation Programming-Fortran (HSPF), 323, 334–335, 337–338, 343, 362
  Illinois Urban Drainage Area Simulator (ILLUDAS), 125, 132, 307–308, 328, 331–333, 349–350

Computer models (*Continued*)
ILUDRAIN, 125, 132, 269, 271, 323, 328, 331–334, 369, 503–511
MIT Catchment Model (MITCAT), 334
Penn State Runoff Model (PSRM), 334–335
Project Formulation-Hydrology, 112
Runoff Quality (RUNQUAL), 334
selection of, 325–328
Stanford Watershed Model, 321, 323
Storage, Treatment, Overflow, Runoff Model (STORM), 323, 334–335
Storm Water Management Model (SWMM), 323, 328, 334–335
TR20, 323, 328, 334, 375
TR55, 334
Concentration times flow method, for nonpoint-source pollution, 232–236, 242
Connecticut, 223
Conservation of mechanical energy, 143–144, 185–188
Constituent rating curve-flow duration curve method, for nonpoint-source pollution, 236–241
Continuous probability (or recurrence interval) criterion, *see also* Single-probability (or recurrence interval) criterion
for detention/retention facilities, 253, 259–261, 269, 273, 287, 380
Continuous simulation, 233, 238, 315, 321–324, 337, 342, 348, 352, 355, 359, 361, 373–375, 382
Control of land use outside of flood-prone areas, overview, 393, 419
Convenience system, 11, 31–34. *See also* Minor system
Conveyance-oriented approach, 1, 25–29. *See also* Storage-oriented approach
Culvert:
analysis of, 184–188
design, 167, 178–184
Cunette (trickle channel), 171

Darcy–Weisbach equation, 167, 186
Death by flooding, 4, 9
Delineation of watersheds, subwatersheds, and subbasins, 77, 79, 81–86
Denver, Colorado, 83–84, 400, 444
Denver Urban Drainage and Flood Control District, 17–19
Design storm, rainfall amount or hyetograph, 80, 98–99, 112–113, 124–125, 129–130, 262–265, 304, 321–324, 361–363
Design year, in master planning, 473
Dikes and flood walls, overview, 393, 402, 404–407
Discharge-probability (frequency) relationship, 63–64, 80, 91–94, 98, 112, 257. *See also* Volume-probability (frequency) relationship
effect of period of record, 92
Disruption of community by flooding, 6, 9
Documentation, in master planning, 479–480
Drag force, exerted by water, 167, 169, 175–177
Durham, North Carolina, 30
Dutch Gap Creek, 396–397, 488

East Branch of Milwaukee River subwatershed, 353
Economic life, 472
Economics of master planning, 493–495
Elm Grove, Wisconsin, 367–369, 484–487
Emergency action plan (EAP), overview, 393, 439, 443
Emergency system, 1, 11, 31–34. *See also* Major system; Overflow system
Encroachment, equal degree of, 163–164
Envelope curves, 87–88
Equilibrium of streams, 35–38
Europe, 15
Experience curves, 88–89

Federal Water Pollution Control Act, 16
Fish kill, 69–70
Flap gate, 406
Flood Boarding and Flood Map (FBFW), 44
Flood Control Act:
of 1917, 17–18
of 1936, 17–18
Flood Disaster Protection Act:
of 1973, 43
of 1977, 43
Flood flow mapping, 10
Flooding zone:
primary, 15
secondary, 15
Flood insurance:
emergency program, 43–45
overview, 42–46, 393, 419–420
regular program, 43–45
subrogation, 46
Flood Insurance Rate Map (FIRM), 44
Flood insurance report, 44
Flood Insurance Studies (FIS), 8, 10, 44, 79, 87
Flood loss (monetary), 5, 18, 42. *See also* Average annual monetary (flood) damage, (AAMD)
direct, 203–205, 212–213

indirect, 203–205, 212–213
intangible, 203–205
private sector, 204–205
public sector, 204–205
tangible, 203–205
typology, 203–204
Floodplain:
delineation of, 40
fringe, 40, 41, 162, 164
regulations, 1, 38–42. *See also* Legal tests, of floodplain regulations; Ordinance
Floodproofing existing and proposed structures and facilities, overview, 393, 420–435
Floodway, 40–41, 77, 140, 141
determination of, 139, 153–154, 162–165
Florida, 437
Flossmoor, Illinois, 464
Flow regulator, 286, 380, 400, 402–403
Fond du Lac, Wisconsin, 395–396, 479, 488
Fort Collins, Colorado, 492

Georgia, 446
Glacial drift, 85
Gravity, operation, 168, 255
Great Britain, 124

Hahns Creek, 250, 266
Hazen's equation, 305–306, 308
High-velocity wave zones, 43
History:
of floodplain management, 19
of National Flood Insurance Program, 42–43
of storage-oriented approach, 29–31
of urban drainage, 15–17
Huff rainfall distribution, 332
Hurricane Agnes, 2, 4–5
Hydraulic analysis and design guidelines, 167–175
Hydrologic cycle:
description of, 53–56
impact of urbanization on, 57–73
meteorologic impacts of urbanization, 67
Hydrologic techniques, accuracy of, 89

Illinois, 98, 99, 129, 332
northeastern, 94–95, 446
Illinois State Water Survey, 331, 333
Impervious area:
directly connected, 125–130, 132
indirectly connected, 125–126
Implementation:
of designed detention/retention facility, 247, 287–290
of master plans, 481–482, 489–493
Indiana:
Association of Cities and Towns, 446
Department of Highways, 290, 491
northwest, 274
Infectious diseases, 15–16
Infiltration system, overview, 393, 412–417
Inlet control device, *see* Flow regulator
Inlet control zone, in sedimentation basin, 299–301, 308
Inspection and maintenance program, overview, 393, 438–441. *See also* Operation and maintenance of stormwater facilities
Interior drainage, 406–407
Inventory, in master planning, 469–471
Iowa, 446

Jackson, Mississippi, 349
James River, 69
Johnstown, Pennsylvania, 4–5

Kansas City, Kansas, 4–5, 14
Kansas City, Missouri, 4–5, 14
Karst topography, 46–47, 85
Kickapoo River, 435
Kinnickinnic River watershed, 14, 220–222, 233–235, 237–240, 354, 356
Kuichling, Emil, 101, 104

Lafayette, Indiana, 98
Lake Winnebago, 479
Land data management system, 470
Land use controls, *see* Control of land use outside of flood-prone areas, overview; Floodplain, regulations; Ordinance
Legal tests, of floodplain regulations, 41–42
Lincoln, Abraham, 462
Log-Person type III method, 91–92, 94

McCarron Lake, 409
Machiavelli, Niccolo, 482
Madison, Wisconsin, 301, 378, 410–411, 480–481
Major system, 11, 21–24, 31–34. *See also* Emergency system; Overflow system
Maning equation, 10, 115, 119, 143, 163, 167, 188–191, 325, 330, 380
roughness coefficients for, 115, 119, 146, 149, 153–154, 156–157, 159–161, 178, 190–191, 320, 329–330, 351
Manitowoc, Wisconsin, 266
Manual, *see* Surface Water Management (SWM) manual, overview
Manual rainfall-runoff methods, 87, 97–133
Manufactured (mobile) home, 209, 433

Menomonee Falls, Wisconsin, 371–372
Menomonee River watershed, 340–341, 352–355, 357–360, 364–366, 371–372, 380, 382–383, 464, 489–490
Metropolitan Sanitary District of Greater Chicago, 259
Miami Conservancy District, 17–18
Milwaukee, Wisconsin, 14, 70, 220, 233, 237–238, 373, 375–377
Minnesota, 8, 161
Minor system (convenience system), 11, 21–24, 31–34
Mississippi, 160
Mississippi River, 10, 18, 435
Modified rational method (MRM), 105–111, 269, 271. *See also* Rational method
  bow method, 111
  rainfall intensity averaging time, 106–108, 110–111
Mt. Pleasant, Wisconsin, 350
Multiple detention/retention facilities, 290–295
Multiple purpose detention/retention facilities:
  concept, 29–31
  examples of, 266–269, 287–288
Multiple purpose-multiple means approach to floodplain management, 9, 17–19

National Flood Insurance Act of 1968, 42
National Flood Insurance Program (NFIP), 1, 10, 42, 43–46, 364, 419, 446
Nationwide Urban Runoff Program (NURP), 6, 70–71, 217, 219
Natural features of landscape in SWM, 34–35
New Hampshire, 36
New Jersey, 94
Nile River, 2
Nonpoint-source pollution, *see also* Concentration times flow method, for nonpoint-source pollution; Constituent rating curve-flow duration curve method, for nonpoint-source pollution; Preliminary screening procedure; Unit load method, for nonpoint-source pollution; Universal soil loss equation for nonpoint-source pollution
  essential features, 67–68, 70–73, 218–219
Normal depth approach, 145, 188, 190–191
North Branch of Chicago River, 362, 364

Oak Creek subwatershed, 353
Oakland, California, 343
Objectives and standards, in master planning, 466–469, 480, 489
Off-line detention/retention facility, 262–265, 271–272. *See also* On-line detention/retention facility
On-line detention/retention facility, 262–263, 269, 271–273. *See also* Off-line detention/retention facility
Operation and maintenance of stormwater facilities, 4, 28, 174–175, 294
Ordinance, *see also* Floodplain, regulations
  building, 40, 419
  health, 39, 419
  land division, 39, 418–419
  sanitary, 34, 419
  zoning, 38–39, 164, 418–419
Orifice:
  coefficients for, 198–200
  equation, 198–199, 277, 312
Outlet control zone in sedimentation basin, 299–301, 304, 310–311
Overflow system, 31–34. *See also* Emergency system; Major system

Pennsylvania, 8
Phasing of master planning, 460–462
Photogrammetry, 155, 161
Physical life, 472
Planning:
  process overview, 453–456, 465
  products overview, 453–454
Plotting position formula, 91–92, 375
Plugging of stormwater facilities, preventing, 171–172
Porter County, Indiana, 252
Prairie du Chien, Wisconsin, 435–436
Preliminary screening procedure, for nonpoint-source pollution, 223–227, 242, 357
Profiles:
  flood stage, 139–141, 145–146, 149, 154–155, 158–162
  streambed, 139–141, 155
  water surface, 142–145
Public:
  communication with, 11–16
  education of and interaction with, 393, 446–448, 462–464, 467, 479
  health risks, 4–5, 15
  interest in SWM, 9, 447, 456–459
  psychological stress, 5
  response to skeptics, 14
  safety risks, 4–5, 9, 162–163, 168–171, 175–177, 294, 427, 429, 446–447, 479
Public Law 92-500 208 studies, 30
Puget Sound, 492

Rapid City, South Dakota, 2, 4, 14, 42
Rational method, 58–60, 100–106, 109, 124, 248, 324, 339, 355, 357, 472. *See also* Modified rational method (MRM)
 rainfall intensity averaging time, 101–111
Regional methods, 77, 87, 93–95, 223, 357
Regulations, *see* Floodplain, regulations; Ordinance
Relocation or demolition of structures and facilities, overview, 393, 435–436
Rend Lake, 240
Reproducibility of hydrologic techniques, 89
Reservation of land for recreation and other open space use, overview, 393, 416–419
Rochester, New York, 101
Root River Canal subwatershed, 353
Roseville, Minnesota, 409–410
Rounding of calculated flows, 89–90

Sedimentation basin–detention/retention system, overview, 393, 410–412
Sedimentation basin overview, 297–313, 393, 408–409
Sedimentation basin–wetland–detention/retention system, overview, 393, 416–417
Sedimentation basin–wetland system, overview, 393, 409–410
Sensitivity studies, 339–340, 349–351
 flood stage profiles, 139, 160–162
 storage volume, 262–265
Settling efficiency, 297–298, 300–301, 304–309
Settling zone, in sedimentation basin, 299–301, 304–311
Single probability (or recurrence interval) criterion, *see also* Continuous probability (or recurrence interval) criterion
 for detention/retention facilities, 253, 259–261, 269, 293, 378–380
Single purpose–multiple means approach, 17–18
Single purpose–single means approach, 9, 17–19
Skokie, Illinois, 29, 190, 299, 380, 400–403, 408
Slope and swale treatment, overview, 393, 436–437
Smith Ditch, 254–255, 287, 361, 363
Soldiers Grove, Wisconsin, 435–436
Southeastern Wisconsin Regional Planning Commission, 17, 19, 327, 337–338, 464, 468
Springfield, Missouri, 69
Stage-discharge relationship:
 detention/retention outlet works, 195–199
 street, 191–192
Statistics, 1, 11–14, 87, 89, 90–93
 risk equations, 12–14
 use of, 11, 14, 323, 373–375, 382
Stokes' law, 306–308
Storage or diversion, overview, 393, 398–403
Storage-oriented approach, 1, 25–31. *See also* Conveyance-oriented approach
Storage zone, in sedimentation basin, 299–301, 309–311
Streambed profile, 139–141
Street:
 flow capacity, 188–192
 storage capacity, 192–195
 sweeping, 224, 226, 438–439
Structural and nonstructural measures:
 definitions, 392
 summary of applicability, 393–394
Structure:
 floodwater entering, 14–15
 hydraulically insignificant, 158–159
 hydraulically significant, 158
Subrogation, 46
Surface Water Management (SWM) manual, overview, 393, 442–446

Target date in master planning, 473
Task Force:
 ASCE, on floodplain regulations, 17–18
 federal, on control policy, 17–18
Tax base, loss of, 436
Tennessee Valley Authority, 17–18
Tibbee River, 160
Time of concentration, 59–60, 86, 100, 101–107, 115, 119, 122, 128, 132, 248, 250–251, 262
Toronto, Canada, 341
Transfer method, 77, 87, 95–97
Trap efficiency, 298, 300, 305–306, 309–310, 408
Trickle channel (cunette), 171
TR (Technical Report) 55 method, 112–124, 248, 269, 271, 273–276, 283, 285, 308, 334, 472

Underwood Creek, 367–368
U.S. government:
 Army Corps of Engineers (COE), 8, 79, 160, 162, 207, 239, 327, 329, 331, 383, 398, 405–406, 435
 Congress, 18
 Department of Commerce, 115
 Department of Transportation, 81, 188, 290

U.S. government (*Continued*)
  Environmental Protection Agency (EPA), 71–72, 79, 223, 357
  Federal Emergency Management Agency (FEMA), 8, 10, 43, 78–79, 87
  Federal Insurance Administration (FIA), 43–44, 46
  Geological Survey (USGS), 79–80, 83, 87, 93, 150, 160, 162, 252, 344, 357
  Housing and Urban Development (HUD), 42, 435
  Hydrologic Engineering Center of the U.S. Army Corps of Engineers, 327–331, 335, 338
  National Technical Information Service (NTIS), 334
  National Weather Service, 80, 373
  Nuclear Regulatory Commission, 405
  Soil Conservation Service (SCS), 79, 81, 98, 100, 112–113, 124, 162, 329, 334
  Water Resources Council, 91
Unit hydrograph, 324
Unit load method for nonpoint-source pollution, 220–223, 242
Universal soil loss equation for nonpoint-source pollution, 227–232, 242, 310
Urban surface water system, components of, 19–25
User fees, 465, 479, 492
Utah, 46
Utility approach, 479, 492

Valparaiso, Indiana, 252, 254, 263, 289, 335, 363, 368, 384, 469, 473, 491
Values:
  aesthetic, 6, 19, 30–31, 34, 164, 170–171, 429, 439–441, 468, 478
  cultural, 6, 19, 34, 164, 438
  ecological, 6, 19, 34, 164, 438
  land, 34–35, 418
  recreational, 6, 19, 30–31, 34–35, 468, 478
Vandalism, preventing, 172–173
Volume-probability (frequency) relationship, 98, 112. *See also* Discharge-probability (frequency) relationship
Vortex, preventing, 172, 312

Washington, D.C., 38, 94
Watersheds, *see also* Delineation of watersheds, subwatersheds, and subbasins
  as basis for master planning, 460–462
Wauwatosa, Wisconsin, 355, 357–360
Weir:
  coefficients, 195, 197–198
  equation, 195–196
  submerged, 197–198, 200
Wisconsin, 69, 94, 232, 242
  southeastern, 59, 66, 88, 114–115, 121, 340–341, 352, 354, 362, 364–366, 371, 380, 462, 489–490